Grafted Biopolymers as Corrosion Inhibitors

WILEY SERIES IN CORROSION

R. Winston Revie, Series Editor

Corrosion Inspection and Monitoring · *Pierre R. Roberge*

Microbiologically Influenced Corrosion · Brenda J. Little and Jason S. Lee

Corrosion Resistance of Aluminum and Magnesium Alloys: Understanding, Performance, and Testing · *Edward Ghali*

Metallurgy and Corrosion Control in Oil and Gas Production · *Robert Heidersbach*

Green Corrosion Inhibitors: Theory and Practice · *V. S. Sastri*

Heterogeneous Electrode Processes and Localized Corrosion · *Yongjun Tan*

Stress Corrosion Cracking of Pipelines · *Y. Frank Cheng*

Corrosion Failures: Theory, Case Studies, and Solutions · *K. Elayaperumal and V. S. Raja*

Challenges in Corrosion: Costs, Causes, Consequences and Control · *V. S. Sastri*

Metallurgy and Corrosion Control in Oil and Gas Production, Second Edition · *Robert Heidersbach*

High Temperature Corrosion: Fundamentals and Engineering · *César A. C. Sequeira*

Grafted Biopolymers as Corrosion Inhibitors: Safety, Sustainability, and Efficiency · *Jeenat Aslam, Chandrabhan Verma and Ruby Aslam*

Grafted Biopolymers as Corrosion Inhibitors

Safety, Sustainability, and Efficiency

Edited by Jeenat Aslam, Chandrabhan Verma, and Ruby Aslam

WILEY

This edition first published 2023

Registered Office
John Wiley & Sons, Inc., 111 River Street, Hoboken, NJ 07030, USA

For details of our global editorial offices, customer services, and more information about Wiley products visit us at www.wiley.com.

Wiley also publishes its books in a variety of electronic formats and by print-on-demand. Some content that appears in standard print versions of this book may not be available in other formats.

Library of Congress Cataloging-in-Publication Data

Hardback ISBN: 9781119881360; ePub ISBN: 9781119881384; ePDF ISBN: 9781119881377; oBook ISBN: 9781119881391

Cover Image: © Vink Fan/Shutterstock
Cover Design: Wiley

Set in 9.5/12.5pt STIXTwoText by Integra Software Services Pvt. Ltd, Pondicherry, India

Contents

About the Editors

Jeenat Aslam, Ph.D., is currently working as an Associate Professor at the Department of Chemistry, College of Science, Taibah University, Yanbu, Al-Madina, Saudi Arabia. She earned her Ph.D. degree in Surface Science/Chemistry from the Aligarh Muslim University, Aligarh, India. Materials & corrosion, nanotechnology, and surface chemistry are the primary areas of her research. Dr. Jeenat has published a number of research and review articles in peer-reviewed international journals like ACS, Wiley, Elsevier, Springer, Taylor & Francis, Bentham Science, and others. She has authored over thirty book chapters and edited more than twenty books for the American Chemical Society, Elsevier, Springer, Wiley, De-Gruyter, and Taylor & Francis.

Chandrabhan Verma, Ph.D., works at the Department of Chemical Engineering, Khalifa University of Science and Technology, P.O. Box 127788, Abu Dhabi, United Arab Emirates. He obtained his Ph.D. in Material Science/Chemistry at the Indian Institute of Technology (Banaras Hindu University) Varanasi, India. He is an American Chemical Society (ACS) member and serves as a reviewer and editorial board member for various internationally recognized ACS, RSC, Elsevier, Wiley, and Springer platforms. Dr. Verma is the Associate Editor-in-Chief of the Organic Chemistry Plus Journal. He is the author of several research and review articles published in ACS, Elsevier, RSC, Wiley, Springer, etc. He has a total citation of more than 9065 with an H-index of 53 and an i-10 index of 142. Dr. Verma has edited many books for the ACS, Elsevier, RSC, and Wiley. Dr. Verma received several awards for his academic achievements.

Ruby Aslam, PhD., is currently a Research Associate fellow under CSIR-HRDG, New Delhi in the Department of Applied Chemistry, Aligarh Muslim University, Aligarh, India. She received her M.Sc., M. Phil., and Ph.D. degrees from the same university. Her main areas of interest in research include the development of stimuli-responsive smart coatings for corrosion detection and protection as well as the assessment of environmentally friendly corrosion inhibitors. She has authored/co-authored several research papers in international peer-reviewed journals of wide readership, including critical reviews and book chapters. She has edited many books for American Chemical Society, Elsevier, Springer, Wiley, De-Gruyter and Taylor & Francis.

List of Contributors

Khamdam Akbarov
Faculty of Chemistry, National University of Uzbekistan, Tashkent Uzbekistan

Ekemini D. Akpan
Centre for Materials Science
College of Science, Engineering, and Technology, University of South Africa, Johannesburg, South Africa

Farhat A. Ansari
Faculty of Pharmaceutical Chemistry
Hygia Institute of Pharmaceutical Education and Research, Uttar Pradesh, India

J. Arun
Centre for waste management – "International Research Centre"
Sathyabama Institute of Science and Technology, Tamil Nadu, India

Jeenat Aslam
Department of Chemistry, College of Science, Taibah University
Al-Madina, Saudi Arabia

Ruby Aslam
Corrosion Research Laboratory
Department of Applied Chemistry
Aligarh Muslim University
Uttar Pradesh, India

Humira Assad
Department of Chemistry
Faculty of Technology and Science
Lovely Professional University
Punjab, India

Rajeshwari Athavale
Department of Chemistry, K.E. T's V. G. Vaze College (Autonomous)
Maharashtra, India

Khasan Berdimuradov
Faculty of industrial Viticulture and Food Production Technology
Shahrisabz branch of Tashkent Institute of Chemical Technology
Shahrisabz, Uzbekistan

Elyor Berdimurodov
Faculty of Chemistry, National University of Uzbekistan, Tashkent Uzbekistan

K. Cial
Corrosion Research Laboratory
Department of Applied Chemistry
Aligarh Muslim University
Uttar Pradesh, India

Lakha V. Chopda
Government Engineering College
Bhuj (Gujarat), India

Fatima Choudhary
Department of Chemistry, K.E. T's
V. G. Vaze College (Autonomous)
Maharashtra, India

Omar Dagdag
Centre for Materials Science
College of Science, Engineering
and Technology
University of South Africa
Johannesburg, South Africa

Walid Daoudi
Laboratory of Molecular Chemistry
Materials and Environment
(LCM2E), Department of Chemistry
Multidisciplinary Faculty of Nador
University Mohamed I, Nador
Morocco

Pragnesh N. Dave
Department of Chemistry, Sardar
Patel University, Vallabh Vidynagar
(Gujarat), India

R. V. Deeksha
Centre for waste management –
"International Research Centre"
Sathyabama Institute of Science and
Technology, Tamil Nadu, India

Manoj Dhameja
Department of Chemistry, Babasaheb
Bhimrao Ambedkar University
Uttar Pradesh, India

Eno E. Ebenso
Centre for Materials Science
College of Science, Engineering
and Technology, University of South
Africa, Johannesburg, South Africa

Abdolreza Farhadian
Department of Polymer & Materials
Chemistry, Faculty of Chemistry and
Petroleum Science, Shahid Beheshti
University, Tehran, Iran

Department of Petroleum Engineering
Kazan Federal University, Kazan
Russian Federation

Omolola E. Fayemi
Department of Chemistry, School
of Chemical and Physical Sciences
and Material Science Innovation
& Modelling (MaSIM) Research
Focus Area, Faculty of Natural and
Agricultural Sciences, North-West
University, Mmabatho, South Africa

Luana Barros Furtado
Federal University of Rio de Janeiro
School of Chemistry, Rio de Janeiro
Athos da Silveira Ramos Avenue
Brazil

Richika Ganjoo
Department of Chemistry, Faculty
of Technology and Science, Lovely
Professional University, Punjab, India

Sailee Gardi
Department of Chemistry, K.E. T's
V. G. Vaze College (Autonomous)
Maharashtra, India

Husnu Gerengi
Corrosion Research Laboratory
Department of Mechanical
Engineering, Duzce University
Duzce, Turkey

Maria José O. C. Guimarães
Federal University of Rio de Janeiro
School of Chemistry, Rio de Janeiro
Athos da Silveira Ramos Avenue
Brazil

Lei Guo
School of Material and Chemical Engineering, Tongren University
Tongren, P. R. China

Preeti Gupta
Department of Chemistry, Babasaheb Bhimrao Ambedkar University, Uttar Pradesh, India

Nellaiah Hariharan
Bangalore Biotech Labs Private Limited (BiOZEEN), Bangalore, India

Rajesh Haldhar
School of Chemical Engineering, Yeungnam University, Gyeongsan
Republic of Korea

Brahim El Ibrahimi
Department of Applied Chemistry, Faculty of Applied Sciences, Ibn Zohr University, Aït Melloul, Morocco

Kundan Jangam
Department of Chemistry, K.E. T's, V. G. Vaze College (Autonomous)
Maharashtra, India

Ali Asghar Javidparvar
School of Metallurgy and Materials Engineering, College of Engineering, University of Tehran, Tehran, Iran

Abduvali Kholikov
Faculty of Chemistry, National University of Uzbekistan, Tashkent
Uzbekistan

Seong-Cheol Kim
School of Chemical Engineering
Yeungnam University, Gyeongsan
Republic of Korea

Ashish Kumar
Department of Chemistry, Faculty of Technology and Science, Lovely Professional University, Punjab, India

NCE, Bihar Engineering University
Department of Science and Technology, Government of Bihar
India

Hariom Kumar
Department of Chemistry, Babasaheb Bhimrao Ambedkar University
Uttar Pradesh, India

Han-Seung Lee
Department of Architectural Engineering, Hanyang University-ERICA, Gyeonggi-do, Republic of Korea

Hassane Lgaz
Innovative Durable Building and Infrastructure Research Center
Center for Creative Convergence Education
Hanyang University ERICA
Gyeonggi do, Korea

Hui-Jing Li
Weihai Marine Organism & Medical Technology Research Institute
Harbin Institute of Technology
Weihai, P. R. China

Sheerin Masroor
Department of Chemistry, A.N. College, Patliputra University
Bihar, India

M. Mobin
Corrosion Research Laboratory
Department of Applied Chemistry
Aligarh Muslim University
Uttar Pradesh, India

Shalini Mohan
Department of Biotechnology
Kalasalingam Academy of Research and Education, Tamil Nadu, India

Paresh More
Department of Chemistry, K.E. T's V. G. Vaze College (Autonomous)
Maharashtra, India

N. Mujafarkani
PG and Research Department of Chemistry, Jamal Mohamed College (Autonomous), Tiruchirappalli
Tamil Nadu, India

Lakshmanan Muthulakshmi
Department of Biotechnology
Kalasalingam Academy of Research and Education, Tamil Nadu, India

Rafaela C. Nascimento
LAQV-REQUIMTE, Instituto de Investigação e Formação Avançada
Universidade de Évora, Évora, Colégio Luís António Verney, Portugal

Lukman O. Olasunkanmi
Department of Chemistry, Faculty of Science, Obafemi Awolowo University
Ile Ife, Nigeria

Department of Chemical Science
University of Johannesburg
Johannesburg, South Africa

Tae Joon Park
Department of Robotics Engineering
Hanyang University, Gyeonggi-do
Korea

M. Parveen
Corrosion Research Laboratory
Department of Applied Chemistry
Aligarh Muslim University
Uttar Pradesh, India

Dwarika Prasad
Department of Chemistry
Shri Guru Ram Rai University
Dehradun, India

P. Priyadharsini
Centre for waste management – "International Research Centre"
Sathyabama Institute of Science and Technology, Tamil Nadu, India

Taiwo W. Quadri
Centre for Material Science, College of Science, Engineering and Technology
University of South Africa
Johannesburg, South Africa

Syed Ali Abdur Rahman
Department of Biotechnology
Sathyabama Institute of Science and Technology, Tamil Nadu, India

Mohamed Rbaa
Laboratory of Organic Chemistry
Catalysis and Environment
Faculty of Sciences, Ibn Tofail University, Kenitra, Morocco

Marziya Rizvi
Department of Chemistry
Babasaheb Bhimrao Ambedkar University, Uttar Pradesh, India

Vandana Saraswat
Department of Chemistry
University Institute of Sciences
Chandigarh University
Mohali, India

Tarun K. Sarkar
Department of Chemistry, IFTM University, Moradabad
Uttar Pradesh, India

Ali Reza Shahmoradi
Department of Chemical Engineering
Shahreza Branch, Islamic Azad University, Shahreza, Iran

Hariom K. Sharma
Engineering Department
University of Technology and Applied Sciences (UTAS)
Dhofar, Sultanate of Oman

Praveen Kumar Sharma
Department of Chemistry, Faculty of Technology and Science, Lovely Professional University, Punjab, India

Rahul Singh
Department of Chemistry, Shri Guru Ram Rai University, Dehradun, India

Karthick Subbiah
Department of Architectural Engineering, Hanyang University-ERICA, Gyeonggi-do
Republic of Korea

Nilufar Tursunova
Faculty of Chemistry, National University of Uzbekistan
Tashkent, Uzbekistan

Dakeshwar Kumar Verma
Department of Chemistry
Government Digvijay Autonomous Postgraduate College
Chhattisgarh, India

Dan-Yang Wang
Weihai Marine Organism & Medical Technology Research Institute
Harbin Institute of Technology
Weihai, P. R. China

Yan-Chao Wu
Weihai Marine Organism & Medical Technology Research Institute
Harbin Institute of Technology
Weihai, P. R. China

Mahendra Yadav
Indian Institute of technology (Indian School of Mines), Dhanbad, India

Ramesh Yamgar
Department of Chemistry
C. S.'s Patkar-Varde College
Maharashtra, India

Preface

Corrosion of metal is a destructive phenomenon that has a significant impact on the anticipated lifetime and use of materials made of metals. Use of corrosion inhibitors is thought to be the most efficient and cost-effective method to block metals against corrosion, especially in acidic conditions, to resolve this type of issue. Studies on "sustainable(green)" corrosion inhibitors, which don't have the negative health effects associated with the organic compounds employed in the past, have been conducted over the past ten years.

In recent times, polymeric biomaterials have received the most important attention in corrosion science. Biomaterials such as natural biopolymers (polysaccharides) and their derivatives are attractive due their affordability, intrinsic non-toxicity, biodegradability, and availability of numerous adsorption sites. These unexpected benefits have led to widespread usage of biopolymers (polysaccharides) and their derivatives for medication delivery, corrosion inhibitors, coating materials, and the removal of hazardous chemicals through adsorption. Though there are various reports on natural biopolymers and their derivatives as corrosion inhibitors. For instance, gums from natural exudates, chitosan, cellulose derivatives, starch and its derivatives, pectin, carrageenan, and alginate. However in order to prevent valuable metals from being damaged by acid solutions, it is still essential to design efficient corrosion inhibitors. Biopolymers (polysaccharides) have been generally studied as corrosion inhibitors because of the presence of a variety of polar functional groups for example OH, COOH, and NH_2 in their arrangement and capability to complex with metals on surfaces. In corrosion inhibition, biopolymers (polysaccharides) characterize a set of chemically stable, biodegradable, and environment-friendly macromolecules with distinctive inhibitory strengths and binding to metal surfaces.

The present book is a collection of major advancements in the field of polymer for the design and testing of the corrosion inhibition effect of sustainable grafted biopolymer corrosion inhibitors. This book explains the synthesis, characterization, and anticorrosive application of some green and environmentally friendly sustainable grafted biopolymers and their derivatives for inhibition of metal corrosion. It has also been explored how their distinct molecular and electrical structures, chemical makeup, and macromolecular weights all have a role in the sorts and ways of protection they offer.

The book is written for scholars in academia and industry, working corrosion engineers and materials science students, and applied and engineering chemistry.

The book is structured into three parts, each of which contains several chapters, in order to condense the detailed explanation of anticorrosive applications of sustainable grafted biopolymer and to offer the reader a sensible and expressive design of the issue.

PART 1 explores the economic and legal issues of corrosion. Topics covered in chapters 1 to 4 are corrosion: basics, economic adverse effects, and its mitigation, corrosion inhibition: past and present developments and future directions, biopolymers as corrosion inhibitors: relative inhibition potential of biopolymers and grafted biopolymers and biopolymers vs. grafted biopolymers: challenges and opportunities

PART 2 discusses an overview of sustainable grafted biopolymers. Topics covered in chapters 5 to 7 are sustainable grafted biopolymers: synthesis and characterizations, sustainable grafted biopolymers: properties and applications, and factors affecting biopolymers grafting.

PART 3 debates sustainable grafted biopolymers as corrosion inhibitors. Topics covered in chapters 8 to 19 are corrosion inhibitors: introduction, classification and selection criteria, chemical, electrochemical, surface characterization, computational techniques for corrosion monitoring, methods of corrosion measurements: chemical, electrochemical, surface and computational, grafted natural exudates gums used as sustainable corrosion inhibitors, grafted pectin as sustainable corrosion inhibitors, grafted chitosan as sustainable corrosion inhibitors, grafted starch used as sustainable corrosion inhibitors, grafted cellulose as sustainable corrosion inhibitors, grafted alginates as sustainable corrosion inhibitors, grafted dextrin as sustainable corrosion inhibitors, grafted biopolymer composites and nanocomposites as sustainable corrosion inhibitors, industrially useful corrosion inhibitors: grafted biopolymers as ideal substitutes.

The goal of this book is to provide the most recent developments in grafted biopolymers for anticorrosive applications. This book is written for a highly diverse group of people who work in chemical engineering, advanced materials research, and other related subjects. Libraries in academic and professional settings, independent research organizations, government agencies, and scientists will all find this book to be an invaluable source of reference information. The chapters' authors and book editors are renowned academic and professional researchers, scientists, and subject matter specialists.

On behalf of John Wiley & Sons, Inc., we thank all contributors for their exceptional and whole-hearted contribution. Invaluable thanks to Mr. Michael Leventhal (Acquisitions Editor), Miss Kelly Labrum (Associate Managing Editor), Miss Elizabeth (Managing Editor), and the Editorial Team at John Wiley & Sons, Inc. for their wholehearted support and help during this project. In the end, all appreciation to John Wiley & Sons, Inc. for publishing the book.

Jeenat Aslam, Chandrabhan Verma & Ruby Aslam
(Editors)

Part 1

Economic and Legal Issue of Corrosion

1

Corrosion

Basics, Economic Adverse Effects, and its Mitigation

*Rahul Singh and Dwarika Prasad**

Department of Chemistry, Shri Guru Ram Rai University, Dehradun - 248001, INDIA
**Corresponding author*

1.1 The Basics of Corrosion

Corrosion is a natural phenomenon that is responsible for the loss of material across the globe, resulting in a loss of approximately 26.1 billion dollars worldwide. Chemically it is expounded as the process of deposition of a layer of oxides or sulfides or chlorides on the surface of materials. Before, corrosion was only studied concerning the degradation of metal surfaces, but nowadays studies also extend to the degradation of plastics and polymers naturally over course of time. The process occurs spontaneously without the requirement of any external factor like catalyst or temperature or energy; where there is moisture there exists corrosion. In fact, in absence of moisture corrosion of steel which results in its cracking is observed, it is mainly due to exposure to di-hydrogen gas which as effect releases methane by reacting with carbon present in steel, categorized as "dry-corrosion." The chemistry of the redox reaction is followed in which one part acts as an anode while another part acts as a cathode. Degradation usually occurs at the anode where oxidation occurs while deposition of oxidation products is usually observed at the cathode where reduction takes place. It is just like a typical galvanic cell. Corrosion is a slow process; it takes days to months and sometimes years depending upon the inhibitor strength that is used. Since corrosion destroys material Fig 1.1 and results in mechanical failure, thus chemical substances are used to inhibitor corrosion or to delay its course of action. Such chemical substances are termed an inhibitor. These inhibitors are broadly categorized into two categories based on their environmental impact: synthetic inhibitor, which is mostly in-organic and causes environmental damage, = and natural inhibitor, which is mostly organic like grafted biopolymer

Grafted Biopolymers as Corrosion Inhibitors: Safety, Sustainability, and Efficiency, First Edition.
Edited by Jeenat Aslam, Chandrabhan Verma, and Ruby Aslam.

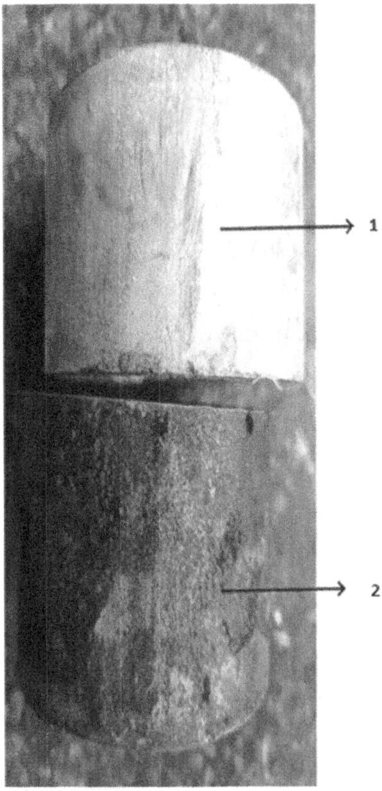

Figure 1.1 An image comparing the surface of a pipe after dipping it into an acid solution with (1) and without (2) mitigator.

and plant extracts used at different concentrations to effectively slow down the rate of corrosion without any hazardous environmental impact. Meanwhile, the research on natural corrosion inhibitors is on the rise, and biopolymers which are polymers produced from natural sources, like DNA/RNA, lipids, collagen, and carbohydrates are under investigation. This research work is majorly undertaken by scholars of chemistry background because of its wide scope. Adverse effects of corrosion spread from the construction sector to the industrial sector to day-to-day life in our houses and in our vehicles, thus the scope of inhibition also extends to wide dimensions. Natural corrosion inhibitors are commonly extracted from plant waste materials. The plant extract is a combination of biopolymer and secondary metabolites. When people separate biopolymers from plant extract and graft from antioxidant materials they act as a very good corrosion inhibitor. The biopolymers are long-chain macromolecules found in living systems because of their complex chemical nature helpful as corrosion inhibitors.

1.2 Corrosion Mitigations

The grafted biopolymer-based corrosion mitigators are good because of some characteristics: 1. The presence of heteroatoms like nitrogen and oxygen which can easily donate their lone pair of electrons to vacant d-orbitals of metal, 2. These mitigators have active π-bonds sites which interact with empty 3d-orbitals of metal, 3. These compounds have large sizes, so grafted biopolymer can cover a large surface area of a metal surface, 4. these have cost-effective because the main constituent biopolymer can easily be extracted from plants, so no problem related to the quality of materials, 5. These are soluble with desired solvents, so no problem related to solubility with different types of cleaning or pickling conditions, 6. A last but very important characteristic of biopolymers is that they are non-toxic and eco-friendly. Low carbon steel (LCS) is a promptly accessible metal combination, which has numerous mechanical properties. In momentum research, mild steel or low-carbon steel has been utilized to test certain properties in corrosive projecting [1, 2]. Nowadays, the current practice for some progressions, for example, cleaning, descaling, and pickling to utilize an acidic climate Fig 1.3, and subsequent disintegration of the metals as low carbon steel is unavoidable [3]. The industrial applications of inhibitors are 1. In the pickling process, 2. In the boiler cleaning process, 3. In the oil well acidization process, 4. In the metallic paint/primer and coating process, 5. In the oil and gas pipelines cleaning process. To secure or protect the metal, the eco-friendly corrosion inhibitors adsorbed on the surface of the metal utilizing pai-electronic frameworks, sulfur, nitrogen, oxygen, and phosphorus [4–7]. This adsorption can occur in two unique manners: physisorption and chemisorption. The physisorption or physical adsorption is reversible and the adsorption enthalpy of approximately 30 kJ/mol, which is low. The chemical adsorption is irreversible and the adsorption enthalpy of approximately 160 kJ/mol, which is high [8] [9]. The physisorption takes place in low temperatures and with the increment of temperature, it decreases [10]. Additionally, it has less activation energy. The chemisorption takes place in high temperatures and with the increment of temperature, it increases [11, 12].

The chemisorption has comparatively higher activation energy. Taking into account that the corrosion process is a natural process, where a metal starts rusting when it comes into direct contact with moisture [13]. It is a combined process of hydrogen evolution (cathodic corrosion) and metal dissolution (anodic corrosion). Therefore, the prevention of steel from the corrosion process is required. Several methods are commercially available to resist steel corrosion but most of those are non-eco-friendly and require a high budget as well [14]. Some crop materials such as organic products, seeds, dry leaves, bark, and peel of some fruits have a non-toxic, non-hazardous, and eco-friendly nature. These materials can be used as corrosion resistance specialists. Its easy availability and economic accessibility make it more

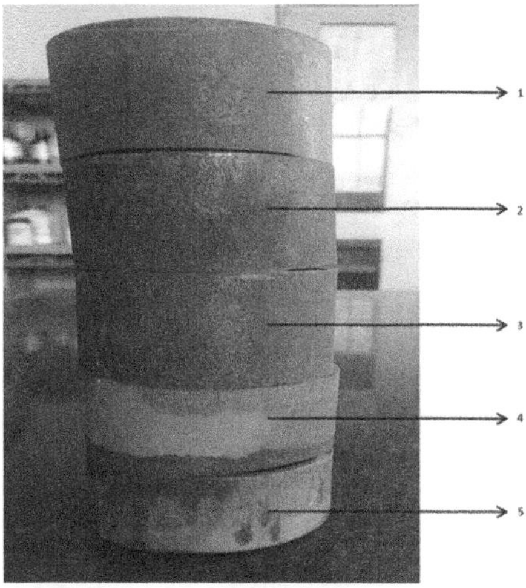

Figure 1.2 The above image shows an increase in the rate of corrosion mitigation from 1 to 5 after the increase in concentration.

favorable [15]. The main types of corrosion mitigators are: 1. Synthetic (organic and inorganic) mitigators, 2. Natural (biopolymer and plant extract) mitigators. Natural biopolymers and their modified grafting biopolymers are mostly non-toxic and eco-friendly. To discard the waste from which we extract grafted bio-polymer, you generally burn it. It may cause air pollution as well. So, we tried to utilize this waste against the corrosion of steel in an acidic medium. It makes this waste a significant component in a monetarily solid environment. Another part of the research is to use the grafted biopolymer as a useful product with a negligible production cost. In this current examination, we utilized a grafted biopolymer as a green corrosion inhibitor with a high inhibitory performance at a low inhibitor concentration Fig 1.2. Even though there are numerous corrosion-resistance materials available, a large portion of them utilize synthetic engineered items that are poisonous and hurtful to the climate [16–19]. Consequently, it is important to foster a harmless to the ecosystem green corrosion inhibitor.

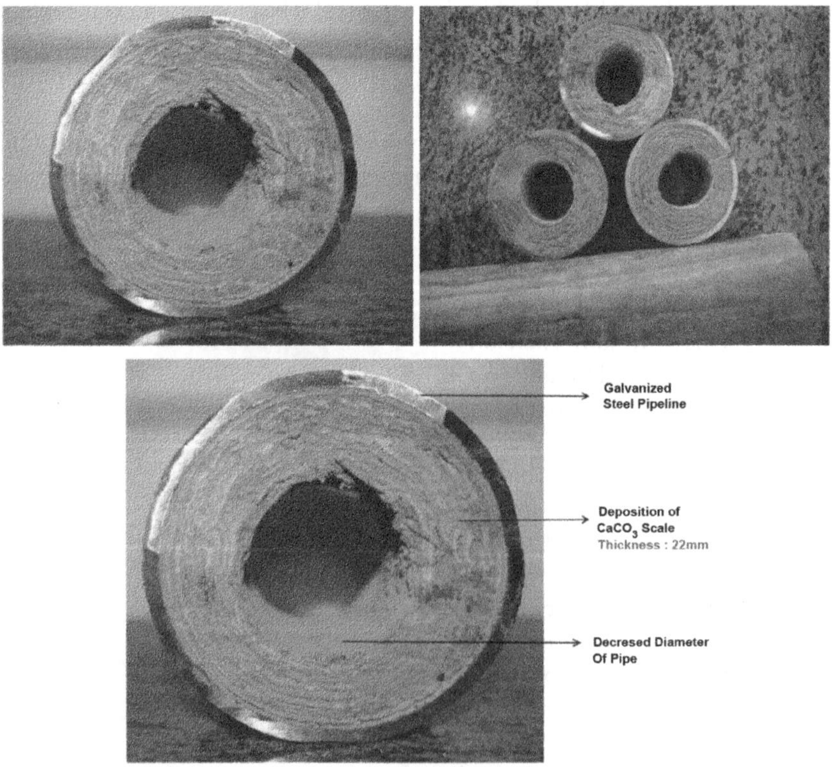

Figure 1.3 Images of a scaled pipeline where required grafting biopolymer as corrosion mitigators during the de-scaling process.

1.3 Corrosion and its Economic Adverse Effects

As per the reports by NACE (National Association of Corrosion Engineers), we are losing $2.5 trillion each year because of corrosion worldwide, which also equals 3.4% of global GDP. As per the country perspective, India lost $1670 billion, China lost $9330 billion, the European Union lost $16950 billion, Germany lost USD 3593 billion, Russia lost $2113 billion, South Korea lost $1198 billion, while Saudi Arabia lost $718 billion due to the corrosion of steel per year [20]. It is a huge economical issue as well. A study by NACE (2013) confirmed that the estimated cost of global corrosion was $ 2.5 trillion (3.4% of the world GDP). As mentioned, this is mostly worth the percentage of corrosion for economic services sectors for all countries. Corrosion costs were 20% for the US, 26% for India, 26% for Japan, 51% for Kuwait, and 20% for the United Kingdom in manufacturing. The global market for corrosion inhibitors was $ 6 billion in 2013, and $ 7.7 billion in 2020 and is estimated to reach about $ 10 billion in 2027 [21]. During the COVID-19 pandemic, consumer behavior has changed across all walks of life. On the other hand, industries will have to restructure their strategies to adapt to the demands of a changing market.

1.4 Conclusion

In the present chapter, the basic definition of corrosion, its economic adverse effect, and the current situations using synthetic and natural corrosion mitigators are reviewed. The chapter starts with a discussion on the review of the corrosion problem; the protection of metals or steels from corrosion inhibition methods has also been discussed. Corrosion protection of steel containing a brief review of synthetic organic substances and grafting biopolymers have been reviewed. This chapter also discuss the economic adverse effect of corrosion on different countries worldwide. Here I have discussed types of corrosion mitigators, out of that biopolymer-based corrosion mitigators are mostly non-toxic and good for our environment. Here I have also discussed good corrosion mitigator properties requirements of industries, which are fulfilled by grafting biopolymer-based corrosion mitigators. High molecular weight biopolymers macro molecules are formed by covalently bonded monomers. Biopolymers are non-toxic, easily available, and eco-friendly, which are alternatives to synthetic mitigators. Grafting biopolymer-based corrosion mitigators are biodegradable macromolecules. Prospect in this area is to go at a molecular level with help of computational studies like DFT and molecular dynamics simulations. Studies of highest occupied and lowest unoccupied molecular orbitals energies and their energies differences, which are useful for understanding molecular adsorption at metal or steel surfaces.

References

1 Verma, D.K. et al. (2021). N–hydroxybenzothioamide derivatives as green and efficient corrosion inhibitors for mild steel: experimental, DFT and MC simulation approach. *Journal of Molecular Structure* 1241: 130648. doi: https://doi.org/10.1016/j.molstruc.2021.130648.

2 Dehghani, A., Bahlakeh, G., Ramezanzadeh, B., and Ramezanzadeh, M. (2019). A combined experimental and theoretical study of green corrosion inhibition of mild steel in HCl solution by aqueous Citrullus lanatus fruit (CLF) extract. *Journal of Molecular Liquids* 279: 603–624. doi: https://doi.org/10.1016/j.molliq.2019.02.010.

3 Majd, M.T., Asaldoust, S., Bahlakeh, G., Ramezanzadeh, B., and Ramezanzadeh, M. (2019). Green method of carbon steel effective corrosion mitigation in 1 M HCl medium protected by Primula vulgaris flower aqueous extract via experimental, atomic-level MC/MD simulation and electronic-level DFT theoretical elucidation. *Journal of Molecular Liquids* 284: 658–674. doi: https://doi.org/10.1016/j.molliq.2019.04.037.

4 Sanaei, Z., Bahlakeh, G., Ramezanzadeh, B., and Ramezanzadeh, M. (2019). Application of green molecules from Chicory aqueous extract for steel corrosion mitigation against chloride ions attack; the experimental examinations and electronic/atomic level computational studies. *Journal of Molecular Liquids* 290: 111176. doi: https://doi.org/10.1016/j.molliq.2019.111176.

5 Tabatabaei Majd, M., Bahlakeh, G., Dehghani, A., Ramezanzadeh, B., and Ramezanzadeh, M. (2019). A green complex film based on the extract of Persian Echium amoenum and zinc nitrate for mild steel protection in saline solution; Electrochemical and surface explorations besides dynamic simulation. *Journal of Molecular Liquids* 291: 111281. doi: https://doi.org/10.1016/j.molliq.2019.111281.

6 Dehghani, A., Bahlakeh, G., and Ramezanzadeh, B. (2019). Green Eucalyptus leaf extract: a potent source of bio-active corrosion inhibitors for mild steel. *Bioelectrochemistry* 130: 107339. doi: https://doi.org/10.1016/j.bioelechem.2019.107339.

7 Dehghani, A., Bahlakeh, G., Ramezanzadeh, B., and Ramezanzadeh, M. (2020). Aloysia citrodora leaves extract corrosion retardation effect on mild-steel in acidic solution: molecular/atomic scales and electrochemical explorations. *Journal of Molecular Liquids* 310: 113221. doi: https://doi.org/10.1016/j.molliq.2020.113221.

8 Ramezanzadeh, M., Bahlakeh, G., and Ramezanzadeh, B. (2020). Green synthesis of reduced graphene oxide nanosheets decorated with zinc-centered metal-organic film for epoxy-ester composite coating reinforcement: DFT-D modeling and experimental explorations. *Journal of the Taiwan Institute of Chemical Engineers* 114: 311–330. doi: https://doi.org/10.1016/j.jtice.2020.09.003.

9 Majd, M.T., Ramezanzadeh, M., Bahlakeh, G., and Ramezanzadeh, B. (2020). Steel corrosion lowering in front of the saline solution by a nitrogen-rich source

of green inhibitors: detailed surface, electrochemical and computational studies. *Construction and Building Materials* 254: 119266. doi: https://doi.org/10.1016/j.conbuildmat.2020.119266.

10 Asfia, M.P., Rezaei, M., and Bahlakeh, G. (2020). Corrosion prevention of AISI 304 stainless steel in hydrochloric acid medium using garlic extract as a green corrosion inhibitor: electrochemical and theoretical studies. *Journal of Molecular Liquids* 315: 113679. doi: https://doi.org/10.1016/j.molliq.2020.113679.

11 Keramatinia, M., Ramezanzadeh, M., Bahlakeh, G., and Ramezanzadeh, B. (2021). Synthesis of a multi-functional zinc-centered nitrogen-rich graphene-like thin film from natural sources on the steel surface for achieving superior anti-corrosion properties. *Corrosion Science* 178: 109077. doi: https://doi.org/10.1016/j.corsci.2020.109077.

12 Mofidabadi, A.H.J., Bahlakeh, G., and Ramezanzadeh, B. (2021). Anti-corrosion performance of the mild steel substrate treated by a novel nanostructure europium oxide-based conversion coating: electrochemical and surface studies. *Colloids and Surfaces A: Physicochemical and Engineering Aspects* 609: 125689. doi: https://doi.org/10.1016/j.colsurfa.2020.125689.

13 Lashgari, S.M., Bahlakeh, G., and Ramezanzadeh, B. (2021). Detailed theoretical DFT computation/molecular simulation and electrochemical explorations of Thymus vulgaris leave extract for effective mild-steel corrosion retardation in HCl solution. *Journal of Molecular Liquids* 335: 115897. doi: https://doi.org/10.1016/j.molliq.2021.115897.

14 Shahini, M.H., Ramezanzadeh, M., Bahlakeh, G., and Ramezanzadeh, B. (2021). Superior inhibition action of the Mish Gush (MG) leaves extract toward mild steel corrosion in HCl solution: theoretical and electrochemical studies. *Journal of Molecular Liquids* 332: 115876. doi: https://doi.org/10.1016/j.molliq.2021.115876.

15 Salmasifar, A., Edraki, M., Alibakhshi, E., Ramezanzadeh, B., and Bahlakeh, G. (2021). Theoretical design coupled with experimental study of the effectiveness of the inhibitive molecules based on Cynara scolymus L extract toward chloride-induced corrosion of steel. *Journal of Molecular Liquids* 332: 115742. doi: https://doi.org/10.1016/j.molliq.2021.115742.

16 Mostafatabar, A.H., Bahlakeh, G., Ramezanzadeh, B., Dehghani, A., and Ramezanzadeh, M. (2021). A comprehensive electronic-scale DFT modeling, atomic-level MC/MD simulation, and electrochemical/surface exploration of active nature-inspired phytochemicals based on Heracleum persicum seeds phytoextract for effective retardation of the acidic-induced c. *Journal of Molecular Liquids* 331: 115764. doi: https://doi.org/10.1016/j.molliq.2021.115764.

17 Naghi Tehrani, M.E.H., Ghahremani, P., Ramezanzadeh, M., Bahlakeh, G., and Ramezanzadeh, B. (2021). Theoretical and experimental assessment of a green corrosion inhibitor extracted from Malva sylvestris. *Journal of Environmental Chemical Engineering* 9 (3): 105256. doi: https://doi.org/10.1016/j.jece.2021.105256.

18 Salmasifar, A., Edraki, M., Alibakhshi, E., Ramezanzadeh, B., and Bahlakeh, G. (2021). Combined electrochemical/surface investigations and computer modeling of the aquatic Artichoke extract molecules corrosion inhibition properties on the mild steel surface immersed in the acidic medium. *Journal of Molecular Liquids* 327: 114856. doi: https://doi.org/10.1016/j.molliq.2020.114856.

19 Lashgari, S.M., Yari, H., Mahdavian, M. et al. (2021). Synthesis of graphene oxide nanosheets decorated by nanoporous zeolite-imidazole (ZIF-67) based metal-organic framework with controlled-release corrosion inhibitor performance: experimental and detailed DFT-D theoretical explorations. *Journal of Hazardous Materials* 404: 124068. doi: https://doi.org/10.1016/j.jhazmat.2020.124068.

20 Haldhar, R., Prasad, D., Saxena, A., and Singh, P. (2018). Valeriana wallichii root extract as a green & sustainable corrosion inhibitor for mild steel in acidic environments: an experimental and theoretical study. *Materials Chemistry Frontiers* 2 (6): 1225–1237. doi: 10.1039/c8qm00120k.

21 Bowman, E. et al. (2016). *International Measures of Prevention, Application, and Economics of Corrosion Technologies Study*, 1e. Houston, Texas, USA: NACE International.

2

Corrosion Inhibition

Past and Present Developments and Future Directions

Lakha V. Chopda[1] *and Pragnesh N. Dave*[2,*]

[1] *Government Engineering College, Bhuj - 370 001 (Gujarat)*
[2] *Department of Chemistry, Sardar Patel University, Vallabh Vidynagar - 388 120 (Gujarat)*
* *Corresponding author*

2.1 Introduction

The corrosion of metal is a natural process as reactive metals tend to go in their most stable form [1]. The form stable form depends on the nature of the corrosive media around the metal. The metals find a huge application in the sphere of human development. The continuous corrosion of metal leads to numerous losses of material. The corrosion of metals is a significant problem worldwide and contributes to a huge economic loss [2] and disturbs the development of any country of the world, hence metal corrosion is considered a serious threat to the economy and society (in the term of deformation metal property in the structure application). The metals at the industry level are frequently in contact with corrosive media, which immediately induces the corrosion of metal. Robust preventive measures are an urgent need to protect precious metals against corrosion [3]. The corrosion inhibitor and coating of metal are the most economic and efficient approaches to enhance the protection efficiency of metal against corrosive media [4, 5]. Various classes of materials such as corrosion inhibitors and coating is reported for the prevention of corrosion [6, 7]. Biopolymers are natural green environment-benign materials that have negligible adverse effects on the environment and are biocompatible [8]. The unique functional groups particularly hydroxyl, carboxylic acid, and amine on the backbone of the biopolymer even make them efficient materials for the prevention of corrosion [9] and help to improve the property of biopolymer by grafting with other polymers [10–12]. The grafting of biomaterials enhanced various properties of biomaterial and makes

Grafted Biopolymers as Corrosion Inhibitors: Safety, Sustainability, and Efficiency, First Edition.
Edited by Jeenat Aslam, Chandrabhan Verma, and Ruby Aslam.

them usable materials in the protection of metals against corrosion. The modified biomaterials by the grating approach have been used to slow metal corrosion in an efficient way. This chapter highlights recent trends of grafting biopolymers and their future perspective for the control of metal corrosion.

2.2 Grafting of Biopolymer

Biopolymers present an important class of polymer as they are cheap, nontoxic, and possess eco-friendly characteristics. The biopolymers are produced from biological sources (cells of living organisms). Renewability, biodegradability, biocompatibility, and inexpensiveness are the main feature of biopolymers that enable biopolymer application in diverse field [13]. Polynucleotides, polypeptides, and polysaccharides are the main category of biopolymers. Polysaccharides are the most abundant biopolymers found in nature that have numerous applications in various fields like drug delivery, food coating, food packaging material, cosmetics, etc. [14, 15]. Grafting is the methodology to alter the property of a biopolymer [16]. The presence of hydroxyl, amine, and carboxylic acid functional groups on the biopolymer is the main active (reactive) site for the grafting polymers and other organic molecules using various methodologies. The polymers are widely grafted over the biopolymer using free radical, controlled radical polymerization (ATRP and RAFT), ROP, and ROMP [10, 11, 17]. The small organic molecule is grafted to the biopolymer by applying simple organic reactions like esterification, amide, imine, and coupling. The grafting of conventional polymers such as PAA, PMA PMMA, PNIPAM, PAN, etc. over the biopolymer gained a significant position in polymer chemistry as they exhibit the properties of synthetic polymer and natural polymer. Polymerization of these polymers grafted over biopolymer is generally carried out by *in situ* polymerization technique using a free radical initiator [10]. In this methodology, free radical initiates polymerization by the formation of free radical species over the biopolymer and monomers used for grafting. The grafting of cyclic monomers such as cyclic ethers, lactones, lactams, carbonates, aziridines, and epoxides over the biopolymer takes place by ROP strategy using Bronsted acid or Lewis acid [18, 19]. This methodology facilitates ring opening of cyclic monomers by the Bronsted acid or Lewis acid and furnished polymerization by nucleophilic addition done by the biopolymers or other nucleophilic agents.

2.3 Grafted Biopolymers for the Corrosion Protection

Biopolymers are cheap, non-toxic, and eco-friendly, hence demand for the application of them for the prevention of corrosion has been dramatically increased. The presence of excellent functionality bearing of heteroatoms over biopolymer makes them potential candidates for the protection of metal against corrosion. The different types of biopolymers have been reported as corrosion inhibitors. Some examples of biopolymers reported as corrosion inhibitors are presented here. Sodium alginate, a promising anion polysaccharide reported for the corrosion protection of API X60 (high-strength carbon steel) in saline media (3.5% NaCl). The corrosion-inhibitive performance of sodium alginate was assessed by the gravimetric and electrochemical methods (OCP, EIS, and EFM). The sodium alginate suppressed the pitting corrosion of API X60 through physisorption over metal the surface [20]. The guar gum is water soluble, non-ionic, non-toxic, biodegradable, and biocompatible biopolymer that has displayed corrosion inhibitive properties towards the carbon steel (CS) in 2M H_3PO_4. The various concentration of guar gum (0.1 to 1.0g/L) at 298–328K has been tested to assess the corrosion effect of guar gum over CS. The electrochemical method showed that a 1.0g/L concentration of guar gum displayed the highest corrosion inhibition efficiency (more than 95%) [21]. The natural polymer is known as iota-carrageenan reported as a corrosion inhibitor for aluminium in 2M HCl. The corrosion inhibition efficiency of ι-carrageenan has been enhanced in the presence of a zwitterionic mediator (pefloxacin mesylate) [22]. The biopolymer pectin exhibited anticorrosion action on 6061 aluminium alloy in an HCl solution. The pectin showed around 80% corrosion inhibition efficiency at 800ppm. concentration [23]. The chitosan and carboxy methyl cellulose (CMC) were reported for the inhibition of corrosion of API 5L X60 steel in a CO_2-saturated 3.5% NaCl solution [24]. The electrochemical techniques EIS and PDP revealed that both inhibitors showed moderate inhibitive performance over API 5L X60 steel in CO_2-saturated 3.5% NaCl solution. The *corynebacterium indologenes* MUT.2 bacterial biopolymer was reported as a corrosion inhibitor for CS in the acidic solution [25]. The biopolymer displayed 58% corrosion inhibition efficiency at 0.5g/L. The mixed types of corrosion inhibition effects have been shown by this biopolymer. The biopolymer (tragacanth gum) was found to effective corrosion inhibitor for carbon steel in 1 M HCl [26]. The corrosion inhibition activity of the inhibitor enhanced as its temperature increased. The dextrin-graft-polyvinyl acetate (Dxt-g-pVAc) was reported as a corrosion inhibitor for mild steel in 15% HCl [27]. The grafting of pVAC over dextrin is accomplished by the ATRP method. The synthesized Dxt-g-pVAc is characterized by FT-IR and FESEM. The investigated EIS and PDP methods showed that the concentration of inhibitor increased (0.025 to 0.15g/L) led to enhance corrosion inhibition efficiency. The inhibitor Dxt-g-pVAc showed more than 90% corrosion inhibition efficiency. The Dxt-g-pVAc exhibited high corrosion inhibition efficiency of 98.39% than DXT (84.56%) at 0.15g/L concentration of inhibitor. The further corrosion inhibition effect of Dxt-g-pVAc was confirmed by thermodynamic and kinetic parameters, FESEM, EDX, and AFM. The value of activation energy (E_a), the entropy of activation (ΔS_{act}), and enthalpy of activation (ΔH_{act}) in the presence of Dxt and Dxt-g-pVAc is higher than without both (blank) which indicates the inhibitive effect of Dxt and Dxt-g-pVAc. The value of the three parameters of Dxt-g-pVAc is more than Dxt. It shows that Dxt-g-pVAc reflected high corrosion inhibition than Dxt.

The surface analysis is the essential additional tool to the verify corrosion inhibition action of inhibitors. The FESEM image in the presence of inhibitor revealed that a smooth surface is obtained in the presence of Dxt and Dxt-g-pVAc. The smoother surface is visualized by Dxt-g-pVAc than Dxt, which also reflects more corrosion inhibition efficiency shown by Dxt-g-pVAc. The EDX analysis indicated that in the presence of Dxt-g-pVAc, the percentage of Fe decreased compared to the presence of Dxt. It confirms that Dxt-g-pVAc is more adsorbed on the surface of the metal. The average roughness (R_a) value of blank and in the presence of Dxt and Dxt-g-pVAc is 891,428, and 51.3 nm respectively. The decrease in the R_a value shows that suppression of corrosion is enhanced. The R_a value of Dxt-g-pVAc is lowered than Dxt and the blank shows the corrosion protection of MS reflected by Dxt-g-pVAc. The mixed-type adsorption (physical and chemical) is shown by Dxt-g-pVAc on the MS surface as the value of ΔG^0_{ads} is −27.98 kJ/mol. The conductive xanthan gum-graft-polyaniline (XG-g-PANI) was prepared and tested for its corrosion inhibition effect for aluminium in 1 M HCl acid by weight loss (WL) and potentiodynamic polarization (PDP) method [28]. The various amount of XG (0.1, 0.2, 0.3, and 0.4 g) was grafted to PANI using 0.4 M of aniline, 1 M HCl, and 0.066 M APS. The resultant material is characterized by FT-IR, UV-Vis, and TGA. The DC electrical conductivity of all materials was evaluated and XG-g-PANI (XG = 0.2 g) showed high DC conductivity (1.58×10^{-1} S/cm) and was selected for corrosion study. The WL and PDP showed that 91.33% and 94.24% corrosion inhibition efficiency is shown by 0.1 g concentration of XG-g-PANI (XG = 0.2 g) which is higher than XG (62.66% and 79.29%) at the same concentration. The corrosion inhibition property of XG-g-PANI (XG = 0.2 g) is also further assessed by SEM and AFM. The R_a value of XG-g-PANI (XG = 0.2 g) is lower than XG and the absence of an inhibitor. The glycerin-grafted starch study as a corrosion inhibitor for C-Mn steel in 1 M HCl [29]. The glycerin was grafted to starch obtained from the maize. The glycerin was grafted over maize starch using only HCl. The grafting of glycerin over starch is confirmed by FT-IR and NMR. The effect of the concentration of gly-g-starch corrosion inhibitor was studied at 25°C by the WL method. The concentration of gly-g-starch increased from 5 mg/L to 300 mg/L contributing to enhancing the corrosion inhibition efficiency. At 200 and 300 mg/L concentrations of inhibitor, 83% and 94% corrosion inhibition efficiency were achieved. The 300 mg/L concentration of glycerin-g-starch showed the highest corrosion inhibition efficiency. The effect of temperature on corrosion inhibition efficiency is evaluated. As the temperature enhanced from 25°C to 50°C did not affect the corrosion inhibition efficiency. The electrochemical method EIS and PDP displayed 91% and 90% corrosion inhibition efficiency at 300 mg/L. The PDP result demonstrated that gly-g-starch showed a mixed type of inhibitor (anodic and cathodic). The value of activation energy (E_a), the entropy of activation (ΔS_{act}), and enthalpy of activation (ΔH_{act}) are lower than blank confirming the corrosion protection property of gly-g-starch. The value of three parameters at 300 mg/L is even low than other concentrations (5, 10, 50, 100, and

200 mg/L). It shows the high corrosion inhibition efficiency of 300 mg/L of gly-g-starch. The ΔG^0_{ads} value of gly-g-starch is between −6.15 to −7.65 kJ/mol suggesting that gly-g-starch is adsorbed over C-Mn steel by physisorption. The grafting of oleic acid over chitosan/graphene oxide (CS/GO) composite and coated over carbon steel [30]. The first graphene oxide (prepared by oxidizing method) and 3 wt% of GO supported over chitosan. The grafting of oleic acid over to chitosan/GO is done by applying a coupling reaction using DCC. The gained material is designated as CS/GO-OA coated over carbon steel (7 x 5 × 0.3 cm). The corrosion resistance of CS/GO-OA was investigated by EIS and PDP method in 3.5 wt % of NaCl for 6 hours. The EIS method demonstrated the high stability of CS/GO-OA over carbon steel. The value of impedance at low frequency is increased in the presence of CS/GO compared to bare metal and the value of impedance at low frequency is enhanced to more than 10^2 folds in the presence of CS/GO-OA. It reveals that OA improves the coating property and enhances the interaction of CS/GO to CS. The PDP method shows that CS/GO-OA significantly decreased the corrosion rate in comparison to bare metal, chitosan, CS/GO, CS-OA, and CS/GO-OA. The value of corrosion rate (mpy) of bare metal is 37.13 and of CS, CS/GO, CS-OA, and CS/GO-OA are 6.61, 3.40, 2.56, and 1.72 mpy, respectively. It shows the pronounced corrosion resistance is shown by CS/GO-OA. The falling in value of oxygen transmission rates values after supporting GO to CS shows that GO effectively prevented the entry of oxygen over the film of CS/GO which ultimately contributed to the high resistance towards the corrosion. The increased water contact angle of CS/GO-OA is an indication of the high barrier property of CS/GO-OA that repelled water over to the metal surface. The water contact angle value of CS/OA is higher than CS/GO reflecting that contribution of OA makes the metal surface more hydrophobic than GO. The corrosion effect of guar gum grafted 2-acrylamido-2-methylpropanesulfonic acid for Cu was metal studied in 3.5 wt% of NaCl [31]. The grafting of 2-acrylamido-2-methylpropanesulfonic acid to guar gum is accomplished by coupling reaction using N, N′-methylenebisacrylamide. The WL method showed that GG-g-AMPS showed 90.3% corrosion inhibition efficiency at 600 mg/L which is higher than 100 and 300 mg/L. The PDP method showed GG-g-AMPS had 95% corrosion inhibition efficiency at 300 mg/L. The SEM and AFM also confirm the corrosion inhibition property of GG-g-AMPS. The stearic acid grafted chitosan was coated over MS [32]. The material was prepared by the coupling reaction using 1-ethyl-3-(3-dimethylaminopropyl) carbodiimide between chitosan and stearic acid. The material is coated over MS by dip coating. The contact angle of stearic acid grafted chitosan was 118° which indicates of hydrophobic nature of the material. The stearic acid grafted chitosan displayed good corrosion protection efficiency towards MS in 0.5 and 1 N HCl than alone chitosan coating. The corrosion inhibitor poly (N-vinyl imidazole) grafted carboxymethyl chitosan was developed [33]. The preparation method involved the first chitosan modified by chloroacetic acid known as carboxymethyl chitosan (CMCh). The 1-Vinylimidazole was grafted over CMCh by the

free radical polymerization method using a potassium persulfate initiator. The corrosion behavior of poly (N-vinyl imidazole) grafted carboxymethyl chitosan (CMCh-g-PVI) was studied for X70 steel in 1 M HCl by EIS and PDP method. The result of both studies showed that the concentration of inhibitor increased from 200 ppm. to 100 ppm. contributing to increasing the corrosion inhibition efficiency as corrosion current density decreased (determined by PDP method) and R_{ct} value increased (determined by EIS method). In the absence of an inhibitor, the value of j_{corr} (corrosion current density) is 499 μA/cm^2 and in the presence of 1000 ppm of inhibitor concentration value of j_{corr} decreased to 47.5 μA/cm^2. The value of R_{ct} in the absence of inhibitor is 89.9 'Ω cm^2 which increased to 1186 'Ω cm^2 in the presence of 1000 ppm concentration of inhibitor. The hydroxyquinoline-*grafted*-Alginate (HQ-g-Alg) was reported as a corrosion inhibitor for MS in 1 M HCl [34]. The inhibitor HQ-g-Alg was prepared by a reaction between sodium alginate and 5-chloromethyl-8-hydroxyquinoline and characterized by NMR (^{1}H &^{13}C) and FT-IR respectively. The corrosion inhibition efficiency towards MS in 1 M HCl increased with enhancing the concentration of inhibitor. The corrosion inhibition efficiency increased from 82.9% to 92/6% as the concentration of HQ-g-Alg increased from 1×10^{-3} to 10×10^{-3} g/L as determined by the PDP method. The EIS method shows the same trend of corrosion inhibition efficiency. At the same concentration of 10×10^{-3} g/L, the EIS method displayed 94.2% corrosion inhibition efficiency. The alginate (Alg) itself showed 87.7% and 89.4% at 10×10^{-3} g/L concentration as evaluated by PDP and EIS methods respectively. The $\Delta G^0{}_{ads}$ is −36.92 kJ/mol and is an indication of a mixed type of adsorption (chemical and physical adsorption). The chitosan-grafted d-glucose showed good corrosion protection of MS in 1 M HCl [35]. The material was prepared between chitosan oligosaccharide (COS) and 1,5-anhydrous-1,2-*O*-isopropylidene-D-glucofuranose in ethanol at 60°C. The corrosion inhibitive action of COS-g-Glu was determined by the EIS and PDP method used to assess the corrosion property of the inhibitor. The three methods revealed that enhancing the concentration of inhibitor led to an increase in the corrosion inhibition efficiency. The COS-g-Glu showed around 97% corrosion inhibition efficiency at a 10^{-3} M concentration of inhibitor. The inhibitor COS-g-Glu displayed the highest corrosion inhibition efficiency than COS (93%), and, 5-anhydro-1,2-*O*-isopropylidene-D-glucofuranose (93%) at 10^{-3} M concentration. The effect of temperature on corrosion inhibition efficiency was studied using the gravimetric (weight loss) method. As the temperature increased from 298 K to 328 K, it contributed to a slight decrease in the corrosion inhibition efficiency. At 298 K, 97% corrosion inhibition efficiency was gained while 90% corrosion inhibition was achieved at 328 K at 10^{-3} M concentration. It shows that higher temperature did not have so much effect on the corrosion inhibition efficiency. The efficiency of COS-g-Glu remained intact at a higher temperature. The value of other parameters (E_a, ΔH_a, and ΔS_a) of COS-g-Glu evaluated at 298 K by the WL method is higher than blank which reflects the high corrosion inhibitive effect of COS-g-Glu. The β-cyclodextrin modified xanthan gum showed 94.74% corrosion inhibition efficiency for L80 steel in 1 M HCl [36]. The PDP results indicated that the inhibitor behaves as a mixed type of inhibitor with a more anodic effect (suppressed metal dissolution). The adsorption mechanism divulged that inhibitor adsorption through the chemisorption mechanism. The unique functions such as hydroxyl, carboxylic acid, and other chemicals interact with the metal surface and contributed to inhibiting the corrosion of metal in 1 M HCl. The same corrosion inhibitor (β-cyclodextrin modified xanthan gum) was used to suppress the corrosion of X80 steel in 1 M H_2SO_4 [37]. The 94.85% corrosion inhibition efficiency was achieved by the β-cyclodextrin modified xanthan gum inhibitor. The inhibitor adsorbed over the metal surface by chemisorption mechanism as assessed by the adsorption experiment.

2.4 Conclusion and Future Prospective

Metal corrosion is a serious issue. The metals immediately undergo corrosion in the presence of a favorable atmosphere (corrosive media) for the attainment of stability. The prevention of metal corrosion is so significant to curb metal corrosion. Metals need robust corrosion protection methods. The coating of metal is a practical available method for the prevention of corrosion. The corrosion inhibitor is an important method that effectively inhibits metal corrosion in the presence of corrosive media. The inorganic, organic, and polymers as corrosion inhibitors are reported for the mitigation of metal corrosion. There are many examples of biopolymers reported for the prevention of metal corrosion in the role of inhibitor. The little work on grafted biopolymer as corrosion inhibitor and coating was reported for curbing metal corrosion. In the literature, the grafting of biopolymers (guar gum, xanthan gum, gum ghatti, chitosan, β-cyclodextrin, cellulose, starch, etc.) by polymer and other organic molecules are reported in multiple for various applications. The application of these materials in the field of corrosion is scant. Explore these materials in the field of corrosion in the role of inhibitor and coating will be the novel direction to nullified metal corrosion.

References

1 Guedes, D., Martins, G.R., Jaramillo, L.Y.A. et al. (2021). Proanthocyanidins with corrosion inhibition activity for AISI 1020 carbon steel under neutral pH conditions of coconut (*Cocos nucifera* L.) husk fibers. *ACS Omega* 6: 6893–6901. doi: https://doi.org/10.1021/acsomega.0c06104.

2 Verma, C., Olasunkanmi, L.O., Ebenso, E.E., and Quraishi, M.A. (2018). Substituents effect on corrosion inhibition performance of organic compounds in aggressive ionic solutions: a review. *Journal of Molecular Liquids* 251: 100–118. doi: https://doi.org/10.1016/j.molliq.2017.12.055.

3 Ates, M. (2016). A review on conducting polymer coatings for corrosion protection. *Journal of Adhesion Science and Technology* 30: 1510–1536. doi: https://doi.org/10.1080/01694243.2016.1150662.

4 Tamalmani, K. and Husin, H. (2020). Review on corrosion inhibitors for oil and gas corrosion issues. *Applied Sciences* 10: 3389. doi: https://doi.org/10.3390/app10103389.

5 Raja, P.B., Ismail, M., Ghoreishiamiri, S. et al. (2016). Reviews on corrosion inhibitors: a short view. *Chemical Engineering Communications* 203: 1145–1156. doi: https://doi.org/10.1080/00986445.2016.1172485.

6 Rani, B.E.A. and Basu, B.B.J. (2012). Green inhibitors for corrosion protection of metals and alloys: an overview. *International Journal of Corrosion* 2012: 1–15. doi: https://doi.org/10.1155/2012/380217.

7 Jain, P., Patidar, B., and Bhawsar, J. (2020). Potential of nanoparticles as a corrosion inhibitor: a review. *Journal of Bio- and Tribo-Corrosion* 6: 43. doi: https://doi.org/10.1007/s40735-020-00335-0.

8 Aaliya, B., Sunooj, K.V., and Lackner, M. (2021). Biopolymer composites: a review. *International Journal of Biobased Plastics* 3: 40–84. doi: https://doi.org/10.1080/24759651.2021.1881214.

9 Umoren, S.A. and Eduok, U.M. (2016). Application of carbohydrate polymers as corrosion inhibitors for metal substrates in different media: a review. *Carbohydrate Polymers* 140: 314–341. doi: https://doi.org/10.1016/j.carbpol.2015.12.038.

10 Roy, D., Semsarilar, M., Guthrie, J.T., and Perrier, S. (2009). Cellulose modification by polymer grafting: a review. *Chemical Society Reviews* 38: 2046. doi: https://doi.org/10.1039/b808639g.

11 Thakur, V.K., Thakur, M.K., and Gupta, R.K. (2013). Development of functionalized cellulosic biopolymers by graft copolymerization. *International Journal of Biological Macromolecules* 62: 44–51. doi: https://doi.org/10.1016/j.ijbiomac.2013.08.026.

12 Kumar, R., Sharma, R.K., and Singh, A.P. (2018). Grafted cellulose: a bio-based polymer for durable applications. *Polymer Bulletin* 75: 2213–2242. doi: https://doi.org/10.1007/s00289-017-2136-6.

13 Biswas, M.C., Jony, B., Nandy, P.K. et al. (2022). Recent advancement of biopolymers and their potential biomedical applications. *Journal of Polymer Environmental* 30: 51–74. doi: https://doi.org/10.1007/s10924-021-02199-y.

14 Zhao, Y., Li, B., Li, C. et al. (2021). Comprehensive review of polysaccharide-based materials in edible packaging: a sustainable approach. *Foods* 10: 1845. doi: https://doi.org/10.3390/foods10081845.

15 Mohammed, A.S.A., Naveed, M., and Jost, N. (2021). Polysaccharides; classification, chemical properties, and future perspective applications in fields of pharmacology and biological medicine (A review of current applications and upcoming potentialities). *Journal of Polymer Environmental* 29: 2359–2371. doi: https://doi.org/10.1007/s10924-021-02052-2.

16 Wohlhauser, S., Delepierre, G., Labet, M. et al. (2018). Grafting polymers *from* cellulose nanocrystals: synthesis, properties, and applications. *Macromolecules* 51: 6157–6189. doi: https://doi.org/10.1021/acs.macromol.8b00733.

17 Messina, M.S., Messina, K.M.M., Bhattacharya, A. et al. (2020). Preparation of biomolecule-polymer conjugates by grafting from using ATRP, RAFT, or ROMP. *Progress in Polymer Science* 100: 101186. https://doi.org/10.1016/j.progpolymsci.2019.101186.

18 Becker, G. and Wurm, F.R. (2018). Functional biodegradable polymers *via* ring-opening polymerization of monomers without protective groups. *Chemical Society Reviews* 47: 7739–7782. doi: https://doi.org/10.1039/C8CS00531A.

19 Koppes, A.N., Keating, K.W., McGregor, A.L. et al. (2016). Robust neurite extension following exogenous electrical stimulation within single-walled carbon nanotube-composite hydrogels. *Acta Biomaterialia* 39: 34–43. doi: https://doi.org/10.1016/j.actbio.2016.05.014.

20 Obot, I.B., Onyeachu, I.B., and Kumar, A.M. (2017). Sodium alginate: a promising biopolymer for corrosion protection of API X60 high-strength carbon steel in saline medium. *Carbohydrate Polymers* 178: 200–208. doi: https://doi.org/10.1016/j.carbpol.2017.09.049.

21 Messali, M., Lgaz, H., Dassanayake, R. et al. (2017). Guar gum as an efficient non-toxic inhibitor of carbon steel corrosion in phosphoric acid medium: electrochemical, surface, DFT and MD simulations studies. *Journal of molecular structure* 1145: 43–54. doi: https://doi.org/10.1016/j.molstruc.2017.05.081.

22 Fares, M.M., Maayta, A.K., and Al-Mustafa, J.A. (2012). Corrosion inhibition of iota-carrageenan natural polymer on aluminum in presence of zwitterion mediator in HCl media. *Corrosion Science* 65: 223–230. doi: https://doi.org/10.1016/j.corsci.2012.08.018.

23 Fathima, H., Pais, M., and Rao, P. (2021). Anticorrosion performance of biopolymer pectin on 6061 aluminum alloy: electrochemical, spectral and theoretical approach. *Journal of Molecular Structure* 1243: 130775. doi: https://doi.org/10.1016/j.molstruc.2021.130775.

24 Umoren, S.A., AlAhmary, A.A., Gasem, Z.M., and Solomon, M.M. (2018). Evaluation of chitosan and carboxymethyl cellulose as eco-friendly corrosion inhibitors for steel. *International Journal of Biological Macromolecules* 117: 1017–1028. doi: https://doi.org/10.1016/j.ijbiomac.2018.06.014.

25 Alipour, A., Bahrami, A., and Saebnoori, E. (2018). Chryseobacterium indologenes MUT.2 bacterial biopolymer as a novel green inhibitor protecting carbon steel corrosion in acidic solution. *Journal of Environmental Chemical Engineering* 6: 4698–4705. https://doi.org/10.1016/j.jece.2018.07.017.

26 Mobin, M., Rizvi, M., Olasunkanmi, L.O., and Ebenso, E.E. (2017). Biopolymer from tragacanth gum as a green corrosion inhibitor for carbon steel in 1 M HCl solution. *ACS Omega* 2: 3997–4008. doi: https://doi.org/10.1021acsomega.7b00436.

27 Biswas, A., Das, D., Lgaz, H. Pal, S. and Nair, U.G. (2019). Biopolymer dextrin and poly (vinyl acetate) based graft copolymer as an efficient corrosion inhibitor for mild steel in hydrochloric acid: electrochemical, surface morphological and theoretical studies. *Journal of Molecular Liquids* 275: 867–878. doi: https://doi.org/10.1016/j.molliq.2018.11.095.

28 Babaladimath, G., Vishalakshi, B., and Nandibewoor, S.T. (2018). Electrical conducting xanthan gum-graft-polyaniline as corrosion inhibitor for aluminum in a hydrochloric acid environment. *Materials Chemistry and Physics* 205: 171–179. doi: https://doi.org/10.1016/j.matchemphys.2017.11.008.

29 Lahrour, S., Benmoussat, A., Bouras, B. et al. (2019). Glycerin-grafted starch as corrosion inhibitor of C-Mn steel in 1 M HCl solution. *Applied Sciences* 9: 4684. doi: https://doi.org/10.3390/app9214684.

30 Fayyad, E.M., Sadasivuni, K.K., Ponnamma, D., and Al-Maadeed, M.A.A. (2016). Oleic acid-grafted chitosan/graphene oxide composite coating for corrosion protection of carbon steel. *Carbohydrate Polymers* 151: 871–878. doi: https://doi.org/10.1016/j.carbpol.2016.06.001.

31 Singh, A., Liu, M., Ituen, E., and Lin, Y. (2020). Anti-corrosive properties of an effective guar gum grafted 2-acrylamido-2-methylpropanesulfonic acid (GG-AMPS) coating on copper in a 3.5% NaCl solution. *Coatings* 10: 241. doi: https://doi.org/10.3390/coatings10030241.

32 Ko, S., Prasad, A.R., Pk, J., and Joseph, A. (2020). Development of self-assembled monolayer of stearic acid grafted chitosan on mild steel and inhibition of corrosion in hydrochloric acid. *Chemical Data Collections* 28: 100402. doi: https://doi.org/10.1016/j.cdc.2020.100402.

33 Eduok, U., Ohaeri, E., and Szpunar, J. (2018). Electrochemical and surface analyses of X70 steel corrosion in simulated acid pickling medium: effect of poly (N-vinyl imidazole) grafted carboxymethyl chitosan additive. *Electrochimica Acta* 278: 302–312. doi: https://doi.org/10.1016/j.electacta.2018.05.060.

34 Fardioui, M., Rbaa, M., Benhiba, F. et al. (2021). Bio-active corrosion inhibitor based on 8-hydroxyquinoline-grafted-alginate: experimental and computational approaches. *Journal of Molecular Liquids* 323: 114615. doi: https://doi.org/10.1016/j.molliq.2020.114615.

35 Rbaa, M., Benhiba, F., Hssisou, R. et al. (2021). Green synthesis of novel carbohydrate polymer chitosan oligosaccharide grafted on d-glucose derivative as a bio-based corrosion inhibitor. *Journal of Molecular Liquids* 322: 114549. doi: https://doi.org/10.1016/j.molliq.2020.114549.

36 Cao, Y., Zou, C., Wang, C. et al. (2021). β-cyclodextrin modified xanthan gum as an eco-friendly corrosion inhibitor for L80 steel in 1 M HCl. *Cellulose* 28: 11133–11152. https://doi.org/10.1007/s10570-021-04240-8.

37 Cao, Y., Zou, C., Wang, C. et al. (2021). Green corrosion inhibitor of β-cyclodextrin modified xanthan gum for X80 steel in 1 M H_2SO_4 at different temperatures. *Journal of Molecular Liquids* 341: 117391. doi: https://doi.org/10.1016/j.molliq.2021.117391.

3

Biopolymers as Corrosion Inhibitors

Relative Inhibition Potential of Biopolymers and Grafted Biopolymers

Rafaela C. Nascimento[1,*], *Luana Barros Furtado*[2], *and Maria José O. C. Guimarães*[2]

[1] *LAQV-REQUIMTE, Instituto de Investigação e Formação Avançada, Universidade de Évora, 7000–671, Évora, Colégio Luís António Verney, Portugal*
[2] *Federal University of Rio de Janeiro, School of Chemistry, 21941–909, Rio de Janeiro, Athos da Silveira Ramos Avenue, Brazil*
* *Corresponding author*

3.1 Introduction

Corrosion, a serious problem in several industrial sectors, generates an annual worldwide cost of trillions of dollars, according to the NACE (National Association of Corrosion Engineers), especially in industrial processes in which the metal or metal alloy is exposed to different acid environments, which are quite corrosive [1–3].

One of the protective methods against corrosion is the use of inhibitors, a common and efficient practice, especially in acid media such as HCl and H_2SO_4 [4–7]. A number of industrial sectors use inhibitors as a protective measure against corrosion, including the oil and gas industry, which uses inhibitors throughout the supply chain, from upstream to downstream [8].

In recent years, heightened ecological awareness has prompted the search for sustainability and the implementation of increasingly rigid environmental regulations, as well as a growing interest in replacing traditional inhibitors, formed primarily by toxic substances that negatively impact human health and the environment, with inhibitors obtained from more ecologically friendly and sustainable natural sources [1, 5, 9–11].

In this respect, the use of biopolymers as ecologically friendly inhibitors has grown in several industrial sectors. Among green inhibitors with significant potential, biopolymers obtained from carbohydrates and their products, structurally modified by

Grafted Biopolymers as Corrosion Inhibitors: Safety, Sustainability, and Efficiency, First Edition. Edited by Jeenat Aslam, Chandrabhan Verma, and Ruby Aslam.

different chemical species, stand out [12]. These biopolymers are readily available, biocompatible, non-toxic, and mainly low-cost, with good performance in acid environments.

3.2 Biopolymers as Corrosion Inhibitors

The significant corrosion inhibition potential of biopolymers is linked to the fact that they can be adsorbed by physical interactions or chemical bonds on metallic surfaces, protecting anodic and cathodic regions, and thereby acting as a mixed inhibitor in the interaction with the metal [5].

Corrosion inhibitor molecules should exhibit certain structural characteristics to enable adsorption onto the metallic surface. Heteroatoms such as nitrogen, oxygen, and sulfur with lone pairs of electrons; aromatic rings; and double and triple bonds are possible adsorption sites due to their high π-electron density. Other important factors that may affect inhibition efficiency are molecule size and steric effects [13, 14]. Fourier transform infrared spectroscopy (FTIR) is widely used to characterize biopolymers and grafted biopolymers. Carbohydrates used as corrosion inhibitors contain hydroxyl groups as well as carbonyl and carboxyl that are naturally present in the backbone. These are possible biopolymer adsorption sites. In addition, grafting or chemically modified biopolymers can enhance adsorption sites with nitrogen (especially amines) or sulfur groups that will prioritize chemical over physical adsorption.

3.3 Structural and Chemical Characteristics of Biopolymers

Carbohydrate-based biopolymers are macromolecules obtained by the polycondensation of one or more monosaccharides, which bind together via glycosidic bonds, forming long linear or branched chains [3]. In nature, these biopolymers are produced by plant and animal cells and can be found in the form of homopolysaccharides or heteropolysaccharides with medium or high molecular mass. The considerable corrosion inhibition potential exhibited by these materials is associated with their macromolecular architecture combined with chemical composition and electronic structure, which enables adequate interaction with different metallic surfaces [12].

Biopolymers such as those derived from cellulose, chitosan, starch, pectin, dextrin, and a wide range of gums, are widely used as corrosion inhibitors [1, 3, 15, 16]. The inhibition potential of these biopolymers is directly related to macrostructural factors associated with the different chemical compositions characteristic of each material, resulting in their high capacity to interact with the metallic

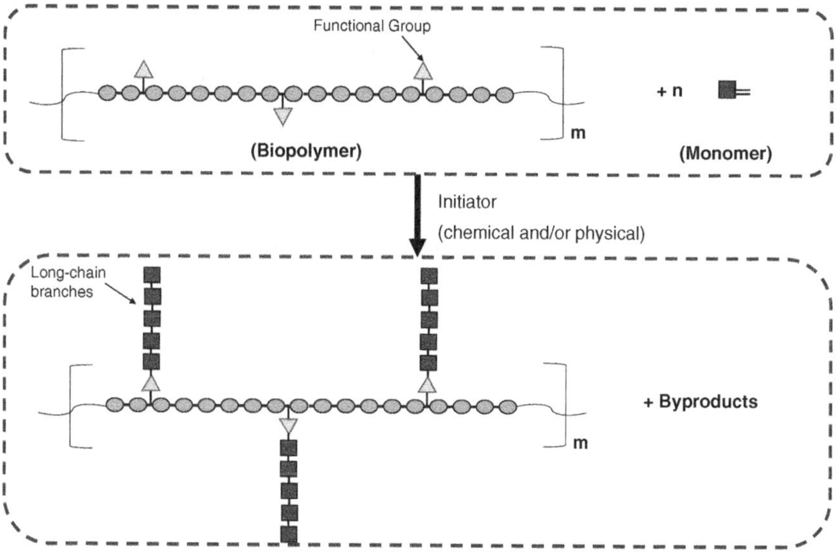

Figure 3.1 Above is a biopolymer graft polymerization reaction scheme.

surface. Among the macrostructural factors are different types of chains (linear or branched), the presence of cyclic rings, or a combination of these factors.

One strategy that has been used to further improve the inhibition efficiency of these biopolymers is the structural modification of the polymer using grafting reactions or insertion of different chemical species, as shown in Figure 3.1. New tailor-made biopolymers can then be produced by strategic selection of chemical compounds [16, 17] or monomers containing heteroatoms that enable good anchoring on the metallic surface [18].

3.4 Grafted Biopolymers

Grafted biopolymers allow the production of novel materials with new characteristics and improved properties. Graft copolymerization is generally carried out by solution technique using organic or aqueous solutions, or mixtures of these. In this process, long-chain branches are inserted into the main biopolymer chain.

The most widely used initiating agents are chemical initiators that produce free radicals or physical agents such as microwave, gamma radiation, and visible ultraviolet radiation. Hybrid initiators formed by the combination of a chemical and physical agent such as microwave radiation have also been used [19]. Among chemical agents, potassium persulfate and the ceric ammonium nitrate in the nitric acid redox system stand out [20]. More recently, biological initiators such as enzymes have been successfully used in the production of grafted biopolymers [21, 22]. The advantage of these initiators is the use of milder synthesis conditions and the integrity of the main chain since the grafting mechanism occurs only through interaction between the enzyme and the end groups of the polymer chain.

The radical mechanism of inserting the polymer chain (long chain branching) into the main biopolymer chain occurs either by abstraction of a hydrogen radical of a functional group, such as the hydroxyl group (OH), or by homolytic breakage of the glycosidic ring, creating start sites for grafting [23, 24]. The formation of long-chain branches from graft sites created in the main chain follows the same conventional polymerization kinetics, with initiation, propagation, and termination steps. In addition to the graft biopolymer, homopolymers may also form during grafting and should be separated from the copolymer grafted by solvent extraction [25, 26].

Grafted biopolymers can be characterized by several techniques since their properties differ from those of the main polymer. FTIR and nuclear magnetic resonance (NMR) are used to determine the functional groups and degree of grafting obtained in synthesis. Molecular mass and molecular mass distribution are calculated by size exclusion chromatography (SEC). The degree of crystallinity can be assessed by differential scanning calorimetry (DSC) and x-ray diffraction (XDR), and thermal stability by thermogravimetric analysis (TGA), respectively. Morphological properties such as miscibility or phase separations are observed by scanning electronic microscopy (SEM) [27–29]. Table 3.1, presents a summary of grafted biopolymers described in the literature as corrosion inhibitors.

3.5 Factors that Influence Grafting Efficiency/Percentage

Grafting efficiency is associated with several factors, such as the nature of the macromolecular chain (macromolecular backbone) and type of monomer to be grafted; initiator agent and its concentration, solubility, nature and operational conditions, including the type of solvent, time and temperature; and the presence of chemical additives such as acids, metal ions, and organic and inorganic salts.

Careful selection of the type of monomer with its specific functional groups is important in expanding and diversifying anchorage sites and increasing the inhibiting potential of these grafted biopolymers. Monomers containing nitrogen heteroatoms, such as acrylamide and its derivatives, caprolactam, anillin, and acrylonitrile are the most widely studied [38, 43, 44]. Acrylic and methacrylic acid, acrylates and methacrylates, vinyl acetate, and some derivatives of methac rylamide with sulfonic groups are other monomers used in graft copolymerization [40, 45, 46]. The grafting process involves the formation of a polymer side chain in the macromolecular biopolymer backbone (Figure 3.2). The grafting compound

Table 3.1 Table 3.1 represents the grafting process variables, post-treatment steps, and characterization techniques.

Backbone	Grafting compound	Temperature (K)	Reaction time	Initiator	Post-treatment	Characterization techniques	References
Dextrin	Poly (vinyl acetate)	338	8 h (in N_2)	Dextrin-bromo macroinitiator, via atom transfer radical polymerization (ATRP)	Precipitation in dichloromethane; vacuum dried at 318 K.	FTIR	[18]
Xanthan Gum	Polyaniline	-	60 W microwave irradiation for 80 s	$(NH_4)_2S_2C_8$	Neutralization – 0.5% NaOH; precipitation in absolute alcohol; Wash with n-methyl-pyrrolidone and acetone; dried in a vacuum oven at 323 K for 24 h.	FTIR	[25]
β-cyclodextrin	Polyacrylamide	323 (N_2 for 45 min, purge)	5 h	Cerium ammonium nitrate	Saturated hydroquinone solution for reaction termination; homopolymer was removed with a solution of formamide and acetic acid (1:1 volume ratio); extraction of the product with acetone; dried at 323 K.	FTIR	[26]
Pectin	Polyacrylamide and polyacrylic acid	318–423	4 h (in N_2)	0.1 mL $K_2S_2O_8$/0.05 mL sodiumbisulphite	Precipitation in acetone; lyophilized	FTIR, TGA, SEM	[27]
Okra mucilage	Polyacrylamide	-	0.5 h (in N_2)	Ceric solution (in 1 N, HNO_3)	Precipitation in isopropanol; separation of homopolymer: N,N-dimethyl formamide, and acetic acid (v/v, 1:1); precipitation in acetone; vacuum dried at 313 K.	FTIR, TGA, SEM, XDR	[28]

(Continued)

Table 3.1 (Continued)

Backbone	Grafting compound	Temperature (K)	Reaction time	Initiator	Post-treatment	Characterization techniques	References
Fenugreek Mucilage	Polyacrylamide	303	24 h (in N_2)	Ceric solution (in 1 N,HNO_3)	Terminated - 0.5 mL of saturated aqueous hydroquinone solution, precipitation in isopropanol; filtered; slurried in acetone; dried in a vacuum oven at 313 K	FTIR, TGA, SEM, XDR	[30]
Guar Gum	Polyacrylamide	-	800 W microwave irradiation for 3 minutes	Ceric ammonium sulfate (CAS)	Filtered and washed with acetone and alcohol-water mixture; dried in an oven for 2 h at 343–353 K	FTIR	[31]
Xanthan gum	Polyacrylamide	348	1 h (in N_2)	$K_2S_2O_8$	Precipitation in acetone	FTIR, GPC	[32]
Cassava starch	sodium allyl sulfonate (SAS) and acryl amide (AA)	328	3.5 h	0.3 g $(NH_4)_2S_2O_8$+0.35 g $NaHSO_3$	Precipitation in acetone; vacuum dried for 24 h at 323 K	-	[33]
Cassava starch	Acrylamide	353	3.5 h (in N_2)	$(NH_4)_2S_2O_8$ and $NaHSO_3$	Precipitation in absolute ethanol; vacuum dried at 323 K for 12 h; purification: reflux in acetone for 12 h; vacuum dried at 323 K for 12 h.	SEM	[34]
Guar gum	Ethyl acrylate	308	90 min	$K_2S_2O_8$	Precipitation in acetone; vacuum dried at 323 K for 24 h.	FTIR	[35]

Gum acacia	Polyacrylamide	343	2 h (15 min in N_2 before the addition of acrylamide and initiator)	$K_2S_2O_8$	Extraction with acetone	FTIR, GPC	[36]
Starch	Glycerin	-	-	-	50% starch solution and 50% glycerine; addition of 1 M HCl until water evaporation; addition of 1 M NaOH for neutralization	^{1}H-NMR, FTIR	[37]
Dextrin	Caprolactam	348	1 h	$K_2S_2O_8$	Dichloromethane to precipitate; washed with absolute ethanol and deionized water; vacuum dried at 333 K.	FTIR	[38]
Chitosan	D-glucose	333	2 h	Not informed	Cooled until 298 K for 30 min; washing the precipitate with absolute ethanol	FTIR, ^{1}H-NMR	[39]
Guar gum	2-acrylamido-2-methylpropanesulfonic acid	343	3 h (in N_2)	$K_2S_2O_8$	Precipitation in acetone; vacuum dried at 323 K for 24 h.	FTIR	[40]
Guar gum	Methyl methacrylate	308	1 h	$K_2S_2O_8$	Precipitation in acetone; vacuum dried at 323 K for 24 h.	FTIR	[41]
Chitosan (CHS)	Glucosyloxyethyl acrylate (GA)	373 (GA) 328 (GA-CHS)	3 h (GA) 72 h (GA-CHS)	1-chloro-2,4-dinitrobenzene	Adjustment of pH 8 with $NaHCO_3$ solution; precipitation in acetone; filtrated under N_2	FTIR, TGA, ^{1}H-NMR, XDR, SEM	[42]

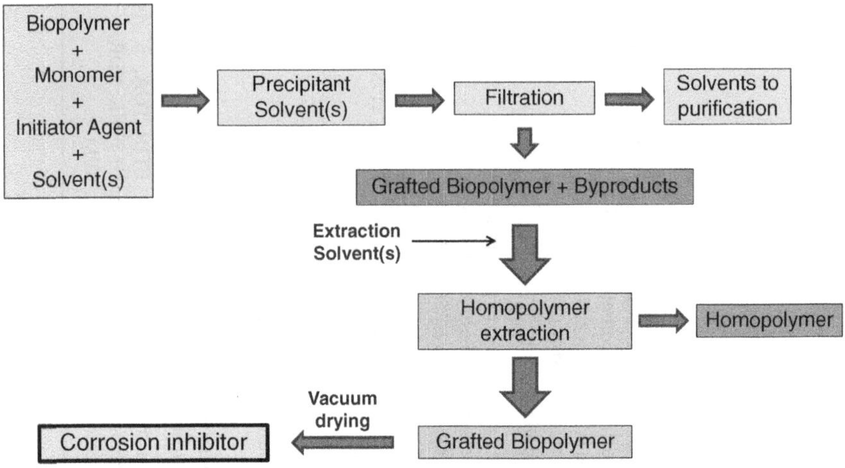

Figure 3.2 Steps of the synthesis of a solution-based graft copolymerization of biopolymers.

should exhibit structural characteristics such as the presence of heteroatoms that might enhance adsorption onto the metallic surface, justifying the grafting process. This process is conducted at specific temperatures and reaction times, using the initiator, as described in Table 3.1. Following grafting, the grafted biopolymer is precipitated by solvents, filtered, purified, and dried, then characterized by the aforementioned techniques.

3.6 Chemical Characteristics of Polysaccharides as Corrosion Inhibitors

Zhang, Ma, Chen, et al. (2020) studied Aloe polysaccharide as a corrosion inhibitor for mild steel in 15% HCl [47]. The FTIR spectrum of the inhibitor indicated the presence of bands at 3295 cm^{-1} related to –OH stretching vibration; 1603 and 1360 cm^{-1} associated with C=O stretching vibration; and at 1076 cm^{-1} corresponding to –C-O-H and –C-O-C stretching vibrations. These results confirm the presence of oxygen heteroatoms in the inhibitor structure.

High oxygen content was also found by Mobin, Ahmad, Basik, et al. (2020) using almond gum to inhibit mild steel corrosion in 1 M HCl [48]. FTIR analysis revealed a band at 3427 cm^{-1} related to –OH stretching; bands at 173 cm^{-1} related to carbonyl group stretching; at 1045 cm^{-1} associated with bending of arabinosyl side chains; and between 800 and 1200 cm^{-1} related to C-O, C-C and C-O-C stretching of the polymer backbone. The same medium was used in another study with *Boswellia serrata* gum as a corrosion inhibitor for low-carbon steel [49].

Bands were reported at 2260 and 3429 cm^{-1} related to –OH stretching vibration; 1645 and 1705 cm^{-1} corresponding to the carboxylate group; and at 1114 cm^{-1} associated with C-O stretching of the ether group.

Other chemical groups were reported applying *Apostichopus japonicus* polysaccharide as a corrosion inhibitor for mild steel in 1 M HCl [50]. FTIR analysis indicated –OH and –NH stretching vibration at 3200–3400 cm^{-1}; C = O stretching vibration of $-CONH_2$ at 1645 cm^{-1}; stretching vibration of –COOH at 1360–1417 cm^{-1}; S=O stretching vibration at 1245 cm^{-1}; and C-O-S stretching vibration at 840 cm^{-1}. These results confirm that this polysaccharide contains heteroatoms such as oxygen, nitrogen, and sulfur. It also has additional adsorption sites, such as nitrogen and sulfur, when compared to Aloe polysaccharide [47].

Moradi, Song, and Xiao (2018) investigated extracellular polymeric substances (EPS) produced by the marine bacterium *vibrio neocaledonicus* sp. as a corrosion inhibitor for carbon steel in artificial seawater and 1 N H_2SO_4 [51]. The FTIR spectrum revealed O-H bond stretching, C-O-C polysaccharide, and COO groups, and P-O-C bonding at 3395.9 cm^{-1}, 1112.7 cm^{-1}, and 615.7 cm^{-1}, respectively. Moreover, the band at 1641.8 cm^{-1} shows the presence of C-O stretching of amide.

The FTIR technique was employed in several studies to confirm the grafting process. A. Biswas, Das, Lgaz, et al. (2019) studied dextrin grafted with poly (vinyl acetate) as a corrosion inhibitor for mild steel in 15% HCl [18]. The FTIR spectrum of dextrin revealed C-O-C symmetric stretching vibration at 1001 cm^{-1}, C-O-C asymmetric stretching at 1145 cm^{-1}, and OH stretching at 3276 cm^{-1}. The grafted polymer exhibited bands at 683 and 1741 cm^{-1} related to C-Br stretching and C=O stretching, respectively, which confirm copolymer formation.

Eduok, Ohaeriand Szpunar, et al. (2020) analyzed glucosyl oxyethyl acrylate graft chitosan (GA-CHS) as a corrosion inhibitor for pipeline steel (API 5 L X70) in chloride-enriched acid solution (1 M HCl) [42]. The grafting process was confirmed by FTIR analysis, identifying the absence of C–H stretching of the $CH_2=CH-$ acrylate double bond group at 810 cm^{-1}. The pipeline steel was severely damaged by the highly corrosive electrolyte, exhibiting uniform and localized corrosion. Adsorption of the grafted polysaccharide, GA-CHS, onto the metallic surface led to significant inhibition of chloride-induced corrosion compared to unmodified chitosan, reducing pit formation. The greater inhibition potential of GA-CHS over unmodified chitosan was attributed to the presence of chemical functional groups such as hydroxyl/amino combined with acrylate functionality.

Singh, Ansari, Quraishi, et al. (2021) grafted guar gum with ethyl acrylate as a corrosion inhibitor for P110 steel in 15% HCl [35]. The FTIR spectrum of guar gum exhibited bands related to –OH stretching at 3462 cm^{-1} and a band at 1639 cm^{-1} associated with the six-membered rings. The grafted biopolymer showed additional bands at 1735 cm^{-1} related to carbonyl, and at 2926 and 2855 cm^{-1} associated with methyl –CH stretching of ethyl acrylate, confirming the grafting process.

The FTIR technique revealed that most of the biopolymers are oxygenated molecules. Some also exhibit nitrogen and sulfur heteroatoms. Thus, the grafting process aims at increasing adsorption sites to promote better surface coverage when compared to the biopolymer.

3.7 Structural Modifications in Biopolymers and Effect of Intensifiers

The structure of Konjac glucomannan (KGM) was modified with amino acids to synthesize polysaccharide esters by Zhang, Yang, Yin, et al. (2018) [52]. The esterification reaction was conducted by mixing the catalyst (p-toluenesulfonic acid) with KGM and the amino acid (L-histidine or L-arginine) in the presence of dimethyl sulfoxide. The corrosion inhibitors were evaluated by weight loss measurements (WL) at 298 K from 100–2000 ppm. The inhibitor obtained from L-histidine showed higher efficiencies (92.4%) in relation to that obtained from L-arginine (89.9%), which may be due to the heterocyclic structure of the imidazole ring, providing better surface coverage.

Gupta, Joshi, Srivastava, et al. (2018) studied chitosan as a corrosion inhibitor for mild steel in 1 M sulfamic acid [53]. The WL measurements revealed maximum efficiency of 73.8% for chitosan (200 ppm), while chitosan (200 ppm) and KI (5 ppm) exhibited 90.3% efficiency. Since the efficiency of the mixture is higher than each compound alone, the synergistic effect is confirmed. Furthermore, the anodic and cathodic branches in potentiodynamic polarization tests (PDP) showed the same characteristic in the absence and presence of the inhibitor, revealing that the corrosive mechanism did not change. Electrochemical impedance spectroscopy (EIS) tests revealed an increase in R_p in the presence of the inhibitor, indicating that the organic compound delayed the charge transfer reaction and, consequently, the corrosive process, by forming an adsorbing film on the metallic surface. Although no explanation was provided for the synergistic effect, other studies [54] reported that iodide ions promote the recharge of the electrical double layer, facilitating the adsorption of the protonated inhibitor.

The synergistic effect of two biopolymers was evaluated by Zhang, Nie, Li et al. (2021) [55]. A polysaccharide mixture composed of chondroitin sulfate derived from pig cartilage and sodium alginate was employed to inhibit mild steel corrosion in 1 MHCl. With respect to WL efficiencies, the mixture performed better than the individual inhibitors (95.18% against 72.78%). Theoretical results indicated possible interactions between these organic inhibitors and the metallic surface. The quantum chemical calculation demonstrated possible bond formation between the inhibitor and the metal would occur by retro-donation and charge transfer between iron atoms and the biopolymer mixture (N and O atoms).

Molecular dynamics (MD) simulations revealed that the interaction and binding energies of the biopolymers follow the same order observed in experimental data. The biopolymer mixture performed better, and the order of efficiency was mixture > chondroitin sulfate > sodium alginate.

Modified alginate was developed to prevent corrosion on mild steel in 1 MHCl [17]. Inhibition efficiency after modifying alginate with 5-chloro-methyl-8-hydroxyquinoline was enhanced compared to the unmodified analog. The 8-hydroxyquinoline-grafted-alginate was prepared via green condensation in water. Surface analysis corroborated corrosion tests, showing a considerable decline in steel dissolution in acidic media for the modified alginate when compared with its analog.

Fares, Maayta, and Al-Mustafa (2012) studied i-Carrageenan and mixtures of the biopolymer with pefloxacin as corrosion inhibitors for aluminum in HCl [56]. The positive charge of $Al\text{-}OH_2^+$ attracts negatively sulfated and negatively charged oxygen heteroatoms of i-Carrageenan. In addition, the adsorption of pefloxacin may favor interaction between i-Carrageenan and the metallic surface through a protonated tertiary amine group in its structure, while its carboxylate group interacts with the metallic surface. The adsorption of pefloxacin onto the positively charged sites can also promote inhibition. These factors contribute to the greater efficiency of the mixture (74.2% in the presence of 400 ppm of pefloxacin + 1600 ppm of i-Carrageenan at 283 K in 2 M HCl) in relation to i-Carrageenan alone (63.3%).

Hasanin and Al Kiey (2020) studied composites of cellulose (NMCC), ethylcellulose (NEC), and carboxymethyl cellulose (NCMC) with niacin (NA) as corrosion inhibitors for copper in 3.5% NaCl [57]. Synthesis consisted of the reaction of equimolar amounts of the cellulosic compound and niacin in boiling water for 3h. TGA revealed higher thermal stability of the inhibitors (NMCC – 500 K, NEC – 506 K, NCMC – 514 K) when compared to NA (473 K). Moreover, the topographic study revealed an irregular shape and size for the NMCC inhibitor; a homogeneous surface texture for NCC, providing better interaction with the metallic surface; and a rough surface and sharp endings for NCMC. Energy-dispersive X-ray spectroscopy analysis (EDX) revealed higher nitrogen content for NEC, which is related to its higher yield. Electrochemical tests were performed in the presence of 100 ppm of the inhibitors. The NEC inhibitor exhibited the highest efficiency by EIS (94.2%), followed by NCMC (81.8%) and NMCC (32.3%). Nyquist plots showed only an increase in arc diameters, without changing the plots, indicating the same corrosive mechanism in the absence and presence of the inhibitors. Nitrogen in the niacin structure of the inhibitors donates electrons to the metallic surface. The π electrons of the aromatic rings of niacin also interact with vacant d orbitals of the metal, promoting inhibition. In addition, NMCC exhibited the lowest efficiency, probably due to the smaller amount of oxygen heteroatom in its structure, providing fewer anchoring points to the substrate.

3.8 Grafted biopolymers Versus Biopolymers

As previously mentioned, the grafting process can enhance corrosion inhibition properties. Several studies have described improvements, especially an increase in adsorption sites in biopolymers compared to non-grafted inhibitors. Roy, Karfa, Adhikari, et al. (2014) analyzed how grafting percentage impacts inhibition efficiency [31]. Polyacrylamide grafted guar gum was obtained by previously dissolving guar gum in water and then adding acrylamide, followed by the initiator (ceric ammonium sulfate). The homogeneous solution was microwave irradiated for 3 min (800 W). Initiator concentration was changed to obtain different percentages of grafted guar gum, maintaining the other factor's constant. A twenty-fold increase in the amount of initiator resulted in 6 different biopolymers from non-grafted guar gum (0%) to 5% polyacrylamide grafted guar gum. It was found that grafting enhances corrosion inhibition compared to non-grafted polymers, but a maximum is achieved with 3% grafting. PDP results showed 71.4% and 75.0% efficiencies for guar gum and polyacrylamide respectively, while the grafted polymers ranged from 84.3% to 91.4%. Increasing grafting up to 3% improves corrosion inhibition but declines thereafter. Nevertheless, all the efficiencies obtained were above their non-grafted counterparts, highlighting that grafting is needed to achieve better performance.

Biswas, Mourya, Mondal, et al. (2018) investigated gum acacia grafted with polyacrylamide (PAM) as a corrosion inhibitor for mild steel in 15% HCl [36]. The authors synthesized different biopolymers by varying the amount of acrylamide from 2 g to 7 g in relation to 1 g of gum acacia (GA), resulting in average molecular weight from 3,260,853 Daltons for gum acacia grafted with 2 g of polyacrylamide (GA-g-2PAM) to 8,705,799 Daltons for gum acacia grafted with 7 g of polyacrylamide (GA-g-7PAM). The WL tests indicated that grafting resulted in higher efficiencies than gum acacia (69.69%), due to the presence of more adsorption sites in the acrylamide structure. In addition, an increase in acrylamide from 2 g to 3 g increased efficiency from 83.42% to 90.49%, which is explained by the rise in the number of adsorption sites, such as –C=O and $-NH_2$. However, a further acrylamide increase to 5 g and 7 g decreased efficiency to 88.90% and 88.29%, respectively. This is due to the steric effects promoted by an increase in the grafted chain. Immersion time between 6 and 24 h was also evaluated for these grafted biopolymers. The results indicated a 25% decrease in efficiency for gum acacia, while for the grafted biopolymers, the decrease was much lower (2–4%). This is due to lower hydrolysis and biodegradation of the grafted biopolymers. The PDP and EIS tests revealed that the plots were similar to the blank, indicating that the corrosive mechanism was not altered by the inhibitor.

Babaladimath, Badalamoole, and Nandibewoor (2018) investigated polyaniline (PANI) grafted onto Xanthan Gum (XG) using microwave-assisted synthesis as a corrosion inhibitor for aluminum in 1MHCl [25]. The XG and XG-g-PANI exhibited WL efficiencies of 62.66% and 91.33%, respectively, at 303 K. The PDP tests indicated that the grafted biopolymer is mixed, and EIS showed that the increase in n from 0.91 to 0.98 with the addition of XG and XG-g-PANI-2, respectively, is related to reduced substrate heterogeneity after adsorption of the grafted biopolymer. In addition, although the XG inhibitor resulted in a smoother surface than that observed by SEM with no inhibitor, pits were also observed. In the presence of the grafted biopolymer, these pits are no longer detected, confirming better surface protection. In this sense, the higher protection provided by XG-g-PANI is explained by the presence of more adsorption sites: the lone pair of electrons in nitrogen may become protonated, adsorbing onto cathodic sites; and p electrons in the PANI chains and the lone pair of electrons in the hydroxyl groups of XG may adsorb onto anodic sites. Adsorption onto both anodic and cathodic sites confirms the mixed behavior of this inhibitor.

Deng, Li, and Du (2020) investigated cassava starch grafted with acrylamide as a corrosion inhibitor for aluminum in 1 M H_3PO_4 [34]. WL tests were performed during 6 h immersion at 223 K, indicating that the grafted biopolymer is more efficient (90.6%) than both acrylamide (45.6%) and cassava starch (32.5%) isolated in the presence of 1.0 g L^{-1} of the inhibitors. The grafted biopolymer showed greater efficiency at higher concentrations; however, from 0.5 g to 1.0 g L^{-1}, there is only a slight increase, due to the better surface coverage. The authors also evaluated a mixture of cassava starch (1.0 g L^{-1}) and acrylamide (1.0 g L^{-1}), reaching 57.6% efficiency. This information indicates that the synergistic effect is not high when the compounds are mixed, meaning that the biopolymer structure must be chemically modified to achieve higher efficiencies. The EIS tests demonstrated that the Nyquist and Bode plots show the same shapes in the absence and presence of the grafted biopolymer, demonstrating that the corrosion mechanism does not change. The Nyquist plots also revealed a first capacitive loop at high frequencies (Al_2O_3 film on the aluminum substrate), a small inductive loop at medium frequencies (adsorption-desorption of intermediates on the aluminum surface), and a second capacitive loop at low frequencies (Al-dissolution). Fathima, Pais, and Rao (2021) also studied the protection of aluminum alloy, employing pectin as a corrosion inhibitor [58]. The EIS tests indicated the same capacitive and inductive arcs, which were related to the same phenomena reported by Deng, Li, and Du (2020) [34].

Biswas, Das, Lgaz, et al. (2019) studied dextrin grafted with poly (vinyl acetate) as a corrosion inhibitor for mild steel in 15% HCl [18]. The PDP tests showed maximum efficiency of 98.39% and 84.56% for the grafted biopolymer and dextrin, respectively, both in the presence of 0.150 g L^{-1} of inhibitor. The E_a values of the grafted biopolymer and dextrin were 29.95 and 35.84 kJ/mol, respectively, indicating the energy barrier for the corrosive process. The higher value for the grafted biopolymer represents this greater adsorption barrier. In addition, the

AFM (Atomic Force Microscopy) study obtained roughness parameters of 891 nm, 428 nm, and 51.3 nm for the blank (absence of inhibitor), the biopolymer, and the grafted biopolymer, respectively. These observations using different techniques indicate higher protection promoted by the grafted biopolymer. The molecular dynamics results show better stability of the grafted biopolymer due to its larger molecular structure and the presence of more sites for adsorption onto the metallic substrate.

Singh, Liu, Ituen, et al. (2020) studied guar gum grafted with 2-acrylamido-2-methylpropanesulfonic acid in 3.5% NaCl as a corrosion inhibitor for copper [40]. The WL measurements revealed that efficiency increased with concentration up to 600 mg L^{-1}, achieving maximum efficiency of 95%. Efficiency increases with immersion time from 95.5% (2 h) to 98.5% (12 h), due to the adsorption of more inhibitor molecules. The grafted biopolymer contains OH, SO_3H, and NH groups, possible anchoring sites on the metallic substrate. In addition, the increase in NaCl concentration from 1% to 11% reduced efficiency from 95% to 55%. This is due to NaCl penetration in the coating, promoting pitting corrosion.

Zheng, Gao, Li, et al. (2021) investigated a grafted biopolymer of chitosan with acrylic acid (NCC) as a corrosion inhibitor for carbon steel in municipal wastewater [59]. A formulation (NHGZ) of the grafter biopolymer with other compounds, such as 2-hydroxyphosphonocarboxylic acid, gluconic acid sodium, and $ZnSO_4$ at a mass ratio of 8:12:25:5, was also studied. NCC showed higher efficiency (54%) than that of chitosan (17%), while the formulation exhibited 97% efficiency. Electrochemical tests were performed from 50–150 mg L^{-1} of NCC and NHGZ. PDP tests showed greater efficiency with an increase in concentration for NCC, revealing that β_c rose from 148.9 mV/cm^2 (50 mgL^{-1}) to 185.4mV/cm^2 (150 mgL^{-1}). This indicates that NCC acts as a cathodic inhibitor. NHGZ tests showed maximum efficiency at 100 mg$L^{-1.}$ Thereafter, the anodic and cathodic parameters increased, reducing inhibition efficiency. The authors associate this finding with the increase in free-conducting particles in the solution, which exacerbates corrosion. The grafted biopolymer interacts with the metallic surface through the conjugated π and unshared electron pairs. However, the protective layer is still insufficiently dense to protect the substrate. Thus, the formulation achieved better results, since Zn^{2+} replaced Fe^{2+} ions, increasing film density, and the 2-hydroxyphosphonocarboxylic acid and gluconic acid sodium compounds allowed the complex formation of organophosphorus acid, once again improving film density.

3.9 Corrosion Tests

3.9.1 Effect of Concentration

The effect of inhibitor concentration is a function of the biopolymer chemical structure as well as the corrosive media. A comparison of mild steel corrosion in 1.0 M HCl (see Table 3.2) shows that inhibitor concentration ranges vary, but maximum efficiency is generally achieved at the maximum concentration. Xanthan gum, with concentrations ranging from 100–1000 ppm, exhibited the lowest efficiency, namely 74.24% at 1000 ppm [60]. Pectin achieved 93.9% efficiency in a similar concentration (around 1000 ppm), despite a larger range (0.0625–2.0 g L^{-1} or 62.5–2000 ppm) being applied [61]. Mixing two biopolymers seems to be another viable approach. Chondroitin sulfate mixed with sodium alginate [55] exhibited the highest efficiency (95.18%) at a lower inhibitor concentration of around 600 ppm (100–600 mg L^{-1}). Alginate chemically modified with 8-hydroxyquinoline was studied [17] using the lowest concentration range for the same steel and acidic media. Concentrations from 1-10 ppm (1×10^{-3} to 10×10^{-3} g L^{-1}) performed best (94.2% efficiency) at the maximum concentration. It is important to highlight that the concentration of 10 ppm for the modified alginate is significantly lower than that used for non-modified polymers (around 1000 ppm).

Gowraraju, Jagadeesan, Ayyasamy, et al. (2017) investigated iota-carrageenan (IC) and inulin (INU) as corrosion inhibitors for mild steel in 0.5 MH_2SO_4 [62]. WL indicated a decrease in corrosion rates (CR) with a rise in inhibitor concentration (1–1000 ppm), achieving maximum efficiencies of 96.67% and 78.16% for IC and INU, respectively, at 1000 ppm. It is important to note that Arukalam, Alaohuru, Ugbo, et al. (2014) reported that the solvent initially diffuses into the polymeric structure, leading to swollen gel [63]. Next, the gel slowly dissolves, forming a true solution, resulting in faster and more efficient adsorption onto the metallic substrate. The IC structure exhibits oxygen and sulfur heteroatoms, and INU-only oxygen heteroatom. This structural difference, associated with the overpowering of polymer-polymer attraction forces by their polymer-solvent counterparts in IC, resulted in easier polymer diffusion, favoring adsorption onto the metallic surface.

Jmiai, El Ibrahimi, Tara, et al. (2018) studied alginate as a corrosion inhibitor for copper in 1 M HCl, observing increased efficiency with a rise in concentration (5×10^{-3} to 10^{-1} mg L^{-1}) [64]. This increase is associated with a larger surface coverage and protection against corrosive media. Zhang, Ma, Chen, et al. (2020) studied Aloe polysaccharide as a corrosion inhibitor for mild steel in 15% HCl, observing the same trend from 200 to 800 mg L^{-1} [47]. However, from 400 mg L^{-1} onwards, the increase in efficiency was lower, indicating the proximity of the substrate saturated with inhibitor molecules. PDP performed by Zhang, Ma, Chen et al. (2020) showed that the inhibitor reduces the anodic dissolution of mild steel and delays the cathodic evolution of hydrogen, acting as a mixed inhibitor [47]. The β_a and β_c parameters did not change significantly in the presence of the inhibitor, which demonstrates that the inhibitor molecules delay the corrosive process by attaching to the reactive sites of the metallic substrate, without changing the dissolution mechanism. This was also observed by other authors [35, 48, 50, 52, 58]. The same trend is revealed by EIS, since the capacitive loops are very similar in the presence and absence of different inhibitor concentrations, indicating the

Table 3.2 Table 3.2 represents information regarding corrosion test details, inhibitor structural characteristics, and types of adsorption.

Media	Material	Backbone	Grafting compound	Inhibitor concentration range	Temperature (K)	Immersion time range	Maximum efficiency (%)	Adsorption-ΔG^0_{ads} (kJ mol^{-1})	References
1 M HCl	Mild steel	Alginate	8-hydroxyquinoline	(1–10) × 10^{-3} g L^{-1}	298	Not informed	94.2% in 10 × 10^{-3} g L^{-1} by EIS	Chemical and physical (36.92)	[17]
15% HCl	Mild steel	Dextrin	Poly (vinyl acetate)	0.025–0.150 g L^{-1}	Not informed	1 h (OCP)	98.39% in 0.150 g L^{-1} by PDP	Chemical and physical (27.98)	[18]
1 M HCl	Aluminum	Xanthan gum	Aniline	-	303	1 h (WL) and 70 min (OCP)	95.32% by EIS	-	[25]
1 M H_2SO_4	X80 steel	β-cyclodextrin	Polyacrylamide	50–200 mg L^{-1}	293–323	36h (WL) and 60 min (OCP)	93.16% in 200 mg L^{-1} at 293 K by EIS	Chemical (56.20–61.70)	[26]
3.5% NaCl	Mild steel	Pectin	Polyacrylamide (PAAm) and polyacrylic acid (PAA)	500–900 ppm (PAAm) 50–500 ppm (PAA)	Not informed	Not informed	89.7% in PAAm 700 ppm; 82.8% in PAA at 400 ppm by EIS	-	[27]
0.5 M H_2SO_4	Mild steel	Okra mucilage	Polyacrylamide	1–100 ppm	298–338	3–72 h (WL)	96.6% in 100 ppm at 298K by EIS	-	[28]
0.5 M H_2SO_4	Mild steel	Fenugreek Mucilage	Polyacrylamide	1–100 ppm	308	24 h (WL)	95.9% in 100 ppm by EIS	Not Informed	[30]
1 M HCl	Mild steel	Guar gum	Polyacrylamide	50–500 ppm	303	1–96 h (WL)	94.4 % by EIS (86% of grafting)	Mostly physical (30.1)	[31]
15% HCl	Mild steel	Xanthan gum	Polyacrylamide	0.05–0.5 g L^{-1}	298–333	6 h (WL)	93.18% in 0.5 g L^{-1} at 298 K by WL	Chemical and physical (24.013)	[32]
1 M HCl	Cold rolled steel	Cassava starch	Sodium allyl sulfonate (SAS) and acryl amide (AA)	5–50 mg L^{-1}	293	6 h (WL)	97.3% in 50 mg L^{-1} by WL	Chemical and physical (32.01)	[33]

1 M H_3PO_4	Aluminum	Cassava starch	Acrylamide	0.1–1.0g L^{-1}	223	6 h (WL)	91.9% in 1.0g L^{-1} by PDP		[34]
15% HCl	P 110	Guar gum	Ethyl acrylate	100–600 mg L^{-1}	308	Not informed	92.61% in 500 mg L^{-1} by EIS	Chemical and physical (24.11)	[35]
15% HCl	Mild steel	Gum acacia	Polyacrylamide	0.05–0.4g L^{-1}	298–333	6–24 h (WL)	94.08% with gum acacia and 3g acrylamide, in 0.4 g L^{-1}, 303K by EIS	Physical and chemical (25.41–27.93)	[36]
1 M HCl	C-Mn steel	Starch	Glycerin	5–300 mg L^{-1}	298–323	2 h (WL) and 30 min (OCP)	94% in 300 mg L^{-1} at 298K by WL	Physical (7.25 at 323 K)	[37]
1 M HCl	J 55 steel	Dextrin	Caprolactam	100–400 mg L^{-1}	298–338	24 h (WL)	83.27% in 300 mg L^{-1} at 298K by WL	Physical and chemical(23.68)-	[38]
1M HCl	Mild steel	Chitosan	D-glucose	10^{-3}–10^{-6} M	Not informed	Not informed	97.6% in 10^{-3} M by PDP	Chemical (46.16)	[39]
3.5% NaCl	Copper	Guar gum	2-acrylamido-2-methylpropane-sulfonic acid	100–600 mg L^{-1}	308–338	24 h (WL), 30 min (OCP)	95% in 600 mg L^{-1} at 308K by WL	-	[40]
3.5% NaCl + CO_2	P110 steel	Guar gum (GG)	Methyl methacrylate (MMA)	10–400 mg L^{-1}	323	Not Informed	96.8% in KI (5 mM) + GG-MMA (300 mg L^{-1}) by WL	Chemical and physical (26.84)	[41]
1.0 M HCl	API 5 L X70 Steel	Chitosan	Glucosyloxyethyl acrylate	100–500 ppm	298	Not informed	90.4% at 500 ppm by PDP	-	[42]
15% HCl	Mild steel	Aloe polysaccharide	-	200–800 mg L^{-1}	303–333	48 h (WL)	96.41% at 800 mg L^{-1} at 303K by EIS	Chemical and physical (25.31)	[47]
1 M HCl	Mild steel	Almond gum	-	10–300 ppm	303–333	6 h (WL) and 30 min (OCP)	96.4% in 300 ppm at 333K by WL	Physical and chemical (31.08 at 333K)	[48]

(Continued)

Table 3.2 (Continued)

Media	Material	Backbone	Grafting compound	Inhibitor concentration range	Tempera-ture (K)	Immersion time range	Maximum efficiency (%)	Adsorption-ΔG^0_{ads} (kJ mol^{-1})	References
1 M HCl	Low carbon steel	*Boswellia serrata* gum	-	10–50 ppm	303–333	6–360 h (WL) and 1 h (OCP)	96.69% in 500 ppm at 303K by EIS	Chemical and physical (26.86 at 333 K)	[49]
1 M HCl	Mild steel	*Apostichopus japonicus* polysaccharide	-	200–800 mg L^{-1}	298	0.5 h (OCP)	95.26% in 600 mg L^{-1} by PDP	Chemical and physical (24.96)	[50]
Artificial seawater and 1 N H_2SO_4	A36 carbon steel	Extracellular polymeric substances (EPS)	-	1–10g L^{-1} (artificial seawater), 200–1200 ppm (H_2SO_4)	Room temperature	Artificial seawater (7 days) and H_2SO_4 (2 h)	94% in artificial seawater in 10 g L^{-1} by PDP; 89.37% in H_2SO_4 in 1000 ppm by PDP	Physical (3.14 in artificial seawater, and 11.59 in H_2SO_4)	[51]
0.5 M HCl	Mild steel	Konjac glucomannan polysaccharide esters	-	100–2000 ppm	298	8 h (WL) and 1 h (OCP)	89.9% (with L-arginine ester) and 92.4 (with L-histidine ester) in 2000 ppm by WL	Chemical and physical (26.36 with L-arginine ester and 29.38 with L-histidine ester)	[52]
1 M Sulfamic acid	Mild steel	Chitosan	-	50–200 ppm, 5 ppm of KI	308–338	3 h (WL)	91.68% in 200 ppm of inhibitor+5 ppm of KI at 308 K by PDP	Chemical and physical (33.42 at 308 K)	[53]
1.0 M HCl	Mild Steel	Chondroitin sulfate	Sodium Alginate	100–600 mg L^{-1}	303–333	48 h	95.18% in 600 mg L^{-1} at 303 K by WL	-	[55]

1–2 M HCl	Aluminium	i-Carrageenan	-	40–1600 ppm	283–313	2 h (WL)	74.2% in 400 ppm of pefloxacin and 1600 ppm of i-Carrageenan at 283K in 2 M HCl by WL	Physical (19.4 at 283 K)	[56]
3.5% NaCl	Copper	Cellulose (NMCC), Ethylcellulose (NEC), and carboxymethyl-cellulose (NCMC) composites with Niacin (NA)	-	10 ppm	298	Not informed	94.7% (NEC), 83.4% (NCMC), and 33.2% (NMCC) by PDP	-	[57]
0.025 N HCl	6061 Aluminium alloy	Pectin	-	0.1–0.8 g L^{-1}	303–323	Not informed	79.79% in 0.8 g L^{-1} at 323 K by PDP	Chemical and physical (25.10 at 323 K)	[58]
Municipal wastewater	Carbon steel	Chitosan	Acrylic acid	25–150 ppm	313	72 h	93.55% for the formulation containing grafted biopolymer in 100 mg L^{-1} by PDP	Chemical and physical (29.18)	[59]
1.0 M HCl	Mild Steel	Xanthan gum	-	100–1000 ppm	303–333	6 h	74.24% in 1000 ppm at 303 K by WL	Mostly physical (22.48)	[60]
1 M HCl	Mild steel	Pectin	-	0.0625–2.0 g L^{-1}	298–333	2 h and 24 h	93.9% in 1.0 g L^{-1} at 298 K by EIS	Predominantly chemisorption rather than physisorption (22.44 to 27.92)	[61]
0.5 M H_2SO_4	Mild steel	Iota-carrageenan (IC) and Inulin (INU)	-	1–1000 ppm	313–333	3 h (WL)	96.62% (IC) and 78.16% (INU) in 1000 ppm at 303 K by WL	Chemical and physical (IC:31.09; INU: 33.03)	[62]

(Continued)

Table 3.2 (Continued)

Media	Material	Backbone	Grafting compound	Inhibitor concentration range	Temperature (K)	Immersion time range	Maximum efficiency (%)	Adsorption-ΔG^0_{ads} (kJ mol^{-1})	References
1 M HCl	Copper	Alginate	-	5×10^{-3}–10^{-1} mg L^{-1}	298–328	4–168 h (WL)	83.06% in 10^{-1} mg L^{-1} at 298K by PDP	Mostly physical (24.20)	[64]
1 M HCl	Mild steel	Chitosan	-	100–500 ppm	303–323	2 h (WL) and 0.5 h (OCP)	98.9% in 500 ppm by PDP	Physical (13.06 at 323 K)	[65]
3.5% NaCl	AZ31 Mg alloy	Chitosan (CHI), dextran (Dex), carboxymethyl cellulose (CMC), sodium alginate (ALG), pectin (PEC), hydroxylethyl cellulose (HEC), and Gum Arabic (GA)	-	0.5–2.0 g L^{-1}	Room temperature	24 h (WL), 30 min (OCP)	ALG+KI+Date palm seed oil: 80.62%; HEC+KI+Date palm seed oil: 85.34%, by PDP	-	[66]
Deoxygenated and CO_2 saturated 3.5 wt% NaCl	Low carbon (API 5L X60 grade)	Chitosan and carboxyl methylcellulose (CMC)	-	25–100 ppm	298–333	1–24 h (EIS)	55% for chitosan and 54% for CMC, in 100 ppm at 298 K by PDP	Physical and chemical (Chitosan: 25.2; CMC: 23.0)	[67]
2% NaCl and 1% acetic acid	Tin	Tomato peel pectin and commercial apple pectin	-	0.2–20.0 g L^{-1}	298	3–24 h (WL)	75.6% (apple pectin) and 73.9% (tomato peel pectin) in 20 g L^{-1} by WL	-	[68]
0.25 M NH_2SO_3H	Zinc	Nanoparticles of maltodextrin	-	10–50 ppm	303–323	Not Informed	93% in 50 ppm at 323 K by PDP	Chemical and physical (25.07 to 31.65)	[69]

15% HCl	Mild Steel	Hydroxyethyl cellulose (HEC)	-	6–50 mM	293–353	6 h	93% in 50 mM at 353 K by EIS	Chemical (52.4)	[70]
1.0 M HCl	Mild Steel	Fenugreek gum	-	100–500 ppm	303	1–36 h	94% in 500 ppm by EIS	Mostly physical (25.5)	[71]
0.5 M HCl	X60 pipeline steel	Pectin	-	50–1000 ppm	298–333	24 h	78.7% in 1000 ppm at 298 K by PDP	Mostly physical (22.13)	[72]
3.5% NaCl	AA5052 aluminum alloy	Konjac glucomannan	-	50–200 ppm	298	Not Informed	94% in 200 ppm by PDP	Mostly physical (27.27)	[73]
15% HCl and 15% H_2SO_4	St37 steel	Gum arabic	-	50–1000 ppm	298 and 333	24 h	92.65% in 15% HCl, 1000 ppm at 298 K by PDP	Mostly physical (23.207 to 26.932)	[74]
1 M HCl	Mild steel	Polydopaminenano-particles (PDA–1, PDA-2, PDA-3)	-	800 ppm	Not Informed	2–48 h	99% for 24 h with PDA-2 by EIS	Not Informed	[75]
0.5–2.0 M HCl	Aluminum	Pectin	-	1–8 g L^{-1}	283–313	1 h	91.3% in 2.0 M HCl, 8 g L^{-1} at 283 K by WL	Physical (6.8 to 18.6)	[76]
Acidic media	Fe (110), Cu (111), Al (111) and Sn (001) surfaces	Alginate	-	Theoretical study	Theoretical study	Theoretical study	Theoretical study	Theoretical study	[77]
1.0 M HCl	Mild Steel	O-fumaryl-chitosan	-	100–500 ppm	303–323	2 h	94.1% in 500 ppm at room temperature by EIS	-	[78]

(Continued)

Table 3.2 (Continued)

Media	Material	Backbone	Grafting compound	Inhibitor concentration range	Temperature (K)	Immersion time range	Maximum efficiency (%)	Adsorption-ΔG^0_{ads} (kJ mol^{-1})	References
1.0 M HCl	Mild Steel	Starch (S), pectin (P) blends	-	Blends S:P (1:1, 1:2, 1:3, 2:1, 3:1)	303–333	Not Informed	88.9% for S:P blend=1:3 at 303 K by EIS	Not Informed	[79]
200 ppm NaCl	Mild Steel	Phosphorylated xanthan gum	-	25–200 ppm	303	7 days	92.54% in 150 ppm by EIS	-	[80]
1.0 M HCl	Carbon Steel	Polysaccharides from *Plantago ovata*	-	200–1000 ppm	303–333	6 h	94.4% in 1000 ppm at 333 K by WL	Chemical and physical (35.6)	[81]
0.5 M H_2SO_4	AISI 1040 Steel	Pectin	-	0.5–5 g L^{-1}	303	1–7 h	83.36% in 5 g L^{-1} for 1 h by WL	Physical (7.05 to 14.24)	[82]
1.0 M HCl	Carbon Steel	Alginate	-	0.5–1.5 g L^{-1}	303	6 h	97.7% in 1.5 g L^{-1} at 303 K by EIS	-	[83]
3.5% NaCl	API X60 steel	Sodium Alginate	-	100–1000 ppm	298 and 343	24 h	87.23% in 1000 ppm at 343 K by WL	-	[84]

same corrosion mechanism. However, in the presence of the inhibitor, the C_{dl} value decreases, due to the decline in the local dielectric constant and/or increased thickness of the electric double-layer capacitor. This demonstrates that the water molecules are replaced by their inhibitor counterparts, preventing metal dissolution [52]. The decrease in C_{dl} also suggests a more compact and thicker inhibitory film on the metallic surface [18, 28, 52]. Other authors also observed better surface coverage with an increase in concentration [28, 49, 51, 52, 58].

Singh, Ansari, Quraishi, et al. (2021) studied guar gum grafted with ethyl acrylate as a corrosion inhibitor for P110 steel in 15% HCl [35]. The WL tests indicated maximum efficiency with 500 mg L^{-1} of the inhibitor since no significant increase was observed at the highest concentration (600 mg L^{-1}). This result shows surface coverage and protection against the acidic media at 500 mg L^{-1}. A further increase in concentration does not provide additional coverage, resulting in the same order of magnitude of efficiency. The PDP tests revealed that the cathodic parameter significantly decreases (100 mg L^{-1}: 77.32 mV/ dec; 500 mg L^{-1}: 38.09 mV/dec) at higher concentrations. This demonstrates that the hydrogen evolution mechanism changed due to barrier/diffusion effects related to the decline in the cathodic transfer coefficient. Other studies corroborated the findings of Singh, Ansari, Quraishi, et al. (2021), namely that efficiency increases as a function of concentration [35]. However, after a certain concentration, no increased efficiency is observed due to the saturation of molecules adsorbed onto the metallic surface [65].

Zhang, Wu, and Li (2021) studied Apostichopus japonicus polysaccharide as a corrosion inhibitor for mild steel in 1 M HCl [50]. The effect of concentration (200–800 mg L^{-1}) was investigated by PDP tests. The results showed maximum inhibition efficiency in the presence of 600 mg L^{-1} of the inhibitor (95.26%) since higher concentrations (800 mg L^{-1}) resulted in decreased efficiency (92.66%). This is explained by the fact that at higher concentrations, micelles were adsorbed onto the metallic surface, leading to weaker adsorption and reduced efficiency. Moradi, Song, and Xiao (2018) observed the same trend [51].

Umoren, Solomon, Madhankumar, et al. (2020) investigated chitosan (CHI), dextran (Dex), carboxymethyl cellulose (CMC), sodium alginate (ALG), pectin (PEC), hydroxyethyl cellulose (HEC), and gum arabic (GA) as a corrosion inhibitor for AZ31Mg alloy in 3.5% NaCl [66]. The EIS tests revealed two capacitive loops at high and medium frequencies and an inductive loop at low frequencies. The capacitive loop at high frequencies is related to the charge transfer and corrosion products/adsorbed inhibitor film resistances. The middle capacitive loop is associated with the diffusion processes through the layer and the inductive loop to the relaxation of the adsorbed corrosion products/inhibitor films. Except for HEC (63.04%, WL) and ALG (56.52%, WL), the other biopolymers increased corrosion in relation to the blank system. Chitosan was the biopolymer that resulted in the highest corrosion rates, likely due to the greater chitosan chelating ability of di and trivalent cations. The other biopolymers may have chelated with Mg^{2+} in the electrolyte rather than on the metallic surface. The two biopolymers that performed best were evaluated from 0.5g to 2.0 g L^{-1}. The EIS results indicated increased efficiency from 0.5 to 1.0 g L^{-1} for both polymers, which may be related to an insufficient number of molecules for surface coverage at the lowest concentration. However, a further rise in concentration from 1.0 to 2.0 g L^{-1} reduced efficiency, associated with molecule aggregation due to the shorter distance between them. In addition, the adsorbed molecules may have interacted with their unadsorbed counterparts, leading to the desorption of the former. The authors also studied a formulation containing potassium iodide (KI), date palm seed oil, and HEC/ALG. The results suggested co-adsorption of the compounds, producing the highest efficiencies (ALG formulation-77.43%; HEC formulation-80.56%).

The articles evaluated studied concentrations from 1–2000 ppm, and the effects of the three main types were observed: (i) increased efficiency with concentration, which is related to better surface coverage; (ii) increased efficiency up to a certain concentration, achieving a stable value, which does not increase at higher concentrations (this is related to the saturation of the metallic surface with inhibitor molecules); and (iii) decreased efficiency after a certain concentration, associated with interaction between adsorbed and unabsorbed molecules. These interactions promote desorption by forming aggregates that do not adsorb as a continuous film on the metallic surface.

3.9.2 Effect of Immersion Time

The investigation of immersion time revealed a trend towards increased efficiency with time, followed by a certain stability. Some compounds needed more time to stabilize due to their chemical composition and adsorption strength onto the metallic surface. After stabilization, most of the compounds showed decreased efficiency at longer exposure times. These are indications that the previous film formation step is related to the diffusion of macromolecules in the corrosive media, which takes more or less time depending on the inhibitor structure. Macromolecular structures tend to stabilize each other rather than interact with metals. This may explain rising efficiencies with time until a certain equilibrium condition. After being adsorbed, biopolymers have shown high stability following long exposure time, such as one week (168 h). However, it is important to note that a highly acidic condition can degrade the biopolymer film, leading to macromolecule saturation on the metallic substrate. This can be avoided by using the optimum immersion time instead of a longer exposure.

Alginate was studied as a corrosion inhibitor for copper in 1 M HCl, evaluating the effect of immersion time by measuring WL from 24 h to 168 h [64]. In the first 24 h, efficiency increases, reaching a maximum of 78.20%. This behavior is related to the formation of a more stable film. However, from 24 h to 60 h, efficiency declines to around 74%, followed by stabilization. This may be related to molecule saturation on the substrate, leading to the rearrangement of the film.

Banerjee, Srivastava, and Singh (2012) investigated Okra mucilage natural polymer grafted with polyacrylamide as a corrosion inhibitor for mild steel in 0.5 M H_2SO_4 [28]. Immersion time was evaluated for both 10 ppm and 100 ppm from 3 h to 72 h. An initial increase in efficiency is observed with immersion time due to greater surface coverage. From 48 h to 72 h, efficiency remains stable, which is related to the stability of surface coverage. However, from 72 h onwards, efficiency declines, since the film starts to degrade in the acidic media.

Mobin, Basik, and Aslam (2018) investigated *Boswellia serrata* gum as a corrosion inhibitor for low-carbon steel in 1 M HCl [49]. WL measurements were conducted from 6 h to 360 h. In the first 72 h, efficiency increased from 91% to 94%, which is associated with the adsorption of more molecules, and from 72 h to 120 h, there was a slight decrease to values above 90%, confirming film stability.

Umoren, AlAhmary, Gasem, et al. (2018) studied chitosan and carboxyl methylcellulose (CMC) as corrosion inhibitors for low-carbon steel in CO_2-saturated 3.5% NaCl [67]. The β-(1–4)-linked D-glucosamine/N-acetyl-D-glucosamine and carboxymethyl-bound glucopyranyl moieties on chitosan, and oxygen and nitrogen heteroatoms on CMC are possible sites for adsorption onto the metallic surface. Chitosan exhibited maximum efficiency of 34%, and CMC 51% in 100 ppm by EIS, after 24 h immersion. The better results in prolonged immersion are related to increased molecular adsorption. CO_2 corrosion involves CO_2 dissolution into the liquid phase followed by hydration to form carbonic acid (H_2CO_3). This species dissociates to form HCO_3^-, CO_2^{3-}, and H^+. The protonated inhibitor can adsorb onto the negatively charged surface and/or the heteroatoms (O in CMC; O and N in chitosan) can donate π-electrons to Fe atoms.

Halambek, Cindrić, and Grassino (2020) evaluated tomato peel pectin and commercial apple pectin as corrosion inhibitors in 2% NaCl and 1% acetic acid for tin, which is applied in the canned food industry [68]. Both types of pectin exhibited maximum efficiency at a maximum concentration (20 g L^{-1}), namely, 75.6% (apple pectin) and 73.9% (tomato peel pectin). This highest concentration was investigated in 3 h to 24 h immersion tests. Inhibition efficiencies increase from 51.8% to 75.6% for apple pectin, and from 37.2% to 73.9% for tomato peel pectin, indicating film formation and better coverage with prolonged times. It is suggested that pectin adsorption occurs via electrostatic interactions on the positively charged tin surface with chemical groups such as carbonyl, carboxylic and ionic carboxyl. In addition, the difference between pectin types is explained by their chemical compositions.

Mobin, Ahmad, Basik, et al. (2020) investigated almond gum as a corrosion inhibitor for mild steel in 1 M HCl [48]. Evaluation from 6 h to 168 h at 303 K by WL measurements revealed the stability of the inhibitor film, maintaining high efficiencies (94.8%) even after longer exposure (168 h). This stability in the acidic media may be related to strong adsorption onto the metallic surface since the evaluation of the temperature effect showed increased efficiency with temperature, which is characteristic of chemical adsorption.

Despite showing high stabilities after long exposure to acidic media, most of the biopolymers studied were tested at low temperatures. The rising temperature could change the time stability of the inhibitor layer, especially for weakly adsorbed molecules.

3.9.3 Effect of Temperature and Inhibition Mechanism

Temperature is an important parameter to correlate with inhibitor efficiencies. Diffusion coefficients, ionic species mobility, steel dissolution, molecular adsorption, and desorption are all temperature dependent. Corrosion rates tend to increase with temperature due to the higher H^+ diffusion towards the metallic surface. The cathodic area is blocked by the adsorbed hydrogen at lower temperatures, but rising temperature disrupts hydrogen adsorption, resulting in a larger exposed cathodic surface area [32].

The effect of rising temperature from 298 K to 328 K on copper corrosion inhibition was evaluated in 1 M HCl with alginate as an inhibitor [64]. Efficiencies decrease from 78.20% (298 K) to 42.54% (328 K) in the presence of the highest alginate concentration of 0.1 mg L^{-1}. It is suggested that alginate molecules desorbed from the metallic surface led to its exposure to the acidic media. Decreased efficiency may be related to the physisorption mechanism, which is weaker than its chemical counterpart. The same reduced efficiency with a rise in temperature was observed by other authors [28, 47, 53]. Mobin, Basik, and Aslam (2018) also associated a decline in efficiency with rising temperatures, depleted adsorption strength, and roughening of the electrode surface, which in turn exacerbate corrosion [49].

Gowraraju, Jagadeesan, Ayyasamy, et al. (2017) investigated iota-carrageenan (IC) and inulin (INU) as corrosion inhibitors for mild steel in 0.5 M H_2SO_4 [62]. WL measurements in the presence of INU revealed an increase in CR with temperature, indicating desorption of the inhibitor molecules, associated with the weaker adsorption promoted by physical interactions with the metallic surface. However, IC molecules contain not only oxygen heteroatoms but also sulfur, which contributes to chemical adsorption. In this case, CR decreases with rising temperatures.

Fathima, Pais, and Rao (2021) investigated pectin as a corrosion inhibitor for aluminum alloy in 0.025 N HCl [58]. The PDP tests indicated greater efficiency from 56.98% to 79.79% at increasing temperatures from 303 K to 323 K in the presence of 0.8 g L^{-1} of inhibitor. The authors suggest that this behavior is related to endothermic chemisorption.

Mobin, Ahmad, Basik, et al. (2020) investigated almond gum as a corrosion inhibitor for mild steel in 1 M HCl [48]. The authors also observed greater efficiency with temperatures from 85.5% (303 K) to 96.4% (333 K) in the presence of 300 ppm of the inhibitor. This biopolymer has a stronger and chemically adsorbed layer on the metallic substrate. Water molecules were also replaced by the macromolecular structures, providing a protective film to avoid the acidic media.

Most literature studies investigated temperatures between 303 K and 338 K related to weaker interaction with the metallic surface. Few articles reported an increase in efficiency with temperature, showing stronger adsorption onto the metallic surface. The adsorption process strongly depends on the presence of chemical groups (adsorption sites on corrosion inhibitors), the interaction mode (planar or lateral) with the substrate, and the surface coverage provided by the inhibitor molecules.

Corrosion inhibition with temperature can be obtained by correlating the corrosion rate and apparent activation energy (E_a) using the Arrhenius equation. Corrosion activation energies (E_a) in the absence and presence of corrosion inhibitors can clarify the predominant adsorption mechanism. A reference value for inhibited systems is provided by E_a obtained in the absence of corrosion inhibitors [10]. There are three possible behaviors: (i) E_a remains the same before and after the addition of the inhibitor; (ii) there is a decrease in E_a for the inhibited system; and (iii) E_a increases after the addition of the inhibitor [69]. When the addition of a corrosion inhibitor to the electrolyte solution increases E_a, it means that the energy barrier for steel dissolution increased in the inhibited system compared to the blank solution [70].

Modified hydroxyethyl cellulose was used to inhibit mild steel corrosion in 15% HCl [70]. The E_a value for the inhibited tests was higher than in the blank test. Moreover, the values obtained for the standard Gibbs free energy of adsorption (ΔG^0_{ads}) represented chemisorption (see Table 3.2). According to the literature, a value of ΔG^0_{ads} around $-20\,\text{kJ mol}^{-1}$ represents physisorption, while those around $-40\,\text{kJ mol}^{-1}$ or more are indicative of chemisorption. Other studies observed similar results for chemisorbed biopolymers such as β-cyclodextrin grafted polyacrylamide [26] and modified chitosan [39].

Pais, George, and Rao (2021) evaluated zinc corrosion in NH_2SO_3H solution using nanoparticles of maltodextrin as a corrosion inhibitor [69]. The E_a values obtained for the blank were higher than those of the inhibited systems, which decline with a rise in inhibitor concentration. Nevertheless, other parameters (negative ΔG^0_{ads} and positive ΔH^0_{ads} values) showed that adsorption occurred and can be either physisorption or chemisorption. Most biopolymers exhibited similar adsorption ΔG^0_{ads} values (between -20 and $-40\,\text{kJ mol}^{-1}$). However, some authors consider that adsorption is mostly physical because the ΔG^0_{ads} obtained were around $-20\,\text{kJ mol}^{-1}$ [60, 71–74].

The biopolymer adsorption mode is used to support corrosion inhibition and surface analysis. These results, combined with FTIR analysis of chemical groups in the backbone or lateral chains of the biopolymer, show how the organic layer is formed to protect the metallic surface.

Zhang, Ma, Chen, et al. (2020) studied Aloe polysaccharide as a corrosion inhibitor for mild steel in 15% HCl [47]. The positively charged metallic surface can attract Cl^-, leading to negatively charged regions where the protonated inhibitor can electrostatically adsorb, delaying anodic dissolution. The protonated inhibitors can also interact with the cathodic sites. Chemical adsorption occurs through the coordination bond of the free electron pair of oxygen heteroatom with the unoccupied d-orbital of iron atoms. The excess negative charge accumulated on the metal surface can be transferred from the d-orbital of iron to the empty π^* orbital of the inhibitor molecules. Thus, the physical and chemical adsorptions form a protective barrier between the metallic substrate and the bulk aggressive solution. The same inhibition mechanism was proposed by other authors [26, 48, 49, 52].

Deng, Li, and Du (2020) investigated cassava starch grafted with acrylamide as a corrosion inhibitor for aluminum in 1 M H_3PO_4 [34]. The authors provided the same explanation for the inhibition mechanism based on physical interaction and the chemical coordination bond. However, the molecular skeleton of the natural biopolymer (cassava starch) acts as a barrier that prevents the aggressive media from adsorbing onto the metallic surface, while the nitrogen heteroatoms provided by the copolymer interact with the metallic surface. Zhang, Wu, and Li (2021) studied *Apostichopus japonicus* polysaccharide as a corrosion inhibitor for mild steel in 1 M HCl [50]. The authors used the same explanation provided in the second paragraph of this section, further explaining that the oxygen atoms in the polysaccharide structure may coordinate with Fe(II) species, forming stable chelates that could be adsorbed onto the metallic substrate. A summary of the main adsorption modes is illustrated in Figure 3.3 for pectin, starch, and guar gum.

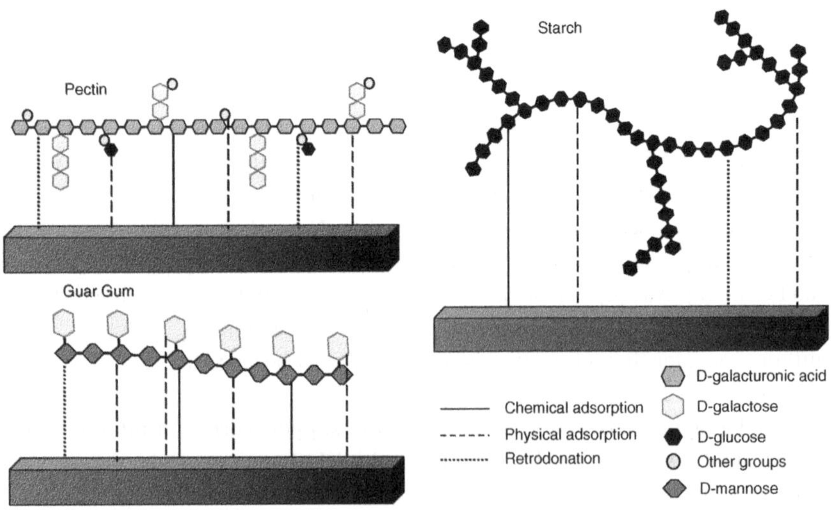

Figure 3.3 Above is a schematic representation of pectin, starch, and guar gum adsorption onto a metallic surface.

3.10 Metallic Alloys and Surface Characterization

Fathima, Pais, and Rao (2021) studied pectin as a corrosion inhibitor for aluminum alloy [58]. The UV-visible absorption spectra of the acidic solution of pectin and the solution with pectin in the presence of the aluminum alloy indicated absorption at 210 nm and 290 nm. These absorptions shift in the presence of the alloy, with no change in spectral shape. This suggests the formation of a complex between Al^{+3} and pectin. The authors also discussed ATR-FTIR spectroscopy, whereby the FTIR spectrum of pectin showed bands at 3342 cm^{-1} related to – OH stretching, at 1742 cm^{-1} associated with C=O stretching, and at 1649 cm^{-1}corresponding to C(=O)–O stretching. The spectrum for the corrosion products exhibited the same bands observed for pectin, albeit at a lower intensity. Intensity decreased and the –OH peak broadened, indicating an interaction between pectin and the metallic surface through the –OH functional groups.

Mobin, Basik, and Aslam (2018) investigated *Boswellia serrata* gum as a corrosion inhibitor for low-carbon steel in 1 M HCl [49]. Most of the biopolymer peaks were present in the adsorption film. The peak shift to lower wavenumbers indicates an interaction between the compounds and the metallic surface. The C-O stretch of ether exhibited the highest shift, from 1114 to 1009 cm^{-1}, confirming that this functional group is an adsorption site.

Mobin, Ahmad, Basik, et al. (2020) investigated almond gum as a corrosion inhibitor for mild steel in 1 M HCl [48]. FTIR analysis of the steel sample revealed bands similar to those observed for the inhibitor, but with lower intensity and shorter wavenumber shifts. The OH stretching vibration observed at 3427 cm^{-1} was shifted to 3393 cm^{-1}, and the carbonyl stretching at 1732 cm^{-1} shifted to 1627 cm^{-1}. These observations indicate the interaction of these functional groups with the metallic surface.

3.11 Computational Studies, in Silico Methods

The inhibition behavior of biopolymers in metallic alloy corrosion has also been investigated using in silico methods, mostly quantum chemical calculations based on Density Functional Theory (DFT) and the Monte Carlo (MC) and Molecular Dynamics (MD) methods [1]. Molecular modeling is a useful tool for correlating biopolymer structure with inhibition efficiency [20, 85, 86]. DFT calculations have been used to determine the energy of molecular orbitals such as the highest occupied (HOMO) and lowest unoccupied molecular orbital (LUMO). The gap between these frontier orbitals may be associated with corrosion inhibition efficiency [12, 87].

However, most computational studies on biopolymers as anti-corrosion agents are limited to the monomer unit of produced biopolymers [77]. Some useful insights into the inhibition process could be obtained if longer chains and the influence of chemical groups in biopolymer chains were studied. Unfortunately, few studies have used this approach in the corrosion field. Since computers are continuously developing, further investigations will likely occur in the coming years.

Oukhrib, Ibrahimi, Oualid, et al. (2020) studied different alginate biopolymer chain lengths to inhibit acid corrosion on Fe (110), Cu (111), Al (111), and Sn (001) surfaces [77]. No experimental results were used by the authors, indicating that continuous progress has been made in this field. The computational results showed the presence of potential metal binding sites interacting with metallic surfaces during the inhibition process. DFT calculations exhibited a trend towards longer alginate fragments that can form chemical bonds with the metal surface because more electrons can be shared through several sites. Monte Carlo simulations revealed that spontaneous adsorption occurred in every simulated system, consisting of alginate fragment/metal, whereby binding energy ($-E_{ads}$) increased progressively with alginate chain length. In addition to potential bonded interactions, alginate can interact via two non-bonded interaction modes. First, the electron-poor sites on metallic surfaces interact with electron-rich atoms of alginate molecules via electrostatic interactions. The second possible mode is that inter-hydrogen bonds can be formed between the hydrogen atoms of alginate molecules and the metal atoms of the studied surfaces. In addition, alginate fragments have shown remarkably preferred adsorption on these surfaces compared with corrosive species in acid media.

Jmiai, El Ibrahimi, Tara, et al. (2018) studied alginate as a corrosion inhibitor for copper in 1 M HCl, using DFT and Monte Carlo calculations to explain the interaction between the biopolymer and the metallic surface [64]. The DFT calculation indicated the presence of the highest negative partial charge in the oxygen atoms, confirming them as electron donors to the copper surface, a low-electron density region, forming coordination bonds. The Monte Carlo simulation showed that the biopolymer adsorbs parallel to the surface. Moreover, adsorption energies rise with the number of monomers, associated with an increase in functional groups for anchoring to the metallic substrate.

Fathima, Pais, and Rao (2021) studied pectin as a corrosion inhibitor for aluminum alloy [58]. Mulliken charges were calculated to understand the chemical interaction between pectin and the metallic surface. Coordination bonds are associated with the chemical interaction between the inhibitor and the substrate, which is influenced by electron density on the heteroatom. Mulliken charges indicate the probable electron distribution and density. For pectin, the charges of all oxygen atoms were between – 0.5071 and – 0.6298, demonstrating the trend toward coordination bond formation.

Singh, Ansari, Quraishi, et al. (2021) studied guar gum grafted with ethyl acrylate as a corrosion inhibitor for P110 steel in 15% HCl [35]. The molecular dynamics simulation obtained adsorption energies of -323.52 and $-180.37\,kJ\,mol^{-1}$ for the grafted biopolymer and biopolymer, respectively. The more negative value for the grafted biopolymer shows a more stable interaction between the inhibitor and the metallic surface, which explains the higher inhibition efficiency for the grafted biopolymer.

3.12 Gaps and Future Trends

Some gaps were identified in the biopolymer and grafted biopolymer scenario. Most experimental studies are carried out at low temperatures (below 333 K). Future research should address ways to improve inhibitor efficiency in more aggressive environments such as highly acidic and high-temperature conditions and devise adequate grafting strategies. Devising a proper grafting strategy is important for maintaining high efficiency in more aggressive conditions. In addition, these macromolecules might have higher mobility, thereby raising the temperature, a hypothesis that should also be investigated in future studies. However, extreme conditions could degrade biopolymers, which is also important to evaluate.

The influence of grafting degree and grafting compound structures associated with the steric effects and inhibition performance need to be thoroughly investigated. These issues will define the point at which the grafting structure changes from an efficiency enhancer to a steric hindrance issue.

Since the presence of the homopolymer has yet to be discussed, the grafted biopolymers need to be better characterized in future research. Homopolymers can interfere with inhibition performance, resulting in inaccurate data interpretation. Thus, the reported efficiency would be exclusively related to the copolymer if the percentage of homopolymers in the copolymer structure was established. Further purification after grafting could be desirable in some cases.

Finally, theoretical studies should address not only the monomer structure but also larger molecules to predict the most promising copolymers in grafting compounds. Theoretical studies with macromolecules can be conducted to corroborate experimental tests, providing new insights on adsorption modes and favorable interactions with metallic surfaces.

Acknowledgments

The work was supported through the project UIDB/50006/2020 | UIDP/50006/2020, funded by FCT/MCTES through national funds.

References

1 Verma, C., Ebenso, E.E., Quraishi, M.A., and Hussain, C.M. (2021). *Materials Advanced* 2: 3806–3850.

2 Benarioua, M., Mihi, A., Bouzeghaia, N., and Naoun, M. (2019). *Egyptian Journal of Petroleum* 28: 155–159.

3 Gruca, A. and Greczek-Stachura, M. (2019). *Inżynieria Materiałowa* 1: 19–24. doi: 10.15199/28.2019.1.4.

4 Ouici, H., Tourabi, M., Benali, O. et al. (2017). *Journal of Electroanalytical Chemistry* 803: 125–134.

5 Yadav, M., Goel, G., Hatton, F.L. et al. (2021). *Current Research in Green and Sustainable Chemistry* 4: 100153.

6 Mourya, P., Banerjee, S., and Singh, M.M. (2014). *Corrosion Science* 85: 352–363.

7 Mobin, M. and Rizvi, M. (2017). *Carbohydrate Polymers* 156: 202–214.

8 Olajire, A.A. (2017). *Journal of Molecular Liquids* 248: 775–808.

9 Furtado, L.B., Nascimento, R.C., Guimarães, M.J.O.C. et al. (2021). *Sustainable Chemistry and Pharmacy* 19: 100353.

10 Furtado, L.B., Rocha, J.C., Gomes, J.A.C.P. et al. (2020). *Materials Chemistry and Physics* 256.

11 Furtado, L.B., Nascimento, R.C., Seidl, P.R. et al. (2019). *Journal of Molecular Liquids* 284: 393–404.

12 Umoren, S.A. and Eduok, U.M. (2016). *Carbohydrate Polymers* 140: 314–341.

13 Cherrak, K., Belghiti, M.E., Berrissoul, A. et al. (2020). *Surfaces and Interfaces* 20: 100578.

14 Monticelli, C. (2018). *Surface Science and Electrochemistry. Encyclopedia of Interfacial Chemistry*, Elsevier. 164–171.

15 Inamuddin Ahamed, M.I., Luqman, M., and Altalhi, T. (2021). *Materials Research Foundations* 107: doi: 10.21741/9781644901496.

16 Rbaa, M., Benhiba, F., Dohare, P. et al. (2020). *Chemical Data Collections* 27: 100394.

17 Fardioui, M., Rbaa, M., Benhiba, F. et al. (2021). *Journal of Molecular Liquids* 323: 114615.

18 Biswas, A., Das, D., Lgaz, H., Pal, S. and Nair, U.G. (2019). *Journal of Molecular Liquids* 275: 867–878.

19 Sosnik, A., Gotelli, G., and Abraham, G.A. (2011). *Progress in Polymer Science* 36: 1050–1078.

20 Vega-Hernández, M.A., Cano-Díaz, G.S., Vivaldo-Lima, E. et al. (2021). *Processes* 9: 37.

21 Slagman, S., Zuilhof, H., and Franssen, M.C.R (2017). *ChemBioChem* 19: 288–311.

22 Cannatelli, M.D. and Ragauskas, A.J. (2016). *Applied Microbiology and Biotechnology* 100: 8685–8691.

23 Roy, D., Semsarilar, M., Guthrie, J.T., and Perrier, S. (2009). *Chemical Society Reviews* 38: 2046–2064.

24 Fakhru'l-Razi, A., Qudsieh, I.Y., Yunus, W.M., Ahmad, M.Z., and Rahman, J. (2001). *Journal of Applied Polymer Science* 82: 1375–1381.

25 Babaladimath, G., Badalamoole, V., and Nandibewoor, S.T. (2018). *Materials Chemistry and Physics* 205: 171–179.

26 Wang, C., Zou, C., and Cao, Y. (2021). *Journal of Molecular Structure* 1228: 129737.

27 Geethanjali, R., Ali Fathima Sabirneeza, A., Subhashini, S. (2014). Water-Soluble andBiodegradable Pectin-Grafted Polyacrylamide and Pectin-Grafted Polyacrylic Acid: Electrochemical Investigation of Corrosion-Inhibition Behaviour on Mild Steel in 3.5% NaCl Media. *Indian Journal of Materials Science* 2014: 9. https://doi.org/10.1155/2014/356075.

28 Banerjee, S., Srivastava, V., and Singh, M.M. (2012). *Corrosion Science* 59: 35–41.

29 Lele, V.V., Kumari, S., and Niju, H. (2018). *Starch* 70: 1700133.

30 Srivastava, V., Banerjee, S., and Singh, M.M. (2010). *Journal of Applied Polymer Science* 116: 810–816.

31 Roy, P., Karfa, P., Adhikari, U., and Sukul, D. (2014). *Corrosion Science* 88: 246–253.

32 Biswas, A., Pal, S., and Udayabhanu, G. (2015). *Applications of Surface Science* 353: 173–183.

33 Li, X., Deng, S., Lin, T., Xie, X. and Du, G. (2020). *Journal of Materials Research and Technology* 9 (2): 2196–2207.

34 Deng, S., Li, X., and Du, G. (2021). *Chinese Journal of Chemical Engineering* 37: 222–231

35 Singh, A., Ansari, K.R., Quraishi, M.A. et al. (2021). *International Journal of Hydrogen Energy* 46: 9452–9465.

36 Biswas, A., Mourya, P., Mondal, D., Pal, s., Udayabhanu, G. (2018). *Journal of Molecular Liquids* 251: 470–479.

37 Lahrour, S., Benmoussat, A., Bouras, B. et al. (2019). *Applied Sciences* 9: 4684.

38 Liu, M., Xia, D., Singh, A., and Lin, Y. (2021). *Processes* 9: 1642.

39 Rbaa, M., Benhiba, F., Hssisou, R. et al. (2021). *Journal of Molecular Liquids* 322: 114549.

40 Singh, A., Liu, M., Ituen, E., and Lin, Y. (2020). *Coatings* 10: 241.

41 Singh, A., Ansari, K.R., and Quraishi, M.A. (2020). *International Journal of Hydrogen Energy* 45: 25398–25408.

42 Eduok, U., Ohaeri, E., Szpunar, J., and Akpan, I. (2020). *Journal of Molecular Liquids* 315: 113772.

43 Bhatt, N., Gupta, P.K., and Naithani, S. (2011). *Cellulose Chemistry and Technology* 45: 321–327.

44 Lv, X., Song, W., Ti, Y. et al. (2013). *Carbohydrate Polymers* 92: 388–393.

45 Taghizadeh, M.T., and Mehrdad, A. (2006). *Iranian Journal of Chemistry and Chemical Engineering* 25: 1–12.

46 Zhang, J., Zhang, S., Yuan, K., and Wang, Y. (2007). *Journal of Macromolecular Science - Pure and Applied Chemistry* 44: 881–885.

47 Zhang, W., Ma, Y., Chen, L. et al. (2020). *Journal of Molecular Liquids* 307: 112950.

48 Mobin, M., Ahmad, I., Basik, M., Murmu, M., and Banerjee, P. (2020). *Sustainable Chemistry and Pharmacy* 18: 100337.

49 Mobin, M., Basik, M., and Aslam, J. (2018). *Journal of Molecular Liquids* 263: 174–186.

50 Zhang, W., Wu, Y.-C., and Li, H.-J. (2021). *Journal of Molecular Liquids* 321: 114923.

51 Moradi, M., Song, Z., and Xiao, T. (2018). *Journal of Materials Science and Technology* 34: 2447–2457.

52 Zhang, K., Yang, W., Yin, X. et al. (2018). *Carbohydrate Polymers* 181: 191–199.

53 Gupta, N.K., Joshi, P.G., Srivastava, V., and Quraishi, M.A. (2018). *International Journal of Biological Macromolecules* 106: 704–711.

54 Jokar, M., Farahani, T.S., Ramezanzadeh, B., and Taiwan Inst, J. (2016). *Chemical Engineer* 63: 436.

55 Zhang, W., Nie, B., Li, H. et al. (2021). *Carbohydrate Polymers* 260: 117842.

56 Fares, M.M., Maayta, A.K., and Al-Mustafa, J.A. (2012). *Corrosion Science* 65: 223–230.

57 Hasanin, M.S. and Al Kiey, S.A. (2020). *International Journal of Biological Macromolecules* 161: 345–354.

58 Fathima, H., Pais, M., and Rao, P. (2021). *Journal of Molecular Structure* 1243: 130775.

59 Zheng, Y., Gao, Y., Li, H. et al. (2021). *Journal of Molecular Liquids* 324: 114679.

60 Mobin, M. and Rizvi, M. (2016). *Carbohydrate Polymers* 136: 384–393.

61 Abou-Elseoud, W.S., Abdel-karim, A.M., Hassan, E.A., and Hassan, M.L. (2021). *Carbohydrate Polymer Technologies and Applications* 2: 100072.

62 Gowraraju, N.D., Jagadeesan, S., Ayyasamy, K. et al. (2017). *Journal of Molecular Liquids* 232: 9–19.

63 Arukalam, I.O., Alaohuru, C.O., Ugbo, K.N. et al. (2014). *International Journal of Advanced Science and Technology* 3: 5–16.

64 Jmiai, A., El Ibrahimi, B., Tara, A. et al. (2018). *Journal of Molecular Structure* 1157: 408–417.

65 Sangeetha, Y., Meenakshi, S., and Sairam Sundaram, C. (2015). *International Journal of Biological Macromolecules* 72: 1244–1249.

66 Umoren, S.A., Solomon, M.M., Madhankumar, A., and Obot, I.B. (2020). *Carbohydrate Polymers* 230: 115466.

67 Umoren, S.A., AlAhmary, A.A., Gasem, Z.M., and Solomon, M.M. (2018). *International Journal of Biological Macromolecules* 117: 1017–1028.

68 Halambek, J., Cindrić, I., and Grassino, A.N. (2020). *Carbohydrate Polymers* 234: 115940.

69 Pais, M., George, S.D., and Rao, P. (2021). *Surfaces and Interfaces* 26: 101418.

70 Farhadian, A., Kashani, S.A., Rahimi, A. et al. (2021). *Journal of Molecular Liquids* 338: 116607.

71 Lgaz, H., Chung, I., Salghi, R. et al. (2019). *Applications of Surface Science* 463: 647–658.

72 Umoren, S.A., Obot, I.B., Madhankumar, A., and Gasem, Z.M. (2015). *Carbohydrate Polymers* 124: 280–291.

73 Zhang, K., Yang, W., Xu, B. et al. (2018). *Journal of Colloid and Interface Science* 517: 52–60.

74 Solomon, M.M., Gerengi, H., Umoren, S.A. et al. (2018). *Carbohydrate Polymers* 181: 43–55.

75 Habibiyan, A., Ramezanzadeh, B., Mahdavian, M., and Kasaeian, M. (2020). *Journal of Molecular Liquids* 298: 111974.

76 Fares, M.M., Maayta, A.K., and Al-Qudah, M.M. (2012). *Corrosion Science* 60: 112–117.

77 Oukhrib, R., Ibrahimi, B.E., Oualid, H.A. et al. (2020). *Journal of Molecular Liquids* 312: 113479.

78 Sangeetha, Y., Meenakshi, S., and Sundaram, C.S. (2016). *Carbohydrate Polymers* 136: 38–45.

79 Sushmitha, Y. and Rao, P. (2019). *Surfaces and Interfaces* 16: 67–75.

80 Vimal Kumar, K. and Rao, B.V.A. (2019). *Materials Today: Proceedings* 15: 155–165.

81 Mobin, M. and Rizvi, M. (2017). *Carbohydrate Polymers* 160: 172–183.

82 Prabhu, D., Sharma, S., Prabhu, P.R., Jomy, J. and Sadanand, R.V. (2021). *Journal of the Iranian Chemical Society*. doi: 10.1007/s13738-021-02368-9.

83 Nadi, I., Belattmania, Z., Sabour, B., et al. (2019). *International Journal of Biological Macromolecules* 141: 137–149.

84 Obot, I.B., Onyeachu, I.B., and Kumar, A.M. (2017). *Carbohydrate Polymers* 178: 200–208.

85 Diaconescu, R., Grigoriu, A.-M., and Luca, C. (2007). *Cellulose Chemistry and Technology* 41: 385–390.

86 Luca, C., Grigoriu, A.-M., Diaconescu, R., and Secula, M. (2011). *Revista de Chimie* 62: 1033–1038.

87 Rbaaa, M., Fardioui, M., Verma, C. et al. (2020). *International Journal of Biological Macromolecules* 155: 645–655.

4

Biopolymers vs. Grafted Biopolymers

Challenges and Opportunities

N. Mujafarkani

PG and Research Department of Chemistry, Jamal Mohamed College (Autonomous), Affiliated to Bharathidasan University, Tiruchirappalli-620 020, Tamilnadu, India

4.1 Introduction

4.1.1 Biopolymers

The term biopolymers means natural polymers which are collected from natural sources such as microorganisms and plants. Biopolymers are easily degradable and therefore, environment friendly. They find applications in a variety of industries, from the food industry to engineering, packaging, and biomedical. Biopolymers are promising materials due to their unique properties such as durability, biocompatibility, and non-toxicity.

Examples of biopolymers are carbohydrates, starch, cellulose, proteins, DNA, RNA, collagen, and lipids.

4.2 Classification

Biopolymers have been classified into the following types based on the:

- type
- origin
- monomeric units

Grafted Biopolymers as Corrosion Inhibitors: Safety, Sustainability, and Efficiency, First Edition.
Edited by Jeenat Aslam, Chandrabhan Verma, and Ruby Aslam.

Based on type

- Carbohydrate-based polymers such as starch or sucrose are being used as a key for manufacturing products
- Starch can be active as a natural polymer. It is made of glucose and found in plant tissues.
- Cellulose is made of glucose derived from natural sources like cotton. It is used for packaging.
- Degradable polymers can be prepared from synthetic materials derived from petroleum.

Based on origin

- Natural biopolymers are biosynthesized by living organisms.
- Synthetic biopolymers are made of renewable materials such as permeable polylactic acid.
- Microbial biopolymers are made by microorganisms.

Based on monomeric units

- Polysaccharides: Starch and cellulose.
- Proteins are made of amino acids. Example.: Collagen, casein, ferritin
- Polynucleotides: Deoxyribonucleic acid (DNA) and ribonucleic acid (RNA)

4.3 Opportunities

Biopolymers are made from natural sources. Biopolymers are much needed in the future because they are the solution for a greener and more sustainable environment. They are renewable, compostable, and emit greenhouse gases. Biopolymers gain prominence and they are high in cytosine, starch, cellulose, and protein.

Biopolymers slowly replace conventional polymers by their properties. These resources are enormously used in several industries such as pharmaceuticals, electronics, food, and agriculture, thus generating huge volumes of waste. Newly, biopolymer-based products have fascinated interest in developing alternate technologies for the removal of toxic metal ions and dyes from wastewater, owing to their comparability, eco-friendliness, economical value, non-toxic, high thermal stability, ease of operation, and excellent sorption capacities.

Chitosan has the property of removing metals from water, which is used for water purification. Additionally, biopolymers belong to one such class of inhibitors to corrosion studies which are naturally available, cheap, eco-friendly, and biodegradable. Phosphorylated chitin also acts as an excellent anti-corrosive agent to protect mild steel from corrosion.

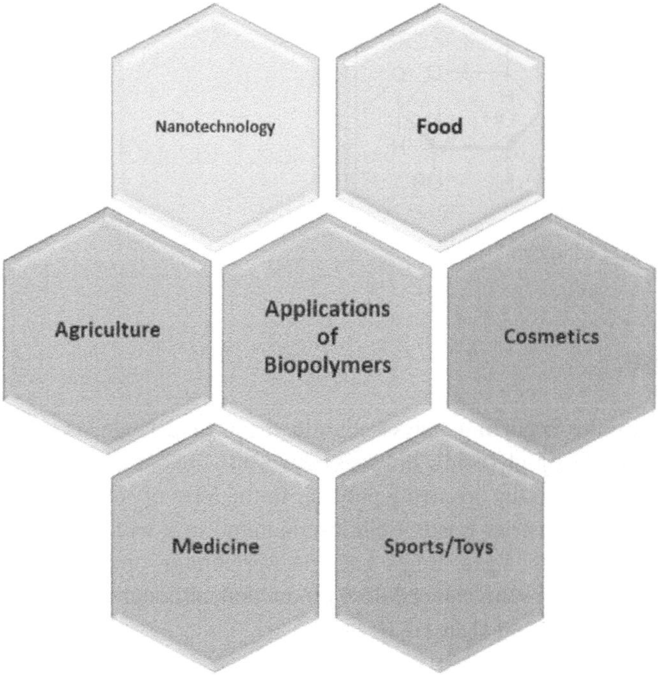

Figure 4.1 Schematic diagram for the applications of biopolymers.

4.4 Ecological Applications of Biopolymers

- Biopolymers are carbon neutral and can be renewed. They are stable since made from living materials.
- Biopolymers can reduce the amount of CO_2 in the atmosphere and reduce carbon emissions. This occurs because the biodegradability of these chemical compounds releases CO_2 and is reabsorbed by crops grown in their place.
- Biopolymers are biodegradable, which means they are less likely to pollute the environment. This is one of the most important benefits of this chemical compound. Though, the products made from this compound are not biodegradable.
- Reduces the dependence of these chemical compounds on non-renewable fossil fuels. These are simply compostable and reduce air pollution. This significantly decreases the harm caused using plastics to the environment. Continuing the use of biopolymers will reduce the use of fossil fuels.

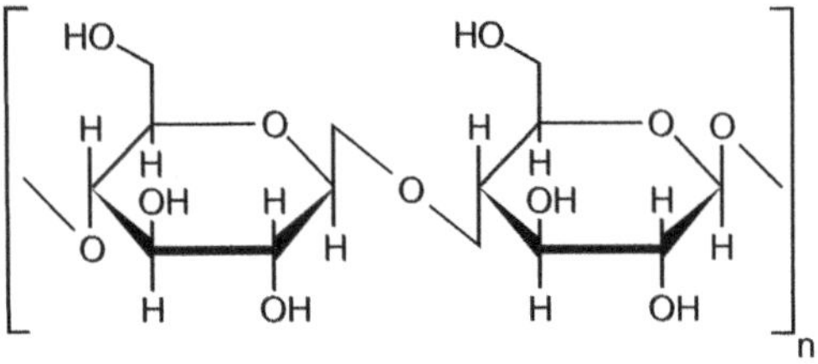

Figure 4.2 Structure of cellulose.

4.5 Cellulose

Cellulose is a naturally occurring polysaccharide made of β-D-glucose. The glucose units are joined by β-glycosidic linkages between C-1 of one unit and C-4 of the next unit. It is a naturally occurring polymer in the form of cotton, wood, and jute. It is an agro-industrial waste. Cellulose is found in a wide variety of organisms, including plants, algae, fungi, bacteria, vertebrates, and tunicates. Vegetable origin is the main source of cellulose production, although cellulose is also synthesized by bacteria and algae [1–3].

4.6 Cotton

Cotton is about 95% cellulose. Cotton fabrics withstand heat and have a high moisture absorption capacity. Because of these properties, cotton fabrics possess wearing comfort in summer. The chemical treatment of cotton with a concentrated alkali solution is called *mercerization*. The treatment lowers the degree of crystallinity, and the mercerized cotton is more amorphous than the native cotton. This improves the luster and moisture absorption capacity.

4.7 Regenerated Cellulose

Cellulose is usually dissolved in alkali and the resulting soda cellulose is treated with carbon disulfide to form an unstable compound *Cellulose xanthate*. This is dissolved in dilute alkali when pure cellulose is reprecipitated by hydrolysis. This form of cellulose is known as *regenerated cellulose*. Regenerated cellulose in the form of fiber is called *rayon* while as a film, it is known as *cellophane*.

4.8 Rayon

Rayon (also known as Viscose or Artificial silk) is in fact, regenerated cellulose. Cellulose obtained from wood pulp is dissolved in strong NaOH when alkali or soda cellulose is formed. This treatment with CS_2 will produce *Cellulose xanthate.* It is dissolved in dilute NaOH. The solution thus obtained is known as a viscose solution.

$$\underset{\text{Cellulose}}{R-OH+NaOH} \rightarrow \underset{\text{Soda cellulose}}{RONa+H_2O}$$

$$\underset{\text{Soda cellulose}}{R-ONa+CS_2} \rightarrow \underset{\text{Cellulose xanthate}}{R-O-C(S)-SNa}$$

The viscous solution is allowed to ripen for a few days. Cellulose xanthate is hydrolyzed, and cellulose is regenerated as shown:

$$R-O-C(S)-SNa+H_2O \rightarrow R-O-C-SH+NaOH \xrightarrow{-CS_2} \underset{\text{Regenerated cellulose}}{R-OH}$$

If the viscose solution is injected into a bath of sulphuric acid in fine jets, the cellulose gets regenerated in the form of filaments (Rayon). Rayon was almost exclusively used for making tire cords. However, this market disappeared with the increased use of nylon.

4.9 Cellophane

Cellophane, a transparent film, is also regenerated cellulose. In the manufacture of cellophane, the viscose solution is extruded into a thin film and then immersed in a bath of ammonium sulfate, sodium sulfate, and sulphuric acid. The xanthate group is removed, and cellulose is precipitated. The film is plasticized with glycerol and dried.

4.10 Cellulose Derivatives

4.10.1 Cellulose Acetate

Cellulose acetate is a cellulosic plastic made by the action of acetic anhydride on cellulose. Wood pulp forms the raw material for cellulose acetate. Sulphuric acid is used as the catalyst.

$$R-OH+(CH_3CO)_2O \rightarrow ROCOCH_3+CH_3COOH$$

Complete acetylation takes place to form primary cellulose acetate (cellulose triacetate). Partial saponification of the primary cellulose acetate leads to the formation of secondary cellulose acetate (cellulose diacetate) from which acetate fiber is produced.

Cellulose acetate is used for making non-inflammable safe photographic and cinematographic films under the name pyroxylin plastic. Thin films of cellulose acetate are used in wrapping and packaging.

4.10.2 Cellulose Nitrate

Cellulose nitrate plastic (celluloid) is produced by nitrating cellulose with a mixture of conc. HNO_3 and conc. H_2SO_4. At elevated temperatures, the polymer chain gets degraded. Hence, nitration is usually done at 20°C. The polymer is then plasticized with camphor and used.

$$\underset{\text{Cellulose}}{R-OH} + HONO_2 \rightarrow \underset{\text{Cellulose nitrate}}{RONO_2} + H_2O$$

Depending upon the degree of nitration, cellulose nitrate has different applications:

Nitrogen content of Cellulose nitrate	Applications
10.5–11%	Celluloid plastic
11–12%	Lacquer
12–12.3%	Photographic and cinematographic film
> 12.3%	Explosive

4.11 Starch

Starch is a naturally occurring biodegradable polymer that is readily available from agricultural processes. It is biodegradable, renewable, and of economical cost. It is found in abundance in rice, wheat, potatoes, corn, beans, and peas [4].

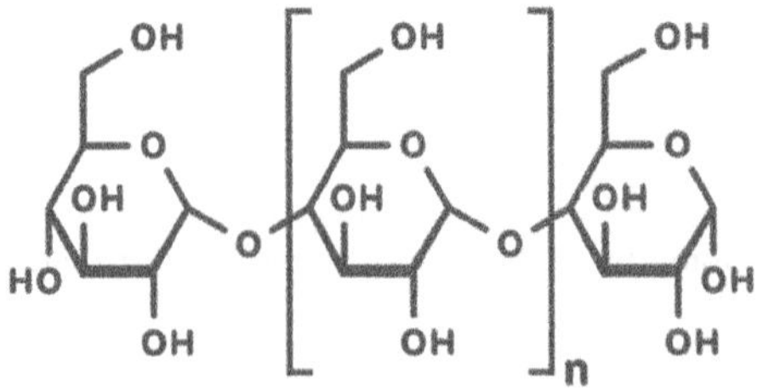

Figure 4.3 Structure of Amylose.

Starch contains a mixture of polysaccharides such as amylose (20%) and amylopectin (80%). Starch particles vary in size, shape, and particle size distribution. The ratio of amylose-amylopectin depends on the botanical origin [5].

Amylose is a straight-chain component of starch. It is made of

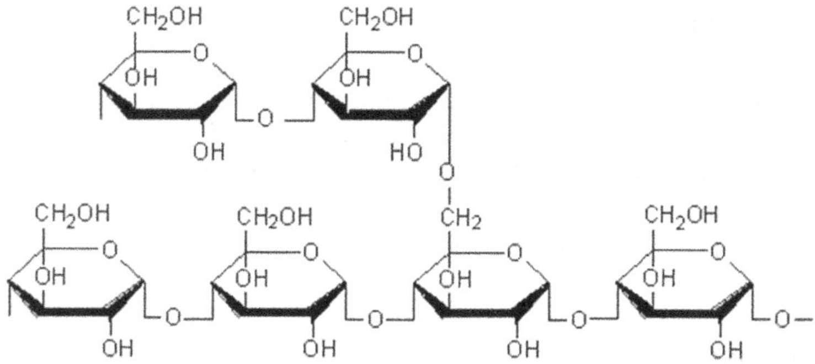

Figure 4.4 Structure of Amylopectin.

glucopyranose units linked by α-(1,4)-glycosidic linkages. It contains α-D-glucose units joined by x-glycosidic linkages between C-1 and C-4 units. Amylose is linear whereas amylopectin is a highly branched polymer.

Amylopectin is an extremely branched chain component of starch with short α-(1,4)-glycosidic chains linked by α-(1,6)-glycosidic branching points [6]. In addition, many side chains are attached to the basic chain by α-glycosidic linkages at C-1 and C-6 atoms. It is a good thickening agent and high viscosity, which produces very weak gels with poor mechanical properties [7].

Starch is a natural polymer that occurs in the form of hydrogels, films, and sponges. Natural and modified starch has been widely used in the chemical, pharmaceutical, environmental, and food industries due to its cheap cost, bio-decomposition, and biocompatibility. Furthermore, starch is used as a plant food and for making pastes [8, 9].

4.12 Silk Fibroin

Silk fibroin is a protein polymer derived from silkworms. Silk fibroin is commonly applied in fashion textiles and medical sewing. Silk is a polypeptide obtained from amino acids such as glycine, alanine, serine, and tyrosine. Garments made of silk will give a soft and warm feel.

4.13 Wool

Wool is a complex polypeptide containing about 20 α-amino acids. It is noted for its high moisture absorption, wrinkle resistance, and good insulating properties. It finds extensive use in suitings, blankets, carpets, and felts. So far, no artificial fiber has equaled it in resilience and flame resistance.

4.14 Proteins

Proteins are complex nitrogen-containing organic compounds with high molecular weight. They are synthesized from animal and plant sources, which produce amino acids in hydrolysis (enzymes). Thus, proteins are polymeric amides that are made of long chains of numerous amino acids bound together by peptide bonds. Proteins are among the most important types of substances present in living cells.

4.15 Collagen

The human body is rich in collagen protein (30%). Collagen is derived from amino acids made of carbon, hydrogen, and oxygen. Collagen consists of important amino acids such as glycine, proline, hydroxyproline, and arginine. Collagen strengthens and protects our various body organs. It is an important component of human skin and nails. Gelatin is another prime protein derived from collagen. The human body comprises sixteen different types of collagens. Due to the biodegradability and biocompatibility of collagen and gelatine, it is considered a good product for use in many fields, including pharmaceuticals and cosmetics [10].

4.16 Chitosan

Chitosan is a primary polysaccharide obtained from chitin. Chitin is considered a very important polysaccharide in the world, after cellulose. Chitosan is also found in the cell walls of yeast, fungi, and other organisms in lower plants and animals [11]. Chitosan is a linear polysaccharide composed of approximately distributed β-linked D-glucosamine and N-acetyl-D-glucosamine. It has been used to encapsulate probiotics. It is biodegradable non-toxic and biocompatible. This biopolymer exhibits biological activity, chemical activity, immune stimulation, enzyme biodegradation, mycotoxin, or epithelial infiltration, and is extensively used as a biological agent due to its good behavior and appropriateness to the human body [12]. Owing to their hemostatic character and hastening wound therapeutic effect, chitosan and chitin are mostly used in the treatment of burns, ulcers, and wounds. Chitosan is also used in tissue renewal and remodeling due to its biodegradability and cell attractiveness [13].

4.17 Challenges of Biopolymers

Biopolymers derived from renewable sources offer some drawbacks, despite the promising trends in compatibility such as less machine-driven properties, fast decay, high hydrophilic capacity, and poor mechanical properties that make

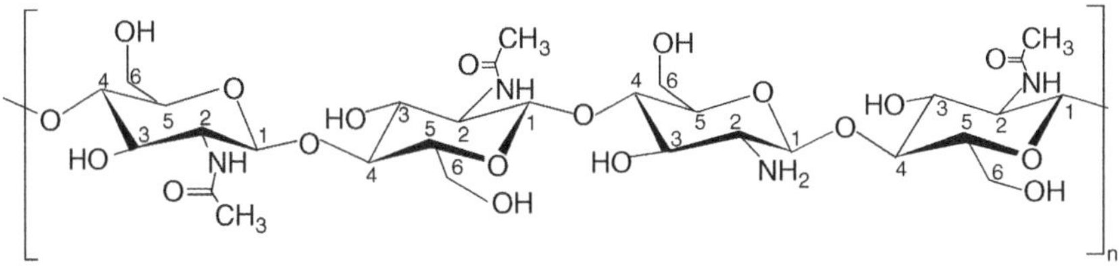

Figure 4.5 Structure of Chitin.

their use impossible, especially in moist atmospheres [14, 15]. In this context, various opinions are arising regarding the adequacy of biodegradable polymers in industries [16]. The challenge will be to progress novel mechanisms for the effectual and economical use of biopolymers. Chitosan is considered to be one of the most effective substances for absorbing contaminants in water purification systems. Chitosan contains amino and hydroxy functional groups. Owing to the presence of the functional groups permits the removal of heavy metal ions, organic compounds, and dyes from an aqueous solution [17–19]. In addition, these functional groups are subject to change, which improves absorption capacity and individuality [20].

4.18 Grafted Biopolymers

4.18.1 Opportunities

Grafting polymers consist of branch molecules. In grafted polymers, the main chain consists of one monomer unit and the branched chains form another monomer unit.

Graft copolymers containing a single branch are known as *miktoarm* star copolymers. The branches of grafted polymers may have different chemical structures or compositions in homo or copolymers [21]. Graft copolymerization is an attractive method for the surface functionalization of biopolymers and can be initiated by chemical methods, radiation techniques, and other systems. Synthetic polymers, biopolymers, or metal surfaces are considered backbones for grafting polymers.

There are three methods used to make graft polymer such as

- grafting to
- grafting from and
- grafting through [22].

Grafted biopolymers tend to enhance the nature of compatible, sustainable, flexible, recyclable, resistance to moisture, and soluble in an aqueous solution. Biopolymers modified by blending show significant improvement in the resistance

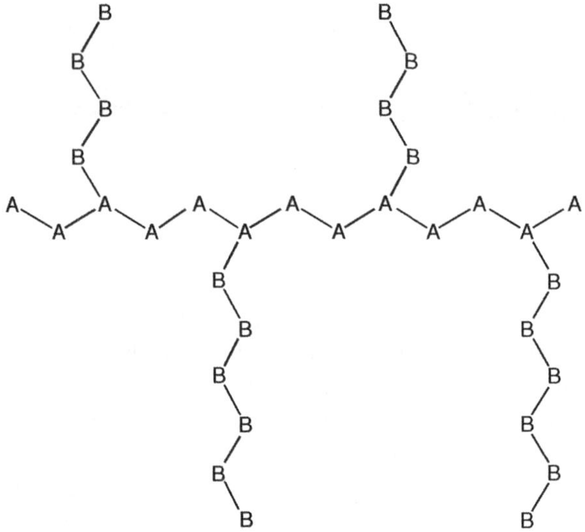

Figure 4.6 Structure of a grafting copolymer.

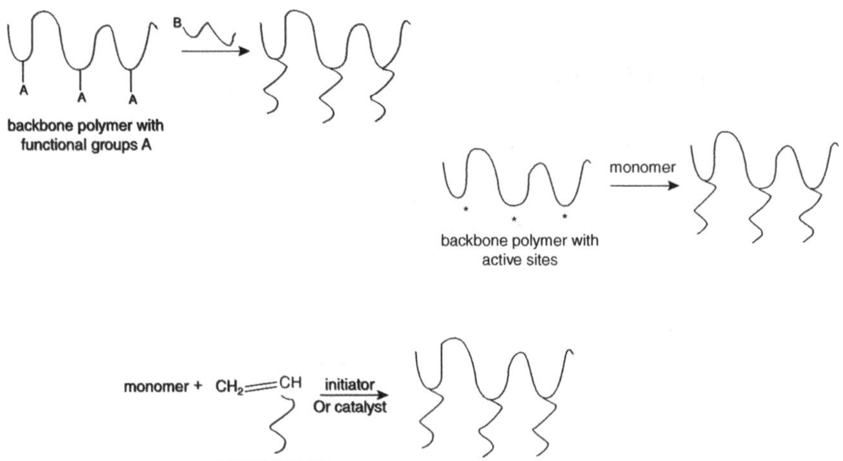

Figure 4.7 Structure of grafting methods of polymers.

of non-flexible polymers. Biopolymer systems with particles of one or more dimensions at the nanometer scale are called bio-nano composites, a special type of material with fire resistance, unique thermal stability, and optical and mechanical properties. Bio-nanocomposites are used efficiently in controlled drug delivery and food packaging [23].

Grafting is one of the finest ways to change the physical and chemical properties of cellulose. The possible uses and current developments of numerous cellulose graft biopolymers complete in the fields of controlled drug delivery, absorption of harmful and non-toxic dyes from wastes, absorption of toxic metal ions from wastewater, conversion of electrolyte, electrodes, and separators of the modern age lithium-ion batteries and fabrication of innovative food packaging materials. Some natural polymers and their modified nature polymers are ecological, non-toxic, and adsorbent. Modified biopolymers are extensively used as anti-corrosive and coating materials [24, 25].

Dextrin is derived from a polysaccharide, which is eco-friendly and has a certain anti-corrosive effect. Dextrin-grafted caprolactam copolymer (Dxt-g-CPL) has a better inhibitory effect in the part of corrosion [26]. The renewed emphasis on the use of biological-based monomers and biopolymers has given impetus to research efforts for the development of improved materials with important contents of biopolymers as a potential way to reduce (synthetic) polymer waste and disposal problems. These trends are aimed at research focusing on polymer grafting, which includes the grafting of polymers to cellulose, chitin/chitosan, or polysaccharides [27–31].

4.19 Challenges

Polymer grafting makes polymers superior to their synthetic and natural counterpart. The challenge is to get constituents that have properties like synthetics [32]. To accomplish this purpose, various methods of recognizing changes in environmental polymers are examined. Carbon Nano Structures can lead to the development of attractive substances that are suitable for a variety of medicinal applications with attractive biopolymers [33].

4.20 Conclusion

In this chapter, the theoretical aspects of the opportunities and challenges of biopolymers and grafted biopolymers are discussed. It considers features such as corrosion inhibitors, removal of toxic/heavy metal ions in the aqueous solution, absorption of harmful and non-toxic dyes from wastes, drug delivery, sustainability, and properties that make possible, their applicability & compatibility with biopolymers and grafted biopolymers. Particularly, many natural polysaccharides and their modified polymers (starch, cellulose, chitosan, etc.) have been extensively used as anti-corrosive and coating materials. Dextrin-grafted caprolactam copolymer act as an anti-corrosive agent to evade corrosiveness.

References

1 Klemm, D., Schumann, D., Kramer, F. et al. (2006). Nanocelluloses as innovative polymers in research and application. *Polysaccharides II* 205: 49.

2 Henriksson, M. and Berglund, L.A. (2007). Structure and properties of cellulose nanocomposite films containing melamine formaldehyde. *Journal of Applied Polymer Science* 106: 2817.

3 Nogi, M., Iwamoto, S., Nakagaito, A.N., and Yano, H. (2009). Optically transparent nanofiber paper. In: *Advanced Materials*, 21 (ed. X. Wang, H. Fu, A. Peng, et al.), 1595. Weinheim: Wiley.

4 Queiroz, V.M., Kling, I.C.S., Eltom, A.E. et al. (2020). Corn starch films as a long-term drug delivery system for chlorhexidine gluconate. *Materials Science and Engineering C* 112: 110852.

5 Qi, X. and Tester, R.F. (2019). Starch granules as active guest molecules or microorganism delivery systems. *Food Chemistry* 271: 182.

6 Chen, J., Wang, Y., Liu, J., and Xu, X. (2020). Preparation, characterization, physicochemical property and potential application of porous starch: a review. *International Journal of Biological Macromolecules* 148: 1169.

7 Šárka, E. and Dvořáček, V. (2017). New processing and applications of waxy starch (A review). *Journal Food Engineering* 206: 77.

8 Hemamalini, T. and Giri Dev, V.R. (2018). A comprehensive review on electrospinning of starch polymer for biomedical applications. *International Journal of Biological Macromolecules* 106: 712.

9 Wang, B., Sui, J., Yu, B. et al. (2021). Physicochemical properties and antibacterial activity of corn starch-based films incorporated with *Zanthoxylum bungeanum* essential oil. *Carbohydrate Polymers* 254: 117314.

10 Sionkowska, A., Wisniewski, M., Kaczmarek, H. et al. (2006). The influence of UV irradiation on the surface composition of collagen/PVP blended films. *Applied Surface Science* 253: 1970.

11 Rinaudo, M. (2006). Chitin and chitosan: properties and applications. *Progress in Polymer Science* 31 (7): 603.

12 Peña, J., Izquierdo-Barba, I., Martínez, A., and Vallet-Regí, M. (2006). New method to obtain chitosan/apatite materials at room temperature. *Solid State Science* 8: 513.

13 Singla, A.K. and Chitosan:, C.M. (2001). Some pharmaceutical and biological aspects-An update. *Journal of Pharmacy and Pharmacology* 53: 1047.

14 Demirgöz, D., Elvira, C., Mano, J.F. et al. (2000). Chemical modification of starch-based biodegradable polymeric blends: effects on water uptake, degradation behavior and mechanical properties. *Polymer Degradation and Stability* 70: 161.

15 Shankar, S. and Rahim, J. (2019). Effect of types of zinc oxide nanoparticles on structural, mechanical, and antibacterial properties of poly(lactide)/ poly(butylene adipate-co-terephthalate) composite films. *Food Packaging and Shelf Life* 21: 100327.

16 Meraldo, A. (2016). Introduction to bio-based polymers. In: *Multilayer Flexible Packaging*, 2e (ed. J.R. Wagner Jr), 47. Rochester: William Andrew.

17 Vakili, M., Rafatullah, M., Salamatinia, B. et al. (2014). Application of chitosan and its derivatives as adsorbents for dye removal from water and wastewater: a review. *Carbohydrate Polymers* 113: 115.

18 Boamah, P.O., Huang, Y., Hua, M. et al. (2015). Sorption of heavy metal ions onto carboxylate chitosan derivatives-A mini-review. *Ecotoxicology and Environmental Safety* 116: 113.

19 Tran, V.S., Ngo, H.H., Guo, W. et al. (2015). Typical low-cost biosorbents for adsorptive removal of specific organic pollutants from water. *Bioresource Technology* 182: 353.

20 Kyzas, G.Z. and Bikiaris, D.N. (2015). Recent modifications of chitosan for adsorption applications: a critical and systematic review. *Marine Drugs* 13: 312.

21 Hadjichristidis, N., Pitsikalis, M., Iatrou, H. et al. (2010). Graft copolymers. In: *Encyclopedia of Polymer Science and Technology*, 2e. (ed. K. Matyjaszewski), 1–38. John Wiley & Sons: Hoboken, NJ, USA. ISBN 978-047-144-026-0.

22 Abetz, V. (et 2005). *Encyclopedia of Polymer Science and Technology* (ed. W. Aktualisiert), Hoboken, N.J.: Wiley–Interscience.

23 Pathania, D., Sharma, G., and Kumar, A. (2017). *Modified Biopolymers: Challenges & Opportunities*. Nova Science Publishers Inc. ISBN: 978-1536121162.

24 Zhang, H., Wang, D., Wang, F. et al. (2015). Corrosion inhibition of mild steel in hydrochloric acid solution by quaternary ammonium salt derivatives of corn stalk polysaccharide (QAPS). *Desalination* 372: 57.

25 Carneiro, J., Tedim, J., and Ferreira, M.G.S. (2015). Chitosan as a smart coating for corrosion protection of aluminum alloy 2024: a review. *Progress Organic Coatings* 89: 348.

26 Liu, M. and Xia, D. (2021). Ambrish Singh and Yuanhua Lin, analysis of the Anti-corrosion performance of Dextrin and its graft copolymer on J55 steel in acid solution. *Processes* 9: 1642.

27 Roy, D., Semsarilar, M., Guthrie, J.T., and Perrier, S. (2009). Cellulose modification by polymer grafting: a review. *Chemical Society Reviews* 38: 2046.

28 Wohlhauser, S., Delepierre, G., Labet, M. et al. (2018). Grafting polymers from cellulose nanocrystals: synthesis, properties, and applications. *Macromolecules* 51: 6157.

29 Jenkins, D.W. and Hudson, S.M. (2001). Review of vinyl graft copolymerization featuring recent advances toward controlled radical-based reactions and illustrated with chitin/chitosan trunk polymers. *Chemical Reviews* 101: 3245.

30 Thakur, V.K. and Thakur, M.K. (2014). Recent advances in graft copolymerization and applications of Chitosan: a review. *ACS Sustainable Chemistry & Engineering* 2: 2637.
31 Kaur, L. and Gupta, G.D. (2017). A review on microwave-assisted grafting of polymers. *International Journal of Pharmaceutical Sciences and Research* 8: 422.
32 Kashyap, P.L., Xiang, X., and Heiden, P. (2015). Chitosan nanoparticle-based delivery systems for sustainable agriculture. *International Journal of Biological Macromolecules* 77: 36.
33 Kasalkova, N.S., Žáková, P., Stibor, I. et al. (2019). Carbon nanostructures grafted biopolymers for medical applications. *Materials Technology* 34 (7): 376.

Part 2

Overview of Sustainable Grafted Biopolymers

5

Sustainable Grafted Biopolymers

Synthesis and Characterizations

Omar Dagdag[1,*], *Rajesh Haldhar*[2], *Sheerin Masroor*[3], *Seong-Cheol Kim*[2], *Elyor Berdimurodov*[4], *Ekemini D. Akpan*[1], *and Eno E. Ebenso*[1,*]

[1] *Centre for Materials Science, College of Science, Engineering, and Technology, University of South Africa, Johannesburg 1710, South Africa*
[2] *School of Chemical Engineering, Yeungnam University, Gyeongsan 38541, Republic of Korea*
[3] *Department of Chemistry, A.N. College, Patliputra University, Patna, 800013, Bihar, India*
[4] *Faculty of Chemistry, National University of Uzbekistan, Tashkent, 100034, Uzbekistan*
[*] *Corresponding authors*

5.1 Introduction

Biopolymers are divided into two groups according to their origin: synthetic and natural biopolymers. Polyesters, polysaccharides, and proteins are three classes of native biopolymers. For example, collagen, keratin, and elastin gluten protein are biopolymers while chitin, cellulose, and chitosan are polysaccharide biopolymers. In polysaccharide biopolymers, glycosidic bonds are bonded to monomeric saccharides [1, 2]. Biopolymers are polymeric compounds derived from living organisms, including monomers such as monosaccharides, nucleic acids, and amino acids [3, 4]. There are two ways to produce biopolymers: monomer synthesis and polymerization. Figure 5.1 shows the production of biopolymers. In particular, the design of microbial biopolymers requires C, N, salts, and other minerals.

The surface structure and morphology of biopolymers can be determined by various microscopic methods, such as AFM [5], SEM [6], TEM [7], and XDR [8] analysis. The functional changes of the biopolymers can be observed by Fourier Transform Infrared (FTIR) spectroscopic analysis. Other methods are used to describe the properties of DSC biopolymers, gravimetric methods, determination of the rate of erosion or swelling, TGA analysis, and XPS analysis [9].

Grafted Biopolymers as Corrosion Inhibitors: Safety, Sustainability, and Efficiency, First Edition.
Edited by Jeenat Aslam, Chandrabhan Verma, and Ruby Aslam.

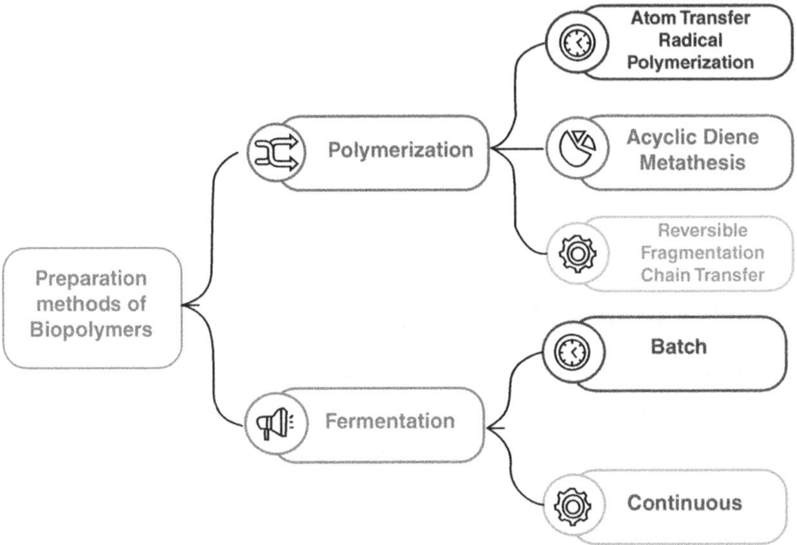

Figure 5.1 Preparation methods of Biopolymers. [3] / Reproduced with permission of Elsevier.

5.2 Grafted Biopolymers: Synthesis and Characterizations

5.2.1 Grafted Polysaccharides

Macromolecular properties (heat capacity, biodegradability, water or oil solubility, and chemical or physical characteristics) of base polysaccharides, such as chitin, starch, and cellulose were modified with the various organic compounds. In their native form, polysaccharides are too strong for degradation [10], but their safety is limited because they promote biodegradation. On the other hand, synthetic polymers are easy to operate, but the strength of the wool is limited. With the introduction of polymeric materials, natural polysaccharides [11] can be transformed into matrices that can be easily processed into materials suitable for various applications. The use of chemicals is one of the best ways to create a harmonious relationship between synthetic and natural materials to create new products and blends. This includes the incorporation of polymer chains, usually monomers, into older polymeric materials.

Some vinyl monomers such as acrylamide [12], N-tert-butyl acrylamide [13], methyl methacrylate [14], acrylonitrile [15], and methacrylamide [16] are inoculated into various polysaccharides to optimize for commercial use. Modified vinyl [10] improves the flocculation properties of polysaccharides and increases the susceptibility to branched bacteria. Guar-*graft*-poly(acrylamide) has been described for the release of pH-sensitive compounds [17]. In addition to protective derivatives, many graft copolymers are based on polysaccharides that help regulate drug release [18].

5.2.1.1 Characterization of Polysaccharide Graft Copolymers

Linked polysaccharides provide different macromolecules for different applications. After the synthesis of polymeric compounds, polysaccharides usually exhibit a plastic layer that is solvent-degradable, and insoluble in various liquids [19]. Such products pose cleaning problems, and such devices can be found in permanent or non-permanent models or modified models. Polymer separation and material identification may require alternative techniques to determine the copolymer. Many copolymer problems can be solved by the general analysis of synthetic polymers [20].

Solvent injection, in which a polymer is injected into a polysaccharide, commonly referred to as a polysaccharide, and there are several ways to determine the increase in activity, including the percentage by weight or the number of grafted wings.

Copolymer separation is important in determining crop yield and should be used with caution. The solubility of polymer materials and the nature of the solvent impact these properties [21] and require a long period of extraction to eliminate the copolymer and/or the complete support from the copolymer surface. However, due to the duration of the test, the gel could not be removed by contact with the copolymers, which interfered with removal and caused errors in the determination of the rate of incorporation. However, laser and ultrasound treatments are ineffective because they can destroy the joint ligaments, leading to a reduction in mass for injection [22].

5.2.2 Modification of Polysaccharides by Microwaves

Modification is an important way to improve the performance of both natural and synthetic polymers [23]. The recycling process provides synthetic materials from a number of popular polymers, including rubber [24], starch, cellulose/cellulose derivatives [25], and alginate [26]. Polysaccharides contained different polar functional groups, such as hydroxy, uronic acid, carboxylic acid, and amino can be electronically loaded or transported depending on the polysaccharide. Most hydrophilic and water-soluble polysaccharides exist in their natural form, while other polysaccharides, such as cellulose and chitosan, are insoluble in water and therefore require special degradation conditions. Although water is primarily an aqueous solution, it degrades many polysaccharides and their derivatives.

In a modification of polysaccharides by microwaves, polysaccharides are used as solvents, as well as their solvents, depending on their solubility in water. However, it is sometimes difficult to destroy soluble polysaccharides to obtain such solutions.

Due to the diversity of polysaccharide compounds, the temperature seems to be higher due to the reduced density of the shell (brown color), which increases the bonding. Another option is to use it as a screen that animators fill with excipients before heating it in the microwave. In the case of dispersion, when the polysaccharide is dispersed with other dispersants under mixed or dry conditions, the local modification of the polysaccharide components is greatly hindered by the heating of the insulator.

Polysaccharides appear to differ in that they can be used in the microwave depending on their function and properties. In traditional heating systems, high levels of turbidity are sufficient because oxygen disinfects and increases heat. IBD produces stable free radicals that do not interfere with the reaction. An important advantage of using a magnetron with polysaccharide separation is that, unlike conventional methods, the reaction is isolated in atmospheric conditions.

To date, most polysaccharides have been affected by microwave radiation (Figures 5.2a and Figures 5.2b) and the impact of microwave radiation on the performances of polymer materials has been investigated. Several methods have been used, including the use of microwaves, to determine polysaccharides in open or closed vessels, and solutions (uniform or heterogeneous) environments.

The most effective popular way is microwave radiation in the presence of pathogenic and/or catalytic and commercial crosslinkers [27]. One lesser-known technique is the use of microwave radiation without any free radical initiator [28]. Both methods of bonding with microwaves involve the use of the substrate of microwaves as a solvent, which may form a solvent, but is useful in microwave ovens [29]. Although polysaccharides changed significantly and rapidly as a result of the first two changes, the yield was generally higher than in microwave culture [30]. Microwave ovens are widely used in polysaccharide implants, but more research is needed to understand the application of the process described above and to understand the importance of the use of polysaccharides in purifying the contents or macromolecules.

5.2.2.1 Gum Acacia

Gum acacia (GA) is a low egg protein arabinogalactan complex and is classified as an arabinogalactan protein complex [32]. Thanks to the combination of conductive polymers such as polyaniline (PANI) and the adaptive matrix GA, it is ideal for manufacturing due to its electrical conductivity, chemical resistance to additives, and thermal stability [33]. The microwave synthesis methodology with ammonium persulfate (APS) catalyst was used to obtain GA [27] as oxidizing agent/starter during disease states. The reaction mixture is the same because all the solvents are dissolved in the aqueous phase. Polyethylene glue based on acacia glue (GA-g-PANI) has electrical properties. The pH change depends on the degree of adhesion and the pH of the material.

5.2.2.2 Alginate

Sodium alginate can be dissolved in water as a polysaccharide [26] with a chain line of d-manuronic acid associated with β (1 → 4) and α – (1 → 4) residue of guluronic acid residues (Figure 5.2a (A)). It is widely used in the food industry for increasing viscosity and as an emulsifier, anti-digestive tablet, dental implant, and textile industry in the printing house. A new wave of alginate [34] (SG-g-PAM) with sodium polyacrylamide, a commercially available composite, was obtained under the same conditions using microwave ovens. Microwave injection was initiated because no external seeds were used for this seed, unlike the gum acacia PANI for which the microwave injection was made.

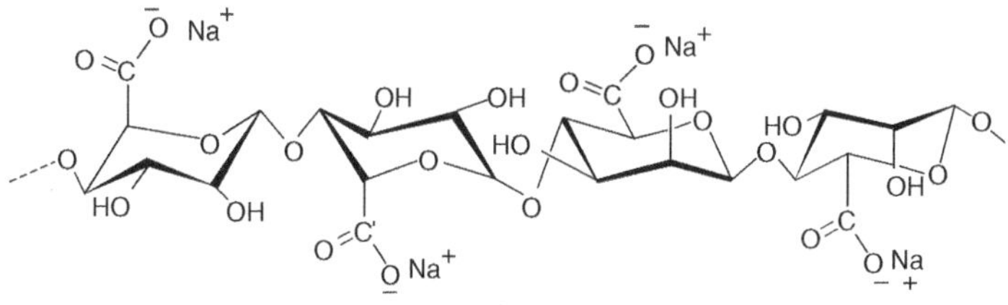

(A)

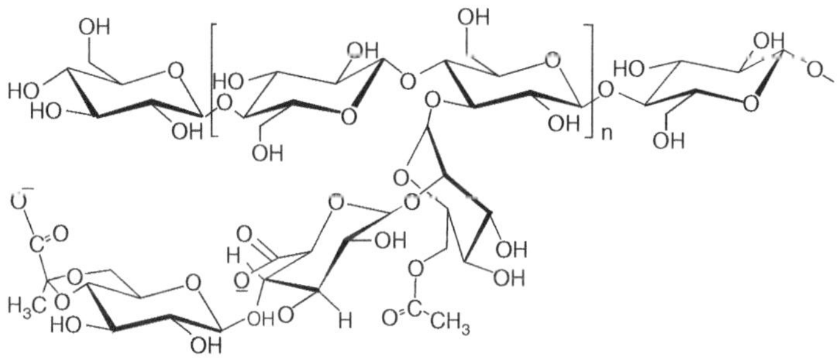

(B)

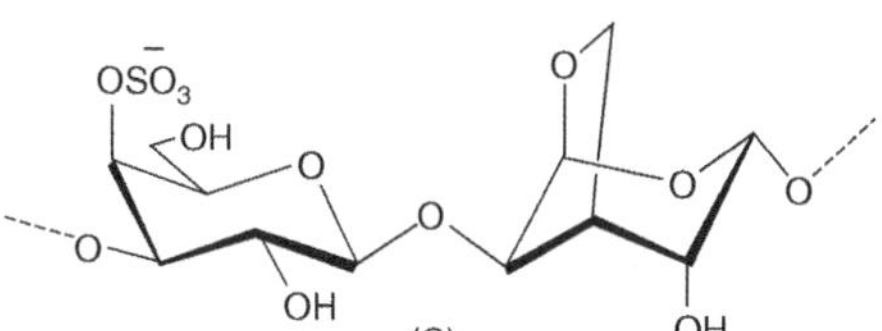

(C)

Figure 5.2 Structures of some "charged" polysaccharides that have been used for grafting modification under microwaves. [31] / Reproduced with permission of Elsevier.

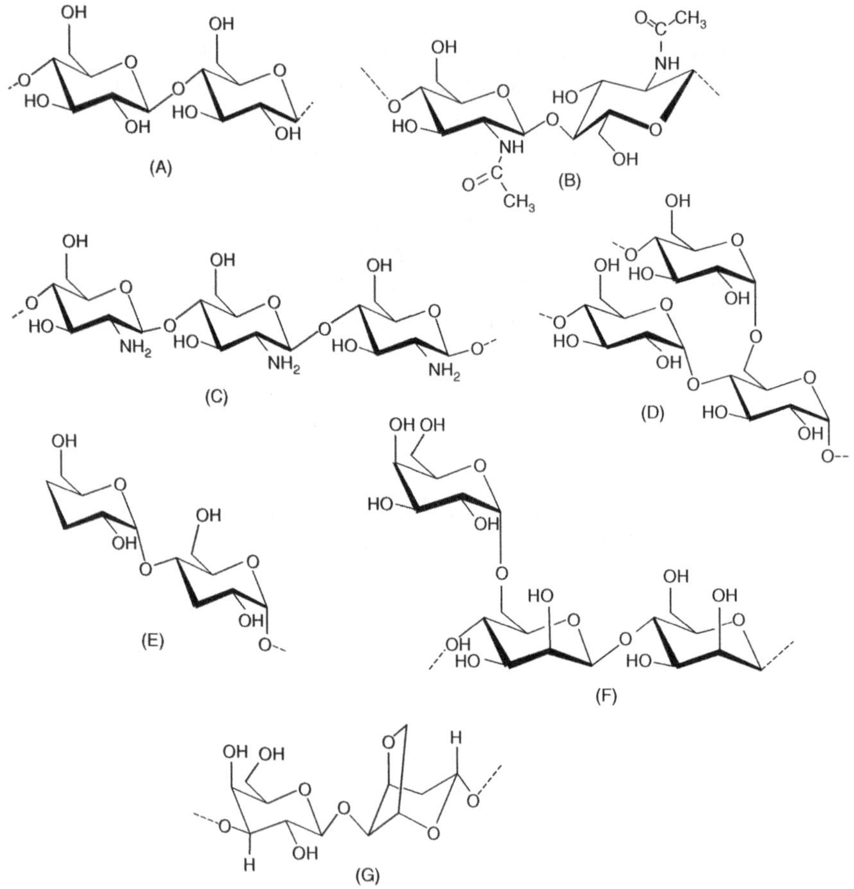

Figure 5.2 (Cont'd)

5.2.2.3 Cellulose

Cellulose, a solid polymer of natural origin, is a compound of the basic formula ($C_6H_{10}O_5$) n and consists of a chain of hundreds of thousands and thousands of D-glucose units combined with β (1 → 4) [35] (Figure 5.2b (A)). There are several scientific studies of cellulose conversion in various commercial applications [36]. The cellulose-based materials were used as superabsorbent polymer resin for heavy ions [37]. The heavy metal ions are environmentally hazardous. This issue was effectively solved by creating green materials based on cellulose superabsorbent polymers. Cellulose is necessary for water and dispersed in an aqueous solution of vinyl monomer and modified by software under various conditions in the presence of a redox initiator.

5.2.3 Grafted Chitosan Derivatives

5.2.3.1 Phosphorylated Chitosan

Phosphorylation of chitosan was performed using phosphorus anhydride with methanesulfonic acid as the solvent to obtain a water-soluble material with good solubility at room temperature. Methanesulfonic acid also acts as a carrier during the reaction (Figure 5.3) [38]. Similarly, phosphorylated chitosan was prepared by reacting chitosan with phosphoric acid at 150°C (Figure 5.4) [38]. Mono-(2-methacryloxyethyl)phosphoric acid copolymerization experiments were also performed on phosphorylated chitosan. These products have strong zwitterionic properties and strong antibacterial properties (Figure 5.5) [38].

In another paper [39], the simultaneous reaction of chitosan with phosphoric acid and formaldehyde acts as an antibiotic in the formation of water-soluble N-mono- and N-diphosphonomethylene compounds (Figure 5.6).

5.2.3.1.1 N-Phthaloylated Chitosan

N-phthaloylchitosan-based polymers are more active than N, O-phthaloylchitosan derivatives [41]. A strongly deacetylated chitosan polymer, which reacts with

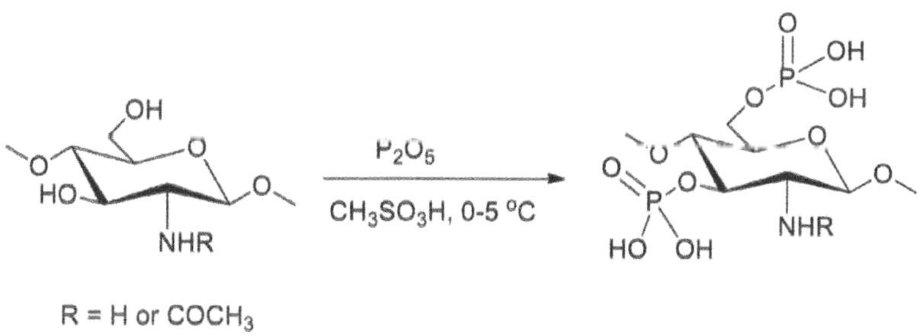

Figure 5.3 Phosphorylation of chitosan using phosphorus pentaoxide. [40] / Reproduced with permission of Elsevier.

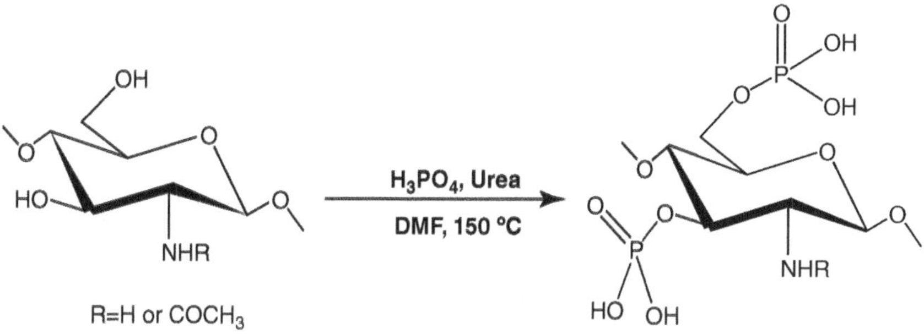

Figure 5.4 A Phosphorylation of chitosan using orthophosphoric acid [40] / Reproduced with permission of Elsevier.

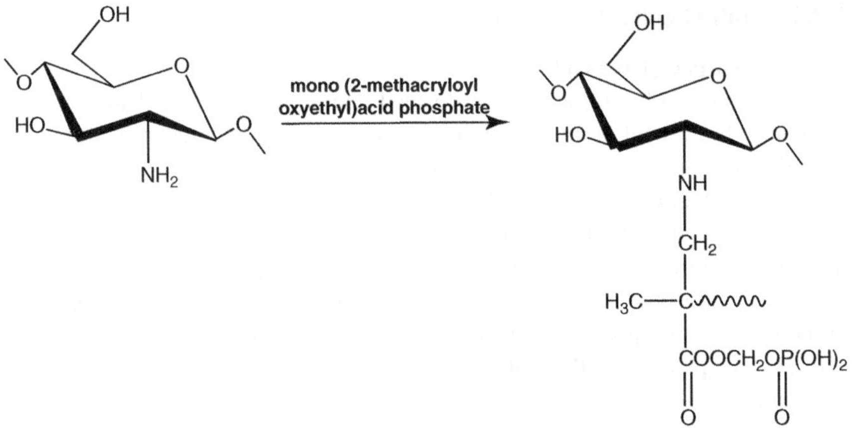

Figure 5.5 Phosphorylation of chitosan using mono-(2-methacryloyl oxyethyl) phosphoric acid. [40] / Reproduced with permission of Elsevier.

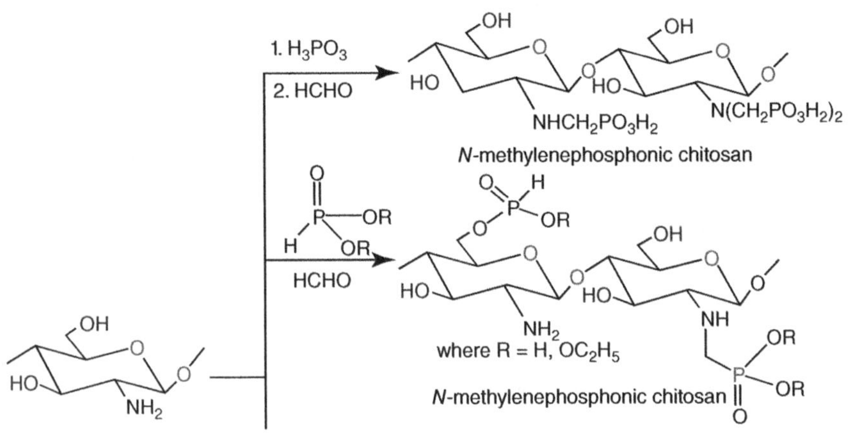

Figure 5.6 A functionalization of chitosan by phosphorous acid and formaldehyde. [40] / Reproduced with permission of Elsevier.

phthalic anhydride, forms an N-phthaloylchitosan derivative, a polar and soluble organic solvent (Figure 5.7) [42].

Huh et al [43] grafted two chitosan compounds: chitosan-containing quaternized C-PET and polyethylene terephthalate (C-PET) to produce a bacterial resistance tissue product that inhibits the rise of bacterial populations. The quaternized C-PET and C-PET have been shown to have high antimicrobial activity, ranging from 75% to 86%.

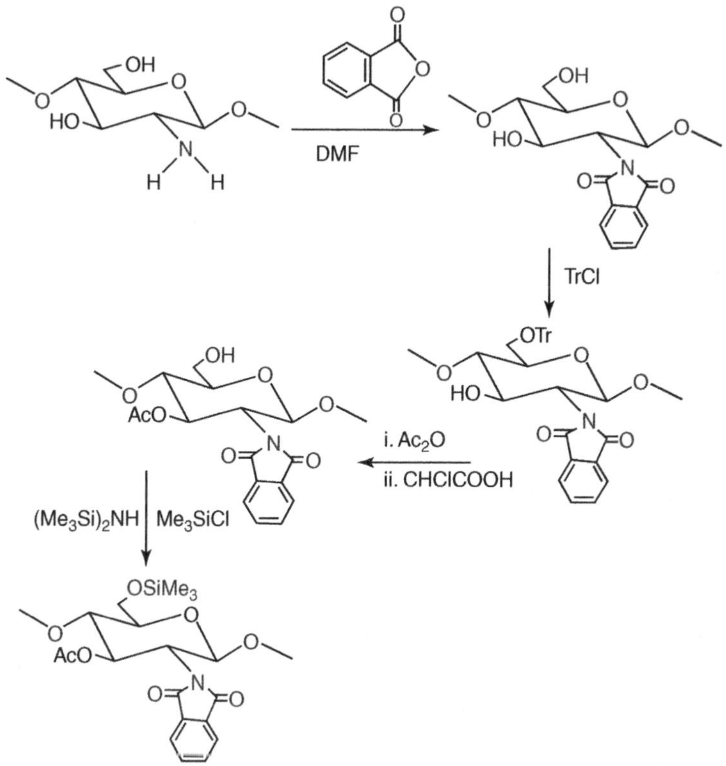

Figure 5.7 A diagram of the N-phthaloyl chitosan derivative. [40] / Reproduced with permission of Elsevier.

5.2.4 Crosslinked Chitosan Derivatives

Nagireddi et al. [44] described the potential of a glutaraldehyde-chitosan (GCC) copolymer to form and release lead(II) ions in a galvanic solution. GCC adsorbent (Figure 5.8) was obtained from the 17:1 (glutaraldehyde and chitosan).

Chitosan-derived biopolymers work better than non-volatile biopolymers. Chitosan-EDTA-modified biopolymer is used as a protective matrix in the fight against antibiotics. The drug is released temporarily due to the ionic interaction of dionic biopolymers such as 1,8-diamino-octane or lysine. Grafting ethylenediaminetetraacetic acid (EDTA) with chitosan biopolymer (Figure 5.9) [31] increases the antimicrobial activity of chelating magnesium ions, which stabilize the cell membrane of extracellular Gram-negative bacteria.

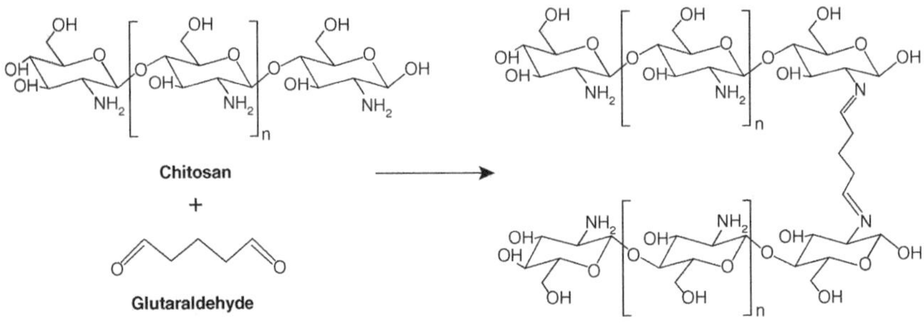

Figure 5.8 An image of the process of Chitosan Grafting by glutaraldehyde. [40] / Reproduced with permission of Elsevier.

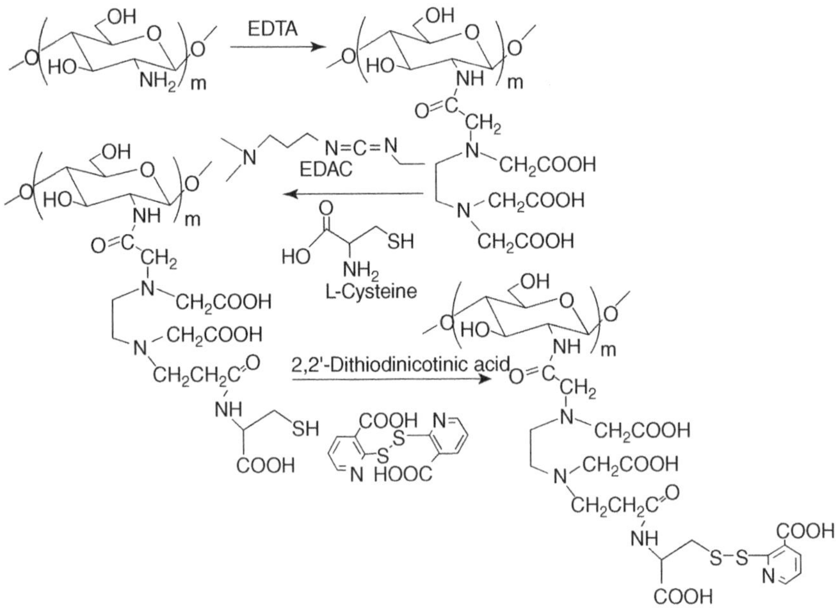

Figure 5.9 A diagram of Ethylene Diamine Tetraacetic Acid (EDTA) grafting on chitosan biopolymer. [40] / Reproduced with permission of Elsevier.

5.2.5 Chitosan Methacrylate Derivatives

Chitosan (CN) -methacrylate (M) [45] was obtained by refining CN and M polymers in a ratio of 1:0.25. The synthesis process was done on a CN/M polymer, resulting in dynamic polymerization with various monomers (Figure 5.10), for example, 2-acrylamido-2-methyl-1-propanesulfonic acid, methacrylamide, and 1-vinyl imidazole. The polymer of the virus is more soluble than biomethylchitosan methacrylate.

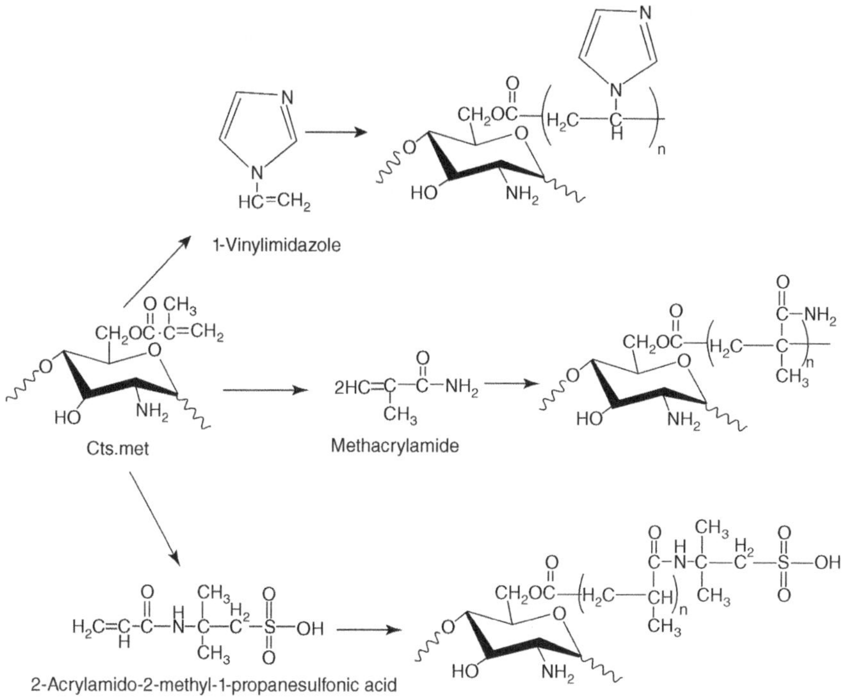

Figure 5.10 A diagram of the process of esterification of a Chitosan and Methacrylic acid. [40] / Reproduced with permission of Elsevier.

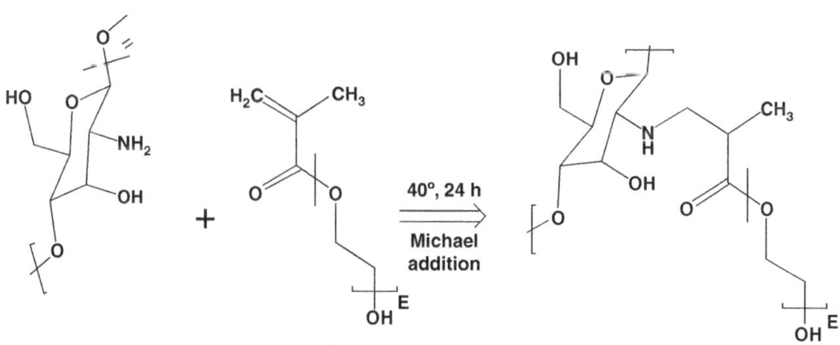

Figure 5.11 A representation of the formation of A Chitosan grafted-poly(ethylene glycol) methacrylate copolymer. [40] / Reproduced with permission of Elsevier.

First, Savin et al. [46] prepared water-soluble chitosan polymethacrylate (ethylene glycol) obtained by the Michael impurity reaction to form non-toxic microparticles/nanoparticles (MNPs) (Figure 5.11). Modification of chitosan improves solubility in aqueous media.

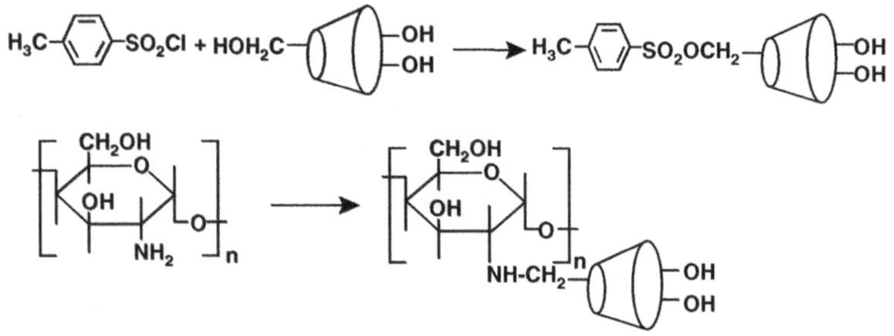

Figure 5.12 The rafting of β-CD on chitosan in the presence of a 2-hydroxypropyl spacer. [40] / Reproduced with permission of Elsevier.

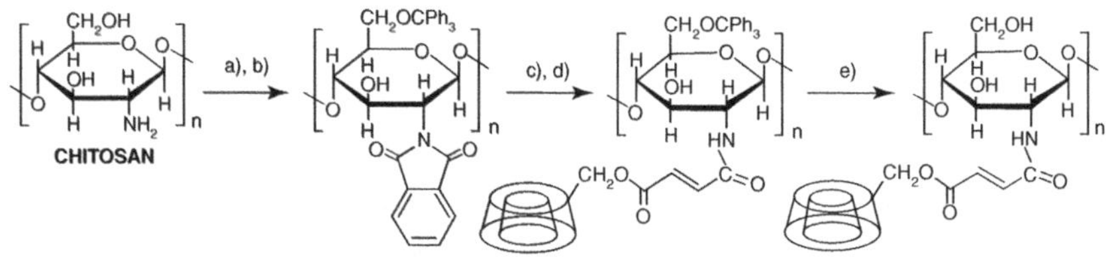

Figure 5.13 Grafting of β-CD on chitosan in the presence of sugar and maleic spacer. [40] / Reproduced with permission of Elsevier.

5.2.6 Targeted Chitosan Modification

1,6-Hexamethylene diisocyanate was prepared by cyclodextrin modification with chitosan. This new modification is for efficient adsorbent for cholesterol [47]. The spacers were a β-CD graft consisting of a 2-hydroxypropyl group and a chitosan-activated epoxide derived from chitosan (Figure 5.12) [48], which reduced glucose and malignancies (Figure 5.13) [49].

5.3 Conclusions

Synthesis and characterizations of grafted biopolymers were discussed and reviewed. The modern trends in the synthesis of grafted biopolymers were discussed. It was found that the hydroxyl functional groups of biopolymers are grafted centers. These polymers can be reacted with the various chemical compounds (organic compounds, polymers, nano-compounds, and other targeted materials) by the hydroxyl functional modification centers. The grafted biopolymers were characterized by the spectroscopic methods.

References

1 Kartik, A., Akhil, D., Lakshmi, D. et al. (2021). A critical review on the production of biopolymers from algae biomass and their applications. *Bioresource Technology* 329: 124868.
2 Han, X., Zheng, Y., Munro, C.J., Ji, Y., and Braunschweig, A.B. (2015). Carbohydrate nanotechnology: hierarchical assembly using nature's other information-carrying biopolymers. *Current Opinion in Biotechnology* 34: 41–47.
3 Yaashikaa, P., Kumar, P.S., and Karishma, S. (2022). Review on biopolymers and composites–Evolving material as adsorbents in removal of environmental pollutants. *Environmental Research* 212: 113114.
4 Khademian, E., Salehi, E., Sanaeepur, H., Galiano, F., and Figoli, A. (2020). A systematic review on carbohydrate biopolymers for adsorptive remediation of copper ions from aqueous environments-part A: classification and modification strategies. *Science of the Total Environment* 738: 139829.
5 Venkateshaiah, A., Padil, V.V., Nagalakshmaiah, M. et al. (2020). Microscopic techniques for the analysis of micro and nanostructures of biopolymers and their derivatives. *Polymers* 12: 512.
6 Fernando, M.S., Wimalasiri, A., Dziemidowicz, K. et al. (2021). Biopolymer-based nanohydroxyapatite composites for the removal of fluoride, lead, cadmium, and arsenic from water. *ACS omega* 6: 8517–8530.
7 Raval, N.P., Shah, P.U., and Shah, N.K. (2016). Nanoparticles loaded biopolymer as an effective adsorbent for adsorptive removal of malachite green from aqueous solution. *Water Conservation Science and Engineering* 1: 69–81.
8 Saravanan, D. and Sudha, P. (2014). Batch adsorption studies for the removal of copper from wastewater using natural biopolymer. *International Journal of ChemTech Research* 6: 3496–3508.
9 Udayakumar, G.P., Muthusamy, S., Selvaganesh, B. et al. (2021). Biopolymers and composites: properties, characterization and their applications in food, medical and pharmaceutical industries. *Journal of Environmental Chemical Engineering* 9: 105322.
10 Singh, R., Tripathy, T., Karmakar, G. et al. (2000). Novel biodegradable flocculants based on polysaccharides. *Current Science-Bangalore* 78: 798–803.
11 Gref, R., Rodrigues, J., and Couvreur, P. (2002). Polysaccharides grafted with polyesters: novel amphiphilic copolymers for biomedical applications. *Macromolecules* 35: 9861–9867.
12 Singh, V., Tiwari, A., Tripathi, D.N., and Sanghi, R. (2004). Microwave-assisted synthesis of guar-g-polyacrylamide. *Carbohydrate Polymers* 58: 1–6.
13 Fares, M.M., El-faqeeh, A.S., and Osman, M.E. (2003). Graft copolymerization onto starch–I. Synthesis and optimization of starch grafted with N-tert-butylacrylamide copolymer and its hydrogels. *Journal of Polymer Research* 10: 119–125.

14 Sharma, B.R., Kumar, V., and Soni, P. (2003). Ce (IV)-ion initiated graft copolymerization of methyl methacrylate onto guar gum. *Journal of Macromolecular Science, Part A* 40: 49–60.

15 Ikhuoria, E., Folayan, A., and Okieimen, F. (2010). Studies in the graft copolymerization of acrylonitrile onto cassava starch by ceric ion induced initiation. *International Journal of Biotechnology and Molecular Biology Research* 1: 010–014.

16 Celik, M. (2006). Preparation and characterization of starch-g-polymetha crylamide copolymers. *Journal of Polymer Research* 13: 427–432.

17 Soppimath, K.S., Kulkarni, A.R., and Aminabhavi, T.M. (2001). Chemically modified polyacrylamide-g-guar gum-based crosslinked anionic microgels as pH-sensitive drug delivery systems: preparation and characterization. *Journal of Controlled Release* 75: 331–345.

18 Maiti, S., Ranjit, S., and Sa, B. (2010). Polysaccharide-based graft copolymers in controlled drug delivery. *International Journal of PharmTech Research* 2: 1350–1358.

19 Singh, V., Sharma, A.K., and Maurya, S. (2009). Efficient cadmium (II) removal from aqueous solution using microwave synthesized guar gum-graft-poly (ethyl acrylate). *Industrial & Engineering Chemistry Research* 48: 4688–4696.

20 Sibilia, J.P. (1996). *A Guide to Materials Characterization and Chemical Analysis*. John Wiley & Sons.

21 Tomasik, P. and Schilling, C.H. (2004). Chemical modification of starch. *Advances in Carbohydrate Chemistry and Biochemistry* 59: 175–403.

22 Kargin, V.A. (1963). Solid-state properties of graft copolymers. *Journal of Polymer Science Part C: Polymer Symposia* 4 (3): 1601–1632.

23 Kabanov, V.Y. and Kudryavtsev, V. (2003). Modification of polymers by radiation graft polymerization (State of the art and trends). *High Energy Chemistry* 37: 1–5.

24 Pandey, P.K., Srivastava, A., Tripathy, J., and Behari, K. (2006). Graft copolymerization of acrylic acid onto guar gum initiated by vanadium (V)–mercaptosuccinic acid redox pair. *Carbohydrate Polymers* 65: 414–420.

25 Zampano, G., Bertoldo, M., and Bronco, S. (2009). Poly (ethyl acrylate) surface-initiated ATRP grafting from wood pulp cellulose fibers. *Carbohydrate Polymers* 75: 22–31.

26 Sand, A., Yadav, M., Mishra, D.K., and Behari, K. (2010). Modification of alginate by grafting of N-vinyl-2-pyrrolidone and studies of physicochemical properties in terms of swelling capacity, metal-ion uptake, and flocculation. *Carbohydrate Polymers* 80: 1147–1154.

27 Tiwari, A. and Singh, V. (2008). Microwave-induced synthesis of electrical conducting gum acacia-graft-polyaniline. *Carbohydrate Polymers* 74: 427–434.

28 Singh, V., Singh, S.K., and Maurya, S. (2010). Microwave-induced poly (acrylic acid) modification of Cassia javanica seed gum for efficient Hg (II) removal from solution. *Chemical Engineering Journal* 160: 129–137.

29 Polshettiwar, V. and Varma, R.S. (2008). Aqueous microwave chemistry: a clean and green synthetic tool for rapid drug discovery. *Chemical Society Reviews* 37: 1546–1557.

30 Singh, V., Tiwari, A., Pandey, S., and Singh, S.K. (2006). Microwave-accelerated Synthesis and Characterization of Potato Starch-g-poly (acrylamide). *Starch-Stärke* 58: 536–543.

31 Singh, V., Kumar, P., and Sanghi, R. (2012). Use of microwave irradiation in the grafting modification of the polysaccharides–A review. *Progress in Polymer Science* 37: 340–364.

32 Dror, Y., Cohen, Y., and Yerushalmi-Rozen, R. (2006). Structure of gum arabic in aqueous solution. *Journal of Polymer Science Part B: Polymer Physics* 44: 3265–3271.

33 Tiwari, A. (2007). Gum Arabic-graft-polyaniline: an electrically active redox biomaterial for sensor applications. *Journal of Macromolecular Science, Part A: Pure and Applied Chemistry* 44: 735–745.

34 Sen, G., Singh, R.P., and Pal, S. (2010). Microwave-initiated synthesis of polyacrylamide grafted sodium alginate: synthesis and characterization. *Journal of Applied Polymer Science* 115: 63–71.

35 Krässig, H.A. (1993). *Cellulose: Structure, Accessibility, and Reactivity*. Gordon and Breach Science Publ.

36 Roy, D., Semsarilar, M., Guthrie, J.T., and Perrier, S. (2009). Cellulose modification by polymer grafting: a review. *Chemical Society Reviews* 38: 2046–2064.

37 Bao-Xiu, Z., Peng, W., Tong, Z., Chun-yun, C. and Jing, S. (2006). Preparation and adsorption performance of a cellulosic-adsorbent resin for copper (II). *Journal of Applied Polymer Science* 99: 2951–2956.

38 Jayakumar, R., Selvamurugan, N., Nair, S.K.V., Tokura, S. and Tamura, H. (2008). Preparative methods of phosphorylated chitin and chitosan—An overview. *International Journal of Biological Macromolecules* 43: 221–225.

39 Heras, A., Rodriguez, N., Ramos, V., and Agullo, E. (2001). N-methylene phosphonic chitosan: a novel soluble derivative. *Carbohydrate Polymers* 44: 1–8.

40 Negm, N.A., Hefni, H.H., Abd-Elaal, A.A., Badr, E.A., and Abou Kana, M.T. (2020). Advancement on modification of chitosan biopolymer and its potential applications. *International Journal of Biological Macromolecules* 152: 681–702.

41 Kurita, K., Ikeda, H., Shimojoh, M., and Yang, J. (2007). N-phthaloylated chitosan as an essential precursor for controlled chemical modifications of chitosan: synthesis and evaluation. *Polymer Journal* 39: 945–952.

42 Dutta, P.K., Dutta, J., and Tripathi, V. (2004). Chitin and chitosan: chemistry, properties, and applications.

43 Huh, M.W., Kang, I.K., Lee, D.H. et al. (2001). Surface characterization and antibacterial activity of chitosan-grafted poly (ethylene terephthalate) prepared by plasma glow discharge. *Journal of Applied Polymer Science* 81: 2769–2778.

44 Nagireddi, S., Katiyar, V., and Uppaluri, R. (2017). Pd (II) adsorption characteristics of glutaraldehyde cross-linked chitosan copolymer resin. *International Journal of Biological Macromolecules* 94: 72–84.

45 Cankaya, N. (2019). Grafting of chitosan: structural, thermal and antimicrobial properties. *Journal of the Chemical Society of Pakistan* 41: 240.

46 Savin, C.-L., Popa, M., Delaite, C. et al. (2019). Chitosan grafted-poly (ethylene glycol) methacrylate nanoparticles as carrier for controlled release of bevacizumab. *Materials Science and Engineering: C* 98: 843–860.

47 Chen, C.-Y., Chen, -C.-C., and Chung, Y.-C. (2007). Removal of phthalate esters by α-cyclodextrin-linked chitosan bead. *Bioresource Technology* 98: 2578–2583.

48 Zhang, X., Wang, Y., and Yi, Y. (2004). Synthesis and characterization of grafting β-cyclodextrin with chitosan. *Journal of Applied Polymer Science* 94: 860–864.

49 Aime, S., Gianolio, E., Uggeri, F. et al. (2006). New paramagnetic supramolecular adducts for MRI applications based on non-covalent interactions between Gd (III)-complexes and β-or γ-cyclodextrin units anchored to chitosan. *Journal of Inorganic Biochemistry* 100: 931–938.

6

Sustainable Grafted Biopolymers

Properties and Applications

Paresh More[1,], Kundan Jangam[1], Sailee Gardi[1], Rajeshwari Athavale[1], Fatima Choudhary[1], and Ramesh Yamgar[2]*

[1] *Department of Chemistry, K.E. T's, V. G. Vaze College Autonomous, Mithagar Road, Mulund (E), Mumbai-400081, Maharashtra, India*
[2] *Department of Chemistry, C. S.'s Patkar-Varde College, Goregaon (W), Mumbai 400 062, Maharashtra, India*
[*] *Corresponding author*

6.1 Introduction

Recently researchers are working very diligently to provide a sustainable environment and have replaced synthetic polymers by using natural biopolymers. Using advanced technologies, composites of biopolymers and grafted biopolymers can be synthesized. These functionalized materials can be used for various applications due to their at-par performance as compared to synthetic polymers. The use of grafted biopolymers has rapidly increased over the last many years due to their versatile applications in day-to-day life. Natural polysaccharides have enormous advantages over synthetic polymers since they are easily available, nontoxic, and cost-effective. However, there are a few disadvantages of the natural polysaccharides such as, they are sensitive to temperature, easily contaminated by microbes, their viscosity gets lowered during storage, and are extremely sensitive to highly stressful conditions [1]. These limitations of natural polysaccharides can be overcome by modifying them, e.g., cross-linking [2], etherification [3], derivatization of functional groups [4–11], and graft polymerization with various techniques. As compared to other methods reported in the literature as well as discussed above, graft polymerization is the best method.

Graft polymerization involves covalently bonded monomers and various polymerized side chains are fixed on the basic polymer chain, also called the main chain or the

Grafted Biopolymers as Corrosion Inhibitors: Safety, Sustainability, and Efficiency, First Edition.
Edited by Jeenat Aslam, Chandrabhan Verma, and Ruby Aslam.

backbone. A variety of functional groups are added to the polymer, and it is grafted not only to make them attractive but also to make them useful for various applications. Graft polymers and graft copolymers are the same. Graft copolymers possess a minimum of two different types of monomer units distinctly different from the main polymer chain. In graft polymerization, one or more than one kind of monomer is involved. This will result in the formation of a homopolymer, or copolymer as depicted in Figure 6.1. It is noteworthy that the chains on the sides are different in structure compared to the main chain and the chains are individually connected.

For many decades, graft polymers have been used in several applications. One such important application is as impact-resistant materials such as elastomer, thermoplastic, compatibilizers, or as emulsifiers in the alloys or the synthesis of stable blends. Further, the graft polymer finds numerous applications in textiles, cable, electrolyte membranes, automobiles, laminates, adhesives, commodity plastics, separation, purification and in the biomedical sciences, etc. The most famous example of a grafted polymer is polystyrene, which is prepared using polybutadiene as a grafted chain on the polystyrene backbone.

There are two approaches "grafting to" or "grafting from" by which the grafting of polymers can be carried out. Highly functionalized monomers are made to react with the main chain (backbone) to result in the grafted polymer this approach is the "grafting to" approach. The substrate is treated with a suitable method and immobilized initiators are generated followed by polymerization. Using this technique, a very high-density polymer can be grafted. Numerous methods were used by the researchers to synthesize graft copolymerization which includes photochemical, chemical, photo-irradiation, enzymatic grafting, plasma-induced, electron beam radiation and gamma radiation, etc. Grafted polymer is a very hot topic as the material can be tailored and used for specific applications using the required monomers and the basic units. In this review article, we are going to discuss grafted biopolymers, their properties and applications.

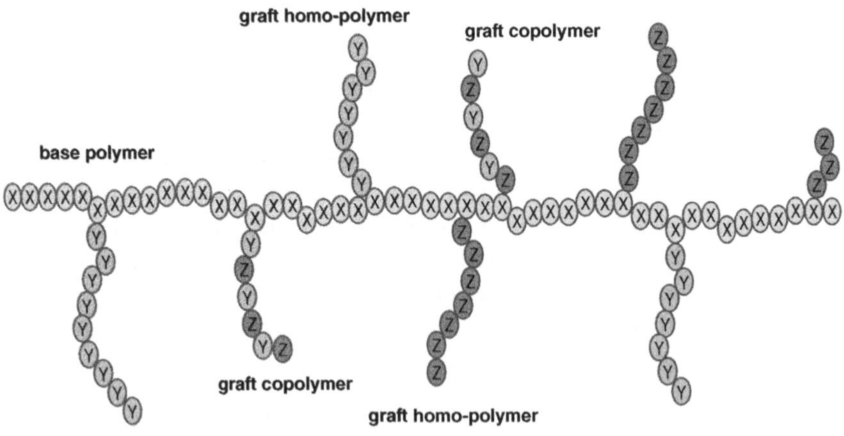

Figure 6.1 The formation of a homopolymer or a copolymer.

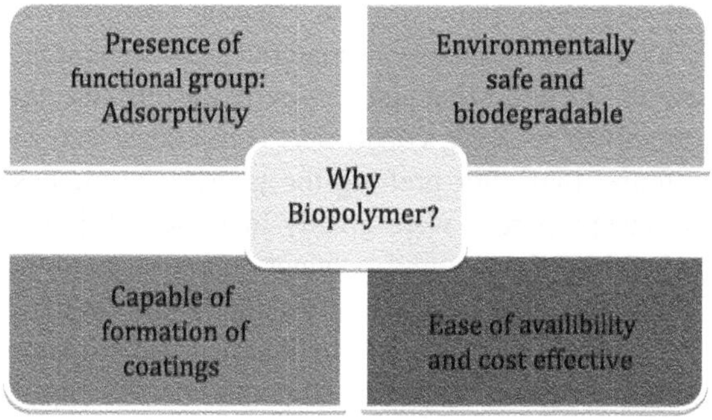

Figure 6.2 The different roles of biopolymers.

This study focuses on a detailed review of some of the excluded biopolymers, grafted biopolymers their properties, and applications. Also, the review includes recent studies on biopolymers which have been reviewed previously. The review explores current trends in research, proposed mechanism, different types of tests, and advanced approaches to give brief insights into the properties of biopolymers and their applications. The need for biopolymers and grafted biopolymers is depicted in Figure 6.2.

6.2 Properties

A few physical properties of biopolymers, biopolymer composites, and grafted biopolymers are the melting point, boiling points, density, viscosity, and shape. The water molecule interacts with the biopolymer and alters the internal structure which results in converting biopolymer moisture sensitivity. The physical and mechanical properties are weakened by plunging numbers of hydrogen bonds. This reduction in the number of hydrogen bonds is in turn caused by moisture absorption by biopolymers as was reported by a group of researchers. This hydrogen bond system can be manipulated by controlling the moisture in the system, the physical and mechanical properties can be tuned as per our requirement [12, 13]. Thermal conductivity and thermal stability are the major contributing parameters that determine thermal properties. Parameters such as stability, thermal degradation rate, melting temperature, degree of crystallinity, crystallization temperature, and glass transition temperature are very crucial because previously mentioned parameters play a significant role in the thermal analysis of biomaterials. The above parameters can be tuned by grafting the biopolymer with a suitable chain. Analytical techniques such as DSC and TGA can help in determining these parameters [14].

Due to intramolecular and intermolecular hydrogen bonding properties of materials changes drastically. Lignin is a thermoplastic material and founds applications in thermosetting materials. We know that lignin is thermally unstable; the thermal properties can be modified by using copolymer, composites, and by grafting biopolymer. It was found that after grafting the physical properties, chemical properties and thermal properties were greatly increased [15]. The mechanical properties of grafted biopolymers and polymer composites can be tuned by applying external conditions. The external factors such as heating and cooling rates, applied pressure, temperature, and deformation are responsible to make the polymer as per our requirement. The tensile strength and the impact strength increase when biopolymers are grafted. Many mechanical properties contribute to the construction of an integrated biopolymer material, and techniques such as Poisson's ratio, rheology, Young's modulus, and elongation are mandatory industrial requirements.

6.3 Applications

Biopolymers and grafted biopolymers are found to be one of the most commercialized materials due to their versatile applications. As the material is biodegradable it is in great demand in packaging film material which has applications in the food, medical, and pharmaceutical industries. These polymers found application in bone tissue engineering, cell-based transplantation as well as in gene therapy through the products such as dialysis, wound dressing material, three-dimensional Scaff holds, artificial skin, and implanted medical devices for biomolecule delivery [16–18].

Biopolymers such as silk, chitosan, keratin, elastin, and collagen were grafted with man-made polymers to get an extraordinary material. The strength, quality, and durability of natural biopolymers are increased by grafting. The current trend in the pharmaceutical and medical industries are linked with normal biopolymers along with synthetic and grafted biopolymers.

6.3.1 In Food Industry

Proteins or polysaccharides have important properties for changing the flow characteristics of a liquid by carrying out different actions such as binding, emulsifying, combining, chelating, encapsulating, and stabilizing which are very helpful in producing emulsions, gels, stabilizers, films as well as thickeners. They can be eaten as additives or as edible films because they have a strong structure because of their fat, their fragmentation, their water-holding capacity, as well as appearance and texture [19].

6.3.1.1 In Food Packaging Material

Non-perishable petrochemical packaging materials were widely used around the world. This has drawn a lot of attention to environmental safety, which led to the interest in the construction of a new compostable packaging material when its sources were based on plants. Advanced nanocomposite technology is integrated with biopolymers to improve mechanical properties to prevent defects in natural polymer packaging materials. The composite properties of nano-sized pores have shown effective resistance to gas as well as water. This development has triggered the production of effective, intelligent food packaging based on various technologies based on encapsulation [20, 21]. Plant-based polysaccharides such as cellulose, chitin, starch, pectin, chitosan, and biopolymers based on proteins like soy protein, wheat gluten, corn zein, gelatin, and milk proteins are processed, treated, distributed, and dried to produce a variety of packaging materials. Additional biopolymers such as PHA, PLA, and other biopolymer compounds, such as starch alloys with PLA, PHB, PCL, PVAs, PHAs, EVOH, and PVOH were tested in it [22]. Biopolymer nanocomposites embedded in nano-scale filters can bring about broad enhancements in mechanical, physical, antimicrobial, gas, and thermal properties. Synthesis of food packing compounds can be increased by improving the blocking properties of oxygen, the vapor of water, and carbon dioxide, along with flavors [23, 24]. Biopolymers namely PLA along with starch have shown broad properties such as bio-nano composites as a combination of nano-size additives namely silver, zinc oxide, and montmorillonite. Despite the lack of toxicology research done on synthetic nanofillers, such compounds have been seen to exhibit the production of oxygen and antimicrobial properties as well as common properties of packaging materials, which are important in increasing the food products' shelf life [25]. To enhances the antimicrobial properties of decaying polymers, extracted through plants, agricultural waste, and other essential oils are seen to stop the spread of many Gram-negative and Gram-positive bacteria. Matrices of PLA, alginate films, and PCL are combined with effective additives of natural sources that help liberate bioactive to prevent the growth of bacteria on food products. This results in extended shelf life. Biodegradable polymers are bonded with compatible materials and have packaging applications [26, 27].

6.3.1.2 Edible Films

Food products are coated using edible films that are in turn made by using biopolymers. Edible films, having been made from biopolymers could be consumed with food because they are prepared with polysaccharides and proteins. They have the important function of enhancing shelf life, food quality, and functional properties [28]. Edible films can be used to develop antimicrobial properties to prevent further contamination and to release substances of active nature gradually. Native resources of antimicrobials such as essential oils and natural acids are accepted as Generally Recommended as Safe, abbreviated as GRAS, and attached to biopolymers made from gums, seaweed, cellulose, starch, and their compounds. Edible films prepared by proteins and polysaccharides appear to be transparent due to hydrocolloids [29]. Whey protein isolate-based edible biofilms were synthesized by adding calcium ion and glycerol that showed upgraded effects on their formation, thickness, along with their evaporation properties. In addition, relatively thin layers formed by whey protein aggregates can withstand UV light [30]. Edible nanocomposite films based on mango puree infused with cellulose nanofibers have significantly increased the temperature change of glass and solid structures. Among the various other biopolymers used, carrageenan and alginates have shown remarkable inhibitory properties [31, 32]. Recently, Poly-γ-glutamic acid (γ-PGA) formulated using the species of Bacillus subtilis gene showed immunomodulatory effects on the immune system. The reason for this was the high molecular weight (HMW) compounds that are used in food processing and food packaging industries [33].

6.3.2 In Pharmaceutical Industry

Biopolymers, composites of biopolymer, and synthetic biopolymers are used in medical systems because of their growing knowledge and interest gained from elaborate research efforts to facilitate biological processing, decomposition, and biodiversity. Biopolymers and biopolymer composites are sold primarily as organs, scaffolds of tissue, and delivery systems also as input devices to carry out controlled delivery to the selected target of genes, drugs, etc. [34].

6.3.2.1 Drug Delivery Systems

Biopolymers are adapted to maintain the demand for protective drug delivery systems and their diversity in structure, medical, and applications of physiological nature. Contributions have been made in the pharmaceutical industry at various levels by assisting in the manufacture of drugs and drug delivery programs. Class of polysaccharides of a protective structure that are heavily used for their ability to form conjugates containing proteins and lipids present in cell walls. Ordinary biopolymers such as cellulose, chitosan, collagen, gelatin, fibroin, and starch may form suspensions for the transport of various molecules. Various preparation methods such as electrospraying, Supercritical fluid extraction, microemulsion, and freeze-drying, were employed to carry molecules in the nose, teeth, ocular, and other systems [35, 36]. Biopolymers have become a much-needed carrier of drugs due to their durability, decay, and low toxic nature in the pharmaceutical delivery system. Biopolymers such as nanoparticles have been tested for degradation of their enzyme by studying pH efficiency for further information on stability. The biopolymer used for drug delivery is often referred to as the excipient material, being the main focus of the present medical industry. Protein-based biopolymers have been considered for their preparation method in relation to their mechanism of drug release to determine their use. The need for a target drug delivery system (TDDS) using the polymeric drug delivery system with improved pharmacokinetics was sought [37]. Intra-articular delivery was done using elastin-like polypeptides (ELPs) that can switch phases to release the drug over time at body temperature. These ELPs can move away from the shared space over a period of time making it useful to combine them with protein drugs to represent ELPs as synthetic proteins in the plant environment [38]. Albumin microspheres are produced using chemical stabilization and heat to transport the drug. A chemical-based modification was done by adding amino acids and cationization on the albumin-based biopolymer having specific tissue specifications. Because of various methods used in the synthesis and reinforcement of the albumin microspheres carrying the drug, bioactive compounds and drug release in a controlled manner were achieved [39].

The delivery of the albumin protein was done by bacterium nanocellulose (BNC). The function along with the integrity of the loaded protein remained the same, thus proving compliance and hydrophilicity. Controlled release kinetics cemented the use of the BNC in the field of targeted drug delivery in the future [40]. Environmental-based biopolymers are widely available which makes the continuation of nanogels and hydrogels an efficient and cost-effective process. Many methods of linking polymerization techniques such as extrusion, precipitation, mixing, along with spray drying were used to form hydrogels. In one particular biopolymer study, dextrin was transformed into a hydrogel in a distinct manner that showed improved drug release and mechanical properties with better gel having stabilizing properties. Because of its non-immunogenic nature along with biocompatibility, it could be converted into nanogels that can be used in treating cancer [41, 42].

6.3.2.2 Medical Implants

Various biopolymers like PLA and Chitosan were recognized in many applications in the pharmaceutical industry because of their compatibility as implanted medical equipment. For example, a common biopolymer found in mammals, Collagen is applied in the medicinal industry as bone marrow, bone scaffold along with cardiovascular transplantation. Chitosan is used as a plant in cardiology as a heart valve, and as a contact lens in the field of ophthalmology along with anaesthesia surgery [43]. Gelatin is applied in various fields as grafts in cardiology, the orthopedic field to replace bone, in dermatology along with 3D biomatrices. Polyhydroxy alkenoates (PHA) are widely applied as implants in arterial cartilage, esophagus, and nerves. As mentioned above chitosan compounds are used to make peripheral nerves, bioartificial liver, bone regeneration, and bone scaffolds, while hyaluronic acid extracts are used in otolaryngology to improve tissue in cartilages and vocal folds. Similarly, PHB has also been used as an anchor for cell culture and as a surgical implant [44, 45].

Protein-based biopolymers, polysaccharides, micro-polymers, and their compounds are created using various methods such as electrospinning, bioprinting of 3D kind, casting, freeze-drying as well as leaching. Above mentioned compounds are actively being applied in medical implants such as membranes, stents, genetically modified vehicles, drugs, cells, and growth factor delivery systems [34].

6.3.2.3 Wound Healing

Because of their amazing properties in addition to their decay and consistency, biopolymers are applied on the skin to form structures showing similar positioning. The best way to close the wound is to connect it automatically but because of the lack of donor sites in the body, chitosan-based clothing is made. The introduction of synthetic garments has been developed to improve the wound by offering a different approach to the causative agent of the wound as well as the method associated with healing. Biopolymer compounds were designed in the form of hydrogels and the formulation was treated by therapeutic agents such as antiseptics, cytokines, growth factors, antioxidants, antimicrobials, and other inhibitory agents to stimulate and speed up the process of wound healing (Figure 6.6). Synthetic coatings use both polymers as well as biopolymers such as PVC, alginate, and hydrocolloids in different types of sprays, films, foam, as well as gel [46].

The healing of wounds is delayed by diabetes mellitus that can be increased by using biopolymers such as starch, cellulose, as well as toxic metal nanoparticles containing alginate. These biopolymers include nanoparticles that have proven anti-inflammatory properties to improve the healing process of diabetes mellitus caused by wounds [47]. Among other biopolymers, chitosan possesses antifungal and bactericidal properties that make it a wound-binding substance having properties needed for healing, regeneration, and wound healing. Like chitosan, chitin was also used in various types of fibers and films in wound treatment. Because of its hemostatic effect, painkilling and antibacterial properties along with enhanced crystallinity, and hydrophobicity, as well as the ability to easily absorb compounds into selected organic acids aid in the synthesis and erythrocytes synthesis in target wounds [48, 49]. Hydrogels are produced with different concentrations using the stimulant-responsive poly (N-vinyl-2-pyrrolidone), chitosan, and (PAA) poly acrylic acid in a neutralized condition that demonstrates the biocompatibility required for controlled drug delivery [50]. Films of Alginate coated with cerium (III) along with the solution of cerium (III) chitosan are designed for the coating of wounds that showed efficient inhibition of Gram-negative and Gram-positive bacteria with flexibility and elasticity [51].

6.3.3 For Sustainable Development

Presently, growing interest in research in a sustainable environment has returned to polymers made from natural biopolymers. They can be divided into chemical-composed biopolymers and natural and microbial biopolymers. Through the use of technological advances, biopolymers and their compounds are identified in the market because of their compact performance with polymers which are based on their performance, structure, and use [52].

Metals are electropositive and hence tend to corrode easily in water, acidic, basic, and saline conditions. They are used in almost every equipment of machines in various industries. Major industries which face problems of corrosion are oil industries where metals are in constant contact with acidic conditions, similarly in the case of water treatment plants use pipes that have more chance to corrode easily, and it affects the economy and the environment considerably [52–54]. Corrosion has a tremendous impact on human life, on nature as well as on the wealth of the country. Hence, scientists are working very hard to understand the process of corrosion and then exploring new approaches so that corrosion can be minimized. Every year India has a loss of around 6 Lakh Crore due to corrosion and the loss in the world's GDP is 3.4 % due to corrosion. To overcome corrosion and to protect metals from corrosion, numerous methods have been developed such as alloying [55–57], coating [58–60], inhibitor [61–63], etc. Rust suppression inhibitors are the simplest and most effective method as they are most effective even if they are very small and in a very rusty environment [64, 65].

Inhibitors penetrate the metal surface by forming a protective oxide film to counteract corrosion and thus reduce corrosion. Further inhibitors can combine chemically, physically, or both; the combination reacts at the surface which results in obstruction of the active sites and the anodic, cathodic, or both reaction rates fall, and corrosion is minimized. During chemical interactions, charge transfers, and charge-sharing functions are performed between the inhibitor and outer cover. This strong bond connects the barrier to the surface with electrostatic force and the contact is loosened with increasing temperature. This interaction is potent compared to physical interaction [66–70].

Different environments for a substrate such as copper [71], aluminium [72], steel [73], magnesium [74], etc., have been reported for the enhancement in corrosion mitigation of inhibitors [75–77]. Also, it protects against corrosion, and toxicity and affects the human body, which was the reason for their limited usage. These inhibitors, if compared with their different functions, are better alternatives for toxic inhibitors which gives an environmentally-friendly motive for their generation [68, 78–80]. The literature studies suggest that many green inhibitors can be obtained from a source such as biopolymers, surfactants, ionic liquids, drugs, rare earth metals, amino acids, etc. [81, 79, 82–85]. The advancement in the field of nanotechnology also helped in improving the corrosion inhibition of metals, especially by using biopolymers. They can be used in innovative approaches for coatings [86] as well as corrosion inhibitors [87].

There are ample examples of biopolymers that are biodegradable such as polysaccharides, nucleic acids, and proteins; mainly obtained from Plants and living organisms including fungi, bacteria, and animals [88, 89]. The structure of biopolymers contains Oxygen and Nitrogen atoms, making them biodegradable [90].

Biopolymers are known for their use in tissue engineering [91], coating [92], composite [93], nanocontainer [94], etc. Their potential as corrosion inhibitors is an important reason for employing them. Presence of heteroatom in the structure, induces adsorption property, one of the important properties that an inhibitor should have for corrosion inhibition [95]

Biopolymers are polymers produced by or derived from living organisms. There is a large classification of biopolymers ranging from plants to animals to small microbes. Plants, microbes, animals, agricultural wastes, and fossils are natural sources of biopolymers. Synthetically manufactured biopolymers are made from biological monomer units such as amino acids, sugars, nucleotides, and natural fats. Chitin, hyaluronic acid, and chitosan occupy most of the animal polysaccharides distribution whereas cellulose, starch, and pectin have the lion's share of plant biopolymers. They possess properties such as non-toxicity, biodegradability, ease of availability, resistance to temperature, moisture, etc. One of them is the corrosion inhibition property, although some organic compounds have this property due to their toxicity, not considered a safe coating for metals that carry oils or water. Biopolymers provide a non-toxic option for corrosion resistive coatings. By grafting, the properties of these biopolymers can be enhanced to broaden the scope of application in various fields. The process of grafting the biopolymers improves the properties of the original biopolymers. Chemical and physical properties are greatly enhanced when synthetic polymer, chemical molecules, metals, etc are grafted on biopolymer.

This study focuses on the detailed review of some of the excluded biopolymers as inhibitors in existing reviews. Also, the review includes recent studies on biopolymers which have been reviewed previously. The review explores current trends in research, proposed mechanisms, different types of tests, and advanced approaches to give brief insights into the properties of biopolymers as corrosion inhibitors.

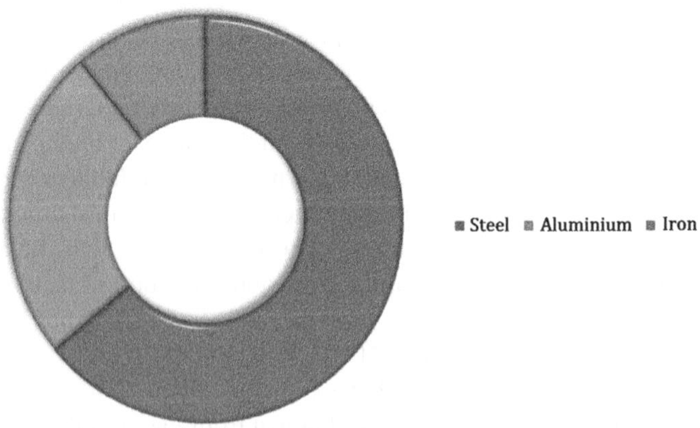

Figure 6.3 The use of grafted biopolymers for corrosion inhibitors.

6.4 Application of Grafted Biopolymers in Corrosion Inhibition

6.4.1 Copper

The absorption property of grafted biopolymers is one important factor that makes them suitable for non-toxic, biodegradable, cost-effective corrosion inhibitors. Observation has been made that if atoms such as nitrogen, oxygen, and sulfur contain single pairs, the presence of double or triple bonds that provide pi-electrons are present then they play an important role in the absorption process. Biopolymers such as sodium alginate, chitosan, hydroxypropyl cellulose, and poly-aspartic acid provide a variety of functional groups, especially electron-donating groups. They can interact with the orbital of metal which promotes bond formation, and adsorption onto the metal surfaces. Guar gum-infused 2-acrylamide-2-methyl propane sulfonic acid contains various functional groups such as hydroxylimine, and hydrogensulfate groups at different positions which contain oxygen, and nitrogen atoms. These atoms have lone pairs which interrelate with d orbitals of copper and help to make a protective layer. The presence of electron donating groups helps to bond formation and adsorption of grafted biopolymer on the outside of copper. GG-AMPS being environmental-friendly, cost-effective, and easily available, is a greener option for corrosion inhibitors of copper, even in a high concentration of the corrosive medium. Rust-resistant properties have been tested with standard weight loss methods and chemical reactions, which ensure GG-AMPS reduction capacity of up to 3.5% NaCl (corrosion media) [96]. Sodium alginate is a hydrophilic natural polymer obtained from marine brown algae. It has been reported that 8-hydroxyquinoline-grafted-alginate has corrosion inhibition properties for copper in acidic media. Qualities such as non-toxicity, biodegradability, and multiple absorption sites make it ideal for green corrosion inhibitors of copper [97].

6.4.2 Steel

It has many varying compositions such as mild steel (contains 0.02–0.03%S, 0.4–0.5%Mn, 0.03–0.08%P, 0.1–0.2%C, and remaining is Fe), carbons steel, X70 steel, cold rolled steel and many more. They are known for their strength and light weightlessness. Thus, used in many industries as equipment. One of them is the oil industry, they use steel pipes for plugging oil wells. In this process, metal acquires deposits such as $CaCO_3$ and $CaSO_4$ which reduces the efficiency of steel. To avoid this problem corrosion inhibitors are applied. Polyaspartic acid is a biodegradable, water-soluble polymerized amino acid. When grafted with tryptophan (PASPTR) (poly aspartic acid) and copolymer grafted with (tryptophan) improves the efficiency of inhibition [98], as reported in the inhibition efficiencies of PASPTR against $CaSO_4$ and $CaCO_3$ scales. The presence of Nitrogen atoms in PASPTR promotes adsorption over the steel surface. The performance of PASPTR was evaluated against $CaSO_4$ and $CaCO_3$ by the method based on static scale inhibition, showing inhibition efficiency of 96 and 90% respectively. PASPTR exhibited good thermal tolerance, stable inhibition efficiency with long reaction times, good chemical stability, great adsorption properties, and lattice distortion properties.

Chia et.al synthesized cysteamine-modified PASP with S atom (PASP-S), in response to ring opening reaction of poly succinimide with cysteamine [99]. In addition, performance tests to prevent corrosion of PASP-S and PASP soft metal in an acidic medium of 0.5 M H_2SO_4 were performed using electrochemical impedance, weight reduction, as well as potentiodynamic polarization methods. Results showed that PASP-S, a mixed inhibitor, significantly improved inhibitory activity in the soft metal area with a high efficiency of 93.9% (weight loss method) at very low concentrations of 100 mgL^{-1} to 298 K compared to I-Unchanged PASP. It was concluded that the presence of a binding thiol group acting on PASP-S played a key role in improving the effectiveness of prevention.

6.4.3 Pectin

Pectin is an acid heteropolysaccharide found in plant cell walls. R. Geethanjali et al. prepared it by mixing it with polyacrylamide abbreviated as (Pec-g-PAAm) along with (Pec-g-PAA) which is polyacrylic acid, to be tested to prevent corrosion of the soft metal in the middle [100]. polymers were compostable and natural. As a non-corrosive, quantitative, electrochemical method was chosen to monitor the corrosion process. The temperature degradation profile has shown that the bonded polymers are suitable for use in areas with high salt temperatures. The electrochemical method concluded that polyacrylic acid linked showed a process of corrosion by closing the cathodic sites, thus acting as a cathodic inhibitor. The polyacrylic acid-containing polymer has poor performance in preventing corrosion of cool metal as compared to graft polymer with polyacrylamide.

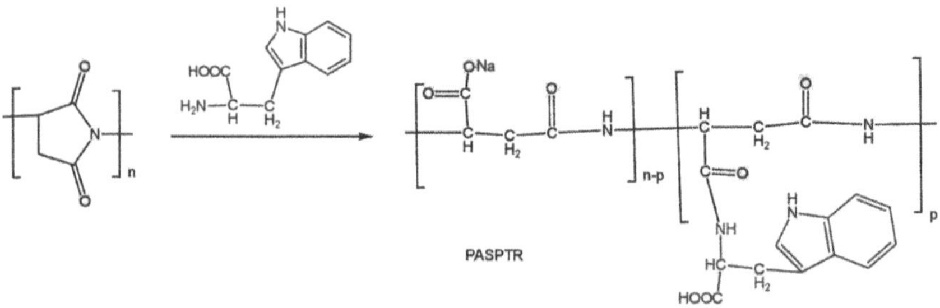

Figure 6.4 An image of a Polyaspartic acid–tryptophan grafted copolymer. [98] / Reproduced with permission of John Wiley & Sons.

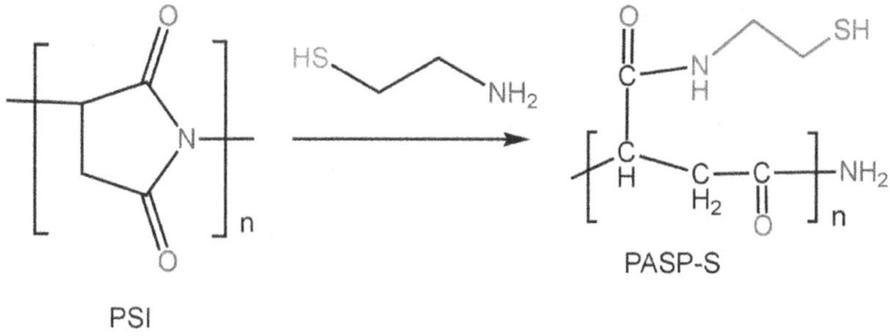

Figure 6.5 A Cysteamine modified poly aspartic acid. [99] / Reproduced with permission of Elsevier.

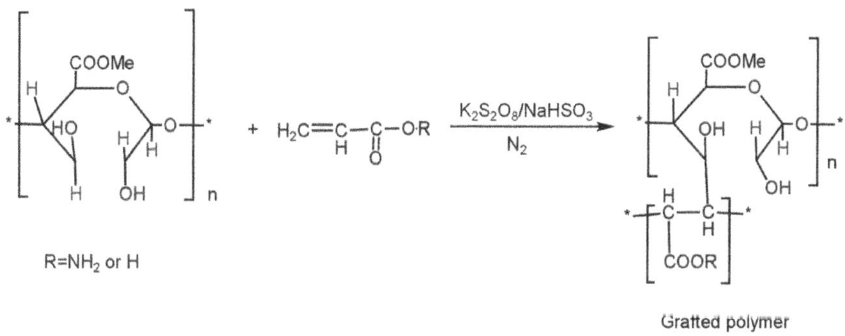

Figure 6.6 The Synthesis of Pec-g-PAA and Pec-g-PAAm. [100] / Hindawi / Licensed under CC BY 3.0.

6.4.4 Chitosan

Chitosan is a biopolymer found in chitin N-deacetylation with anti-bacterial activity, and biocompatibility and shows an anti-corrosion effect on metal in acid sources. Unfortunately, its use as a corrosion inhibitor has been banned due to its low solvability in water in presence of neutral acid. Cai Wang et.al, employed modified chitosan-oligosaccharide (MCO) which was prepared by grafting tryptophan with heterocyclic structure CO to enhance its inhibition property [101].

Sodium silicate (SS) has been reported to act as a green barrier to prevent the corrosion of steel metal. Wang et al. investigate the effect of combining MCO and SS in different formulations to maximize its use as a corrosion inhibitor in adverse conditions (high temperature / long immersion / high rotation speed) [101]. The results proved that the MCO + SS alloy exhibits high corrosion resistance to protect the carbon steel. K.O. Shamsheera et al. reported chitosan polymer composed of 1-ethyl-3- (3-dimethyl aminopropyl) carbodiimide conducted a coupling reaction with stearic acid techniques [102]. It was used to test the protection of corrosion in Mild Steel at 0.5 and 1N HCl by using methods like non-electrochemical and electrochemical. K.O. Shamsheera et al reported that stearic acid films attached to chitosan (CSA) provided better protection against corrosion in MS than blank CS film.

M. Rbaa et al. reported a long-lasting, non-toxic, and corrosive inhibitor of mild steel (COS-g-Glu) biopolymer prepared by gluing glucose moiety in Chitosan [103].

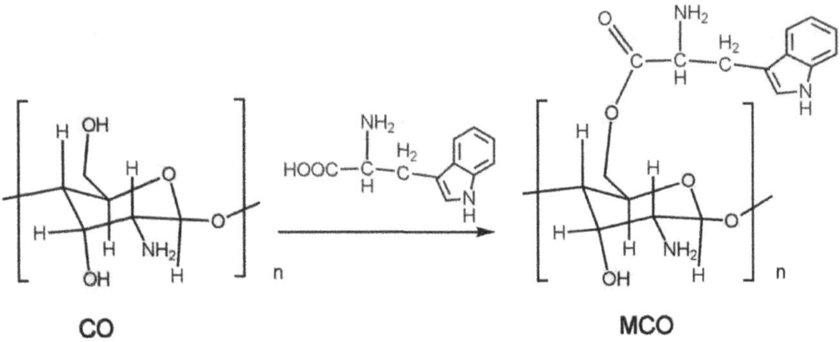

Figure 6.7 The synthesis of a modified chitosan-oligosaccharide. [101] / Reproduced with permission of Elsevier.

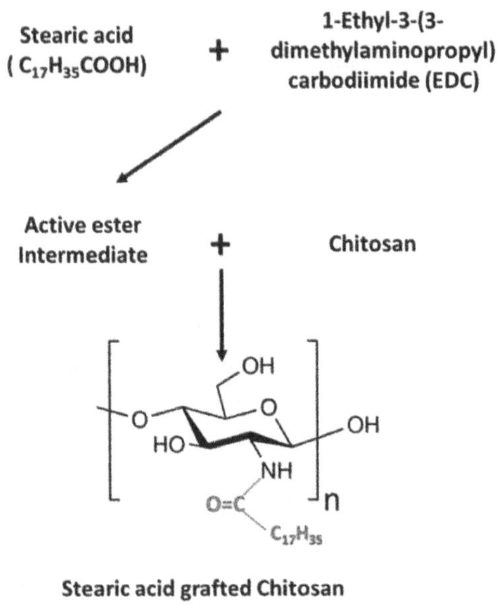

Figure 6.8 Synthesis of a Stearic acid grafted Chitosan.

Corrosion protection was tested in an acidic environment (1.0 M HCl). Electrochemical tests have provided evidence of improved inhibitory efficacy by increasing inhibitor concentration and reaching a high rate of 97% at 10^{-3} M.C.G. Cui et.al, reported the effects of chitosan oligosaccharides (BHC and PHC) as green corrosion inhibitors. The performance of BHC and PHC was analyzed at P110 iron concentration at 3.5wt.% NaCl CO_2-saturated solution at 80°C was tested using gravimetric measurements and electrochemical impedance spectroscopy (EIS) [104]. The results provided 85.70% and 88.59% inhibitory efficacy, in BHC and PHC concentrations of 100 mg / L respectively. BHC and PHC are mix-type inhibitors. Surface analysis tests BHC also showed greater inhibition effectiveness than PHC when tested in a non-abrasive method, EIS. U. Eduok et al. activated, water-soluble poly (N-vinyl imidazole) attached to carboxymethyl chitosan composite (CMCh-g-PVI) embedded in chitosan, used as a corrosion inhibitor for API X70 in 1 M HCl. The CMCh-g-PVI compound behaved as a composite inhibitor and significantly reduced X70 iron corrosion compared to the equivalent concentrations of chitosan and carboxymethyl chitosan [105].

Fayyad et al. introduced an eco-friendly corrosion-resistant oxide film for CS and CS/GO (chitosan / graphene oxide film) functionalized with oleic acid to treat the corrosion of carbon steel [106]. The impact of using CS/GO nanocoating as a protective layer was assessed by electrochemical impedance spectroscopy and potentiodynamic polarization measurements. Experimental outcomes have shown that GOs improve the effects of CS decay. Shamsheera KO et al. modified chitosan film by adding stearic acid [102]. The film was used to prevent Mild metal corrosion at 0.5 and 1 N HCl using electrochemical and non-electrochemical methods. The results showed that stearic acid films attached to chitosan (CSA) provided significant protection against MS rust at 0.5 and 1 N HCl over the unmodified CS film. Majidi et al. made films of graphene oxide-chitosan (GO-CS) and graphene oxide-chitosan- ZnO (GO-CS-ZnO) and tested their anti-corrosion properties [107]. Rust prevention results showed higher performance of GO-CS hybrids than pure GO. In addition, the GO-CS-ZnO ternary nano-hybrids have produced a very high anti-corrosion function. GO rust efficiency (42.35%) improved to 83.81% and 85.61% of GO-CS and GO-CS-ZnO nanocomposites by 500 ppm conc.

6.4.5 Mucilage

S. Banerjee et al. produced a green inhibitor by binding polyacrylamide with Okra mucilage, polysaccharide (O-g-PAM) and was tested as an inhibitor of corrosion of mild iron in 0.5 M H_2SO_4 cases [108]. Blocking power was tested using gravimetric and electrochemical techniques. Gravimetric analysis reported that polyacrylamide coupling with Okra mucilage increased the effectiveness of inhibition to 91.8% which was only 72.5% of PAM. An increase in the concentration of O-g-PAM from 1 to 100 ppm weight by volume also exhibited an increase in inhibition efficiency from 65.7% to 91.8%, however it decreases with the increase in immersion time and temperature beyond 72 h.

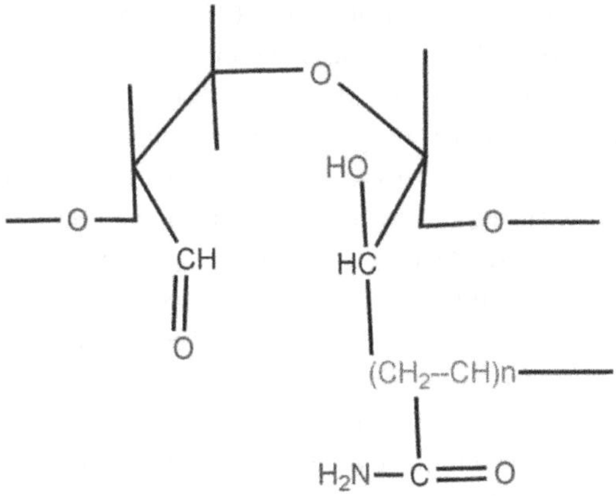

Figure 6.9 Polyacrylamide-grafted copolymer of Okra mucilage (O-g-PAM). [108] / Reproduced with permission of Elsevier.

Mishra et al. produced a very pure graft copolymer of Polyacrylamide-grafted-fenugreek mucilage (Fen-g-PAM) by attaching acrylamide (AM) to fenugreek mucilage (Fen) using the ceric ion polymerization process [109]. Srivastava et al. checked the use of Fen-g-PAM as a corrosion inhibitor on steel at 0.5 M H_2SO_4 for weight loss, potentiodynamic polarization, and electrochemical impedance spectroscopy [110]. Fen-g-PAM, a cathodic inhibitor, has shown an Inhibition efficiency of 78% at low concentrations such as 1ppm and an efficiency of up to 96% at 100 ppm.

6.4.6 Chitin

Chitin is not just a second abundant polysaccharide along with cellulose but is nontoxic and decaying. V. Kumar et.al, converted chitin into phosphorylated chitin (P-chitin or PCT) which increases melting in an aqueous environment and complex with ions [111]. He also tested its protective effect on the corrosion of soft metal in a chloride-neutral environment. PCT exhibited a high efficiency of 73% concentration at 150 ppm 94.36% inhibition efficacy was observed in the synergistic formulation of 100 ppm Zn^{2+}+200ppm PCT.

6.4.7 k-Carrageenan

L. Muthulakshmi et al. grafted k-Carrageenan, a gel-based sea-based biopolymer in the form of sol-gel [112]. The coat was applied to the (SS) 304 stainless steel of medical grade and was tested below the natural electrolyte content in the medium of 3.5 wt% NaCl. Evaluation of the effectiveness of inhibition is done using electrochemical techniques.

The results reported that the combined SS 304 acts as a promising corrosion inhibitor effective up to 65% in 227 h and up to 97% in 24 hours.

6.4.8 Starch

S. Lahrour et.al, examined the effect of corrosion inhibition on C-Mn steel which has been often subject to corrosion [113]. Starch, a bio-copolymer acquired from maize starch, was grafted with glycerin to prepare a green corrosion inhibitor. The assessment was done by weight-loss and electrochemical techniques. The best outcome, 98.07% was reported at the Optimum temperature of 323 K, with an inhibitor concentration of 300 mg/l. M. Bello et.al tested two inhibitors derived from cassava starch for corrosion of, AS (activated starch) changed by implementing gelatinization and CMS (carboxymethylated starch) with two degrees of substitution (DS) [114]. They were evaluated by electrochemical impedance spectroscopy (EIS) on carbon steel in a basic 200 mgL^{-1} NaCl solution. M. Bello et. al concluded that activated starch (AS) had good inhibitive properties but the corrosion inhibition act of carboxymethylated starch (CMS) improved with an increase in the degree of substitution (DS).

Xianghong Li et.al, introduced ternary graft copolymer of cassava starch consisting of cassava starch having sodium allyl sulfonate-acrylamide graft copolymer (CS-SAS-AAGC) made from cassava starch (CS) altered by fusion of two monomers of acrylamide (AA) and sodium allyl sulfonate (SAS) simultaneously [115]. The inhibition showing CS-SAS-AAGC on cold rolled steel (CRS) in 1.0 M HCl solution was experimentally examined by electrochemical and weight loss techniques. The result shows that CS-SAS-AAGC exhibits 97.2% inhibitory efficacy at 50 mg / L, and its inhibitory potential is stronger than that of CS, SAS, or AA. CS-SAS-AAGC behaves as a composite inhibitor.

6.4.9 Cellulose

I.O. Arukalam et.al investigated the impeding effect of hydroxyethyl cellulose (HEC) as a corrosion inhibitor for mild steel in 0.5 M H_2SO_4 solution using techniques such as potentiodynamic polarization, and electrochemical impedance spectroscopy (EIS) techniques [116]. A rise in inhibition efficiency was reported with increased concentration and temperature. Potentiodynamic polarization measurements concluded that HEC is a composite inhibitor because it gives both cathodic and anodic reactions.

Farhadian et.al introduced hydroxyethyl cellulose (HEC) as a dynamic inhibitor in diminishing mild steel (MS) corrosion [117]. It was evaluated using methods like weight loss, electrochemical impedance spectroscopy (EIS), and potentiodynamic polarization (PDP) techniques. The incorporation of 1% of polyurethane polymer into the CHEC structure enhanced its inhibition efficiency in the acidic solution, at high temperatures. CHEC functioned as a composite inhibitor, having 93% efficiency at 800C.

6.4.10 Alginate

A green corrosion suppressor, by grafting 8-hydroxyquinoline on Alginate (HQ-g-Alg) was developed by Fardioui et al. for evaluating its performance on mild steel in the acidic medium of 1 M HCl solution. The study deduced that the modified alginate behaves as a mixed inhibitor and the suppression efficiency of the modified alginate holds more potential to perform as an inhibitor than the unmodified analog [97].

Alzahrani et.al analyzed the performance of (Butyl-2-methylprop-2-enoate)-grafted alginate/Fe3O4 composite and tri-iron tetraoxide (Fe_3O_4) nanoparticles (NPs) as a corrosion-resistant coat for carbon steel [118]. Electrochemical studies showed that the quality of coating varied between the 2 combinations, with PBMA-gft-Alg/Fe_3O_4 nanocomposite being more effective than Fe_3O_4NPs coating in the optimum conditions 150^0C and immersion time of 2 hrs.

6.4.11 Dextrin

Dextrin is a group of carbohydrates having low molecular weight. Biswas et al. constructed a dextrin-derived graft copolymer (Dxt-g-pVAc) and applied it as a corrosion suppressor for steel in a 15% HCl medium. It was found from electrochemical measurements that the assembled graft copolymer exhibited good anti-corrosion efficiency (98.39%) in comparison to native dextrin (84.56 %) [119].

6.4.12 Polyacrylamide

Azzam et al. worked on three inhibitors, polyacrylamide (PACM), poly(2-methoxy-aniline) (free amino groups of PACM, grafted by the oxidation of 2-methoxyaniline), Polyacrylamide-graft- P2-MeOANI Grafting polyacrylamide onto poly (2-methoxyaniline). Further investigation was done for testing the application and compared it with polyacrylamide and poly (2-methoxyaniline) [120]. The suppressing effect of the three polymers on the mild steel corrosion in the 1.0 M HCl. A rise in temperature caused enhanced inhibition performance. Poly 2-methoxy-aniline proved to be a very successful corrosion inhibitor at all temperatures.

6.4.13 Glucomannan

Luo et al. tested the mixture of the bis quaternary ammonium salt (BQAS) and glucomannan (GL) as a green eco-friendly inhibitor as a corrosion barrier of mild steel by electrochemical measurements [121]. It was found that the mixture of

BQAS and GL acts as a mixed inhibitor and provides effective corrosion inhibition to 98%. EIS studies also suggest that the inhibition efficiency of the mixture rises with a surge in the immersion time and inhibitor concentration, whereas decreases with the temperature.

6.4.14 Gum

A. Singh et al. synthesized a mixed-type corrosion inhibitor from Guar gum and ethyl acrylate (GG-EEA) [96]. The composite of natural type was employed as an inhibitor of corrosion to lessen the evolution of hydrogen and corrosion protection of P110 steel in presence of acid. The techniques such as (EIS) electrochemical impedance spectroscopy, weight loss, and potentiodynamic polarization (PDP) were used to find corrosion inhibition results. The corrosion inhibition performance of GG is 77.5%, and of GG-EEA is 92.3%. J. Balaji et.al, investigated the sol-gel coating of Chitosan-doped-hybrid/TiO_2 nanocomposite for the corrosion resistance of aluminum metal in an alkaline medium which was achieved through (SAMs) self-assembled monolayer method [122]. Condensation and hydrolysis of tetraethoxysilane (TEOS), 3-glycidoxypropyltrimethoxy silane (GPTMS), and titanium (IV) isopropoxide (TIP) in acidic solution produced Organic-inorganic hybrid sold. The efficiency was increased by adding chitosan to nano-TiO_2/Hy-based sol-gel significantly in the protection of aluminium in the environment of a harsh kind. The corrosion current density was calculated by using Potentiodynamic Polarization measurements. The presence of Si-O-Ti, Al-O-Ti, and Si-O-Si bonds was confirmed by the result of FTIR analysis. The electrochemical measurements results proved the chitosan nano-TiO_2/doped-Hy sol-gel coating was very effective in suppressing the anodic dissolution process of aluminum to a greater manner compared to nano-TiO_2/Hy. It was concluded that the chitosan fused nano-TiO_2/Hy layer forms a protective coating with adhesive properties and provides better corrosion resistance for the aluminum metal in a neutral medium.

T. Sugama et al, examined the reduction in corrosion of aluminum surfaces by using coatings of Poly(itaconic acid)-modified chitosan. The man-made poly(itaconic acid) (PIA) polymer was fused with the linear CS chains and water-insoluble chitosan (CS) was made by crosslinking the CS chains [123]. Environmentally safe water-based coatings were applied to the aluminum (Al) substrate by following a dip-withdrawing method of the simple kind. Three factors held an important part in reducing the corrosion rate of the Al: the first was the conformation of polymer which contains the bonds of hydrophobic amide, along with a minimum amount of hydrophilic unreacted NH_2 in CS and COOH in PIA; next because of the enhancement in the degree of crosslinking and grafting caused few chances of the coating films to moisture; and third was the development in the making of the–COO–Al linkage because of the networking at the interface between the Al substrate and the grafted CS. The proper proportions of PIA to CS give rise to the PIA-infused and films of -crosslinked CS polymer that included all factors and it also showed less conductivity of ions, which imparts 694 hours of salt-spray resistance for the coated Al panels.

A bioinspired macromolecule was investigated which consisted of grafted Chitosan-cinnamaldehyde Schiff base by D.S. Chauhan et al, which can be employed as a corrosion suppressor for the industry of oils and gas [124]. Microwave irradiation was used to make Schiff base in a single step between chitosan and cinnamaldehyde (Cinn-Cht) and characterization was done by using techniques of spectroscopy. Suppression of corrosion on carbon steel was done in acidic (15% HCl) by this Schiff base. The electrochemical impedance measurements, weight loss tests, and polarization studies give information about the corrosion evaluation. The dosage of the inhibitor increased and reached a peak value of 85.16% at 400 mg/L concentration. The inhibition of corrosion was also increased. Langmuir isotherm was followed in the inhibitor adsorption. The corrosive solution made using Potassium iodide (KI) was incorporated to make more improvements in the corrosion inhibition performance. This shows an increase in the efficiency of inhibition to 92.45% at a concentration of 10 mM. Many studies on a surface like SEM analyses indicated the adsorption of the inhibitor along with the formation of protection film on the steel surface. Mainly the protonated form of inhibitor adsorbs on the outer part of the metal this was shown by the DFT which performed the computational studies. The neutral inhibitor showed less adsorption energy than the inhibitor in the protonated form; this was shown by the studies of Monte Carlo simulation.

G. Babaladimath et al. investigated Electrical conducting Xanthan Gum which was grafted by polyaniline as a corrosion suppressor for aluminium in HCl environment biopolymer [125]. Microwave radiation was used for grafting which was carried out by polymerization of aniline in an acidic medium. The FTIR, UV-Vis, and TGA data analysis confirmed the formation of the graft copolymer. DC conductivity measurements were applied to see the enhancement in conductivity of XG on grafting with PANI. Various methods such as weight loss, AC impedance, potentiodynamic polarization, and surface studies were used to check the corrosion inhibitor property of the two-i. e XG-g-PANI and XG in a harsh acid environment on an aluminium surface. Both species were found to be excellent inhibitors.

Cassava starch, a graft copolymer as an effective inhibitor for corrosion on aluminium in phosphoric acid was investigated by S. Deng et al. For this study, grafting of acrylamide (AA) with cassava starch (CS) was done which gives rise to cassava starch-acrylamide graft copolymer (CS-AAGC). 1060 aluminium was examined with an efficient inhibitor for 1.0 M H_3PO_4 media [126]. Weight loss and electrochemical methods give an idea about the adsorption behavior pattern and electrochemical mechanism of CS- AAGC. Characterization of the protected aluminium surface was done by AFM, SEM, XPS, and contact angle measurements. A drastic decrease was seen in the degree of corrosion and surface roughness of aluminium after the addition of CS-AAGC to the media. This was confirmed by SEM and AFM drastically decreased. Results confirmed that the inhibitive ability of CS-AAGC was better as compared to CS, AA, or CS/AA mixture, and the maximum inhibition efficacy was 90.6% at 20 °C at 1.0 g·L^{-1} CS-AAGC concentration. CS-AAGC mainly retards the anodic reaction because it acts as a mixed inhibitor. CS-AAGC increases polarization resistance and EIS shows three-time constants.

X. Li et al, investigated the Cassava starch-sodium allyl sulfonate-acrylamide graft copolymer as a successful inhibitor of aluminium corrosion in an HCl solution. cassava starch (CS) helped in, in situ polymerization of sodium allyl sulfonate (SAS) and acrylamide (AA) [115].

Electrochemical and weight loss techniques were employed to test the performance of inhibition by immersing aluminium in an acidic solution. 95% inhibition efficiency was seen at low concentrations of inhibitors at 100 mg/L, CS-SAS-AAGC. The inhibitive performance was much higher when compared to the SAS, CS, AA, CS/ SAS, CS/AA, or mixture of CS/AA/SAS. Langmuir adsorption isotherm was obeyed in CS-SAS-AAGC adsorption on the aluminium outer part. It was found that CS-SAS-AAGC mainly inhibits the cathodic reaction by behaving as a mixing-type inhibitor. Two-time constants were seen in electrochemical impedance spectroscopy (EIS) i.e., a large loop at increased frequencies was followed by low frequencies with a big inductive one. Contact angle images and scanning electron microscope (SEM)gave an idea about the protection of aluminium surfaces by CS-SAS-AAGC. The adsorption mechanism was analyzed using quantum chemical calculations and molecular dynamic (MD) simulation. Grafted Biopolymers with their efficiency are depicted in Table 6.1.

Table 6.1 Grafted Biopolymers with their efficiency.

Metal	Biopolymers	Grafted with	Efficiency	Reference
Copper	Guar gum	2-acrylamide-2-methylpropanesulfonic acid	95%	[96]
	Sodium Alginate	8-hydroxyquinoline	89%	[97]
Steel	Poly aspartic acid	Tryptophan	90–96%	[98]
		Cysteamine	93.9%	[99]
	Chitosan	Tryptophan	80%	[98]
	Pectin	Polyacrylic acid	89%	[100]
		MCO and SS	96–98%	[101]
		1-ethyl-3-(3-dimethylaminopropyl) carbodiimide	83–89%	[102]

(Continued)

Table 6.1 (Continued)

Metal	Biopolymers	Grafted with	Efficiency	Reference
		Stearic acid	-	[102]
		GO-oleic acid	-	[102]
		Glucose	97%	[103]
		GO-ZnO	85.61%	[107]
	PHC		85.70%	[104]
	BHC		88.59%	[104]
	carboxymethyl chitosan composite	Poly (N-vinyl imidazole)	93%	[105]
	Mucilage	Polyacrylamide	91.8%	[108]
	Fenugreek mucilage	Acrylamide	96%	[109]
	K-carrageenan	SS 304	97%	[112]
	maize starch	Glycerin	98.07%	[113]
	cassava starch	Sodium allyl sulfonate-acryl amide	97.2%	[115]
	glucomannan	Bis quaternary ammonium salt	98%	[121]
	Guar gum	Ethyl acrylate	92.3%	[96]
Aluminium	Chitosan	Cinnamaldehyde	92.45%	[122]
		Hybrid/TiO2 nanocomposite	-	[122]
		Poly (itaconic acid) (PIA)	-	[123]
	Xanthan Gum	Polyaniline	95.32%	[125]
	cassava starch	Acrylamide (AA)	90.6%	[126]
		Sodium allyl sulfonate-acrylamide	95%	[115]

6.5 Future Scope

1) All the grafted biopolymers will replace biopolymers and synthetic polymers due to their sustainability, economic viability, green approach, and versatile applications.
2) Aluminium due to its properties like high corrosion resistivity and inexpensiveness used for domestic as well as industrial purposes. Aluminium metal matrix

composites with SiC nanoparticles are an improved version of the metal that possesses the high thermal capacity, conductivity, damping capacity, etc., for which it is used in electronic packaging, aeronautics, aviation, automobiles, etc. But this modification does not reinforce the corrosive layer of aluminium that is Aluminium oxide. Biopolymers are a greener option to add a corrosive inhibitive layer but still, there is very less work has been reported about grafted corrosion inhibitors, especially for aluminium in an acidic medium.

3) Lignin which is obtained from wood, sugar cane, etc. is eco-friendly and effective against corrosion making it ideal to be used as coatings. Grafted lignin variants may lead to significant corrosion-resistant metal coatings.
4) Synergism is the addition of corrosive material to biopolymer and obtains an improved corrosion-resistant coating. This method was applied in fewer biopolymers, thus having the potential to explore grafted biopolymers with synergism.
5) One of the major users of metal is the construction industry. Metal bars which are used for building pillars, frameworks, etc are susceptible to corrosion. To reduce corrosion, the galvanization method is used however the use of biopolymers which is non-toxic not been reported in this field.

6.6 Conclusion

For several years researchers are extensively working very hard to develop highly effective methods to synthesize grafted polymers with versatile applications following green chemistry principles. Numerous substrates are used to replace the toxic inhibitors out of which biopolymers are the best as they are easily available, environment friendly, and highly effective. In this review report recent investigations on biopolymers, grafted biopolymers, their properties, and their applications are covered in detail. Numerous biopolymers such as pectin, chitosan, mucilage, starch, cellulose, alginate, dextrin, glucomannan, and gum were explored. The biopolymers were selected such that they reduce corrosion of respective metals in specific conditions. The corrosion mainly occurs in harsh conditions such as acidic, saline, high humidity, etc. Almost every biopolymer studied here is capable of corrosion inhibition in an acid medium. Especially steel and its alloys have been used in machinery for oil extraction industries which mainly operated in acidic conditions thus poly aspartic acid, chitosan, pectin, PHC, BHC, carboxymethyl chitosan composite, mucilage, fenugreek mucilage, K-carrageenan, maize starch, cassava starch, glucomannan, guar gum are used. The electrochemical technique has been used to modify different biopolymers for effective grafting. Grafting induces various functional groups which enhance the adsorptive as well as corrosion mitigation properties of biopolymers. Existing corrosion inhibitors have heterocyclic organic compounds which are toxic to the environment. Thus, Biopolymers are a greener alternative for corrosion inhibition.

References

1 Gupta, S., Sharma, P., and Soni, P.L. (2004). Carboxymethylation of Cassia occidentalis seed gum. *Journal of Applied Polymer Science* 94: 1606–1611.

2 Bastos, D.C., Santos, A.E.F., Silva, M.L.V., and Simao, R.A. (2009). Hydrophobic corn starch thermoplastic films produced by plasma treatment. *Ultramicroscopy* 109: 1089–1093.

3 Edgar, K.J., Buchanan, C.M., Debenham, J.S. et al. (2001). Advances in cellulose ester performance and application. *Progress in Polymer Science* 26: 1605–1688.

4 Battaerd, H.A.J. and Treglar, G.W. (1967). *Polysaccharide Based Graft Copolymers*. New York, NY: Wiley.

5 Fares, M.M., El-faqeeh, A.S., and Osman, M.E. (2003). Graft copolymerization onto starchel. Synthesis and optimization of starch grafted with N-tert-butylacrylamide copolymer and its hydrogels. *Journal of Polymer Research* 10: 119–125.

6 Haag, A.P., Maier, R.M., Combie, J., and Geesey, G.G. (2004). Bacterially derived biopolymers as wood adhesives. *International Journal of Adhesion and Adhesives* 24: 495–502.

7 Jenkins, D. and Hudson, S. (2001). Review of vinyl graft copolymerization featuring recent advances toward controlled radical-based reactions and illustrated with chitin/ chitosan trunk polymers. *Chemical Reviews* 101: 3245–3273.

8 Kaith, B.S. and Kalia, S. (2008b). Microwave-enhanced synthesis of flax-g-poly(MMA) for use in phenolic composites as reinforcement. *E-Journal of Chemistry* 5: 163–168.

9 Moreira, A.C.E., Oliveira, M.G., and Soares, B.G. (1997). Thermal degradation studies of poly(EVAL-g-methylmethacrylate). *Polymer Degradation and Stability* 58: 181–185.

10 Reis, A.V., Guilherme, M.R., Cavalcanti, O.A. et al. (2007). Synthesis and characterization of pH-responsive hydrogels based on chemically modified Arabic gum polysaccharide. *Polymer* 47: 2023–2029.

11 Shi, R., Zhang, Z., Liu, Q. et al. (2007). Characterization of citric acid/glycerol co-plasticized thermoplastic starch prepared by melt blending. *Carbohydrate Polymers* 69: 748–755.

12 Hubbell, D.S. and Cooper, S.L. (1977). The physical properties and morphology of polyε-caprolactone polymer blends. *Journal of Applied Polymer Science* 21: 3035–3061.

13 Kulasinski, K., Guyer, R., Keten, S. et al. (2015). Impact of moisture adsorption on the structure and physical properties of amorphous biopolymers. *Macromolecules* 48: 2793–2800.

14 Gan, P.G., Sam, S.T., Bin Abdullah, M.F., and Omar, M.F. (2020). Thermal properties of nanocellulose-reinforced composites: a review. *Journal of Applied Polymer Science* 137: 48544.

15 Sen, S., Patil, S., and Argyropoulos, D.S. (2015). Thermal properties of lignin in copolymers, blends, and composites: a review. *Green Chemistry* 17: 4862–4887.

16 Cheung, H.Y., Lau, K.T., Tao, X.M. et al. (2008). A potential material for tissue engineering: silkworm silk/PLA biocomposite. *Composites Part B: Engineering* 39 (6): 1026–1033.

17 Chaabouni, E., Gassara, F., and Brar, S.K. (2014). Biopolymers Synthesis and Application. In: *Biotransformation Waste Biomass into High Value Biochem* (ed. C. Emna, F. Gassara, and S.K. Brar), 415–443. New York, NY: Springer, New York. ISBN 978-1-4614-8004-4.

18 Sionkowska, A. (2011). Current research on the blends of natural and synthetic polymers as new biomaterials: a review. *Progress in Polymer Science* 36: 1254–1276. doi: 10.1016/j.progpolymsci.2011.05.003.

19 Chronakis, I.S., and Kasapis, S. (1995). Food applications of biopolymer—theory, and practice. *Developments in Food Science* 37: 75–109.

20 Fabra, M.J., Lopez-Rubio, A., and Lagaron, J.M. (2014). 15-Biopolymers for food packaging applications. *Smart Polymers and their Applications* doi: 10.1533/9780857097026.2.476.

21 Rhim, J.W. and Ng, P.K.W. (2007). Natural biopolymer-based nanocomposite films for packaging applications. *Critical Reviews in Food Science and Nutrition* 47: 411–433.

22 Udayakumar, G.P., Kirthikaa, G.B., Muthusamy, S. et al. (2020). Comparison and evaluation of electrospun nanofiber membrane for the clarification of grape juice. In: *Sustainable Development in Energy and Environment* (ed. I.G. Murthy), 77–92. Springer.

23 Ghanbarzadeh, B., Oleyaei, S.A., and Almasi, H. (2015). Nanostructured materials utilized in biopolymer-based plastics for food packaging applications. *Critical Reviews in Food Science and Nutrition* 55: 1699–1723.

24 Tang, X.Z., Kumar, P., Alavi, S., and Sandeep, K.P. (2012). Recent advances in biopolymers and biopolymer-based nanocomposites for food packaging materials. *Critical Reviews in Food Science and Nutrition* 52: 426–442.

25 Othman, S.H. (2014). Bio-nanocomposite materials for food packaging applications: types of biopolymer and nano-sized filler. *Agriculture and Agricultural Science Procedia* 2: 296–303.

26 Muthuraj, R., Misra, M., and Mohanty, A.K. (2018). Biodegradable compatibilized polymer blends for packaging applications: a literature review. *Journal of Applied Polymer Science* 135 (24): 1–35.

27 Vald́es, A., Mellinas, A.C., Ramos, M. et al. (2014). Natural additives and agricultural wastes in biopolymer formulations for food packaging. *Frontiers in Chemistry* 2: 1–10.

28 Han, J.H. (2013 May 1). Chitosan-based edible films and coatings: a review. *Materials Science and Engineering: C* 33 (4): 1819–1841.

29 Campos, C.A., Gerschenson, L.N., and Flores, S.K. (2011). Development of edible films and coatings with antimicrobial activity. *Food and Bioprocess Technology* 4: 849–875.

30 Fang, Y., Tung, M.A., Britt, I.J. et al. (2002). Tensile and barrier properties of edible films made from whey proteins. *Journal of Food Science* 67: 188–193.

31 Tavassoli-Kafrani, E., Shekarchizadeh, H., and Masoudpour-Behabadi, M. (2016). Development of edible films and coatings from alginates and carrageenans. *Carbohydrate polymers* 137: 360–374.

32 Azeredo, H.M.C., Mattoso, L.H.C., Wood, D. et al. (2009). Nanocomposite edible films from mango puree reinforced with cellulose nanofibers. *Journal of Food Science* 74: 31–35.

33 Sung, M.H., Park, C., Kim, C.J. et al. (2005). Natural and edible biopolymer poly-γ-glutamic acid: synthesis, production, and applications. *Chemical Record* 5: 352–366.

34 Bin Park, S., Lih, E., Park, K.S. et al. (2017). Biopolymer-based functional composites for medical applications. *Progress in Polymer Science* 68: 77–105.

35 Jacob, J., Haponiuk, J.T., Thomas, S., and Gopi, S. (2018). Biopolymer based nanomaterials in drug delivery systems: a review. *Materials Today Chemistry* 9: 43–55.

36 Singh, A.V. (2011). Biopolymers in drug delivery: a review. *Pharmacology Online* 674: 666–674.

37 Gopi, S., Amalraj, A., Sukumaran, N.P. et al. (2018). Biopolymers and their composites for drug delivery: a brief review. *Macromolecular Symposia* 380: 1800114.

38 Betre, H., Liu, W., Zalutsky, M.R. et al. (2006). A thermally responsive biopolymer for intra-articular drug delivery. *Journal of Controlled Release: Official Journal of the Controlled Release Society* 115: 175–182.

39 Patil, G.V. (2003). Biopolymer albumin for diagnosis and in drug delivery. *Drug Development Research* 58: 219–247.

40 Müller, A., Ni, Z., Hessler, N. et al. (2013). The biopolymer bacterial nanocellulose as drug delivery system: investigation of drug loading and release using the model protein albumin. *Journal of Pharmaceutical Sciences* 102: 579–592.

41 Das, D. and Pal, S. (2015). Modified biopolymer-dextrin based crosslinked hydrogels: application in controlled drug delivery. *RSC Advances* 5: 25014–25050.

42 Oh, J.K., Lee, D.I., and Park, J.M. (2009). Biopolymer-based microgels/nanogels for drug delivery applications. *Progress in Polymer Science*34: 1261–1282.

43 Rebelo, R., Fernandes, M., and Fangueiro, R. (2017). Biopolymers in medical implants: a brief review. *Procedia Engineering* 200: 236–243.

44 Bhatt, R. and Jaffe, M. (2015). Biopolymers in medical implants. In: *Excipient Applications in Formulation Design and Drug Delivery* (ed. A. Narang and S. Boddu), 311–348. Springer.

45 Bonartsev, A.P., Bonartseva, G.A., Shaitan, K.V., and Kirpichnikov, M.P. (2011). Poly (3- hydroxybutyrate) and poly (3-hydroxybutyrate)-based biopolymer systems. *Biochemistry (Moscow) Supplement Series B: Biomedical Chemistry* 5: 10–21.

46 Demir, A. and Cevher, E. (2011). Biopolymers as wound healing materials: challenges and new strategies. *Biomaterials Applications for Nanomedicine.* London Intech.

47 Vijayakumar, V., Samal, S.K., Mohanty, S., and Nayak, S.K. (2019). Recent advancements in biopolymer and metal nanoparticle-based materials in diabetic wound healing management. *International Journal of Biological Macromolecules* 122: 137–148.

48 Bano, I., Arshad, M., Yasin, T. et al. (2017). Chitosan: a potential biopolymer for wound management. *International Journal of Biological Macromolecules* 102: 380–383.

49 Singh, R., Shitiz, K., and Singh, A. (2017). Chitin and chitosan: biopolymers for wound management. *International Wound Journal* 14: 1276–1289.

50 Rasool, A., Ata, S., and Islam, A. (2019). Stimuli-responsive biopolymer (chitosan) based blend hydrogels for wound healing application. *Carbohydrate Polymers* 203: 423–429.

51 Kaygusuz, H., Torlak, E., Akın-Evingür, G. et al. (2017). Antimicrobial cerium ion-chitosan crosslinked alginate biopolymer films: a novel and potential wound dressing. *International Journal of Biological Macromolecules* 105: 1161–1165.

52 Ramezanzadeh, B., Mehdipour, M., and Arman, S. (2014). *Progress in Color, Colorants and Coatings* 8: 69–86.

53 Ontiveros-Rosales, M., Olivares-Xometl, O., Likhanova, N. et al. (2017). *Research on Chemical Intermediate* 43: 641–660.

54 Atta, A., El-Mahdy, G., Allohedan, H., and Abdullah, M. (2016). *International Journal of Electrochemical Science* 11: 882–898.

55 Lutton Cwalina, K., Demarest, C., Gerard, A., and Scully, J. (2019). *Current Opinion in Solid State and Materials Science* 23: 129–141.

56 Zhang, M., Hector, L., Guo, Y. et al. (2019). *Computational Materials Science* 165: 154–166.

57 Kim, Y., Park, I., An, B. et al. (2020). *Materials Chemistry and Physics* 241: 122275.

58 Zhang, J., Zhang, W., Wei, L. et al. (2019). *Materials Engineering* 304: 1900374.

59 Ranjandish Laleh, R., Savaloni, H., Abdi, F., and Abdi, Y. (2019). *Progress in Organic Coatings* 127: 300–307.

60 Manoj, A., Ramachandran, R., and Menezes, P. (2020). *International Journal of Advanced Manufacturing Technology* 106: 2119–2131.

61 Kowsari, E., Arman, S., Shahini, M. et al. (2016). *Corrosion Science* 112: 73–85.

62 Solomon, M., Umoren, S., Quraishi, M., and Salman, M. (2019). *Journal of Colloid and Interface Science* 551: 47–60.

63 Ye, Y., Yang, D., Chen, H. et al. (2020). *Journal of Hazardous Materials* 381: 121019.

64 Hu, P., Wu, Z., Wang, J. et al. (2020). *Green Energy and Environment* 5: 214–222.

65 Cao, S., Liu, D., Ding, H. et al. (2020). Towards understanding corrosion inhibition of sulfonate/carboxylate functionalized ionic liquids: an experimental and theoretical study. *Journal of Colloid and Interface Science* 579: 315–329.

66 Kamali Ardakani, E., Kowsari, E., and Ehsani, A. (2020). Imidazolium-derived polymeric ionic liquid as a green inhibitor for corrosion inhibition of mild steel in 1.0 M HCl: an experimental and computational study. *Colloids and Surfaces A: Physicochemical and Engineering Aspects* 586: 124195.

67 Arthur, D., Jonathan, A., Ameh, P., and Anya, C. (2013). A review on the assessment of polymeric materials used as corrosion inhibitors of metals and alloys. *International Journal of Industrial Chemistry* 4: 2.

68 Umoren, S. and Solomon, M. (2015). Effect of halide ions on the corrosion inhibition efficiency of different organic species – a review. *International Journal of Industrial Chemistry* 21: 81–100.

69 Yahya, S., Othman, N., Daud, A. et al. (2015). The influence of temperature on the inhibition of carbon steel corrosion in acidic lignin. *Anti-Corrosion Methods and Materials* 62: 301–306.

70 McCafferty, E. (2010). Societal aspects of corrosion. In: *Introduction to Corrosion Science*. Springer: New York, NY. https://doi.org/10.1007/978-1-4419-0455-3_1.

71 Li, H., Zhang, S., Tan, B. et al. (2020). Investigation of losartan potassium as an eco-friendly corrosion inhibitor for copper in 0.5 M H_2SO_4. *Journal of Molecular Liquid* 305: 112789.

72 Mohammadi, I., Shahrabi, T., Mahdavian, M., and Izadi, M. (2020). Sodium diethyldithiocarbamate as a novel corrosion inhibitor to mitigate corrosion of 2024-T3 aluminum alloy in 3.5 wt% NaCl solution. *Journal of Molecular Liquid* 307: 112965.

73 Obot, I., Umoren, S., and Ankah, N. (2019). Pyrazine derivatives as green oil field corrosion inhibitors for steel. *Journal of Molecular Liquid* 277: 749–761.

74 Su, H., Liu, Y., Gao, X. et al. (2019). Corrosion inhibition of magnesium alloy in NaCl solution by ionic liquid: synthesis, electrochemical and theoretical studies. *Journal of Alloys and Compounds* 791: 681–689.

75 Liu, Y., Song, Z., Wang, W. et al. (2019). Effect of ginger extract as a green inhibitor on chloride-induced corrosion of carbon steel in simulated concrete pore solutions. *Journal of Cleaner Production* 214: 298–307.

76 Dehghani, A., Bahlakeh, G., Ramezanzadeh, B., and Ramezanzadeh, M. (2019). A combined experimental and theoretical study of green corrosion inhibition of mild steel in HCl solution by aqueous *Citrullus lanatus* fruit (CLF) extract. *Journal of Molecular Liquid* 279: 603–624.

77 Parthipan, P., Narenkumar, J., Elumalai, P. et al. (2017). Neem extract as a green inhibitor for microbiologically influenced corrosion of carbon steel API 5LX in a hypersaline environment. *Journal of Molecular Liquid* 240: 121–127.

78 Deyab, M.A. (2020). Understanding the anti-corrosion mechanism and performance of ionic liquids in desalination, petroleum, pickling, de-scaling, and acid cleaning applications. *Journal of Molecular Liquid* 309: 113107.

79 Kesavan, D., Gopiraman, M., and Sulochana, N. (2012). Green inhibitors for corrosion of metals. *Chemical Science Review and Letters* 1 (1): 1–8.

80 Petrovićmihajlović, M. and Antonijević, M. (2015). Copper corrosion inhibitors. Period 2008-2014. A review. *International Journal of Electrochemical Science* 10: 1027–1053.

81 Verma, C., Ebenso, E.E., and Quraishi, M.A. (2017). Ionic liquids as green and sustainable corrosion inhibitors for metals and alloys: an overview. *Journal of Molecular Liquid* 233: 403–414.

82 Gece, G. (2011). Drugs: a review of promising novel corrosion inhibitors. *Corrosion Science* 53: 3873–3898.

83 Popoola, L. (2019). Organic green corrosion inhibitors (OGCIs): a critical review. *Corrosion Reviews* 37: 71–102.

84 Verma, C., Ebenso, E., Bahadur, I., and Quraishi, M. (2018). An overview of plant extracts as environmentally sustainable and green corrosion inhibitors for metals and alloys in aggressively corrosive media. *Journal of Molecular Liquid* 266: 577–590.

85 Hamadi, L., Mansouri, S., Oulmi, K., and Kareche, A. (2018). The use of amino acids as corrosion inhibitors for metals: a review. *Egyptian Journal of Petroleum* 27: 1157–1165.

86 Mekonnen, T., Haile, T., and Ly, M. (2021). Hydrophobic functionalization of cellulose nanocrystals for enhanced corrosion resistance of polyurethane nanocomposite coatings. *Applications of surface science* 540: 148299.

87 Solomon, M., Gerengi, H., and Umoren, S. (2017). Carboxymethyl cellulose/silver nanoparticles composite: synthesis, characterization, and application as a benign corrosion inhibitor for St37 steel in 15% H2SO4 medium. *ACS Applied Materials and Interfaces* 9: 6376–6389.

88 Akeem Azeez, M. and Olatunde, O. (2018). Recent developments on the application of carbohydrate polymers. *IOSR Journal of Applied Chemistry (IOSR-JAC)* 11: 68–80.

89 Machado, G., Santos, F., Lourega, R. et al. (2020). Biopolymers from lignocellulosic biomass. *Biorefining Technology* Wiley, 125–158. https://doi.org/10.1002/9781119568858.

90 Khademian, E., Salehi, E., Sanaeepur, H. et al. (2020). A systematic review on carbohydrate biopolymers for adsorptive remediation of copper ions from aqueous environments-part A: classification and modification strategies. *Science of the Total Environment* 738: 139829.

91 Mokhtarzadeh, A., Alibakhshi, A., Hejazi, M. et al. (2016). Bacterial-derived biopolymers: advanced natural nanomaterials for drug delivery and tissue engineering. *TrAC – Trends in Analytical Chemistry* 82: 367–384.

92 Amiri, S. and Rahimi, A. (2014). Preparation of supramolecular corrosion-inhibiting nanocontainers for self-protective hybrid nanocomposite coatings. *Journal of Polymer Research* 21: 1–8.

93 Shi, S. and Su, C. (2020). Electrochemical behavior of hydroxypropyl methylcellulose acetate succinate as a novel biopolymeric anticorrosion coating. *Materials Chemistry and Physics* 248: 122929.

94 Thakur, V., Singha, A., and Thakur, M. (2012). Biopolymers based green composites: mechanical, thermal and physico-chemical characterization. *Journal of Polymers and the Environment* 20: 412–421.

95 Rao, S. (2019 October 5). Photocatalytic recovery of H_2 from H_2S containing wastewater: surface and interface control of photo-excitons in $Cu_2S@TiO_2$ core-shell nanostructures. *Applied Catalysis B: Environmental* 254: 174–185.

96 Singh, A. et al. (2021). Chemically modified guar gum and ethyl acrylate composite as a new corrosion inhibitor for a reduction in hydrogen evolution and tubular steel corrosion protection in an acidic environment. *International Journal of Hydrogen Energy* 46 (14): 9452–9465.

97 Fardioui, M. et al. (2021). Bio-active corrosion inhibitor based on 8-hydroxyquinoline-grafted-alginate: experimental and computational approaches. *Journal of Molecular Liquids* 323: 114615.

98 Sun, X. et al. (2015). Poly (aspartic acid)–tryptophan grafted copolymer and its scale-inhibition performance. *Journal of Applied Polymer Science* 132 (45): 42739.

99 Chai, C. et al. (2019). Cysteamine modified polyaspartic acid as a new class of green corrosion inhibitor for mild steel in sulfuric acid medium: synthesis, electrochemical, surface study and theoretical calculation. *Progress in Organic Coatings* 129: 159–170.

100 Geethanjali, R. et al. (2014). Water-soluble and biodegradable pectin-grafted polyacrylamide and pectin-grafted polyacrylic acid: electrochemical investigation of corrosion-inhibition behavior on mild steel in 3.5% NaCl media. *Indian Journal of Materials Science* 214: 356075.

101 Wang, C. et al. (2019). Modified chitosan-oligosaccharide and sodium silicate as an efficient sustainable inhibitor for carbon steel against chloride-induced corrosion. *Journal of Cleaner Production* 238: 117823.

102 Shamsheera, K.O. et al. (2020). *Chemical Data Collection* 28: 100402.

103 Rbaa, M. et al. (2021). *Journal of Molecular Liquids* 322: 114549.

104 Cui, C.G. et al. (2019). Chitosan oligosaccharide derivatives as green corrosion inhibitors for P110 steel in a carbon-dioxide-saturated chloride solution. *Carbohydrate Polymers* 203: 386–395.

105 Eduok, U. et al. (2018). Electrochemical and surface analyses of X70 steel corrosion in simulated acid pickling medium: effect of poly (*N*-vinyl imidazole) grafted carboxymethyl chitosan additive. *Electrochimica Acta* 278: 302–312.

106 Fayyad, E.M. et al. (2016). Oleic acid-grafted chitosan/graphene oxide composite coating for corrosion protection of carbon steel. *Carbohydrate Polymers* 151: 871–878.

107 Majidi, H.J. et al. (2020). *International Journal of Biological Macromolecules* 148: 1190–1200.

108 Banerjee, S. et al. (2012). Chemically modified natural polysaccharide as green corrosion inhibitor for mild steel in acidic medium. *Corrosion Science* 59: 35–41.

109 Mishra, A. et al. (2006). Biodegradable graft copolymers of fenugreek mucilage and polyacrylamide: a renewable reservoir to biomaterials. *Carbohydrate Polymers* 65 (1): 58–63.

110 Srivastava, V. et al. (2009). Inhibitive effect of polyacrylamide grafted with fenugreek mucilage on corrosion of mild steel in 0.5 M H_2SO_4 at 35°C. *Journal of Applied Polymer Science* 116 (2): 810–816.

111 Kumar, V. and Rao, A. (2017). Chemically modified biopolymer as an eco-friendly corrosion inhibitor for mild steel in a neutral chloride environment. *New Journal of Chemistry* 41: 6278–6289.

112 Muthulakshmi, L. et al. (2021). The benefits of *k*-Carrageenan-gelatin hybrid composite coating on the medical grade stainless steel (SS304) used as an anticorrosive barrier. *Materials Chemistry and Physics* 258: 123909.

113 Lahrour, S. (2019). Glycerin-grafted starch as corrosion inhibitor of C-Mn steel in 1 M HCl solution. *Applied Sciences* 9: 4684.

114 Belloa, M. et al. (2010). Modified cassava starches as corrosion inhibitors of carbon steel: an electrochemical and morphological approach. *Carbohydrate Polymers* 82 (3): 561–568.

115 Li, X. et al. (2020). Cassava starch ternary graft copolymer as a corrosion inhibitor for steel in HCl solution. *Journal of Materials Research and Technology* 9 (2): 2196–2207.

116 Arukalam, I.O. et al. (2015). Inhibition of mild steel corrosion in a sulfuric acid medium by hydroxyethyl cellulose. *Chemical Engineering Communication* 202: 112–122.

117 Farhadian, A. et al. (2018). Easy synthesis, characterization, and cell cytotoxicity of green nano carbon dots using hydrothermal carbonization of Gum Tragacanth and chitosan bio-polymers for bioimaging. *Journal of Molecular Liquids* 259: 284–290.

118 Alzahrani, E. et al. (2021). *Arabian Journal of Chemistry* 14 (5): 103100.

119 Biswas, A. et al. (2019). Biopolymer dextrin and *poly* (vinyl acetate) based graft copolymer as an efficient corrosion inhibitor for mild steel in hydrochloric acid: electrochemical, surface morphological and theoretical studies. *Journal of Molecular Liquids* 275: 867–878.

120 Azzam, M.S. et al. (2018). Control the corrosion of mild steel using synthesized polymers based on polyacrylamide. *Egyptian Journal of Petroleum* 27 (4): 897–910.

121 Luo, X. et al. (2017). Corrosion inhibition of mild steel in simulated seawater solution by a green eco-friendly mixture of glucomannan (GL) and bisquaternary ammonium salt (BQAS). *Corrosion Science* 125: 139–151.

122 Balaji, J. et al. (2017). Chitosan-doped-hybrid/TiO_2 nanocomposite-based sol-gel coating for the corrosion resistance of aluminum metal in 3.5% NaCl medium. *International Journal of Biological Macromolecules* 104 (Pt B): 1730–1739.

123 Sugama, T. et al. (1996). Polyorganosiloxane-grafted potato starch coatings for protecting the aluminum from corrosion. *Thin Solid Films* 289: 39–48.

124 Chauhan, D.S. et al. (2020). *International Journal of Biological Macromolecules* 158: 127–158.

125 Babaladimath, G. et al. (2018). Electrical conducting Xanthan Gum-graft-polyaniline as corrosion inhibitor for aluminum in a hydrochloric acid environment. *Materials Chemistry and Physics* 205: 171–179.

126 Deng, S. et al. (2021). An efficient corrosion inhibitor of cassava starch graft copolymer for aluminum in phosphoric acid. *Chinese Journal of Chemical Engineering* 37: 222–231.

7

Factors Affecting Biopolymers Grafting

Marziya Rizvi[1,], Preeti Gupta[1], Hariom Kumar[1], Manoj Dhameja[1], and Husnu Gerengi[2]*

[1] Department of Chemistry, Babasaheb Bhimrao Ambedkar University, Lucknow 226003, Uttar Pradesh, India
[2] Corrosion Research Laboratory, Department of Mechanical Engineering, Duzce University, Duzce 81620, Turkey
* Corresponding author

7.1 Introduction

Grafting biopolymers is of immense importance in the field of polymer engineering. On a broader view, biopolymer grafting process, along with trends in the field, should be researched. This mends the differences between the synthesis principles, characterization techniques, and different applications of polymer grafting. It is expected that the development of innovation for the grafting of biopolymers will render more applicable strategies in the field of polymer engineering.

Graft copolymers are macromolecular-chain polymeric structures with one or more units of a block(s), which are connected to the main polymeric chain. In the case of these graft copolymers, the chemical properties of the polymeric backbone branches as well as structural properties, like molecular numbers and weights, spatial distributions of polymeric chain branches, etc., are the basic parameters to reveal various important performances. Graft copolymerization of various natural polysaccharides has been investigated to modify their structures to make them attractive biomaterials since native polysaccharides are mostly not suitable because of their substantial swelling and poor stability in the biological environment [1, 2]. Numerous modifications of natural polysaccharides through graft-copolymerization procedures have been already studied concerning "grafting-from" (i.e., growth of polymeric chains from the initiating sites on polysaccharidic backbone) and "grafting-to" (i.e., coupling of preformed polymeric chains to the polysaccharide) methods [3]. Among these two methods, "grafting-from" is the most familiar method produced through a variety of chemical-based or high-energy irradiation-based graft-copolymerization procedures. Recently, several vinyl monomers, like acrylamide (AM), methacrylamide (MA), methyl methacrylate (MMA), N-acrylonitrile, tert-butylacrylamide, N-poly vinylpyrrolidone (NPVP), etc., have been already grafted to numerous plant-derived

Grafted Biopolymers as Corrosion Inhibitors: Safety, Sustainability, and Efficiency, First Edition.
Edited by Jeenat Aslam, Chandrabhan Verma, and Ruby Aslam.

polysaccharides for optimizing the potential properties [4, 5]. The characteristics of these graft copolymers of plant polysaccharides are extremely reliant on the characteristics and inherent structural features of plant polysaccharides, the type and characteristics of the grafting monomers, grafting efficiency, and grafting ratio. By virtue of the favorable intrinsic characteristics of the structures and properties of the plant-derived polysaccharides, plant polysaccharide-g-copolymer has been employed in several industrial fields, especially in chemical engineering, dyeing, biomaterials, drug delivery, foods, agriculture, paper-making, wastewater treatment, etc [3].

Innovation in this class of materials is still needed, for example, Hebeish and Guthrei investigated the synthesis of cellulose derivatives by different methods, such as vinyl graft copolymerization onto cellulose, radiation-induced grafting, and grafting by chemical activation of cellulose [6]. It is suggested that the synthesis procedure may have a significant effect on the final characteristics of the grafting of biopolymers, making them suitable for specific applications. In the chapter, several of the many variables that control grafting generally will be discussed, including the nature of the backbone, monomer, solvent, initiator, additives, temperature, etc.

7.2 Nature of the Backbone

As grafting involves the covalent attachment of a monomer to a pre-formed polymeric backbone, the nature of the backbone (viz. physical nature and chemical composition) will play a significant role in the processing. Ng et al. [7] found that owing to the immense size of the polymeric chain, bonding between the amino residues, insolubility, the cysteine linkages and intramolecular H-bonding in wool are the crucial factors that are responsible for shaping and setting characteristics, thus making cellulose, resistant to grafting reactions in water. Moreover, oxidative reactions are initiated, and free radicals are formed in the presence of UV light, leading ultimately to grafting if monomers are present.

$$\mathrm{CeO-H} \xrightarrow{h\nu} \mathrm{CeO^{\cdot}} + \mathrm{H^{\cdot}} \tag{7.1}$$

$$\mathrm{CeO} + \mathrm{nM} \rightarrow \mathrm{CeO} - \mathrm{M_n} - \mathrm{M^{\cdot}} \tag{7.2}$$

CeO-H refers to cellulose. Cellulose has a highly structured and bulky architecture. In the presence of an appropriate solvent, swelling of the backbone may take place which enhances the mobility of radicals generated in the monomer (e.g., by irradiation) to active sites on the substrate backbone that affect grafting significantly. A similar situation is observed for the wool. When wool absorbs UV light, degradation takes place at the cystine disulfide bonds and led to the formation of the oxidized SO_3 and the reduced -SH species. This photochemical cleavage of wool's cysteine cross-links near the fibrous surface. This results in soluble proteins being released from the irradiated fibers. Furthermore, at the same time created a number of bonding sites in wool architecture where grafting may occur. Ibrahem and Nada [8] reported that with an increase in the degree of substitution, crystallinity decreases thereby affecting the grafting of acrylamide on acetylated wood pulp. As the crystallinity decreases, it becomes less ordered and facilitates the grafting reaction. The solvent and amorphous fraction can also play a role. In the case where styrene grafts to polyethylene, the viscosity in the amorphous region increases with the addition of MeOH–H_2SO_4 or MeOH along with the monomer, thus increasing the grafting rate [9]. Clark et al. [10] disclosed the effect of the microstructural characteristics of the backbone on the course of both grafting and cross-linking reactions. In the presence of maleic anhydride or peroxide, both the grafting and crosslinking events were boosted for polyethylene (PE) with high levels of terminal unsaturation to get reacted. Moreover, to eliminate undesirable side reactions, when styrene was added as a co-monomer, cross-linking was still observed for PE with a high concentration of terminal unsaturation. The results were attributed to low reactivity between the allylic radical generated on the PE backbone and styrene is believed to be responsible for increased cross-linking. There are numerous reports regarding the role of chemical composition in grafting. For example, since lignin is a good scavenger of radicals, the presence of lignin (phenolic-OH) in straw affects the grafting of 2-methyl vinyl pyridine [11]. A similar phenomenon has also been witnessed in ethyl acrylate grafted to a sisal fiber system (sisal fiber contains 8% lignin). It was found that the grafting rate is higher when NaOH is used as a lignin remover, but the reverse has also been reported, i.e., in the presence of lignin, the grafting yield increases if the backbone is ozonized and grafted using Fe^{2+}–H_2O_2 as the initiator. In this case, ozone oxidizes lignin, thereby resulting in the formation of the carboxylic group in the lignin structure that favors the free radical formation which ultimately influences grafting [12]. This phenomenon has also been seen where acrylonitrile grafts on pulp utilize the xanthation method. In cases where lignin is present in the cellulose structure, chain transfer may occur to lignin from the OH radical, giving rise to less reactive lignin radical [13]. The grafting also influenced the presence of functional groups in the backbone. Styrene is grafted relatively with high efficiency on cellulose acetate *p*-nitrobenzoate. These results indicate that the pendant aromatic nitro group is more effective in obtaining a graft co-polymer [14]. The replacement of -OH by -SH groups in a cellulose substrate increases the level of grafting. This enhancement is initiated by Ce^{4+} ion which occurs by H-abstraction from C-atoms having -OH groups. However, in the case of MMA grafting on holo cellulose (comprising a mixture of the hemicellulose and α-cellulose), the mode of initiation does not the H-abstraction, and -SH group is associated with a marked decrease in the level of grafting [15]. In the case of vinylidene chloride grafting on

PVA, the vicinal -OH groups in PVA influence head-to-head monomer incorporation [16]. Rao and Rao [17] studied the rates of grafting of acrylic acid on different backbones. They explained the trends observed based on chain transfer of the growing polymer chain in the ordered nylon-6 > polyester ≈ PP fiber. The grafting gets also influenced by the treatment with chemical agents. Ghosh et al. [18, 19] observed good grafting percentage and efficiency in the case of diethylene triamine-modified oxycellulose. This might be due to complex formation, which initiates grafting. Annealing of the backbone also influences grafting, e.g., annealed PVA gave a higher degree of grafting than untreated backbone with vinyl monomers [20].

7.3 Effect of Monomer

Together with the nature of the backbone, the reactivity of the monomer is also an important factor in grafting. The reactivity of monomers depends upon various factors, viz. steric and polar nature, the concentration of monomers, and swellability of the backbone in the presence of the monomers. Recently, Naguib et al. [21] calculated the reactivity ratio of vinyl imidazole and acrylic acid by different techniques [22–25], with results given in Table 7.1. It was observed that maximum grafting on polypropylene films by γ-irradiation was achieved for a monomer composition with 40% acrylic acid and 60% vinyl imidazole. However, it is also observed that the monomer reactivity ratios for the grafting process are completely different from the values observed for conventional solution polymerization, e.g., for mixtures of monomers acrylonitrile/styrene, acrylamide/styrene [26].

In addition, the graft co-polymerization of mixed monomers is a more complex process than the individual monomers grafting owing to the synergistic effects resulting from the differences in monomer reactivity ratios. Nurkeeva et al. [27] studied the grafting of vinyl ether of monoethanolamine (VEMEA) in the presence of the more active monomer, vinyl ether of ethylene glycol (VEEG).

Monomer reactivity ratios (r^1 and r^2) for the conventional solution co-polymerization of VEEG and VEMEA are found to be equal to 1.2 and 0.7, respectively. The addition of the more active VEEG to the binary mixture enhances the grafting

Table 7.1 Reactivity ratios of vinyl imidazole and acrylic acid. [Reproduced with permission from [21] © Elsevier].

Fineman-Ross		Kelen-Tudos		Non-linear least square	
r^1	r^2	r^1	r^2	r^1	r^2
0.121	1.125	0.121	1.126	0.122	1.126

r^1, reactivity ratio of vinyl imidazole and r^2, reactivity ratio of acrylic acid of both monomers increasing the rate of co-polymerization. The difference in the grafting of ethyl acrylate EA (60.8%) and vinyl acetate VA (2.6%) on wool can be explained based on the nature of these monomers. The ethyl acrylate is highly reactive to free radicals and since vinyl acetate acts as an electron-donating monomer, it is extremely susceptible to monomer concentration [28]. Thus, the percentage of grafting of ethyl acrylate is higher because the loss of ethyl acrylate in side reactions is minimal. On the other hand, being less reactive to radicals, vinyl acetate is reduced in side reactions. Nagaty et al. [29] also observed differences in grafting ethyl acrylate, methyl methacrylate MMA and acrylonitrile AN on insoluble starch. In this case, the reactivity order was found to be AN> E ≈MMA. Here, the grafted polymethyl acrylate forms gel over the starch granules, acting as a barrier to monomer diffusion to the vicinity of starch. The order of the monomers on wool in terms of grafting is MA (methyl acrylate)>EA>MMA>VA> A(acrylic acid) [30]. The reactivity of the first three monomers is explained by steric considerations. Thus, MMA, being a highly crowded monomer that forms a complex with Ce4þ less readily and thereby affords minimum grafting. In contrast, VA is susceptible to monomer transfer reaction and tends to terminate the growing grafted chain by that process, which results in poor grafting efficiency. Since AA and its polymer are soluble in water, AA tends to undergo homo polymerization preferentially, resulting in poor grafting efficiency. Bhattacharya et al. [31] have compared grafting for substituted acrylamides. They revealed that the grafting order on cellulose acetate is acrylamide>methyl acrylamide>*N, N* dimethyl acrylamide. The methyl group in methyl acrylamide may reduce the mobility of the monomer that suppresses the grafting. The stability of tertiary (I) polymer radical may also be a reason for the low grafting with methyl acrylamide, whereas polymer radical from acrylamide is secondary (II) (Figure 7.1). The secondary radicals are more reactive than the tertiary.

With *N,N*-dimethyl acrylamide, two methyl groups play a remarkable role in the extent of grafting. Due to the steric effect of the two-methyl groups, the easy approach of the monomer to the backbone is maximally hindered, and thus the extent of grafting is the least. This phenomenon was also observed by earlier workers in the case of substituted acrylates [32, 33]. The grafting order on

$$\sim CH_2-\dot{C}(CH_3)-CONH_2 \quad \textbf{(I)} \qquad \sim CH_2-\dot{C}H-CONH_2 \quad \textbf{(II)}$$

Figure 7.1 Tertiary (I) polymer radical and secondary (II) polymer radical from acrylamide.

cellulose by means of a Ce^{4+} initiation is methyl acrylate>ethyl acrylate>butyl acrylate>methyl methacrylate. The explanation of reactivity can be offered in terms of steric and polar effects. As seen previously, it was also proposed that grafting depends upon the stability of the radical. The polymer radical that is formed in the case of methyl methacrylate is relatively stable, whereas in case of methyl methacrylate, which is the most reactive, the corresponding polymer radical is probably stable. It has also been described by Dworjanyn and Garnett [34] that during the radiation grafting, certain substituents activate whereas others deactivate the monomers, e.g., relative to styrene, o-methyl styrene strongly activates the cellulose while 2-vinyl pyridine is strongly deactivated. In general, the grafting efficiency will also depend on the concentration of the monomer. For example, during comparison of the grafting efficiency observed for 4-vinyl pyridine and methyl acrylate monomers, Kaur et al. [35] explained the phenomena in terms of the monomer's polymerizability and the solubility ($k_p/k_t^{1/2}$ values 213×10^{-3} and 7×10^{-3} for MA and 4-vinyl pyridine, respectively). The higher value for MA is responsible for more homo-polymer formation than for grafting as compared to 4-vinyl pyridine. In addition, 4-vinyl pyridine, being more soluble than MA in an aqueous medium results in higher grafting in an aqueous environment owing to its greater accessibility to the active sites. It is often reported that up to a certain limit, the grafting efficiency increases with monomer concentration and then decreases with a further increase in the monomer concentration [36]. This behavior may be attributed to an initial increase of the monomer concentration in proximity to the backbone. After a certain limit, the increase in monomer concentration accelerates the homo-polymerization reaction rather than grafting.

7.4 Effects of Solvent

In grafting mechanisms, the solvent acts as a carrier through which monomers are transported to the close vicinity of the backbone. The choice of the solvent depends upon several parameters, including the solubility of monomer in the solvent, the miscibility of the solvents if more than one is used, the swelling properties of the backbone, the generation of free radicals in the presence of the solvent, etc. The nature of the solvent and the polymer decides the solubility of the monomer, e.g., alcohols are useful solvents for grafting styrene [37–39]. The reason behind this is that alcohols can swell the backbone effectively and can dissolve the styrene so that the monomer can easily diffuse in the cellulosic structure. However, the extent of grafting decreases progressively when the alcohol is changed from methanol to ethanol to isopropanol and *t*-butanol. This decrease in grafting may be due to the gradually decreased swelling properties of the alcohol, known to be corroborated by the bulkiness of the alcohol molecules. A similar trend for the alcohols (methanol>ethanol>propanol>butanol>pentanol) in the case of grafting of methyl methacrylate on nylon-6 was also observed by Lenka [40]. A similar observation was also reported for styrene grafting on PVC by the irradiation technique [34]. In this case, it was found that the radiolysis fragments of methanol, particularly H-atoms would also support the grafting by H-abstraction reactions that creates the grafting sites in the backbone polymer. It is observed that, unlike styrene, acrylamide is not grafted at all from pure alcohol media. This may be clarified by the following reactions. The effect of solvent would primarily loosen the network of the polymer so that a grafting reaction can occur [41]. In a solution, the efficiency of grafting will depend upon the relative reactivity of monomer and solvent in competing for the radicals induced on the base polymer.

$$C^{\cdot} + X \rightarrow CX^{\cdot} \tag{7.3}$$

$$C^{\cdot} + CH_3OH \rightarrow CH + CH_2OH \tag{7.4}$$

where $C^{\cdot}$ represents the peroxy radical of the base polymer, formed under the conditions of the experiment and X denotes the monomer styrene or acrylamide. The H-abstraction rate constant follows the order isopropanol>ethanol>methanol>*t*-butanol. When reaction (a) is operative, the desired graft co-polymer will be obtained, whereas in the case of reaction (b) the formation of a homopolymer occurs rather than a graft. This can apprehend as polymer backbone radical abstracts H-atom from alcohol and produces $^{\cdot}CH_2OH$, leading to homopolymerization. For example, it appears that in grafting onto cellulose acetate, reaction (a) is faster when styrene is the monomer, but the reverse is probably true, with reaction (b) faster if acrylamide is the monomer. Therefore, alcohols are suitable solvents for grafting styrene, but not acrylamide, on cellulose acetate. Conversely, when water is added gradually, grafting begins to be noticeable. On increasing water for the acrylamide in the alcohol system, the extent of grafting increases. When the grafting is just initiated, as the alcohol is changed from *t*-butanol to isopropanol, the limiting fraction *f* of water (in the alcohol-water mixture) varies from 30% to 70%.

The addition of water to alcohol in the grafting medium suppresses the grafting of styrene. Incidentally, although cellulose acetate has a greater affinity for water than for MeOH, grafting from the alcohol-water mixture is affected by the decreased solubility of styrene in the solvent [39]. Nevertheless, when water is slowly added to alcohol, the grafting process is characterized by a detrimental effect up to a certain limit. Beyond these limits, however, with the gradual increase of water content, the extent of grafting increases, attaining a maximum and then decreasing again. The appearance of such maxima can be explained by means of Trommsdorff type effect. Since water is a poorer solvent for polystyrene, the growing homopolymer chains become sufficiently immobilized, as a result of which their rates of collisions with the growing graft chains are decreased and consequently termination is inhibited. Such a decrease in termination rate unaccompanied by a change in initiation rate leads to a higher steady concentration of radicals and hence to a higher overall rate of graft co-polymerization. A subsequent decrease in grafting on further addition of water is associated with the decreasing solubility of styrene in the solvent mixture. Diffusion of the monomer controls chain growth and chain termination in the internal structures of the polymer [42]. The solubility parameters d of the solvents should be close to the polymeric backbone so that the necessary chemical energy to disrupt intermolecular cohesive forces between polymer chains and permit chain mobility. In general, the formation of radicals that can initiate grafting may be suppressed by the chemical nature of polyethylene terephthalate, and in addition, ordering and the high crystallinity of the amorphous regions retard the diffusion of monomer. Considering PET, an $(AB)_x$ alternating copolymer, where A is the semirigid aromatic segment -CO-C_6H_4- with a d value of 9.8 and B is a flexible aliphatic ester -COO- CH2-CH2- with a value of 12.1, where solubility parameters are close to those of PET, e.g. DCE (9.0), Py (10.61), DMSO (12.93). These solvents promote diffusion and incorporation of the monomer and thereby subsequently grafting takes place. The inclusion of the solvents in the films increased with temperature in the medium and treatment time. As wetting of the polymer by the solvents is an important criterion, the surface tension data gives useful information regarding the key role in grafting. In the presence of tetrahydrofuran and alcohol, the high graft yield can be attributed to the low surface tensions of those solvents, which may significantly improve the wetting of monomers towards PTFE [43, 44]. Similarly, owing to similar equilibrium swelling of polypropylene fiber in both solvents (benzene and toluene), the grafting of methacrylic acid on polypropylene fibers was reported to be the same in benzene and toluene [45]. In both solvents, the homopolymer yields are also almost similar. As compared to other solvents like benzene or toluene, the significant feature of methacrylic acid grafting on polypropylene fiber is that chlorobenzene gives a higher grafting yield. This may be due to the comparatively higher degree of swelling of polypropylene fiber by chlorobenzene. The swelling behavior of polypropylene (PP) fibers in benzene–methanol mixture decreases continuously and almost linearly with increasing methanol content in the solvent mixture. No swelling is observed in the methanol solvent. In this system, polymethyl methacrylate precipitates out from the reaction mixture. Thus, the viscosity of the reaction medium is not affected. Hence, the information on the swelling of PP fibers in the reaction medium can be utilized to regulate the graft levels. The presence of water is essential for the grafting off collagen. Experimentally observed that initially, the water adsorbs on the surface of collagen, and then grafting sites are formed directly between hydrated collagen and Bu_3B most likely via the complex [46] (Figure 7.2).

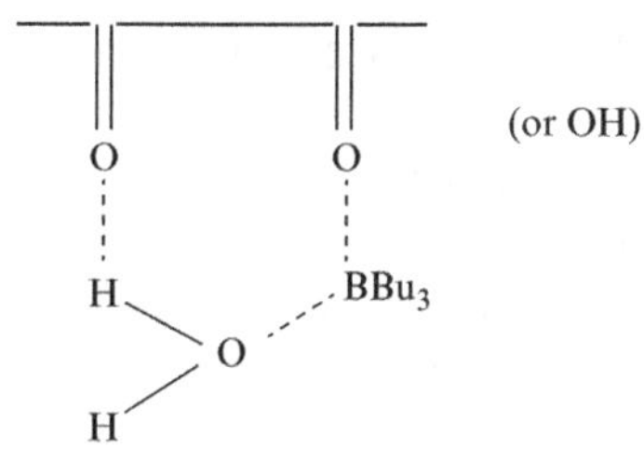

Figure 7.2 Structure of the complex between collagen and Bu_3B [46]. [Reproduced with permission from [46] © Wiley].

The generation of free radicals of solvent is very important for the grafting mechanism, besides the free radical formation monomer, and on the backbone. Consequently, the proper choice of solvent is important. The graft copolymer (i.e., AN on starch) had more grafted chains when prepared in aqueous organic solvent systems than preparation in water alone [47]. In the presence of CH_3OH, the more frequent grafting could have resulted from chain transfer to starch via free radicals derived from CH_3OH. The latter has approximately the same effect as water in the grafting concentration of polystyrene on starch. But ethyl alcohol was less effective as a promoter of homograft copolymers. In starch, methyl alcohol permeates starch structure effectively and as a result, free radicals derived from methanol can easily disrupt H-bonds. In the case of acrylonitrile grafting on isotactic polypropylene, PP-OOH may undergo swelling in the presence of water that facilitates the accessibility of the monomer to the active sites [48]. As water has zero chain transfer constant, the side reaction involving chain transfer is minimal in water. Thus, water has been proven an excellent medium for grafting. Also, as compared to grafting indioxane, grafting is prominent in a water medium. This can be understood in terms of H-bonding. PP-OOH is capable of forming H-bonded complexes I and II with water and dioxane, respectively (see Figure 7.3).

Complex(II) being fairly stable blocks some -OOH groups of PPOOH resulting in a decrease in grafting. In the presence of methanol, grafting takes place without the accelerating effect. This is realized by the fact that various chain transfer and H-abstractions are accelerated in the presence of methanol leading to a decrease in grafting. The reactivity order for different solvents is water>bulk>dioxane>methanol.

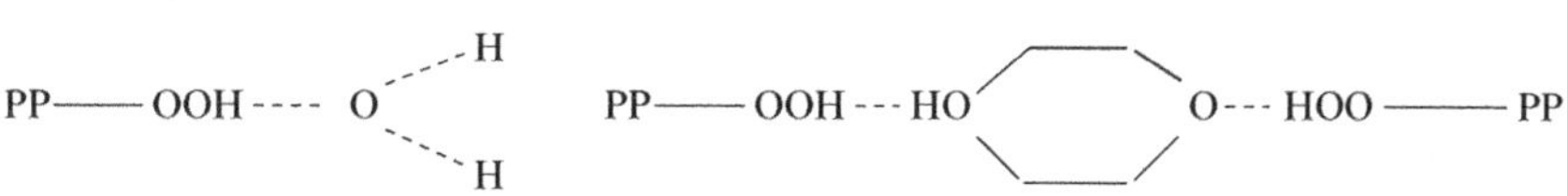

(I) (II)

Figure 7.3 H-bonded complex of polypropylene with water and dioxane. [Reproduced with permission from [48] © Wiley].

7.5 Effect of Initiator

Apart from the radiation technique, all chemical grafting reactions require an initiator, and its nature, solubility, concentration as well as function need to be deliberated. There are numerous kinds of initiators: (Fe^{2+}–H_2O_2), AIBN, $K_2S_2O_8$, etc. The nature of the initiator has a profound effect on grafting. For example, as discussed above, resonance stabilization is exhibited by AIBN. No such resonance stabilization exists with conventional peroxide initiators. Therefore, higher grafting yield should be procured with peroxide initiators than with AIBN [49]. In another example, AIBN gives poor grafting and $K_2S_2O_8$ is unsuitable as an initiator in the grafting of HEMA on cellulose since that degrades the cellulose chain. The rate of grafting is dependent upon the backbone polymer, the monomer as well as the concentration of the initiator [50, 51]. There are various empirical relationships regarding the dependence of the grafting efficiency on the initiator concentration [52, 53]. It is specious from the observations that once a certain initiator concentration is reached, the conversion of grafted monomer could not increase even if the higher levels of initiator [54, 55]. Sanli and Pulet [42] explained the behavior using Bz_2O_2in the case of acrylamide grafting on PET (polyethylene terephthalate). Increasing the Bz_2O_2 concentration to a certain limit enhanced the grafting which implies that the secondary free-radical species ($C_6H_5^{\cdot}$) and/or the primary free-radical species ($C_6H_5COO^{\cdot}$) are formed by the dissociation of Bz_2O_2 in the polymer system that participates directly in the initiation of grafting. Above a certain limit, the abundance of these radicals leads to their participation and the termination of the growing polymer, as well as to the combination of phenoxo and phenyl radicals, thereby giving rise to decreased grafting [55]. However, it is found that the maximum graft efficiency occurs at a particular initiator concentration which in turn also depends on the grafting system [56, 57]. At higher concentrations, Ce^{4+} participates in the termination of growing grafted chains as follows, based on an example of grafting onto wool [56]:

$$WM_1^{\cdot} + Ce^{4+} \rightarrow WM_1 + Ce^{3+} + H^+ \qquad (7.5)$$

$$WM_2^{\cdot} + Ce^{4+} \rightarrow WM_2 + Ce^{3+} + H^+ \qquad (7.6)$$

where WH refers to wool; M_1 and M_2 are the two monomers. Another prime factor is the solubility of the initiator in the grafting medium. Ideally, to initiate the grafting reaction through monomers, the initiator must be fully soluble. Nakamura et al. [58] reported an interesting experimental observation regarding vinyl acetate, styrene, methacrylic acid, and methyl methacrylate grafting on sericin (obtained by scouring silk fiber). It has been displayed that water-soluble initiators were superior to water-insoluble initiators in obtaining graft co-polymers having homogeneous molecular weight distribution for the grafting of water-soluble monomers (e.g., methacrylic acid) to the sericin.

7.6 Role of Additives on Grafting

The extent of graft co-polymerization or grafting yield significantly depends on the presence of additives such as acids, inorganic salts, and metal ions. Thus, the reaction between the backbone and the monomer must compete with any reactions between additives and the monomer. Although to augment the grafting efficiency, some additives may enhance the monomer/backbone reaction. If the reaction between the monomer and the additive is dominant, the reverse will be spot-on. In some grafting processes, the role of acid additives is typically important [9, 59, 60]. However, the nature of the acids is also taken into consideration. Among the mineral acids, only sulfuric acid is effective, dependent, of course, on the nature of the polymer backbone. The use of nitric acid is limited, for example, during irradiation it tends to degrade cellulose. Hydrochloric acid has been found to exert a negative effect on grafting efficiency, as chlorine tends to be incorporated into the polymer structure [61]. The enhancement of the grafting efficiency by acid mechanism has been the subject of considerable work. In the case of radiation grafting, the acid enhancement is attributed to two on-board predominant factors: the extent to which the grafting monomer is soluble in the bulk solution and the radiolytic yield of H-atoms. Enhanced grafting (e.g., of styrene to polyolefins) due to the effects of thermalized electron captures reaction leads to increased G(H), the number of molecules formed per 100 eV of absorbed energy, and hence more sites for grafting are available by H-atom abstraction reactions. There is an equilibrium concentration of monomer absorbed in any grafting system. As the grafting proceeds, owing to the changed composition of the grafted polymer, the grafting region may continually change. Thus, in grafting styrene to cellulose, during the initial part of the reactions, the grafting region is cellulosic, however, as the reaction proceeds, an increased polystyrene component is observed in the grafting region. Thus, the chemical structure of the region at the specific time of grafting decides the degree to which the monomer is absorbed by the grafting region. The addition of alkali acids can also affect the nature of the backbone, the initiator as well as solvent thereby can influence the grafting. Zaharan and Zhody [62] revealed that the presence of sulfuric acid or alkali controls the grafting yield when styrene and ethyl acrylate are co-grafted on sisal fiber. Due to the alkali treatment, the increase in crystallinity will result in a reduction in the sorption capacity of the fiber. As a result, during the grafting process, the amount of monomer solution sorbed in the fiber will be reduced. This accounts for the decrease in the grafting efficiency for sisal fibers subjected to alkali treatment. By contrast, better grafting results onto cellulose are obtained when the fibers are subjected to the combined treatment and fibrillation due to the intracrystallite swelling by the acid facilitating the subsequent penetration of NaOH solution. Moreover, an increase in the ordering of the fibers in addition to an increase in the crystalline regions may result from the combined treatment. As compared to that of the fibers subjected to the alkali treatment alone, these effects are reflected in the slight decrease in the grafting yield of fibers subjected to the combined treatment. For grafting, the solvent structure is considered to be an important point of view, and the lower molecular weight alcohols are seen as efficient reagents to be deliberated [43]. In the presence of an acid, the radiolytically produced H-atoms (reactions(a) and (b)) abstracted H-atoms from the base polymer (PH) (reaction (c)) results in yielding additional grafting sites [44]

$$CH_3OH + H^+ \rightarrow CH_3OH_2^+ \tag{7.7}$$

$$CH_3OH_2^+ + e \rightarrow CH_3OH + H^+ \tag{7.8}$$

$$PH + H^\cdot \rightarrow= P^\cdot + H_2 \tag{7.9}$$

where PH refers to a polymer. Misra et al. [63, 64] studied the effect of sulphuric acid in the mechanism for the grafting of methyl acrylate onto cellulose in the presence of Ce^{4+} as initiator. In the presence of sulfuric acid, maximum grafting takes place. In an aqueous medium, initiator Ce^{4+} is believed to combine with water in the following manner

$$Ce^{4+} + H_2O \leftrightarrow \left[CeOH^{+3}\right] + H^+ \tag{7.10}$$

$$2[CeOH^{+3}] \leftrightarrow + \left[Ce - O - Ce\right]^{+6} H_2O \tag{7.11}$$

Thus, Ce^{4+} exists as $[Ce^{4+}]$, $[CeOH^{+3}]$, and $[Ce\text{-}OCe]^{+6}$ in an aqueous solution. The concentration of these species varies with the acid concentration in the manner described. From Equation 7.10

$$\left[CeOH\right]^{+3} = k_1 Ce^{4+} / \left[H^+\right] \tag{7.12}$$

where $[Ce^{4+}]$ = total concentration of the ceric ion. From the above, it is clear that with an increase in $[H_2SO_4]$, the equilibrium shifts towards the formation of more and more of $[CeOH]^{3+}$ and $[Ce]^{4+}$. $[Ce^{4+}]$ that facilitates the formation of a complex with the base polymer. Having smaller sizes, these species facilitate the formation of a complex between the polymer (e.g., cellulose) and Ce^{4+} ion, increasing the percentage of grafting. Beyond an optimum $[H_2SO_4]$, considerable $[Ce^{4+}]$ and $[CeOH]^{+3}$ are formed, which results in a decrease in grafting efficiency due to the acceleration of the termination of growing grafted chains. The same behavior was also observed for the grafting of poly(methyl methacrylate) onto wool in the presence of nitric acid [56, 65]. In the presence of inorganic salt (e.g., $LiClO_4$ or $LiNO_3$), the enhancement of grafting is also been established by partitioning phenomena [66, 67]. Owing to the overall monomer partitioning effect, metal salts such as $LiClO_4$ are more efficient than acids in enhancing photographs. This partitioning behavior may be interpreted as an example of the salting out technique employed in solvent extraction, except that here one phase is a solid. With the addition of salts, the monomer solubility decreases, inducing the increased partitioning of monomers into the substrate. The net result of this driving force is equilibrium monomer concentrations within the substrate and higher rates of monomer diffusion. It has also been shown that in promoting UV photochemical grafting, $LiClO_4$ is more efficient than $LiNO_3$ [68]. Synergistic effects in the simultaneous use of these additives have been reported [67, 69]. Generally, grafting efficiency enhances in the presence of a metal ion (e.g., Fe^{2+}, Cu^{2+}) and Mohr's salt that preferentially reduces homopolymer formation. Partitioning phenomena also appeared to control the efficiency of this process. This can be followed visually in the grafting system when copper salts are used in the case of polypropylene[71]. When a natural polymer (e.g., cellulose) is used as the backbone, a high proportion of Cu^{2+} ions are partitioned from the bulk solution to the vicinity of the backbone. The grafting gets reduced due to the relatively high concentration of Cu^{2+} near the backbone polymer. Because the depletion of Cu^{2+} in the bulk solution reduces the scavenging of monomer radicals in the monomer solution, and therefore the efficiency of undesired homo-polymerization is increased. In the case of styrene grafting onto cellulose, though initially, it behaves in its pure form, as grafting proceeds, cellulose is progressively enriched with polystyrene, causing the grafting process to resemble that for polypropylene. Kubota and Hata [70] found that hydroquinone influences the distribution of MAA grafted chains in polyethylene film, though it does not affect graft efficiency in the system. In the grafting of methyl acrylate and vinyl acetate on cellulose acetate, the addition of NaCl or $NaNO_3$ and their role has also been studied. The presence of NaCl or $NaNO_3$ affected the graft co-polymerization by enhancing the oxidation of cellulose by the transition metal ions (viz. Ce^{4+}). The transition metal ions initiate the formation of free radicals for grafting, but it left the homopolymerization almost unaffected [71]. A comparison of different inorganic salts was studied by Lenka et al. [72, 73] showing that the grafting of methyl methacrylate onto nylon-6 was affected in the order $CuSO_4$, KCl, NaCl, NaBr, $LiNO_3$, NaF, $MnSO_4$. The addition of either $CuSO_4$ or Na-lauryl sulfate surfactant was found to separately suppress the formation of homopolymer. Therefore, increased efficiency of grafting in the case of methyl methacrylate grafting on nylon-6 is visualized. Apart from the inorganic salt effects and the acid enhancement, partitioning phenomena seems to probably involve when using organic inclusion compounds like urea [74] and multifunctional acrylates [75–77] that also may increase the grafting efficiency [78]. The role of semiconductors (e.g., colloidal Fe_2O_3) in grafting has also been studied [79]. The colloidal Fe_2O_3 acts as a photosensitizer, which absorbs photons of bandgap energy, promoting the electrons to the conduction band, which are then scavenged by methyl viologen. The valence band holes can abstract H-atoms from the backbone (e.g., of cellulose acetate), thus creating active sites onto which acrylamide was grafted. The grafting mechanism is as follows:

$$\alpha Fe_2O_3 \rightarrow h^+ + e \tag{7.13}$$

$$e + MV^{2+} \rightarrow MV^+ \tag{7.14}$$

$$h^+ + Cell - H \rightarrow Cell \tag{7.15}$$

$$Cell + M \rightarrow Cell - M(acrylamide) \tag{7.16}$$

The effects of complexing agents, such as EDTA, potassium fluoride, and ascorbic acid on the grafting efficiency of PEA [poly(ethyl acrylate)] onto cellulose have also been studied [80]. These additives were found to reduce the grafting of PEA on cellulose. However, the decrease in the percentage of grafting was much less, and grafting occurred without homopolymer formation when KF as a complexing agent was used. When the grafting is initiated with Fenton's reagent, the primary radical $OH^{\cdot}$ responsible for the generation of active sites on cellulose gets destroyed by a reaction with Fe^{3+} formed during the reaction of Fe^{2+} and H_2O_2. Consequently, by the use of suitable agents (e.g., ascorbic acid, KF, or EDTA), grafting may be favored by the complexation of Fe^{3+}. It is also observed, however, that a further increase in the concentration of KF considerably reduces the grafting of ethyl acrylate. Similar behavior was also seen in the presence of Fenton's reagent during the grafting of vinyl acetate. This may indicate that whereas at a lower concentration KF reduces Fe^{3+} by complex formation, promoting grafting. At higher concentrations of KF, oxidized KF to elemental fluorine derived by the oxidation of KF by H_2O_2 may add to a vinyl monomer. This results in a consequent decrease in the percent grafting. Thus, the addition of KF did not improve the percentage of grafting of VA to cellulose. The effect of amines upon ceric ion-initiated grafting of PMA or poly(methyl acrylate) onto wool [81] has been explained by assuming a complex formation between the ceric ion and wool. Amines are known to form complexes with ceric ions in the following manner:

$$Ce^{4+} + RNH_2 \leftrightarrow Complex \rightarrow Ce^{+3} + H^+ + RNH^{\cdot} \tag{7.17}$$

The ceric amine complex decomposes to give free-radical species. This free-radical species at lower concentrations generate more active sites on wool by H-abstraction. However, there exists a critical concentration of amines that promotes grafting. With a further increase in concentration, the percentage of grafting decreases owing to the termination of growing grafted chain-by-chain transfer with the amine:

$$W - (M)_n^{\cdot} + RNH_2 \xrightarrow{Ktr} RNH_2 \rightarrow W(M)_n - H + RNH^{\cdot} \tag{7.18}$$

The complex formation between Ce^{4+} and the amine will be determined primarily by three factors: the nucleophilicity, the basicity, and the steric requirements of the amine [82]. In the case of grafting poly(methyl acrylate), the reactivity of amines followed the order: triethylamine>diethylamine>*n*-butylamine>triethanolamine>*N*,*N*-dimethylaniline. It was observed that with an increase in the basicity of the amines, the grafting percentage increases linearly. This would be expected because the complexation of ceric ions with amines should be enhanced by the increased basicity of the amine. The exceptional behavior in the case of triethylamine might be because with triethylamine both the nucleophilic reactivity of the amine to form a complex with Ce^{4+}, and the stability of the complex between Ce^{4+} and the amines is dependent on the ability of various substituents in the amines to increase electron density at nitrogen. Because of its three ethyl groups, triethyl amine is more nucleophilic than diethylamine, which will facilitate the formation of a complex with Ce^{4+}. Because of increased crowding, the triethylamine-Ce^{4+} complex readily undergoes decomposition to yield free-radical species. That thereby abstracting H-atom creates more active sites on wool to initiate grafting. The grafting efficiency generally increases as the size of the amines increases. This indicates that owing to their large size, only at a higher concentration of the ion, tertiary amines will form complexes with Ce^{4+}. The oxygen atom at the b-position in triethanolamine reduces the electron density in nitrogen and hence does not aid in the formation of a complex. Complex formation between Ce^{4+} and dimethylaniline is not favored because of the electron delocalization in the amine. This will reduce and may completely suppress the grafting event (see Figure 7.4). In the case of grafting poly(vinyl acetate) onto wool, the effect of amines has been studied for triethyl and triethanolamine [83].

In the presence of vinyl acetate, complex formation (Figure 7.4) between triethylamine and Ce^{4+} is inhibited, and hence on the addition of the amine, no acceleration in grafting occurs. As discussed earlier, the oxygen reduces the electron density of the nitrogen with triethanolamine, and complex formation between triethylamine and Ce^{4+} is not favored. In addition, the complex formation between triethanolamine and

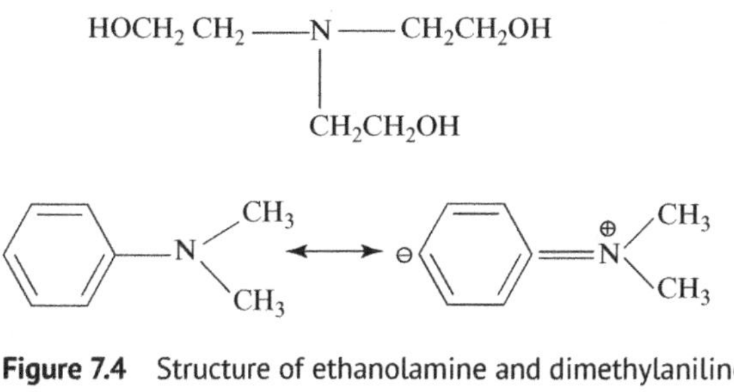

Figure 7.4 Structure of ethanolamine and dimethylaniline. [85] / John Wiley & Sons.

Ce^{4+} is also inhibited by the size factor. Thus, neither the grafting nor homopolymerization is favored in the presence of triethanolamine. The reactivity of various amines onto wool towards the graft copolymerization [84] followed the order of diethylamine, dipropylamine, ammonia,triethylamine, Triethanolamine, and pyridine. Though DEA (diethylamine) is as nucleophilic as dipropylamine (DPA), only DEA enhances the grafting rate tremendously. However, no accelerating effect upon grafting efficiency is observed in the presence of DPA. This may be explained based on the steric factor, such that DEA, having a smaller steric requirement than DPA, therefore easily forms a complex with Ce^{4+}. The substantial difference in behavior between DPA and DEA can also be explained by the steric factor. Ammonia, having a smaller steric requirement than TEA, forms a complex with Ce^{4+} more easily than TEA. With triethanolamine and pyridine, all three factors (nucleophilicity, basicity, and steric size) are responsible for giving a low efficiency of grafting. L-Threonine bears -NH_2, –COOH and-OH groups that are capable of forming complexes with Ce_4 [65]. In general, with increasing L-threonine concentration, the efficiency of grafting methyl acrylate on cellulose was found to increase, but at higher concentrations, the complex formation was reduced. This may be through the competition by an enhanced rate of abstraction of H-atom from L-threonine by the growing grafted chains:

$$R_{poly}O(M_n)^{\cdot} + RNH_2 \rightarrow R_{poly}O(M_n) - H + R - NH^{\cdot} \tag{7.19}$$

$$R_{poly}O(M_n)^{\cdot} + ROH \rightarrow R_{poly}O(M_n) - H + R - O^{\cdot} \tag{7.20}$$

$$R_{poly}O(M_n)^{\cdot} + RCOOH \rightarrow R_{poly}O(M_n) - H + R - COO^{\cdot} \tag{7.21}$$

Grafting of methyl acrylate on cellulose was suppressed by 5-hydroxy tryptamine and 5-hydroxytryptophan additives. The phenolic OH groups present in these additives inhibit both grafting and polymerization. This may indicate that the mode of action of radio-protecting agents accounts for a free-radical mechanism.

7.7 Effects of Temperature

The kinetics of graft co-polymerization is highly affected by the temperature. In general, with an increase in temperature, grafting yield increases until a limit is attained. One factor in this can be the faster monomeric diffusion processes in the backbone increases with increasing temperature, facilitating grafting, as observed by Dilli et al. [36]. In the case of grafting MMA on silk, with increasing temperature, the graft yield increases significantly due to the greater swelling of silk. Thus, the corresponding enhanced rate of diffusion of the monomers in the vicinity of silk will result [85]. However, Sun et al. [36] explained this behavior as with the increase in temperature, the thermal decomposition rate of the initiator and the initiator efficiency in producing free radicals on base polymer also increases, resulting in an increased concentration of polymer macroradicals, and thus enhanced the graft polymerization. Increasing the temperature initially enhances the grafting yield and facilitates the decomposition of peroxide. However, with an increase in temperature in the case of acrylamide grafting on cellulose acetate, the grafting yields subsequently decrease, as reported by Maldas [86]. The initial increase in grafting is due to the decomposition of peroxides formed as a result of irradiation of the base polymer in air, making the requisite radicals available for grafting. Conversely, due to the increased molecular motion with increased temperature, resulting in increased radical decay led to the subsequent decrease. An interesting reflection is that for a temperature near the glass transition temperature, the maximum graft yield occurs [42, 87]. For the temperatures below the Tg (glass transition temperature), the radicals formed in the polymer chains cannot react, owing to the reduced diffusion of the monomer. Whereas for T above Tg, the number of radicals available for grafting will decrease with increasing temperature. With increasing temperature, the combination of monomer radicals results in lower graft yield [88].

7.8 Conclusion

The increasing environmental awareness continuously propels the exploration and exploitation of eco-friendly raw materials like plant-derived materials as substitutions for synthetic materials. Among various plant-derived materials, plant polysaccharides have long been exploited in different fields for their inherent advantages of being widely abundant, inexpensive, nontoxic, naturally renewable, and biodegradable. However, pristine plant polysaccharides often demonstrate several unsatisfactory characteristics, and eventually, their applications are limited. Graft copolymerization is the most attractive and potential method, which could be employed to improve various properties and widen the uses of plant polysaccharides. Indeed, graft copolymerization facilitates the introduction of functional groups onto polysaccharide backbones to control the aggregation state of molecular chains, polymer charges, rheological properties, hydrophobic potential, complex formation ability, etc. The current chapter presents the current advancements in graft copolymerization of various plant polysaccharides and their uses with a brief description of the concept and methods of graft copolymerization. Graft copolymerization onto different plant polysaccharides is found to be significantly controlled by initiator nature, concentrations of initiator, concentrations of monomer, reaction temperature, and reaction time. The characteristics of grafted polymers have been customized by the nature of side chains including their molecular structures and molecular numbers. The published literature indicated that grafted plant polysaccharides are promising materials with wide-ranging applications in industrial processes such as flocculants, decolorizing agents, thickeners, adsorbents, drug delivery carriers, electrical conductors, etc. The vast applications of plant polysaccharide-based graft copolymers suggest that these biomaterials have a bright and commercial future in various industrial fields. Therefore, it is reasonable to expect that grafted biopolymers continue to revolutionize the advancements of green technologies soon.

References

1 Thakur, V.K., Thakur, M.K., and Gupta, R.K. (2013). Graft copolymers from cellulose: synthesis, characterization, and evaluation. *Carbohydrate Polymers* 97: 18–25.

2 Thakur, V.K., Thakur, M.K., and Gupta, R.K. (2014). Graft copolymers of natural fibers for green composites. *Carbohydrate Polymers* 104 (87–93).

3 Wang, A. and Wang, W. (2013). Gum-g-Copolymers: synthesis, Properties, and applications. In: *Polysaccharide Based Graft Copolymers* (ed. S. Kalia and M. Sabaa), Berlin, Heidelberg: Springer. https:doi.org/10.1007/978-3-642-36566-9_5.

4 Celik, M. (2006). Preparation and characterization of starch-g-polymethacrylamide copolymers. *Journal of Polymer Research* 13 (427–432).

5 Sharma, R.K. (2011). Lalita, Synthesis and characterization of graft copolymers of N-Vinyl- 2- Pyrrolidone onto guar gum for sorption of Fe $^{2+}$ and Cr^{6+} ions. *Carbohydrate Polymers* 83: 1929–1936.

6 Hebeish, A. and Guthrei, J.T. (1981). *The Chemistry and Technology of Cellulosic Copolymers*. Berlin: Springer-Verlag. https:doi.org/10.1007/978-3-642-67707-6.

7 Ng, L.-T., Garnett, J.L., Zilic, E., and Nguyen, D. (2001). Effect of monomer structure on radiation grafting of charge transfer complexes to synthetic and naturally occurring polymers. *Radiation Physics and Chemistry* 62: 89–98.

8 Ibrahem, A.A. and Nada, A.M.A. (1985). Grafting of acylamide onto cotton linters. *Acta Polymerica* 36 (6): 320–322.

9 Chappas, W.J. and Silverman, J. (1979). The effect of acid on the radiation-induced grafting of styrene to polyethylene. *Radiation Physics and Chemistry* 14: 847–852.

10 Clark, D.C., Baker, W.E., and Whitney, R.A. (2000). Peroxide-initiated comonomer grafting of styrene and maleic anhydride onto polyethylene: effect of polyethylene microstructure. *Journal of Applied Polymer Science* 79 (1): 96–107.

11 Tyuganova, M.A., Galbraikh, L.S., Ulmasove, A.A., Tsarevskaya, I.Y., and Khidoyator, A.A. (1985). Use of rice straw as cellulosic raw material for ion exchanger production. *Cell Chemical Technology* 19 (5): 557–568.

12 Kokta, B.V., Valade, J.L., and Daneault, C. (1981). Modification of mechanical and thermo mechanical pulps by grafting with synthetic polymers. II. Effect of ozonation on polymerization parameters and pulp properties. *Transactions* 7: 5–10.

13 Hornof, V., Kokta, B.V., and Valade, J.L. (1976). The xanthate method of grafting. IV. Grafting of acrylonitrile onto high-yield pulp. *Journal of Applied Polymer Science* 20: 1543–1554.
14 Nakamura, S., Yoshikawa, E., and Matsuzuki, K. (1980). Graft copolymerization of styrene onto cellulose acetate p-nitrobenzoate by chain transfer reaction. *Journal of Applied Polymer Science* 25: 1833–1887.
15 Okieima, E.F. and Idehem, I.K. (1987). Graft copolymerization of acrylonitrile methylmethacrylate and vinyl acetate on bleached holocellulose by use of ceric ions. *Journal of Macromolecular Science Chemistry* 11: 1381–1391.
16 Storey, R.F., Dantiki, D.S., and Goff, L.J. (1988). Ceric ion initiation of vinylidene chloride from poly(vinyl alcohol). *Polymer Materials Science and Engineering* 58: 565–569.
17 Rao, M.H. and Rao, K.N. (1985). Radiation-induced grafting onto synthetic fibers. A new mechanism of crosslinking and grafting of acrylic acid and acrylonitrile to synthetic fibers. *Radiation Physics and Chemistry* 26 (6): 669–672.
18 Ghosh, P. and Bandyopadhyay, A.R. (1986). (Photo) graft copolymerization of methylmethacrylate (MMA) on polyamine-modified oxycellulose in a limited aqueous system. *Journal of Applied Polymer Science* 31 (5): 1499–1507.
19 Ghosh, P. and Ghosh, T.K. (1982). Photograph copolymerization of methylmethacrylate acrylonitrile and acrylamide on oxycellulose using potassium persulfate as the initiator in a limited aqueous system. *Journal of Macromolecular Science Chemistry A* 18 (3): 361–370.
20 Meissner, H. and Meublein, G. (1986). Pfropfreaktionen am polyvinylkohol Mitt: zurPfropfung von vinylmonomeren auf getempertempolyvinylalkohol. *Acta Polymers* 37 (5): 323–324.
21 Naguib, H.F., Aly, R.O., Sabaa, M.W., and Mokhtar, S.M. (2003). Gamma radiation-induced graft copolymerization of vinylimidazolcacrylic acid onto polypropylene films. *Polymer Testing* 22 (7): 825–830.
22 Fineman, M. (1950). Ross, S.D., Linear method for determining monomer reactivity ratios in copolymerization. *Journal of Polymer Science* 5: 259–262.
23 Kelen, T. and Tudos, F. (1974) New improved linear graphical method for determining copolymerization reactivity ratios. *Journal of Macromolecular Science Chemistry* 1–91975.
24 Behnken, D.W. (1964). Estimation of copolymer reactivity ratios: an example of nonlinear estimation. *Journal of Polymer Science* 2 (2): 645–648.
25 Tidwell, P.W. and Mortimer, G. (1965). An improved method of calculating copolymerrization reactivity ratios. *Journal of Polymer Science* 3: 369–378.
26 El-Naggar, A.M., Zhody, M.H., Sahar, S.M., and Allan, E.A. (2001). Reactivity ratios during radiation-induced grafting of comonomer mixtures onto polyester fabrics. *Polymer International* 50: 1082–1088.
27 Nurkeeva, Z.S., Aal, A.A., Khutoryanskiy, V.V., and Mun, G.A., Beksyrgaeva, A.G. (2003). Radiation grafting from binary monomer mixtures I vinyl ether of

monoethanolamine and vinyl ether of ethylene glycol. *Radiation Physics and Chemistry* 67 (6): 717–722.

28 Misra, B.N., Sharma, R.K., and Mehta, I.K. (1982). Grafting onto wool. XV. Graft copolymerization of MA and MMA by use of Mn(acac)3 as initiator. *Journal of Macromolecular Science Chemistry* 3: 489–500.

29 Nagaty, A., Abd-El-Mouti, F., and Mansour, O.Y. (1980). Graft polymerization of vinyl monomers onto starch by use of tetravalent cerium. *European Polymer Journal* 16: 343–346.

30 Misra, B.N., Mehta, I.K., and Dogra, R. (1980). Ceric ion initiated graft copolymerization of vinyl monomers comparison of monomer reactivities. *Journal of Applied Polymer Science* 25 (2): 235–241.

31 Bhattacharya, A., Das, A., and De, A. (1998). Structural Influence on grafting of acrylamide based monomers on cellulose acetate. *Indian Journal of Chemical Technology* 5: 135–138.

32 Varma, D.S. and Narashinan, V. (1972). Thermal behavior of graft copolymers of cotton cellulose and acrylate monomers. *Journal of Applied Polymer Science* 16: 3325–3339.

33 Varma, D.S. and Narashinan, V. (1975). Grafting of formaldehyde crosslinked and cyanoethylated cotton cellulose with acrylate monomers. *Journal of Applied Polymer Science* 19: 29–36.

34 Dworjanyn, P.A. and Garnett, J.L. (1992). Radiation grafting of monomer on plastics and fibers. In: *Rad Processing of Polymers* (ed. A. Singh and J. Silverman), 93. Munich: Hanser Publishers.

35 Kaur, I., Misra, B.N., Gupta, A., and Chauhan, G.S. (1998). Graft copolymerization of 4-vinyl pyridine and methyl acrylate onto polyethylene film by a radiochemical method. *Journal of Applied Polymer Science* 69: 599–610.

36 Sun, T., Xu, P., Liu, Q., Xue, J., Xie, W. (2003). Graft copolymerization of methacrylic acid onto carboxymethyl chitosan. *European Polymer Journal* 39: 189–192.

37 Dilli, S. and Garnett, J.L. (1967). Radiation-induced reactions with cellulose. III. Kinetics of styrene copolymerisation in methanol. *Journal of Applied Polymer Science* 11 (6): 859–870.

38 Yasukawa, T., Sasaki, Y., and Murukami, K. (1973). Kinetics of radiation-induced grafting reactions. II. Cellulose acetate-styrene systems. *Journal of Polymer Science Polymer Chemistry* 11 (10): 2547–2556.

39 Bhattacharyya, S.N. and Maldas, D. (1982). Radiation-induced graft copolymerization of mixtures of styrene and acrylamide onto cellulose acetate. I. Effect of solvents. *Journal of Polymer Science Polymer Chemistry* 20 (4): 939–950.

40 Lenka, S. (1982). Grafting vinyl monomers onto nylon-6. X. Graft copolymerisation of methylmethacrylate onto Nylon 6 using peroxydiphosphate as initiator. *Journal of Applied Polymer Science* 27 (4): 1417–1419.

41 Lee, S., Rengarajan, R., and Parameswara, V.R. (1990). Solid phase graft copolymerization: effect of the interfacial agent. *Journal of Applied Polymer Science* 41: 1891–1894.

42 Sanli, O. and Pulet, E. (1993). Solvent-assisted graft copolymerization of acrylamide on poly(ethylene terephthalate)films using benzoyl peroxide initiator. *Journal of Applied Polymer Science* 47: 1–6.

43 Garnett, J.L. (1979). Grafting. *Radiation Physics and Chemistry* 14: 79–99.

44 Nho, Y., Garnett, J.L., Dworjanya, P.A., and Jin, J.H. (1992). Radiation-induced graft copolymerization of 2-hydroxyethyl methacrylate and styrene onto polytetrafluoroethylene. *Journal of Korean Industrial and Engineering Chemistry* 3: 491–498.

45 Mukherjee, A.K. and Gupta, B.D. (1985). Radiation-induced graft copolymerization of methacrylic acid onto polypropylene fibers. II. Effect of solvents. *Journal of Applied Polymer Science* 30: 2655–2661.

46 Kojima, K., Iguchi, S., Kajima, Y., and Yoshikuni, M. (1983). Grafting of methylmethacrylate onto collagen initiated by tributyl borane. *Journal of Applied Polymer Science* 28: 87–95.

47 Walling, C. (1957). *Free Radicals in Solution*, 285. New York: Wiley.

48 Misra, B.N., Sood, D.S., and Mehta, I.K. (1985). Grafting onto polypropylene. I. Effect of solvents in gamma radiation-induced graft copolymerization of poly (acrylonitrile). *Journal of Polymer Science Polymers Chemistry* 23: 1749–1756.

49 Nishioka, N., Matsumoto, K., and Kosai, K. (1983). Homogeneous graft copolymerization of vinyl monomers onto cellulose in a dimethyl sulfoxide-paraformaldehyde solvent system. II. Characterization of graft copolymers. *Polymers* 15 (2): 153–158.

50 Anbarason, R., Jayasehara, J., Sudha, H., Nirmala, P.V., and Gopalon, A. (2000). Peroxosalts initiated graft copolymerization of o-toluidine onto rayon fiber—a kinetic approach. *International Journal of Polymeric Materials* 48 (2): 199–223.

51 Anbarason, R., Jayasehara, J., Sudha, H., Nirmala, P.V., and Gopalon, A. (2000). Peroxydisulphate initiated graft copolymerization of o-toluidine onto synthetic fibers—a kinetic approach. *Macromolecular Chemistry and Physics* 201: 1869–1876.

52 Celik, M. and Sacak, M. (2000). The rate of grafting and some kinetic parameters of the graft copolymerization of methacrylic acid on poly(ethylene terephthalate) fibers with azobisisobutyronitrile. *Turkey Journal of Chemistry* 24 (3): 269–274.

53 Bhattacharyya, S.N. and Maldas, D. (1983). Graft copolymerization onto cellulosics. *Progress in Polymer Science* 10: 171–270.

54 Khalil, M.I., El Rafie, M.H., Bendak, A., and Hebeish, A. (1982). Graft polymerization of methylmethacrylate onto wool using dimethylaniline/Cu(II) system. *Cellulose Chemistry and Technology* 16 (5): 465–471.

55 Patil, D.R. and Fanta, G.F. (1993). Graft copolymerization of starch with methyl acrylate: an examination of reaction variables. *Journal of Applied Polymer Science* 47: 1765–1772.

56 Sood, D.S., Kishore, J., and Misra, B.N. (1985). Grafting onto wool. XXVII. Graft copolymerization of mixed vinyl monomers by use ofcericamonium nitrate as redox initiator. *Journal of Macromolecular Science Chemistry* 22: 263–268.

57 Kaji, K., Hatada, M., Yoshizawa, I., Kohara, C., and Komal, K. (1989). Preparation of hydrophilic polyethylene foam of open cell type by radiation grafting of acrylic acid. *Journal of Applied Polymer Science* 37: 2153–2164.

58 Nakamura, K., Hirose, S., Hatakeyama, T., and Hatakeyama, H. (1984). Structure and function of cellulose. *Sen'igakkaishi.* 40 (9): 327–331.

59 El Assy, N.B. (1991). Effect of mineral and organic acids on radiation grafting of styrene onto polyethylene. *Journal of Applied Polymer Science* 42: 885–889.

60 Hoffman, A.S. and Ratner, B.D. (1979). The radiation grafting of acrylamide to polymer substrate in the presence of cupric ion. I. A preliminary study. *Radiation Physics and Chemistry* 14: 831–840.

61 Garnett, J.L., Jankiewicz, S.V., and Sangster, D.F. (1990). Mechanistic aspects of the acid and salt effect in radiation grafting. *Radiation Physics and Chemistry* 36 (4): 571–579.

62 Zaharan, A.H. and Zhody, M.H. (1986). Effect of radiation chemical treatment on sisal fibers. I. Radiation-induced grafting of methyl acrylate. *Journal of Applied Polymer Science* 31: 1925–1934.

63 Misra, B.N., Chauhan, G.S., and Rawat, B.R. (1991). Grafting onto wool. XXVIII. Effects of acids on gamma radiation-induced graft copolymerization of methylmethacrylate onto the wool fiber. *Journal of Applied Polymer Science* 42: 3223–3227.

64 Misra, B.N., Mehta, I.K., Rathore, M.P.S., and Lakhanpal, S. (1993). Effect of L(2) threnine, 5-hydroxytryptophane, and 5-hydroxytryptamine on the ceric-ion-initiated grafting of methyl acrylate onto cellulose. *Journal of Applied Polymer Science* 49: 1979–1984.

65 Misra, B.N. and Chandel, P.S. (1977). Grafting onto wool. I. Ceric ioninitiated grafting of poly(methyl acrylate) onto wool. *Journal of Polymer Science Polymer Chemistry* 15: 1545–1554.

66 Garnett, J.L., Jankiewicz, S.V., Long, M.A., and Sangster, D.F. (1985). The role of inorganic salts in accelerating the radiation-induced grafting of styrene to cellulose and polyethylene. *Journal of Polymer Science Polymer Letters* 23: 563–566.

67 Ang, C.H., Garnett, J.L., Levol, R., and Long, M.A. (1983). Accelerated radiation-induced grafting of styrene to polyolefins in the presence of acid and polyfunctional monomers. *Journal of Polymer Science Part C: Polymer Letters* 21: 257–261.

68 Garnett, J.L., Jankiewicz, S.V., and Long, M.A. (1986). Inorganic salts as additives in accelerating the photographing of styrene to cellulose and polyethylene. *Journal of Polymer Science Part C: Polymer Letters* 24: 125–129.

69 Nho, Y.C., Garnett, J.L., and Dworjanyan, P.A. (1992). The role of cationic salts in enhancing the photosensitized grafting of styrene in methanol to polypropylene in the presence of acid additives. *Journal of Polymer Science Part A: Polymer Chemistry* 30: 1219–1221.

70 Kubota, H. and Hata, Y. (1991). Effect of hydroquinone on the location of methacrylic acid-grafted chains introduced into polyethylene film by photographing. *Journal of Applied Polymer Science* 42: 2029–2033.

71 Fernandez, H.J., Casino, I., and Guzman, G.M. (1991). Grafting of vinyl acetate-methyl acrylate mixture onto cellulose. Effects of inorganic salts. *Journal of Applied Polymer Science* 42: 767–778.

72 Lenka, S. (1982). Grafting vinyl monomers onto Nylon 6.0 XI. Graft copolymerisation of methylmethacrylate onto nylon 6 using peroxydisulfate as initiator. *Journal of Applied Polymer Science* 27 (6): 2295–2299.

73 Lenka, S., Nayak, P.L., and Mohanty, A. (1985). Graft copolymerisation onto natural rubber, rubber using a potassium bromate/thiourea redox system. *Angewandte Makromolekulare Chemie* 134: 1–9.

74 Dworjanyan, P.A. and Garnett, J.L. (1988). Synergistic effect of urea with polyfunctional acrylates for enhancing the photographing of styrene to polypropylene. *Journal of Polymer Science Polymer Letters* 26: 135–138.

75 Dworjanyan, P.A. and Garnett, J.L. (1989). The role of multifunctional acrylates in radiation grafting and curing reactions. *Radiation Physics and Chemistry* 33: 429–436.

76 Aug, C.H., Garnett, J.L., Long, J.L., and Levol, R. (1982). Polyfunctional monomers as additives for enhancing the radiation copolymerization of styrene with polyethylene polypropylene and PVC. *Journal of Applied Polymer Science* 27: 4893–4895.

77 Ang, C.H., Garnett, J.L., Levol, R., and Long, M.A. (1983). Novel additives for enhancing UV and radiation grafting of monomers to polymers and use of these copolymers as ion exchange resins. *ACS Symposium Series* 212: 209–223.

78 Dworjanyan, P.A., Garnett, J.L., Khan, M.A. et al. (1993). Novel additives for accelerating radiation grafting and curing reactions. *Radiation Physics and Chemistry* 42 (1–3): 31–40.

79 Chatterjee, S., Sarkar, S., and Bhattacharyya, S.N. (1993). Colloidal ferric oxide: a new photosensitizer for grafting acrylamide onto cellulose acetate films. *Polymers* 34: 1979–1980.

80 Misra, B.N., Dogra, R., and Mehta, I.K. (1980). Grafting onto cellulose. V. Effect of complexing agents on Fenton's Reagent (Fe^{2+}–H_2O_2) initiated grafting of poly (ethyl acrylate). *Journal of Polymer Science Polymer Chemistry* 18: 749–752.

81 Misra, B.N. and Chandel, P.S. (1980). Grafting onto wool. IV. Effect of amines upon ceric ion-initiated grafting of poly(methyl acrylate). *Journal of Polymer Science Polymer Chemistry* 18: 1171–1176.

82 Khetrapal, R.C., Kaul, J., Mehta, I.K., and Misra, B.N. (1984). Grafting onto soluble protein—III. Effect of amines in ceric ion initiated graft copolymerization of ethyl acrylate onto gelatin. *Indian Journal of Chemistry* 23: 983–987.

83 Misra, B.N. and Sood, D.S. (1981). Graft copolymerization of vinyl monomers onto wool by use of TBHP–FAS system as redox initiators. In: *Physicochemical Aspects of Polymer Surfaces* (ed. K.L. Mittal), 881–891. New York: Plenum Press.

84 Misra, B.N. and Mehta, I.K. (1980). Grafting onto wool. Graft copolymerization of methylmethacrylate (MMA) by use of Ce^{4+-} amine system as redox initiator. *Journal of Polymer Science Polymer Chemistry* 18: 1911–1918.

85 Samal, S., Sahu, G., Lenka, S., and Nayak, P.L. (1987). Photoinduced graft copolymerization. XI. Graft copolymerization of methylmethacrylatE onto silk using the isoquinoline-fsulfur dioxide charge transfer complex as the initiator. *Journal of Applied Polymer Science* 33 (5): 1853–1888.

86 Maldas, D. (1983). Radiation-induced grafting of styrene and acrylamide on cellulose acetate from a binary mixture. Thesis, University of Calcutta. 47.

87 Sacak, M. and Pulat, E. (1989). Benzoyl peroxide initiated graft copolymerization of poly(ethylene terephthalate) fibers with acrylamide. *Journal of Applied Polymer Science* 38: 539–546.

88 Schamberg, E. and Hoigne, J. (1970). Radical and radiation-induced grafting of some synthetic high polymers within the temperature range of their glass transition. *Journal of Polymer Science Part A: Polymer Chemistry* 8: 693–698.

Part 3

Sustainable Grafted Biopolymers as Corrosion Inhibitors

8

Corrosion Inhibitors

Introduction, Classification and Selection Criteria

Humira Assad[1], *Richika Ganjoo*[1], *Praveen Kumar Sharma*[1], *and Ashish Kumar*[1,2,*]

[1] *Department of Chemistry, Faculty of Technology and Science, Lovely Professional University, Phagwara, Punjab, India*
[2] *NCE, Bihar Engineering University, Department of Science and Technology, Government of Bihar, Bihar 803108, India*
* *Corresponding author*

Abbreviations

AFM	Atomic Force Microscopy
CIs	Corrosion Inhibitors
CMP	Chemical Mechanical Polishing
CR	Corrosion Rate
EIS	Electrochemical Impedance Spectroscopy
GCIs	Green Corrosion inhibitors
GPH	Glycerophosphate
GDP	Gross Domestic Product
ILs	Ionic Liquids
MS	Mild Steel
MBT	Mercapto benzothiazole
MP	Melting Point
OCIs	Organic Corrosion Inhibitors
VCIs	Volatile Corrosion Inhibitors
VPIs	Vapor Phase Inhibitors
XPS	X-ray Photoelectron Spectroscopy

Grafted Biopolymers as Corrosion Inhibitors: Safety, Sustainability, and Efficiency, First Edition.
Edited by Jeenat Aslam, Chandrabhan Verma, and Ruby Aslam.

8.1 Introduction

Corrosion is the deterioration of metals caused by interaction with corrosive aqueous environments (air, moisture, or soil) and the formation of noble compounds as a consequence of a direct chemical or electrochemical reaction [1, 2]. Corrosion has caused death and damage expenditures in almost every technical profession throughout the years [3]. As a result, it is seen as a potentially catastrophic environmental concern, necessitating ongoing study toward a permanent remedy [4]. Corrosion is an ecological issue that has cost-effective, environmental, and safety implications in a variety of engineering applications, including building construction, chemical, vehicle, mechatronics, metallurgical, and medical. There has been discussion of several types of material corrosion in various conditions [5]. As a result, in addition to the widely used defensive layers and coatings, cathodic/anodic defense, and corrosion inhibitors, there is a need to create new approaches and strategies for dealing with these harmful phenomena [6]. However, various studies in anticorrosion material applications in the above-listed engineering domains demonstrated that utilizing corrosion inhibitors is the most effective and straightforward method of avoiding detrimental metal and alloy deterioration in corrosive conditions [7]. When modest amounts of corrosion inhibitors are introduced to a corrosive medium, they produce a monomolecular layer adsorption surface that keeps the metal from interacting directly with the corrosive agents, minimizing or preventing corrosion [8]. The electrode process, the environment, and the manner of protection have all been used to classify them. Corrosion inhibitors bind to the metal/metal oxide surface (e.g., through physisorption, chemisorption, complexation, or precipitation) and prevent oxygen from reaching the cathode, hydrogen from leaving the cathode, or metal dissolving (anodic inhibitors) [8].

8.2 Chemistry and Adverse Impact of Corrosion

Metal gets corroded by two different types of mechanisms which are high-temperature and aqueous corrosion. In high-temperature corrosion the metals when exposed to hot gases for example in turbines and boilers cause corrosion. Further in the case of aqueous corrosion metals are often subjected to water, soil, chloride, or an electrolyte. The former is an electrochemical process in which there are two main reactions the anodic and the cathodic which work simultaneously and are complementary and in these reactions, the electrons contribute in a very important manner. Wet corrosion operates according to the principles of thermodynamic laws and the kinetics of electrochemistry. Since hot corrosion involves the development of protective layers which are characteristically oxides, the kinetics of the corrosion process are typically more intricate, relying on several variables, for example: adhesion, consistency, porosity, conduction type (ionic or electronic), and film conductivity.

The common reaction for describing the corrosion of a metal, M, is as follows [9]:

$$\mathbf{M + aggressive\ environment \rightarrow corrosion\ products} \tag{8.1}$$

When water with dissolved gases behaves as an electrolyte, corrosion occurs as a result of two electrochemical reactions known as anodic and cathodic reactions.

In the anodic process there takes place the oxidation of the metal and in the cathodic process, there happens the reduction of the metal. Both the anodic as well as reactions of the cathodic site are given below [9]:

The reaction that occurs at the anode in the case of Fe is [9]

$$\mathbf{2Fe \rightarrow 2Fe^{2+} + 4e^-} \tag{8.2}$$

Cathodic Half Reaction (reduction)

Cathodic reactions depend upon the pH of the solution.

- In acidic and deaerated solution

$$\mathbf{H^+ + 2e^- \rightarrow H_2} \tag{8.3}$$

Where electrons are made accessible and then these electrons are taken by the cathodic reaction for example electrons are taken by oxygen and the formation of hydroxide ion takes place.

$$\mathbf{O_2 + 2H_2O + 4e^- \rightarrow 4OH^-} \tag{8.4}$$

- In neutral and deaerated solution

$$\mathbf{2H_2O + 2e^- \rightarrow H_2 + 2OH^-} \tag{8.5}$$

Metal ion formed at the anode combines with hydroxyl ion and form the iron hydroxide, which oxidized further to hydrated ferric oxide.

$$\mathbf{2Fe^{2+} + 4OH^- \rightarrow 2Fe(OH)_2} \tag{8.6}$$

$$\mathbf{Fe(OH)_2 + \frac{1}{2}O_2 + (X-2)H_2O \rightarrow Fe_2O_3\,.\,XH_2O} \tag{8.7}$$

Hydrated ferric oxide (rust) ⥥

A corrosion process demonstrates an electron flow from the anodic part of the metal to the cathodic part where the electrons are accepted. This electron flow takes place the other way around the current. The current is generated in the corrosion cell which is a metal surface and so the metal starts deteriorating.

Industrial shutdowns, product contamination or loss, a squandering of valuable resources, decreased effectiveness, high-cost upkeep, and overpriced overdesign are all consequences of corrosion [10]. The US economy suffers roughly $300 billion per year due to metal corrosion. Corrosion has a far greater impact on the harmless, dependable, and effective functioning of machines and buildings than a mere loss of metal. Even if the quantity of metal damaged is little, catastrophes of several machines and the need for costly alternates may fall. These costs may be mitigated if better corrosion-resistant materials were utilized and the best corrosion-related scientific procedures were implemented [11].

8.3 Corrosion Inhibitors

Chemical inhibitors are used in a variety of ways to reduce the pace at which corrosion occurs. In the oil extraction and processing sectors, inhibitors have long been the first line of defense against corrosion. The issue of corrosion inhibitors has sparked a slew of scientific research. However, the majority of what we know comes from both laboratory and field trials and errors. There are relatively few rules, equations, or ideas that can be used to guide inhibitor development or usage [12]. An inhibitor, by definition, is a chemical compound or mixture of chemicals that, when given in very small amounts to a corrosive environment efficiently inhibits or decreases corrosion without causing substantial reactivity with the environment's components. Corrosion inhibitor concentrations may range from one to fifteen thousand parts per million (0.0001 to 1.5 wt. %) [13]. Inhibitors are essential in closed environmental systems with excellent circulation because they guarantee appropriate and controlled inhibitor doses. Recirculation cooling water systems, oil production, oil purifying, as well as acid pickling of steel components, are all examples of when such circumstances might be reached. Antifreeze for radiators in automobiles is the most well-known use for inhibitors. Inorganic or organic substances may be used as inhibitors, and they are often dissolved in water. Inhibitors often act by adsorbing on the metallic surface and producing a coating to protect the surface. Inhibitors are often dispersed or spread from a solution. Some of them are included in the composition of a protective coating. Inhibitors reduce corrosion by enhancing anodic or cathodic polarization behavior (Tafel slopes), decreasing ion migration or dissemination onto the metallic surface, and raising the metal surface's electrical impedance [14]. The inhibitor molecule's inhibitory effectiveness (E) is determined by the system's characteristics (temperature, pH, time, and composition of the metal, and so on) as well as the inhibitor molecule's structure [15].

8.4 Corrosion Inhibitor Classification

Corrosion inhibitors can be man-made or natural substances. Based on the electrode process, type of corrosive media, and protection mechanism, corrosion can be broadly classified into numerous types as shown in Figure 8.1.

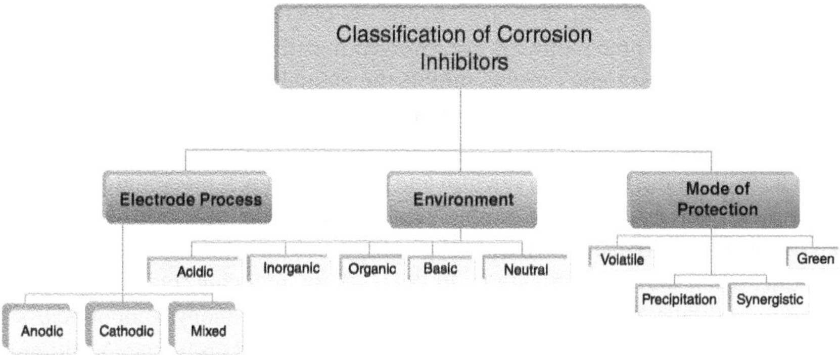

Figure 8.1 Classification of corrosion inhibitors.

8.4.1 Electrode Process-Based

8.4.1.1 Anodic Inhibitors or Passivating Inhibitors

Passivating inhibitors are anodic inhibitors that promote anodic polarization and thereby change the corrosion potential to the cathodic direction. Phosphates, chromates, tungstates, and other ions of transition elements with elevated O_2 concentration act as anodic inhibitors and suppress anodic corrosion by creating a sparingly soluble substance with a freshly generated metal ionic moiety. They adsorb on the metallic outer layer, providing a defensive coating or obstacle that slows down corrosion. Anodic inhibitors form a reedy defensive coating along the anodic terminal, raising their potential and so slowing the corrosion process. While this sort of control is harmed, it may be risky since serious local assault can happen if some parts are left undefended due to inhibitor reduction [10]. Anodic inhibitors include orthophosphates, silicates, and other inorganic inhibitors. Even though anodic inhibitors are commonly utilized, some of them have unfavorable properties. Anodic inhibitors are categorized as harmful as they stimulate corrosion, such as pitting, when employed in extremely low quantities [16]. Passivating inhibitors are divided into two categories oxidizing anions (viz. nitrites, chromates, and nitrates) and non-oxidizing anions (like tung states, phosphates, and molybdates) that passivate metal in the nonexistence and presence of O_2 respectively [16]. When concentrations fall below the required levels, passivation inhibitors may essentially promote pitting and hasten corrosion. As a result, it's critical to keep an eye on the inhibitor concentration.

8.4.1.2 Cathodic Inhibitors

Cathodic inhibitors either lessen the cathodic process or precipitate on cathodic sites selectively; raising surface impedance and limiting the migration of active moieties to these sites. Cathodic inhibitors may act as cathodic precipitates, cathodic poisons, or oxygen scavengers, among other things. Cathodic inhibitors change the corrosion potential from cathodic to anodic. The cations move to the cathode terminals, where they are chemically/electrochemically precipitated, obstructing them. Three pathways are involved in cathodic inhibitors' inhibitory activity [12]. The cathodic poisons inhibit the cathodic reduction procedure by inhibiting the hydrogen recombination as a means of protective emancipation, while the metal's susceptibility to hydrogen-induced cracking is increased. The cathodic precipitates including calcium and magnesium will precipitate as oxides to produce a defensive coating on the metallic surface that functions as a shield and the oxygen scavengers such as As^{3+} and Sb^{3+} on iron solubility in acids create a product when they counter with O_2 in the medium, reducing corrosion [16].

Certain cathodic inhibitors, for example, arsenic and antimony compounds, act by increasing the difficulty of hydrogen recombination and discharge. Additional cathodic inhibitors, for instance, calcium, zinc, or magnesium, may precipitate as oxides on the metallic outer layer, creating a defensive layer. By inhibiting cathodic depolarization induced by oxygen, O_2 scavengers may aid to avoid decomposition. Sodium sulfite (Na_2SO_3) is most likely the most frequent oxygen scavenger used at room temperature. Cathodic inhibitors decrease corrosion by decreasing the electrochemical corrosion cell's reduction reaction rate. This is accomplished by precipitation obstructing the cathodic locations. When cathodic inhibitors delay the cathodic reaction, they are efficient. The elements Bi, As, and Sb are known as cathodic poisons because they slow down the hydrogen reduction process and so slow down overall corrosion. The rate of corrosion will be greatly slowed if oxygen is removed from the corrosive environment [15]. This can be accomplished by vacuum deaeration, boiling of the solution to reduce the amount of dissolved oxygen, and using scavengers of oxygen, like Na_2SO_3 and N_2H_4, to react with oxygen in the mixture and eliminate it [16].

8.4.1.3 Mixed Inhibitors

Mixed inhibitors are inhibitors that hinder both the anodic and cathodic activities participating in corrosion [17]. They are generally film-forming chemicals that form precipitates on the exterior, obstructing both anodic and cathodic spots directly. Anodic inhibitors are often harmful inhibitors, predominantly when their concentrations are too low. On the other hand, cathodic inhibitors are typically considered harmless. Mixed inhibitors are less hazardous than pure anodic inhibitors, and they may not enhance corrosion severity in certain instances. The silicates and phosphates are the most prevalent inhibitors in this group. These inhibitors have the added benefit of controlling both cathodic and anodic corrosion procedures [16].

8.4.2 Based on Environment

8.4.2.1 Acidic Environment Inhibitors

Acid corrosion inhibitors, notably those for Fe and steel, are ubiquitously included in a broad array of industries (like in the petroleum sector). Organic and inorganic acidic corrosion inhibitors are both possible. Because passivating inorganic inhibitors can be detrimental in extreme pH (acidic) and end up causing severe geographic strikes once the unresponsive overlay is completely obliterated, organic inhibitors are being used primarily as acidic inhibitors [18].

8.4.2.2 Inorganic Corrosion Inhibitors

In an acidic environment, substances like arsenic trioxide (As_2O_3) and antimony trioxide (Sb_2O_3) have been testified to act as an inhibitor. The safeguarding in this particular instance is responsible for the lessening of electropositive ions and accumulation on the metallic substrate, as well as the decline of the overvoltage of the primary cathodic depolarization response [19]. It has recently emerged that the addition of heavy metal ions including lead (II), titanium (I), manganese (II), and cadmium (II) inhibits iron deterioration in acids.

8.4.2.3 Organic Corrosion Inhibitors

Organic corrosion inhibitors (OCIs) are extensively employed due to their high performance over a variety of temperatures, interoperability with shielded substances, dissolution rate, and low toxicity [20]. This is a critical concern for researchers, as evidenced by a growing amount of literature. These molecules function as anodic and cathodic inhibitors. OCIs work by adsorbing on the substrate to ensure a defensive barrier that displaces H_2O from the metallic substrate and protects it from deterioration. This is not a mere physical or solely chemical adsorption process. The chemical nature of OCIs, the existence and charge on the outer layer of the compound, the dispersion of charge in the chemical compound, and the category of the belligerent environment (pH and/or electrode potential) all affect adsorption. The electrostatic attraction between the charged metallic surface and the charged inhibitor molecule underpins physisorption. Chemical adsorption is linked to relationships involved in donating and accepting free electron pairs of inhibitor molecules to metal d-orbitals with unoccupied, reduced energy d-orbitals (Figure 8.2).

Effective OCIs should comprise heteroatoms (nitrogen, oxygen, sulfer, and phosphorus) with lone electron pairs and a substituent with π-ēs (aromatic rings and multiple bonds) that can communicate with available orbital d metal and favors adsorption [22]. The basic adsorption-free energy (G°ads) represents the sort of adsorption. The electrostatic interaction is associated with values of up to 20 kJ/mol (physical adsorption). More negative values, less than 40 kJ/mol, directly correlate to chemical adsorption. A negative value represents that both procedures are haphazard. The basic enthalpy of adsorption also provides useful data about the corrosion inhibition framework. Endothermic adsorption (H °ads>0) is associated with chemisorption, whereas exothermic adsorption (H °ads<0) is associated with physical/chemical adsorption [23]. The appearance of heteroatoms with a single pair of electrons, such as N, O, S, or P, as well as π-electrons, improves adsorption [24]. Chemisorption is characterized by the transmission or ability to

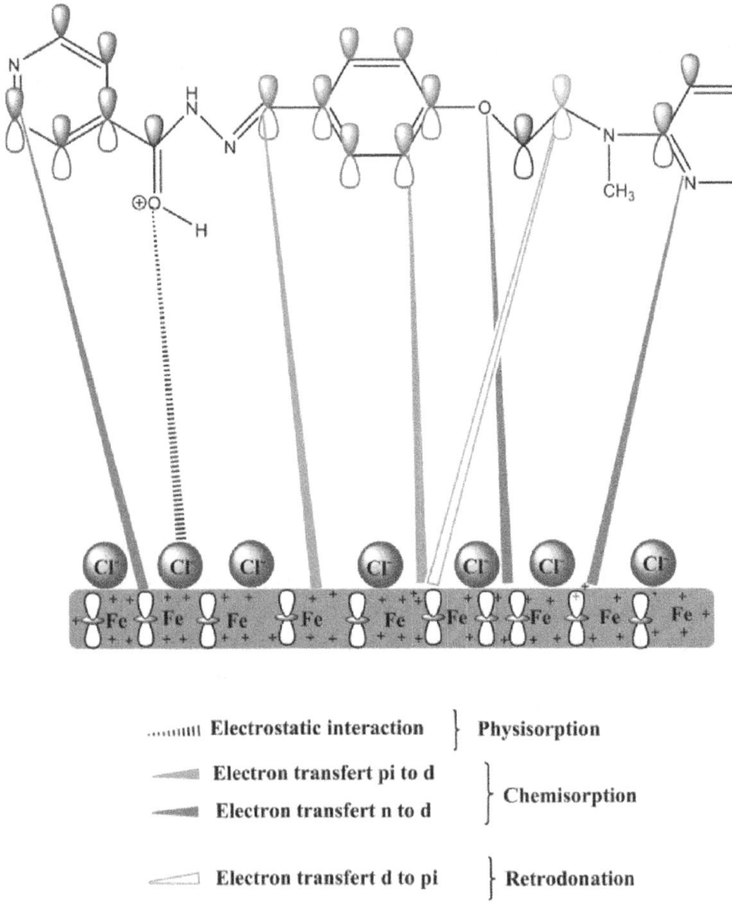

Figure 8.2 Adsorption mechanism of inhibitor molecule on the surface of mild steel in acidic solution. [21] / Springer Nature.

share unrestrained charged particles between the adsorbate molecules and the metallic surface [25]. The sequence of CI is the inverse sequence of these atoms' electronegativity.

$$P > S > N > O$$

In an acidic medium, heteroatoms are protonated, which enhances the interconnections between the inhibitor and the outer layer. The binding of OCIs onto the exterior of a decaying material can be viewed as a substitution procedure between the organic molecule, particularly the aliphatic network in an aqueous environment, and H_2O particles chemisorbed on the outer layer of the metal:

$$\textbf{Organic(sol)} + \textbf{xH}_2\textbf{O(ads)} \leftrightarrow \textbf{Organic(ads)} + \textbf{xH}_2\textbf{O(sol)} \tag{8.8}$$

Where,

Organic(sol) represents the organic moieties dissolved in the aqueous environment,

Organic(ads) represents the organic moieties adsorbed onto the surface of metal substrates,

H_2O(ads) are water molecules that have been adsorbed onto the material's outer layer,

H_2O(sol) are water molecules that are existing in the bulk solution, and x is the size ratio that represents the number of H_2O molecules that one organic inhibitor molecule substitutes.

Because of the repugnance of the non-polar hydrophobic segment of the inhibitor and the polar solution, the aliphatic system affects corrosion prevention. At the metal/water link, the hydrophobic channels provide a barrier-thin coating. The diameter and molecular weight of organic inhibitors also affect inhibition efficiency [26]. The relatively large the molecule, the more effective the inhibition:

$$\mathbf{R_3N > R_2NH > RNH_2}$$

Where, R is a chain of hydrocarbons (say methyl, ethyl, butyl, etc).

The content of corrosion inhibitors has a significant effect on inhibitor effectiveness. Since the adsorption of the inhibitor increases over time adsorbent dose, the corrosion rate decreases.

8.4.2.4 Alkaline Inhibitors

In basic conditions, materials that produce amphoteric oxides are prone to corrosion. In alkaline conditions, numerous organic compounds are commonly used as metal inhibitors. In an alkaline solution, Punitha, R. et al., evaluated the ACTP polymer as a corrosion inhibitor with aluminum metal. The findings demonstrate that the degradation efficiency determined by using the weight loss technique increased with increased corrosion inhibitor concentrations. Polarization studies revealed that ACTP (novel modified polyacrylic acid) and PAA (polyacrylic acid) inhibitors are primarily anodic in existence. Furthermore, an impedance study demonstrated that in the presence of inhibitors, charge transfer resistance (Rct) tends to increase and double-layer capacitance (Cdl) reduces, implying adsorption of the inhibitor onto the surface of aluminum [27]. Moreover, Altaf et al. investigated the anticorrosion effects of various metallic ions (aluminium, lead, manganese, etc.) as dopants and five azoles (benzotriazole, mercaptobenzothiazole (MBT), benzimidazole, mercaptobenzimidazole, and thiadiazole) as an outermost surface on pure copper and 4 brass alloys in borate buffer at pH 10.4. Due to the obvious accessibility of dopant ions, the rate of corrosion (C_R) of all brasses was ascertained to be more gradual than that of pristine Cu (0.132 mm year^{-1}), and MBT demonstrated the maximum corrosion protection efficacy ranging from 91 to 96% for different brasses [28].

8.4.2.5 Neutral Inhibitors

Since the processes in the two main components are distinguishable, inhibitors that operate well enough in acidic solutions don't work well enough in neutral solutions [29–31]. In a pH-neutral NaCl solution, Yang, J. et al., evaluated the corrosion protection of pure Mg encompassing an extreme amount of iron pollutant. The findings demonstrate that adding sodium 2,5 pyridine dicarboxylate, 3-methylsalicylate, and fumarate to a 0.5 wt. percent neutral NaCl solution contributes to high anticorrosion effectiveness of commercial purity Mg. The corrosion performance improves as the intensity of inhibitors rises. The inhibition efficiency of the given inhibitors is greater than 93% after a 2-hour initial stabilization time frame [32]. Moreover, W. Zhuang et al. synthesized Gemini imidazoline surfactants from a diverse array of saturated fats and explored their anticorrosion behavior in NaCl solution for X70 carbon steel utilizing electrochemical impedance spectroscopy (EIS), polarization curves, and molecular dynamics. According to the findings, such surfactants have an outstanding inhibitory impact on carbon steel X70 in NaCl solution, and Gemini imidazoline has a relatively high limited effectiveness in an alkaline medium than in a neutral solution [33]. In neutral solutions, inhibitors engage with the metal's oxide-covered exterior, attempting to prevent oxygen reduction interactions at the cathode. These inhibitors protect the surface of the substrate from abrasive wear. Specific surface-active chelating inhibitors are capable of inhibiting near-neutral remedies [34].

8.4.3 Based on Mode of Protection

8.4.3.1 Precipitation Inhibitors

In essence, these inhibitors are frequently film-forming, such as silicates and phosphates. They are impactful at both anodic and cathodic locations obstructing. They offer some protection on the metallic surface as they hasten. Magnesium and calcium are abundant in hard water. When these salts precipitate on the outer layer of the metal, they form a shielding thin coating. The effectiveness of film-forming inhibitors is affected by many parameters like pH, the composition of water, and temperature.

8.4.3.2 Synergistic Inhibitors

This is a solitary inhibitor that is used in water-cooling applications. The combination of inhibitors (anodic and cathodic) is quite often used to reach greater corrosion-resistant characteristics. Some examples include chromate-phosphates, polyphosphate silicates, zinc-tannins, and zinc-phosphates. Ionic liquids (ILs) also have the important property of continuing to act as a synergistic inhibitor. Even though ILs and organic salts possess this property, they could be utilized as innovative organic inhibitors, preferably with synergistic interactions. Numerous research has been carried out to investigate the use of biologically appropriate anions and cations to generate that could move ahead toward the effectiveness of chromates. Chong, A. L. et al. [35] characterized a category of ILs and organic salts with double features and functions by comprising both anions and cations and expressed effectual inhibition. To create these salts, the imidazolinium cation was coupled with carboxylate anions. Regardless of the type of anion in the salts, these substances were unearthed to have captivating physical properties including simple ionic conduction and synergistic corrosion inhibition on steel specimens. In this research, the influence of pH on the corrosion hindering the effectiveness of the organic salt for mild steel in chloride configurations was investigated. Moreover, Li Feng et al. used electrochemical techniques, weight loss examinations, and SEM to investigate the corrosion-inhibitory activity of thiazolyl-based ionic liquids between anions and cations for copper in hydrochloric acid solution. The findings demonstrate that the inhibitor compounds communicated with copper to form a barrier film that inhibited corrosion through physical and chemical adsorption, and those organic cations and halogen ions inhibited corrosion synergistically. Theoretical calculations and molecular dynamics simulations were used to refine the results [36].

8.4.3.3 Volatile Corrosion Inhibitors

This class of inhibitors is also referred to as vapor phase inhibitors. When the inhibitor compounds in the vapor meet the material surface, the inhibitor adsorbs as shown in Figure 8.3. Moisture then hydrolyzes it, allowing defensive ions to be distributed. Ferrous metal corrosion is inhibited using amines and nitrites.

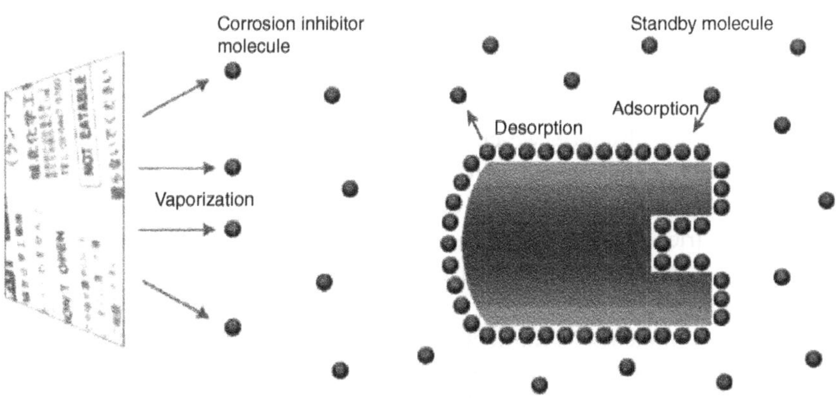

Figure 8.3 Schematic representation of volatile inhibitors. [38] / IntechOpen / Licensed under CC BY 3.0.

Amines can be used to protect abrasive wear in volatile environments. Amines are cationic inhibitors. They are frequently added to HCl-rich settings in the version of a polyhydric alcohol combination at a concentration of 0.2% (glycol). Long-chain aliphatic diamines are used in the production of wells. When a ferrous metal system is involved, neutralizing amines task by changing the ph of the electrolyte, prevents rust establishment [37]. Moreover, Nitrites also have a volatile pattern that could be used. Sodium nitrite is vulnerable to chloride and sulfate-induced localized corrosion in an open recirculating framework. When the pH falls below 5.5–6.0, nitrite loses its inhibitory effect. Nitrite can also be used to impede engine coolant in road vehicles and marine diesel engines. It has been discovered that nitrite can lessen deterioration in diluted seawater. Nitrites in oil-soluble form have been used in the oil and gas sector to impede corrosion caused by water resolved at the tank's underside.

8.4.3.4 Green Corrosion Inhibitors

The present practice in inhibitor use is toward greener, more ecologically responsible chemical compounds. For environmental protection, scientists have enhanced their efforts to investigate the inhibitory potency of natural goods such as peels, seeds, fruit shells, and leaves, which contain various organic substances (e.g. amino acids, alkaloids, flavonoids, pigments, tannins, etc.) that repress metal dissolution reactions and preclude damage to the environment. However, there are no universally accepted definitions of green or "environmentally friendly" corrosion inhibitors. In practice, corrosion inhibition studies have transitioned to consider the personal environment and public health. Researchers have concentrated on using ecologically sustainable chemical compounds to accomplish this goal due to their cost-effectiveness, renewability, and good biocompatibility. such as plant extracts, outdated benign pharmaceutical drugs, and so on, which encompass a high concentration of organic elements. Greener substitutes can be created by reorganizing existing products or investigating novel intermediates for producing environmentally sustainable goods, as well as minimizing effluents in terms of harmful effects and biodegradability. The expenses of corrosion and biomagnification are subdivided. Toxicity can occur even during the fabrication or execution of the substance, which can only be taken into consideration when designing GCIs. As mentioned above amino acids, alkaloids, pigments, and tannins are examples of green alternatives for poisonous materials. Due to their recyclability, eco-friendliness, better rates, and obtainability, extracts of some species of plants and medicinal herbs, and also their derivative materials, have been probed as corrosion inhibitors for many metals and alloys just below a wide variety of environments. N. Karki et al., (2021) investigated the corrosion inhibition of Equisetum hyemale stem extract in 1.0 M H_2SO_4 medium against MS. The studies showed that it is a potent inhibitor and that the IE of EHE is greater than 85% at 1000 ppm, which enhances with the recommended dose and decreases with temperature increase [39]. Furthermore, excerpts can be obtained from any part of a plant. For instance, Al Hasan and co-workers investigated the influence of stem aqueous extracts of bacopa monnieri and lawsoniainermis (henna) on low-carbon steel in a 0.5 M sodium hydroxide solution [40]. The isolate was made from 10 g of bacopa monnieri powder and 5 g of henna and evaluated at concentration levels of 0.5, 1, and 2%. The combined efforts of this stem excerpt resulted in a mixed-type inhibitor with high corrosion inhibition effectiveness. The PP technique generated the highest inhibition performance of about 80%, as per the authors. Moreover, weight loss observations indicate a 65% inhibitory activity. Furthermore, fruit excerpts have also been recommended as corrosion inhibitors. Additionally, Sanaei and co-workers reported the corrosion inhibition effect of the rosa canina fruit aqueous extract on mild steel in a 1 M HCl electrolyte in 2019. EIS, UV–VIS spectroscopy, and SEM imaging were used to assess the extract's characteristics. With an 800 ppm concentration, the highest corrosion inhibition effect of about 86% was acquired. Furthermore, the extract acted as a mixed-type inhibitor, inhibiting both cathodic and anodic reactions. It has been proposed that physically adsorbed molecules (extract) bind to metal in local cathodes, delaying metal dissolution when the cathodic reaction is precluded, whereas chemically adsorbed particles safeguard the anodic regions [41].

8.5 Selection Criteria for Inhibitors

Proper inhibitor selection should be accomplished by correlating the adequate inhibitor chemical composition with the corrosion circumstances and identifying relevant physical properties for the application conditions. When choosing the material characteristics of an inhibitor, the technique of application and system attributes must be taken into account. According to Faltermeier (1999) [42], there are several good suggestions to consider when selecting corrosion inhibitors: Some of the major recommendations are enlisted below:

- The preference for inhibitors begins with the identification of physical parameters. However, the inhibitor's solubility in H_2O or ROH is a major determinant.
- The solvent used to apply the inhibitor ought to be cost-effective, quasi, and non-hazardous.
- The corrosion inhibitor should specifically bind with the metal, forming a polymeric thin coating as a consequence. The polymer sheet serves as a deterrent between the metal and its surroundings, and to prevent additional metal oxidation, the composite material should be thick, tightly packed, and free of void spaces.
- It's also important to make sure the inhibitor gets to all of the metal substrates. It is critical to ensure that all dead ends, coffers, and crevice zones are approached

by the inhibited liquid, especially when first filling a framework. This will be incentivized in many structures by fluid movement in service, but in putatively static systems, it will be attractive to demonstrate a flow pattern for the interim purpose of providing revived delivery of inhibitor [43].
- The constituted inhibitor-metal polymer film should be insoluble in water or organic solvents.
- The corrosion inhibitors should be impactful at specific pH particularly ranging from pH 2 to pH 8.
- The most effective inhibitors would respond in both anodic and cathodic regions.
- The implementation of the inhibitor should not modify the external identity of the metal substrate.
- The inhibitor should not pose a risk to the user.
- The inhibitor should be easily accessible in the basic state from chemical distributors.
- The inhibitor's valuation is a useful metric, but it should still be affordable for a wide range of applications.
- Moreover, when deciding between potential inhibitors, probably the easiest corrosion testing is recommended first to eliminate inappropriate candidates. The ideology of preliminary inspection assessments should be that underperformers are not promoted. An inhibitor that performs badly in initial screening evaluations may perform well in the authentic system; however, the user rarely has the reserves to assess all plausible inhibitors. The inhibitor consumer must use assessment methods that robustly preclude inadequate inhibitors while also excluding a few good inhibitors.

8.6 Mechanism of Corrosion Inhibition

Metal structural damage is inhibited by organic materials through the adsorption of solutes at the metallic surface. As a result, understanding the process of corrosion inhibition requires an understanding of the adsorption concept [44]. In a set of experiments, the theoretical recognition of the connection between the chemical structure of inhibitor materials and their capacity to absorb on the metal substrate (shown in Figure 8.4) and thus inhibit the metal dissolution is geared.

It was discovered that the existence of a heteroatom with unsaturated bonds in a chemical molecule has an inhibitory role and lessens the corrosion process. The distance of the alkyl chain was also developed to increase safeguarding effectiveness. The deposition of ampholytic substances was discovered to be dependent on the characteristics of both hydrophilic and hydrophobic components, as well as their ability to establish clusters in solution. When the lipophilic portion was encircled by 10–12 C-atoms, the corrosion protection of ampholytic substances

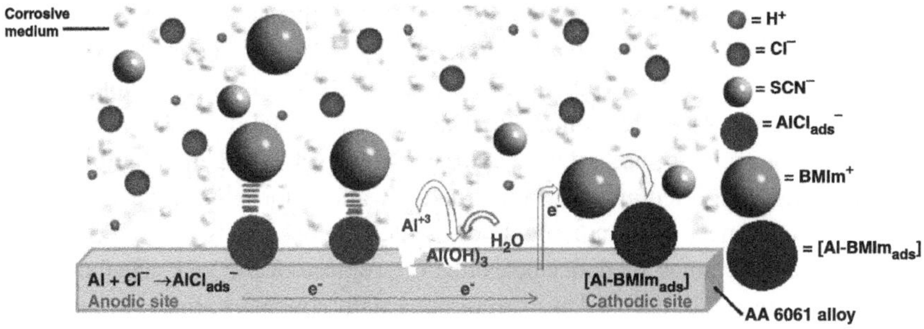

Figure 8.4 Mechanism of corrosion inhibition. [45] / MDPI / Licensed under CC BY 4.0.

was greatest. The alignment of adsorbed organic compounds at the metallic surface is of specific importance, and it is affected by pH and/or electrode potentials. Numerous studies on the CI of Fe have found that molecular orbit communication is related to the inhibition effects of organic molecules. When contrasting the CI of Fe by thiophenol, phenol, and aniline, it was discovered that thiophenol has a good performance resulting from the interaction between the S lone pairs and empty metallic orbits. Organic materials containing one or more polar functions (with N,O, and S atoms) were revealed to be indeed very effectual corrosion inhibitors, as have heterocyclic substances encompassing polar moieties and π-electrons [5, 46]. The polar function is commonly deemed as the active site for the emergence of the adsorption mechanism.

Numerous methods have been used to discover a relationship between inhibitor characteristics and electronic density, electron mobility, electron donor features of various functional moieties, substituent position, and steric factors. Sulfur-comprising materials are usually more efficient CIs for steel in acidic media than nitrogen-containing compounds because Sulfur is an improved electron giving than N and thus adsorbs well on the surface of the metal. When particularly in comparison to other heteroatoms, phosphorus as a heteroatom in a series of samples of organic compounds is conveniently polar and has lesser electronegativity. As a result, it is anticipated that phosphorous substances would be the most useful in decreasing steel dissolution in acidic media. For short-term implementations, scientists have created momentary corrosion protection designed to prevent corrosion of equipment's metal surfaces during transport and storage. There are numerous types of temporary protection measures. Among these, the use of volatile corrosion inhibitors (VCIs) is an efficient and practical method [47]. VCIs are a class of anticorrosive chemicals that extend their corrosion-inhibiting properties to a metal surface through volatilization within an enclosed space. Despite the reality that VCIs are being used for a protracted time to prevent corrosive environments, the exact mechanism of these substances is not perfectly known. It is reported that the main variables held to account for their performance are vapor pressure and interrelations with the metal substrate. The VCIs form an adsorbed monolayer that precludes corrosive entities like moisture, deleterious ions (Cl^- or SO_4^{2-}), and others from penetrating. The adsorbed monolayer may affect the speed of electrochemical processes such as metal dissolution or oxygen reduction. The metal surface is almost entirely made up of anomalous physical grains with varying sizes, orientations, grain structures, and deformities. As a result, VCI adsorption on a metal substrate might not even be homogeneous, and the corrosion rate may be ascertained by deficiencies like the rupture of the VCI safeguard film. However, many inhibitors used in aqueous or partially aqueous settings are described by four broad environmental subgroups:

- Aqueous acid compositions have been used in metal-cleaning operational processes such as pickling to eliminate rust or mill scaling all through metal smelting and production, as well as for post-service intervention of the metal substrate.
- Natural waters, supplying waters, and refrigeration systems water with near-neutral pH (5 to 9).
- Oil production, both overt and covert, and also handling and transportation practices.
- Atmospheric or gaseous corrosion in confined areas, such as mass transit, storage of goods, or any other constricted exercise.

Furthermore, technologically advanced surface methodological approaches, like X-ray photoelectron spectroscopy (XPS), atomic force microscopy (AFM), and so on, help in providing critical insight into the interaction of corrosion inhibitor molecules with that of the metallic surface.

8.7 Industrial Application of Corrosion Inhibition

8.7.1 In Concrete

Corrosion of reinforcing steel in concrete, also known as rebar corrosion, is a difficult and critical concern from the standpoints of mechanical stability and economics [48]. It is guesstimated that countering and resolving corrosion damage in bridges could save $450–550 million annually. The expense can be reduced dramatically by addressing corrosion issues in bridge structures. Corrosion inhibitors are presumably more appealing in terms of price and easy operation. A new bridge is projected to total \$730/m^3. When corrosion inhibitors are being used in combination with relatively impermeable concrete, this cost rises by US\$52/m^3. This price hike is negligible when compared to approximated restoration expenses of more than US$2600/$m^3$ [48]. The elements that influence protection against corrosion of structural bars by concrete are as follows:

- the arrangement of a big fence to the entrance of corrosive species including chloride,
- the establishment of passive oxide owing to the increased pH of the pore solution,
- the limitation of galvanic corrosion by the excellent electrical resistance of the material, and
- the establishment of preventative mineral levels on the reinforcement bars, which precludes the steel from reacting with the surroundings.

Furthermore, as with any framework, the corrosion inhibitor should reduce the corrosion process of steel rebar over time. Other prerequisites include:

- the inhibitor being dissolvable in mixing water and not easily leachable from rebar,
- the inhibitor being congruent with the aqueous concrete transition stage, and the inhibitor being consistent with the aqueous cement stage,
- the inhibitor should not influence characteristics of concrete-like curing time, resilience, or longevity.

Moreover, corrosion attacks can be prevented by interacting with the metal substrate and forming a protective barrier. Calcium nitrite is by far the most frequently used inhibitor in concrete, though organic inhibitors have gained increasing attention in recent years. On a massive scale, calcium nitrite is being used as an inhibitor in concrete. Calcium nitrite protects the steel in concrete from chlorides, has no influence on the properties of concrete, and is widely suitable for commercial use in concrete.

8.7.2 In Cooling Water Systems

Chilling waters are used in a variety of different industry sectors, which include petroleum refining, steel plants, petroleum industry, hydroelectric dams, food central processing units, and chemical processing plants. Carbon steel is usually used within heating systems and recirculation pump chiller dispersion piping due to its low cost. Other components used in heating systems include cast iron copper pipes, naval brasses, copper-nickel alloys, austenitic stainless steels, Inconel 625, Hastelloy C, and titanium. However, deterioration in open cooling water hardware is more prominent and is inextricably linked to the formation of mineral scales, solid acquisition, and microbial ensnarling. As a consequence, comprehensive cooling water augment formulation includes anti-scaling agents, polymeric diluents, and biocides, as well as CIs to protect iron (Fe), copper (Cu), aluminum (Al), and their alloys. To be impactful in cooling water, an inhibitor must satisfy the following requirements:

- the complete material must be protected from damage,
- low concentrations of the inhibitor must be efficacious,
- the procedure must be efficacious under a range of settings like pH, temperature, heat flux, and water quality,
- it need not generate deposits that hamper thermal performance,
- it must have satisfactory toxic effects for release,
- the procedure must reduce the growth of carbonate and sulfate scales, and
- it must counteract microorganism-induced biological processes.

Chromates, nitrates, phosphates, silicates, and other corrosion inhibitors are used to prevent corrosion in cooling water systems. However, each inhibitor molecule operates under specific conditions such as temperature, pH, and so on. Because solitary inhibitor molecules have less efficiency and certain limitations, mixtures of inhibitors usually provide improved inhibition safeguards than comparable levels of individual inhibitors. A multi-component inhibitor system's underlying mechanism is synergistic. For instance, phosphates, as opposed to polyphosphates, are more resistant to degradation over a wider pH and temperature range. Besides that, when merged with Zn or Fe ions, they protect metal surfaces more effectively via chemical adsorption, inevitably forming firm metal materials and adsorbing on particulate matters, resulting in improved anti-scaling and dispersion functionality. Low-phosphorus all-organic synergistic substances, like phosphono-tri carboxylates and fatty amines, as well as phosphono-carboxylates and polyvinyl pyrrolidone, have recently shown remarkable characteristics while possessing a lower impact on the environment. Their potency stems mainly from the formation of a combative layer composed of adsorbed organic chemical substituents coupled with iron oxides or hydroxides [49, 50].

8.7.3 Acid Pickling

Acid treatments are widely used in industrial applications for a variety of purposes including acid pickling, industrial acid hoovering, acid descaling, and petroleum well acidizing. Inhibitors are required to reduce corrosion. The form of acid, intensity of acid, temperature, flow velocity, the existence of solubilized inorganic or organic materials, and category of metals revealed to the acidic medium all influence the choice of an inhibitor. Hydrochloric, sulfuric, nitric, hydrofluoric, citric, formic, and acetic acids are by far the most frequently utilized acids. The inhibitor used is based on the effectiveness of the acid pickling. If the pickling is to remove the mill scale from hot-rolled steel, the inhibitor will be chosen based on the acid concentration, temperature, pickling time, and steel type. In general, pickling with HCl requires 200 g/L of HCl at 60 °C for about 30 minutes, depending on the nature of the steel and the processing parameters of the steel mill. At temperatures up to 90 °C, sulfuric acid at a concentration of 200–300 g/L may be used [48]. These are drastic circumstances that necessitate the use of potent inhibitors. Moreover, pickling inhibitors should comply with the following requirements:

- efficacious inhibition of metal dissolution,
- the dearth of over-pickling procedures,
- impactful pickling at a limited adsorbent dose,
- efficient at extreme temperatures,
- thermal and chemical stability,
- effective inhibition of hydrogen admittance into the material,
- excellent surfactant characteristics, and
- decent frothing characteristics.

However, the basic feature is that the inhibitor minimizes metal dissolution even in the presence of dissolved salts that encourage metal dissolution, including ferrous sulfate.

8.8 Summary

Corrosion is among the most frequent and severe problems encountered in industrial sectors. The problem is unavoidable, and it will have a massive effect as catastrophic events. While attempting to resolve, this problem is inconceivable, taking preventative methods to ensure the surface of the metal from rust and corrosion is much more expensive. And for its low cost and ease of using it, the utilization of corrosion inhibitors is among the most successful tactics for shielding metal surfaces from decomposition. In this regard, this chapter presents an overview of corrosion inhibitors, their classification, and inhibition mechanism as well as literature in which research teams used multiple kinds and procedures of corrosion inhibitors to lessen corrosion in various alloys or metals-based devices. To decrease the occurrence of corrosion in various alloys, various chemical inhibitors were used from time to time. However, emphasis must be laid down on the essence and mixtures of metals displayed, the existence of the corrosive medium, and the operational parameters in terms of flow, temperature, and heat exchange, when selecting inhibitors. Inhibitor concentration levels should also be monitored continuously, and damages should be overcome either by adequate inhibitor additions or by comprehensive fluid resuscitation. Regular inspection should be used whenever conceivable, but keep in mind that outcomes from surveillance equipment, probes, coupons, and so on correspond to the behavior of that specific segment at that specific component of the system. Nonetheless, amidst this prudence, corrosion monitoring in an inhibited system is well common and highly used.

References

1 Sharma, S., Ganjoo, R., Saha, S., Kr. et al. (2021). Experimental and theoretical analysis of baclofen as a potential corrosion inhibitor for mild steel surface in HCl medium. *Journal of Adhesion Science and Technology* 36 (19): 1–26.

2 Ganjoo, R. and Kumar, A. (2022). Current trends in anti-corrosion studies of surfactants on metals and alloys. *Journal of Bio-and Tribo-Corrosion* 8 (1): 1–35.

3 Bashir, S., Lgaz, H., Chung, I.-M., and Kumar, A. (2021). Effective green corrosion inhibition of aluminum using analgin in acidic medium: an experimental and theoretical study. *Chemical Engineering Communications* 208 (8): 1121–1130.

4 Thakur, A. and Kumar, A. (2021). Sustainable inhibitors for corrosion mitigation in aggressively corrosive media: a comprehensive study. *Journal of Bio-and Tribo-Corrosion* 7 (2): 1–48.

5 Assad, H. and Kumar, A. (2021). Understanding functional group effect on corrosion inhibition efficiency of selected organic compounds. *Journal of Molecular Liquids* 344: 117755.

6 Bashir, S., Thakur, A., Lgaz, H., Chung, I.M. and Kumar, A. (2020). Corrosion inhibition performance of acarbose on mild steel corrosion in acidic medium: an experimental and computational study. *Arabian Journal for Science and Engineering* 45 (6): 4773–4783.

7 Bashir, S., Thakur, A., Lgaz, H., Chung, I.M. and Kumar, A. (2020). The corrosion inhibition efficiency of bronopol on aluminum in 0.5 M HCl solution: insights from experimental and quantum chemical studies. *Surfaces and Interfaces* 20: 100542.

8 Ebadi, M., Basirun, W.J., Khaledi, H., and Hapipah Mohd, A. (2012). Corrosion inhibition properties of pyrazolylindolenine compounds on a copper surface in acidic media. *Chemistry Central Journal* 6 (1): 1–10.

9 Revie, R.W. (2008). *Corrosion and Corrosion Control: An Introduction to Corrosion Science and Engineering*. John Wiley & Sons.

10 Schweitzer, P.A. (1989). *Corrosion and Corrosion Protection Handbook*, Vol. 1. CRC Press.

11 Kadhim, A., Al-Amiery, A.A., Alazawi, R., Al-Ghezi, M.K.S., and Abass, R.H. (2021). Corrosion inhibitors. A review. *International Journal of Corrosion and Scale Inhibition* 10 (1): 54–67.

12 Wandelt, K. (2018). *Encyclopedia of Interfacial Chemistry: Surface Science and Electrochemistry*. Elsevier.

13 Lukovits, I., Kalman, E., and Zucchi, F. (2001). *DFT Calculations for Corrosion Inhibition of Ferrous Alloys by Pyrazolopyrimidine Derivatives*. 3–8.

14 Bastidas, D.M., Cano, E., and Mora, E.M. (2005). Volatile corrosion inhibitors: a review. *Anti-Corrosion Methods and Materials* 52 (2).

15 Godinez, L.A., Meas, Y., Ortega-Borges, R., and Corona, A. (2003). Corrosion inhibitors. *Revista de Metalurgia* 39: 140–158.

16 Dariva, C.G., and Galio, A.F. (2014). Corrosion inhibitors–principles, mechanisms and applications. *Developments in corrosion protection* 16: 365–378.

17 Murulana, L.C., Kabanda, M.M., and Ebenso, E.E. (2015). Experimental and theoretical studies on the corrosion inhibition of mild steel by some sulphonamides in aqueous HCl. *RSC Advances* 5 (36): 28743–28761.

18 Goyal, M., Kumar, S., Bahadur, I. Verma, C. and Ebenso, E.E. (2018). Organic corrosion inhibitors for industrial cleaning of ferrous and non-ferrous metals in acidic solutions: a review. *Journal of Molecular Liquids* 256: 565–573.

19 El-Haddad, M.N. (2013). Chitosan is a green inhibitor for copper corrosion in an acidic medium. *International Journal of Biological Macromolecules* 55: 142–149.

20 Fouda, A.S., El-Askalany, A.H., Melouk, A.F., and Elsheikh, N.S. (2020). New synthesized nicotinonitrile derivatives as effective corrosion inhibitors for carbon steel in the acidic environment: electrochemical, surface analysis, and quantum methods. *Journal of Bio-and Tribo-Corrosion* 6 (2): 1–15.

21 Chaouiki, A., Lgaz, H., Rachid Salghi, M. et al. (2020). Inhibitory effect of a new isoniazid derivative as an effective inhibitor for mild steel corrosion in 1.0 M HCl: combined experimental and computational study. *Research on Chemical Intermediates* 46 (6): 2919–2950.

22 Oguzie, E.E., Ying, L., Wang, S.G., and Wang, F. (2011). Understanding corrosion inhibition mechanisms—experimental and theoretical approach. *Rsc Advances* 1 (5): 866–873.

23 Assad, H., Fatma, I., Kumar, A. et al. (2022). An overview of MXene-Based nanomaterials and their potential applications towards hazardous pollutant adsorption. *Chemosphere* 298: 134221.

24 Guo, L., Dong, W., and Zhang, S. (2014). Theoretical challenges in understanding the inhibition mechanism of copper corrosion in acid media in the presence of three triazole derivatives. *RSC Advances* 4 (79): 41956–41967.

25 Guo, L., Chengwei, Q., Zheng, X. et al. (2017). Toward understanding the adsorption mechanism of large-size organic corrosion inhibitors on an Fe (110) surface using the DFTB method. *RSC Advances* 7 (46): 29042–29050.

26 Malik, M.A., Hashim, M.A., Nabi, F. et al. (2011). Anti-corrosion ability of surfactants: a review. *International Journal of Electrochemical Science* 6 (6): 1927–1948.

27 Punitha, R., Kirupha, S.D., Vivek, S., and Ravikumar, L. (2019). Synthesis and corrosion inhibition studies of modified polyacrylic acid bearing triazole moieties on aluminum in alkaline medium. *Journal of Polymer Research* 26 (12): 1–12.

28 Altaf, F., Qureshi, R., Yaqub, A., and Ahmed, S. (2019). Electrochemistry of corrosion mitigation of brasses by azoles in basic medium. *Chemical Papers* 73 (5): 1221–1235.

29 Mistry, B.M., Patel, N.S., Sahoo, S., and Jauhari, S. (2012). Experimental and quantum chemical studies on corrosion inhibition performance of quinoline derivatives for MS in 1N HCl. *Bulletin of Materials Science* 35 (3): 459–469.

30 Khaled, K.F. (2008). Molecular simulation, quantum chemical calculations, and electrochemical studies for inhibition of mild steel by triazoles. *Electrochimica Acta* 53 (9): 3484–3492.

31 Umoren, S.A., Li, Y., and Wang, F.H. (2010). Synergistic effect of iodide ion and polyacrylic acid on corrosion inhibition of iron in H2SO4 investigated by electrochemical techniques. *Corrosion Science* 52 (7): 2422–2429.

32 Wang, Y. and Zuo, Y. (2017). The adsorption and inhibition behavior of two organic inhibitors for carbon steel in simulated concrete pore solution. *Corrosion Science* 118: 24–30.

33 Zhuang, W., Wang, X., Zhu, W. et al. (2021). Imidazoline Gemini surfactants as corrosion inhibitors for carbon steel X70 in NaCl solution. *ACS omega* 6 (8): 5653–5660.

34 Aejitha, S. and Asthuri, P.K. (2014). Geethamani, "Inhibition effect of Antigononleptopus extract on mild steel in sulphuric acid medium. *Indian Journal of Applied Research* 4 (12): 51–53.

35 Chong, A.L., Mardel, J.I., MacFarlane, D.R., Forsyth, M. and Somers, E.A. (2016). Synergistic corrosion inhibition of mild steel in aqueous chloride solutions by an imidazolinium carboxylate salt. *ACS Sustainable Chemistry & Engineering* 4 (3): 1746–1755.

36 Feng, L., Zhang, S., Yao, L. et al. (2019). Synergistic corrosion inhibition effect of thiazolyl-based ionic liquids between anions and cations for copper in HCl solution. *Applied Surface Science* 483: 901–911.

37 Eddy, N. (2011). Green corrosion chemistry and engineering: opportunities and challenges.

38 Palanisamy, G. (2019 August 7). Corrosion inhibitors. *Corrosion Inhibitors*. Available from http://dx.doi.org/10.5772/intechopen.80542.

39 Karki, N., Neupane, S., Chaudhary, Y. et al. (2021). Equisetum hyemale: a new candidate for green corrosion inhibitor family. *International Journal of Corrosion and Scale Inhibition* 10 (1): 206–227.

40 Al Hasan, N.H.J., Alaradi, H.J., Al Mansor, Z.A.K., and Al Shadood, A.H.J. (2019). The dual effect of stem extract of Brahmi (Bacopamonnieri) and Henna as a green corrosion inhibitor for low-carbon steel in 0.5 M NaOH solution. *Case Studies in Construction Materials* 11: e00300.

41 Sanaei, Z., Ramezanzadeh, M., Bahlakeh, G., and Ramezanzadeh, B. (2019). Use of Rosa canina fruit extract as a green corrosion inhibitor for mild steel in 1 M HCl solution: a complementary experimental, molecular dynamics and quantum mechanics investigation. *Journal of Industrial and Engineering Chemistry* 69: 18–31.

42 Faltermeier, R.B. (1999). A corrosion inhibitor test for copper-based artifacts. *Studies in Conservation* 44 (2): 121–128.

43 Mercer, A.D. (1994). Corrosion inhibition: principles and practice. In: *Corrosion*, 2, 17–10. Butterworth-Heinemann.

44 Popova, A., Christov, M., and Zwetanova, A. (2007). Effect of the molecular structure on the inhibitor properties of azoles on mild steel corrosion in 1 M hydrochloric acid. *Corrosion Science* 49 (5): 2131–2143.

45 Wang, X., Huang, A., Lin, D. et al. (2020). Imidazolium-based ionic liquid as efficient corrosion inhibitor for AA 6061 alloy in HCl solution. *Materials* 13 (20): 4672.

46 Bashir, S., Thakur, A., Lgaz, H., Chung, I.M. and Kumar, A. (2019). Computational and experimental studies on Phenylephrine as an anti-corrosion substance of mild steel in an acidic medium. *Journal of Molecular Liquids* 293: 111539.

47 Subramania, A., Sathiya Priya, A.R., and Vasudevan, T. (2006). Diethylamine phosphate as VPI for steel components. *Materials Chemistry and Physics* 100 (1): 193–197.

48 Sastri, V.S. (2012). *Green Corrosion Inhibitors: Theory and Practice*, Vol. 10. John Wiley & Sons.

49 Ochoa, N., Moran, F., Pébère, N., and Tribollet, B. (2005). Influence of flow on the corrosion inhibition of carbon steel by fatty amines in association with phosphonocarboxylic acid salts. *Corrosion Science* 47 (3): 593–604.

50 Ochoa, N., Moran, F., and Nadine, P. (2004). The synergistic effect between phosphonocarboxylic acid salts and fatty amines for the corrosion protection of carbon steel. *Journal of Applied Electrochemistry* 34 (5): 487–493.

9

Methods of Corrosion Measurement

Chemical, Electrochemical, Surface, and Computational

Hassane Lgaz[1], *Karthick Subbiah*[2], *Tae Joon Park*[3], *and Han-Seung Lee*[2,*]

[1] *Innovative Durable Building and Infrastructure Research Center, Center for Creative Convergence Education, Hanyang University ERICA, 55 Hanyangdaehak-ro, Sangrok-gu, Ansan-si, Gyeonggi-do, 15588, Korea*
[2] *Department of Architectural Engineering, Hanyang University-ERICA, 55 Hanyangdaehak-ro, Sangrok-gu, Ansan-si, Gyeonggi-do 15588, Republic of Korea*
[3] *Department of Robotics Engineering, Hanyang University, 55 Hanyangdaehak-ro, Ansan, Gyeonggi-do 15588, Korea*
* *Corresponding author*

9.1 Introductions

Corrosion of metals and alloys is one of the biggest environmental and economic problems. This makes it a highly active research field both fundamentally and practically. Several methods have been developed to detect, monitor, and understand corrosion and corrosion protection processes in response to their serious consequences. Ranging from basic chemical methods such as weight loss to electrochemical and advanced surface analytical techniques, the development of corrosion monitoring techniques is a never-ending task.

Corrosion inhibitors are one of the widely used corrosion control strategies to reduce the corrosion rate of metals and alloys. Adding a small quantity of a corrosion inhibitor to a corrosive solution can significantly increase the metal's life span. The corrosion inhibition process is usually monitored using several chemical and electrochemical techniques. Weight loss is the primary and oldest method to determine the corrosion rate of a metal exposed to uninhibited and inhibited corrosive solutions. However, in-depth analysis of corrosion and corrosion inhibition processes can only be achieved using a destructive and non-destructive electrochemical method such as potentiodynamic polarization curves and electrochemical impedance spectroscopy.

Evaluating the chemical state of the corroded and inhibited metal surface is critical to developing a better understanding of corrosion mechanisms and the effectiveness of any corrosion protection method. To this end, several surface analytical techniques such as electron scanning microscope (SEM), atomic force microscopy (AFM), X-ray photoelectron spectroscopy (XPS), X-ray diffraction (XRD), etc. have been used in corrosion inhibition studies. Besides the techniques mentioned above, computational methods based on Density Functional Theory

Grafted Biopolymers as Corrosion Inhibitors: Safety, Sustainability, and Efficiency, First Edition.
Edited by Jeenat Aslam, Chandrabhan Verma, and Ruby Aslam.

(DFT) or molecular dynamics (MD) simulation have been used to study the adsorption behavior of corrosion inhibitors and their interactions with metal surfaces.

This chapter reports a brief overview of the most used corrosion monitoring techniques. This includes experimental techniques that are used to characterize the behavior of metal/electrolyte interfaces such as open circuit potential (OCP), electrochemical impedance spectroscopy (EIS), linear polarization resistance (LPR), potentiodynamic polarization (PDP) and electrochemical noise analysis (ENA), as well as techniques used to identify the chemical composition or morphology of the metal surface such as XPS, XRD, SEM, and AFM. Additionally, a brief overview is given of theoretical methods used in corrosion inhibition modeling, particularly DFT-based methods and molecular dynamics simulations.

9.2 Corrosion Measurements

Many techniques have been developed and used to monitor metal corrosion in the aqueous environment. The schematic diagram measurement techniques are given in Figure 9.1.

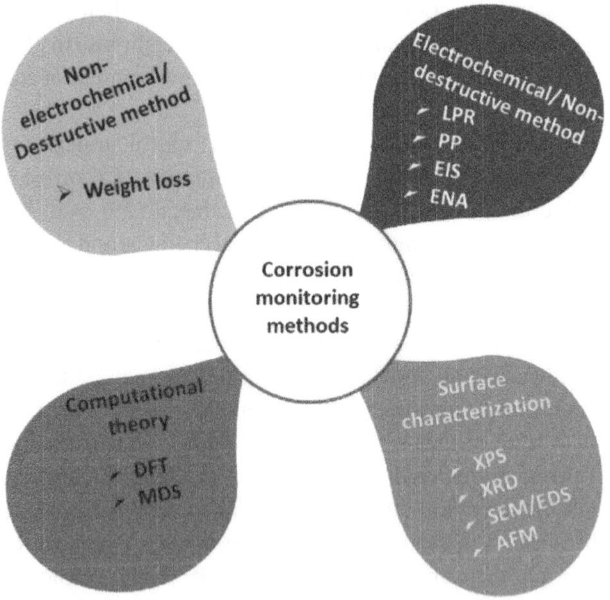

Figure 9.1 Schematic of corrosion monitoring methods. *Source:* Authors.

9.2.1 Non-electrochemical Method for Corrosion Monitoring

9.2.1.1 Weight Loss Methods

Weight loss (WL), commonly known as gravimetric or coupon test, is a fundamental chemical technique for corrosion measurements. It allows a direct measure of a material loss, thus predicting its degradation over time. Before putting metals into service, it is essential to perform preliminary experiments to expect the potential mass loss per exposure times or temperatures in a corrosive environment. Moreover, while in service, the first stage evaluation of the effect of corrosion inhibitors on the metal's degradation behavior is frequently conducted by weight loss tests. The gravimetric tests do not require highly sophisticated instruments, and this simplicity makes it one of the most used corrosion and corrosion inhibition tests. Monitoring the corrosion process of metals is achieved by determining the corrosion rate (CR). Generally, the corrosion rate can be defined as the speed of mass loss from metal exposed to a corrosive medium.

Weight loss experiments can be performed in laboratories or field environments; however, specific requirements are needed for each practice, and these are well detailed in NACE and ASTM recommended standards [1, 2]. However, a particular type of corrosion, such as pitting, requires specific monitoring guidelines as described in ASTM G46-94 [3].

For the determination of corrosion rate, the mass loss, which is defined by mass differential (Δm), is used along with other parameters such as metal density (ρ), time of exposure (t), and sample area (A) using the following equation:

$$CR(mmpy) = K\frac{\Delta m}{\rho At}; \tag{9.1}$$

where K is a conversion parameter used for representing the CR in different units. Δm is the mass loss in grams, t in hours, A is the area in cm^2, and ρ in $g\,cm^{-1}$. The corrosion rate is usually used to determine the corrosion inhibition performance of corrosion inhibitors as a percent of inhibition efficiency as follows:

$$\eta(\%) = \frac{CR_0 - CR_i}{CR_0} \tag{9.2}$$

where CR_0 is the corrosion rate of uninhibited metal, and CR_i is the corrosion rate of the inhibited metal.

Reviewing the literature, weight loss is used as a basic corrosion monitoring technique in most works dealing with corrosion inhibition of metals and alloys. By using TITLE-ABS-KEY ("weight loss", "corrosion inhibition"), a total of 3192 research documents can be found only on the Scopus database between 2000 and 2021, as shown in Figure 9.2. However, despite its large application, the weight loss method has several limitations. For example, while the technique can provide insights into the corrosion rate over time, no information about corrosion and

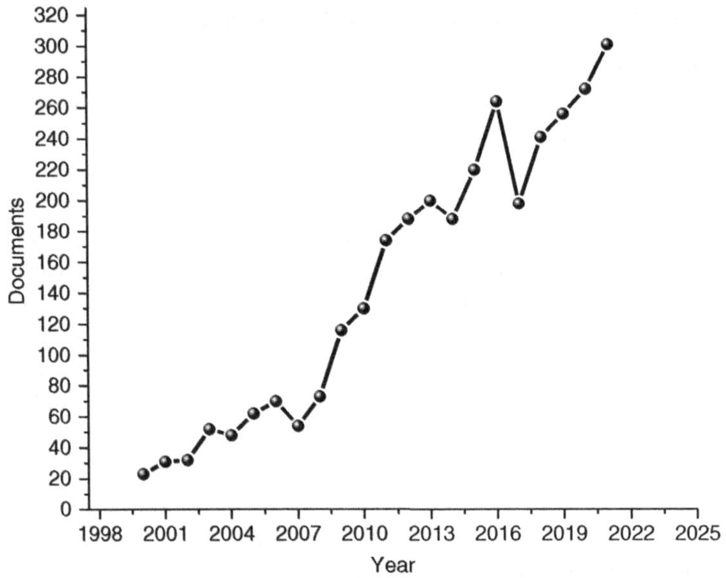

Figure 9.2 Number of published studies on weight loss measurement for the corrosion inhibition of metals based on the Scopus database using TITLE-ABS-KEY ("weight loss", "corrosion inhibition").

corrosion inhibition mechanisms can be derived from WL outcomes. It is usually used in combination with other monitoring techniques such as electrochemical techniques to investigate the corrosion inhibition mechanism and the inhibition performance and/or with surface characterization techniques to identify corrosion products and inhibitor films. Many researchers are using this procedure to investigate the corrosion inhibition performance of many corrosion inhibitors. To cite a few, in 2021, weight loss techniques have been used to monitor the corrosion inhibition performance of several compounds, including Parsley Extract and its synergy with iodide for carbon steel-Q235 in an acidic medium [4], newly synthesized pyran derivative for mild steel in 1 M HCl [5], Allium Jesdianum for mild steel in 1 M HCl solution [6], Anthocyanin for mild steel in HCl media [7], methylcellulose polysaccharide for inhibition of corrosion of magnesium in acidic solutions [8], Carbon Dots for Q235 Steel in 1 M HCl [9], and Isonitrosoacetanilide derivatives for mild steel in 1 mol/L HCl [10], among others [11–28].

9.2.2 Electrochemical Methods for Corrosion Monitoring

By nature, metal corrosion is a combination of electrical and chemical processes. Thus, several corrosion monitoring methods have been developed based on electrochemical principles [29]. Here, electrochemical changes are monitored, which

Table 9.1 Different electrochemical method and advantages.

Method	Advantages
OCP	Non-destructive, can be used for continuous online monitoring in the laboratory and field monitoring.
LPR	Non-destructive, rapid, ideal for screening of inhibitors.
PDP	Destructive, sluggish
EIS	Non-destructive, and when the three-point geometric approach is utilized, it can be quick; beneficial for online monitoring of inhibitor performance in both the lab and in the field.
ENA	Non-destructive methods can be utilized in the laboratory and the field for continuous on-line monitoring of inhibitor performance.

are caused by electron transfer processes due to the corrosion of the material. The most widely used techniques to detect corrosion on the metal surface were linear polarization resistance, open circuit potential (OCP), electrochemical impedance spectroscopy (EIS), linear polarization resistance (LPR), potentiodynamic polarization (PDP), and electrochemical noise analysis (ENA). Different electrochemical methods and their advantages are summarized in Table 9.1.

9.2.2.1 OCP Measurement

A variety of corrosion detection and measurement procedures will provide information on the causes, detection, and rate of corrosion. OCP measurements are the most common way to detect corrosion [30]. The most common method of routine metal examination is OCP monitoring. The potential that any metal generates in interaction with an aqueous medium (e.g., atmospheric moisture, acid rain, natural fluids, aggressive solution, etc.) indicates its tendency to react with it. The metal will develop a potential (mV or V) in an aqueous solution dependent on the concentration of activity species in the aqueous medium. It is one of the two electrode measurement systems. When immersed in a given electrolyte, the OCP can be determined by identifying the potential difference between a suitable reference electrode (saturated calomel electrode (SCE), copper/copper sulfate electrode (CSE), silver/silver chloride electrode, etc.) and the metal's surface [31]. This measurement is made with a high-accuracy, high-impedance voltmeter that can see and record tiny voltages without requiring a lot of current flow.

As described above for metal in aqueous media, the OCP of the entire corrosion reaction is directly related to the change in free energy associated with that reaction by the equations below [32].

$$\Delta G = -nFE \tag{9.3}$$

where ΔG is standard free energy, n and F denote the number of electrons and Faraday constant, and E is equal to Cell potential or OCP. This equation can be used to predict the spontaneous direction of any electrochemical reaction. The value of the free energy change for a given reaction is an essential component, as it reveals whether the reaction is spontaneous or non-spontaneous. For example, the ΔG value is more electronegative, which indicates that the spontaneous corrosion reaction occurs on the metal surface (i.e., the metal anodic dissolution potential (oxidation) should be more active than the cathodic reaction (hydrogen evolution), whereas the ΔG value is positive, suggesting that no spontaneous reaction occurs [32]).

Furthermore, OCP measurements are useful for identifying the anodic and cathodic reactions on the metal surface in the aqueous environment. The OCP of a metal is a valuable metric for assessing current and future corrosion risk and detecting and monitoring the electrochemical reaction that causes corrosion. As a result, this method has become critical for designers of industries, equipment systems, building designers, and maintenance engineers to reduce economic losses due to material corrosion and detection costs, costly and high-risk failures, and repair and replacement, which can be very useful to identify the repair and rehabilitation at the proper time before high corrosion risk. However, while OCP measurement can provide information about corrosion reaction on the metal surface (anodic and cathodic reaction), this will not indicate the corrosion rate of metal in the aqueous mediums.

Besides, measuring the OCP of a metal exposed in an aqueous medium with inhibitors can evaluate the inhibitors' adsorption on the metal surface, which may be used to determine if the anodic, cathodic, or both processes have been reduced [33]. For example, the OCP and time behavior of metals in the aqueous medium with the presence of inhibitors are shown in Figure 9.3. Here it can be seen that the OCP of the metal exposed to an aqueous solution with the presence of an

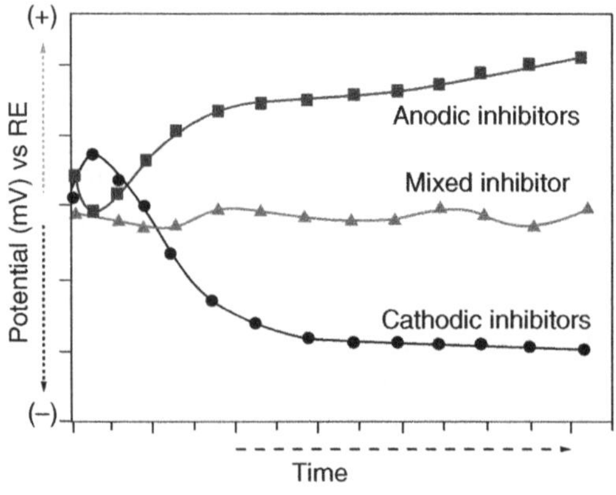

Figure 9.3 OCP measurement of different type of inhibitors. *Source:* Authors.

anodic inhibitor was gradually shifted to the positive direction, which indicates the inhibitor molecules have absorbed on the anodic site of the metal surface to form an oxide layer, causing corrosion resistance. The chromates, nitrates, tungstate, molybdates, benzoic acid, etc., are commonly used as anodic corrosion inhibitors [29]. Whereas the OCP of the metal in the aqueous medium with the presence of cathodic inhibitors (sulfite, bisulfite ions, catalyzed redox reaction by nickel, etc.), the OCP values of metal is shifted to the cathodic region, which due to the inhibitor molecules reduce the cathodic reaction or precipitation on the cathodic site [34–36]. It's worth noting that the OCP versus time curve for the metal exposed to the mixed inhibited system was nearly straight, indicating that steady-state potential had been reached [37]. Mixed inhibitors diminish both the cathodic and anodic reactions. They often create a film on the anodic site, and a precipitated cathodic site on the metal surface, inhibiting both anodic and cathodic sites [38]. However, for mixed inhibitors, this OCP measurement may not be valid, and further electrochemical approaches will be required to confirm this. Some examples of mixed inhibitors include methotrexate, chitosan, allopurinol, activated acyclovir, and enzyme.

9.2.2.2 Linear Polarization Studies

Linear polarization resistance (LPR) is a fast, non-destructive material corrosion testing technology that allows corrosion rates to be evaluated directly and in real-time [39]. Current flow between the anodic and cathodic regions on the metal surface during the corrosion process, although direct measurement of this current flow is unfeasible due to its low magnitude and low external energy requirements. However, altering the potential will result in observable current flow at the corroding surface. The metal is polarized by ±10 or 20 mV relative to its Open Circuit potential (OPC) to accomplish this operation. At the same time, the metals/alloys potential (working electrode) is switched to anodic and cathodic directions by the external energy, a current flow between the working and counter electrodes. The LPR curve was drawn in the potential (E) vs. current density (A/m^2), shown in Figure 9.4.

After using the linear fitting, the slope of the potential versus current curve can be used to evaluate the polarization resistance of the metals/alloys. The following equation can be used to calculate the polarization resistance (R_p).

$$R_p = \frac{\Delta E}{\Delta i} \tag{9.4}$$

$\Delta E/\Delta i$ = slope of the polarization resistance plot, where ΔE is expressed in volts (V) and Δi is expressed in Ampere (A). The slope has units of resistance, hence polarization resistance [31]. Using this polarization resistance, the Stern-Geary equation [40] can compute the corrosion current and corrosion rate.

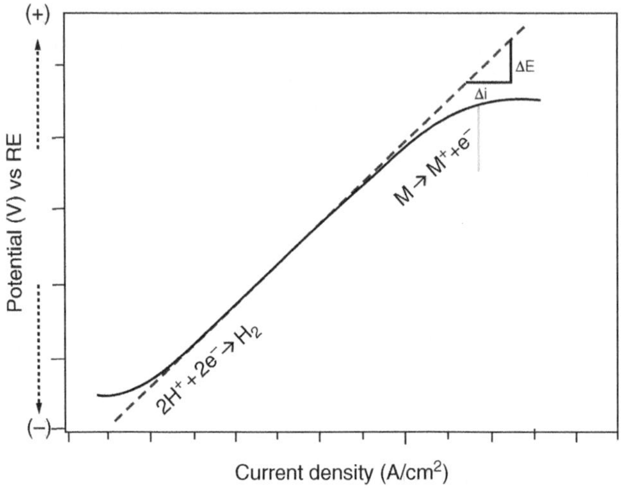

Figure 9.4 LPR measurement methods. *Source:* Authors.

$$I_{corr} = \frac{B}{R_p} \tag{9.5}$$

where B is a Stern-Geary constant, R_p is equal to polarization resistance. However, a B value of 52 mV and 26 mV were considered for the rebar corrosion's passive and active state, respectively [41]. The corrosion current is directly related to the corrosion rate through the following equation:

$$Corrosion\ rate = \frac{K \times I_{corr} \times Eq.W}{Da} \tag{9.6}$$

where K = constant that defines the units for the corrosion rate (1.288×10^5 for mils per year (mpy) and 3.272×10^3 for millimeter per year (mmpy), I_{corr} is a corrosion current, Eq. W is an equivalent weight of metals/alloys, D is the density of metals/alloys, and a is an exposure area.

9.2.2.3 Potentiodynamic Polarization Studies

The potentiodynamic polarization (PDP) test is another commonly used technique for corrosion current and corrosion rate measurement of metal in the aqueous medium [42, 43]. The principle of this technique, from OCP of the metal surface, is shifted to anodic and cathodic using the extensive range of potential (±200 mV), and the response of polarization current with polarization potential was recorded. The polarization curve was drawn in the potential (E) vs. logarithmic of current (log i). The resulting current is plotted on a logarithmic scale as shown in Figure 9.5.

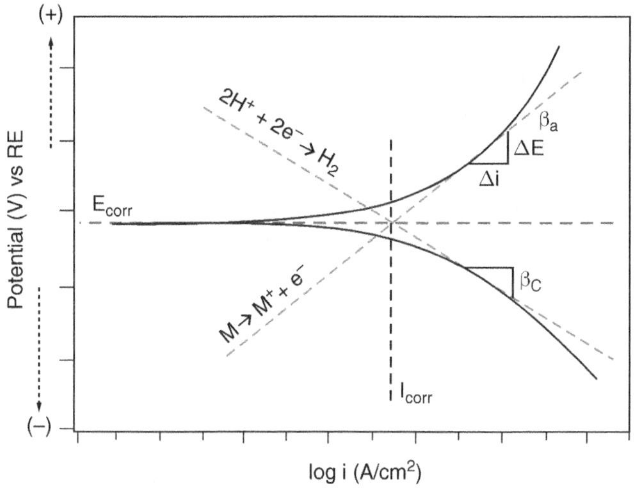

Figure 9.5 Potentiodynamic polarization curve analysis. *Source:* Authors.

The phrase "corrosive potential" (E_{corr}) is used to describe the potential at which no net current flow exists, as estimated by fitting the potential versus current curve. The Butler-Volmer equation expresses the Tafel polarization techniques for measuring the corrosion current [44].

$$i = I_{corr}\left\{\exp\left[S_1(E - E_{corr})\right] - \exp\left[-S_2(E - E_{corr})\right]\right\} \tag{9.7}$$

where I_{corr} is the corrosion current density, S1 and S2 are slop of the anodic (2.303/β_a) and cathodic curve (2.203/β_c), respectively. β_a and β_c are the anodic and cathodic Tafel constant, respectively. E_{corr} is the corrosion potential (V), and E and i represent the potential (V) and current (A) at any time.

In addition, the following equation is used to calculate the Stern-Geary constant using anodic (β_a) and cathodic (β_c) Tafel constant obtained from Tafel polarization methods.

$$B = \frac{\beta_a \beta_c}{2.303\left(\beta_a + \beta_c\right)} \tag{9.8}$$

Tafel constants, labeled β_a, and β_c, must be calculated from the potentiodynamic polarization curve's anodic and cathodic portions. The unit of the Tafel constant is either mV/decade or V/decade. A decade of current is one order of magnitude. A Tafel constant calculation is demonstrated in Figure 9.5. The obtained B values were used to get the corrosion current and examined in Equation (9.5) as well by calculating the corrosion rate using Equation (9.6).

Some examples of polarization curves for metal corrosion in the aqueous solution with inhibitors are presented in Figure 9.6. The curve in this illustration depicts metal anodic and cathodic reactions in the aqueous solution. It can be shown from Figure 9.6a that the shift in the E_{corr} values in a positive direction (> 85 mV) and resistivity behavior of the straight-line curve are recorded at the anodic site, which shows the reduction of anodic reaction/metal dissolution by inhibitor molecules. The anodic type inhibition was described in this way, whereas, mixed-type inhibitors (Figure 9.6b) suppressed both anodic and cathodic reactions as well as changes in E_{corr} values lesser than <85 mV [45, 46]. The E_{corr} values shifted to the negative direction (>85 mV) and the resistivity behavior of straight-line curves formed at the cathodic region (Figure 9.6c). This is because inhibitor molecules reduce the cathodic reaction (oxygen reduction), or because precipitation occurs on the cathodic site [45, 46]. In addition, the I_{corr} values obtained from the curve fitting the Tafel region are utilized to estimate the inhibitor molecules' corrosion inhibition efficiency by the following equation

$$\eta\% = \left[1 - \frac{I_{corr(Inh)}}{R_{ct(0)}}\right] \times 100 \tag{9.9}$$

where $I_{corr(0)}$ and $I_{corr(inh)}$ represent the corrosion current density of metal absence and presence of inhibitors, respectively. In addition to the foregoing, PDP approaches are successfully employed to study metal corrosion and inhibitor corrosion inhibition capabilities in the aqueous medium, as well as to differentiate the type of inhibitors.

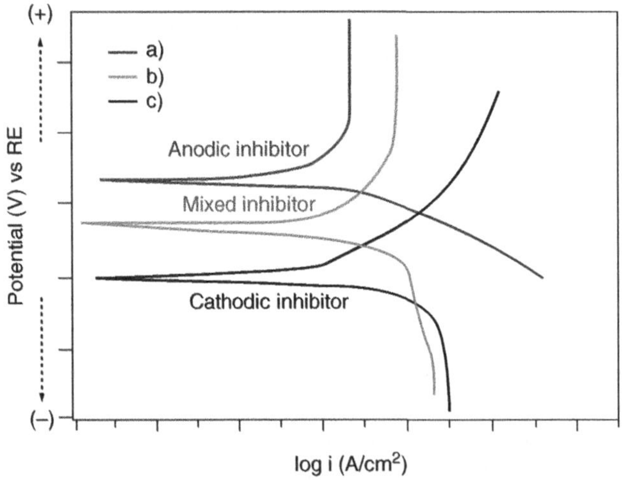

Figure 9.6 Potentiodymic polarization curve for different type of inhibitors. *Source:* Authors.

Both LPR and PDP approaches are based on the administration of either constant fixed current followed by potential monitoring (galvanostatic) or a specified potential followed by current monitoring (PDP) (potentiostatic). The fundamental difference between these two procedures (LPR and PDP) is that the LPR technique requires a change in potential of less than ±20 mV, whilst the PDP technique allows for a shift of potential up to ±200 mV [46]. The interpretation of test data for the computation of corrosion rate is another variation between LPR and PDP. In PDP, the corrosion rate can be estimated by simply substituting Tafel slope values (βa and βc) to obtain the corrosion current, which can then be analyzed in Equations (9.5) and (9.7), followed by computing the corrosion rate using Equations (9.6).

- Under ideal conditions, the Tafel Extrapolation's accuracy is on par with or better than that of traditional weight loss procedures.
- PDP plots can directly indicate the corrosion current, which can be connected to the corrosion rate.
- This approach can be used to measure extremely low corrosion rates
- Tafel plots can be useful for investigations like inhibitor evaluations and alloy comparisons because they can quickly determine corrosion rates.

However, PDP is a destructive approach that alters the electrode surface while employing a larger potential shift (from OCP), therefore it cannot be utilized for continuous corrosion rate monitoring [47].

9.2.2.4 Electrochemical Impedance Spectroscopy (EIS)

In recent years, EIS or AC-impedance has been widely used to characterize material properties. Further, it has been used comprehensively in assessing the corrosion rate of the metal in the aqueous medium, and is the most commonly used non-destructive technique for determining the corrosion of metals/alloys in the aqueous medium [47, 48]. In this technique, a small amplitude of sinusoidal voltage is applied to the metals/alloys, and the resultant current and phase angle is recorded at various frequencies.

The impedance is defined as the ratio of the voltage and current [29].

$$R = \frac{E}{I} \tag{9.10}$$

A small excitation signal usually measures electrochemical impedance. The response of the current to the sinusoidal potential in the linear system will be a sin ewave with a similar frequency but transferred in phase, as given in Figure 9.7.

The excitation signal, the potential sine wave is defined by Equation (9.9)

$$E_t = E_0 \sin(\omega t) \tag{9.11}$$

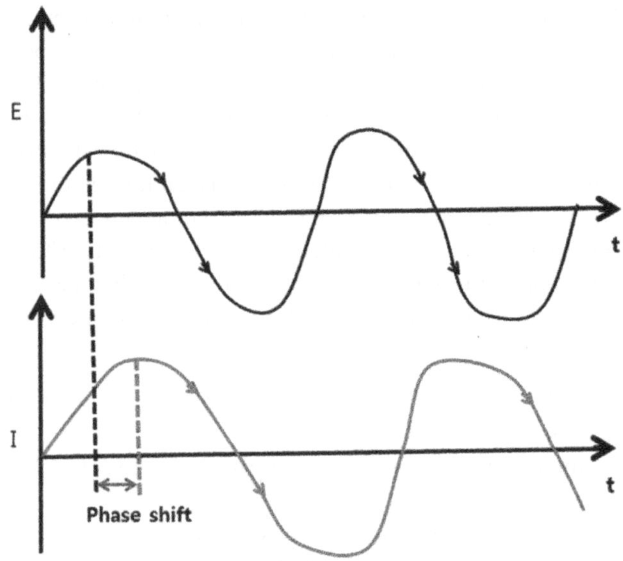

Figure 9.7 Potential and current response in linear system. *Source:* Authors.

where E_t is the potential at time t, E_0 is the signal's amplitude, and ω is the radial frequency.

The relationship between radial frequency ω and frequency f is:

$$\omega = 2\pi f \tag{9.12}$$

In a linear system, the response signal and the current sine wave are expressed by the equation:

$$I_t = I_0 sin(\omega t + \varphi) \tag{9.13}$$

where, I(t) = is the current at time t, I_0 = maximum amplitude, ω = frequency in radians per second = 2πf (where f = frequency in Hertz), t= time, ϕ = phase shift in radians

The rewritten Ohm's Law allows us to calculate the impedance of the system as:

$$Z = \frac{E_t}{I_t} = \frac{E_0 sin(\omega t)}{I_0 sin(\omega t + \varphi)} = Z_0 \frac{sin(\omega t)}{sin(\omega t + \varphi)} \tag{9.14}$$

The impedance, Z, can now be described based on the magnitude of Z_0 and θ is a phase shift.

Equation (9.15) with Euler's relationship given by:

$$e^{j\theta} = cos\theta + jsin\theta \tag{9.15}$$

where, $j=(-1)^{1/2}$

The potential and current response can be:

$$E_t = E_0 e^{j\omega t} \tag{9.16}$$

$$I_t = I_0 e^{j(\omega t-\theta)} \tag{9.17}$$

So, the impedance Z:

$$z(\omega) = \frac{E_t}{I_t} = \frac{E_0 e^{j\omega t}}{I_0 e^{j(\omega t-\theta)}} = Z_0 e^{j\theta} \tag{9.18}$$

$$z(\omega) = Z_0 (cos\theta + jsin\theta) \tag{9.19}$$

$$ReZ = Z' = Z_{real} = Z_0 cos\theta$$

$$Z'' = Z_{imag} = Z_0 sin\theta$$

The response impedance plot is drawn from both the real and imaginary parts. The real part (Z) is plotted on the x-axis and the imaginary part (Z″) is plotted on the y-axis and gives a semicircle plot. This plot is called the Nyquist plot as shown in Figure 9.8. The modified Randles cell analyzes the variation of the impedance with frequency from the graph, which consists of solution resistance (R_S), a double layer capacitor (C_{dl}), and a charge transfer resistance (R_{ct}) (or polarization resistance). The charge-transfer resistance is in parallel with the double-layer capacitance [31]. The equivalent circuit for a basic Randles cell is shown in Figure 9.8. Most of the equivalent circuit elements in the model are common electrical elements such as resistors, capacitors/constant phase elements, and inductors. The equivalent electrical circuit can be useful to determine the charge transfer resistance of metals/alloys in the aqueous medium [49]. The Nyquist plot gives a semi-circle; the diameter of the semicircle is R_S+R_{ct}. R_s is the ohmic

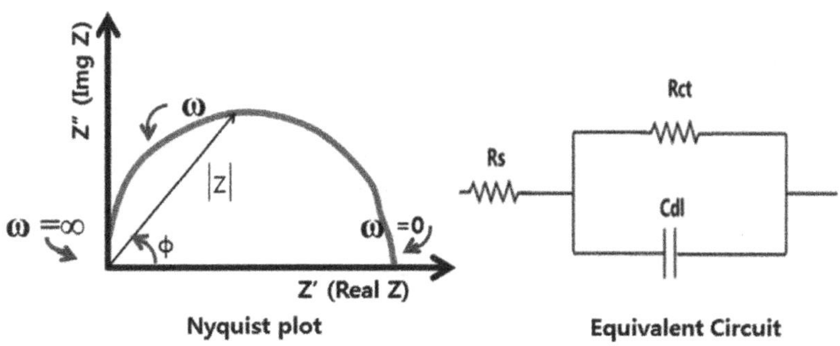

Figure 9.8 Nyquist plot with equivalent circuit. *Source:* Authors.

resistance of the solution between the reference electrode and the metal surface. At the maximum point on the semicircle, the frequency f_{max} can be found, and the double-layer capacitance value is given by [50]:

$$C_{dl} = \frac{1}{2\pi R_{ct} f_{max}} \tag{9.20}$$

Another common illustration is the bode plot, in which the impedance value and phase angle are plotted against frequency. That means Z (impedance), phase (degree), and their corresponding frequency was plotted in the primary y-axis, secondary y-axis, and x-axis, respectively, which is shown in Figure 9.9. This Bode graphic presentation is a valuable additional way to calculate the resistance of solution (R_s), charge transfer resistance (R_{ct}), and double-layer capacitance (C_{dl}).

Furthermore, EIS approaches have been utilized to investigate metal/alloy corrosion and corrosion inhibition in aqueous media in the presence of inhibitors. EIS spectra were analyzed using an appropriate equivalent circuit for the optimal fitting of the plots, to ensure that the employed different circuit provides the best fit for common dissimilar Nyquist plots for metals exposed in an aqueous medium with inhibitors, which is illustrated in Figure 9.10. The equivalent circuit model parameters in Figure 9.10 can be described as follows: Rs denotes the resistance of solution, R_f represents the passive layer resistance, and their corresponding constant phase elements CPE_f (due to inhibitor molecules adsorption/formation oxide layer). The charge resistance of metals and their corresponding constant phase element are denoted as R_{CT} and CPE_{CT}. For perfect fitting or in the case of admittance (Y_0), CPE has been used instead of a pure capacitor. Z_w denotes the

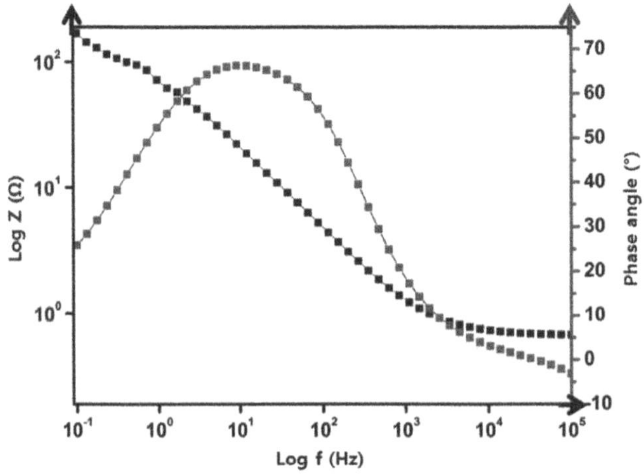

Figure 9.9 Bode and phase angle plots. *Source:* Authors.

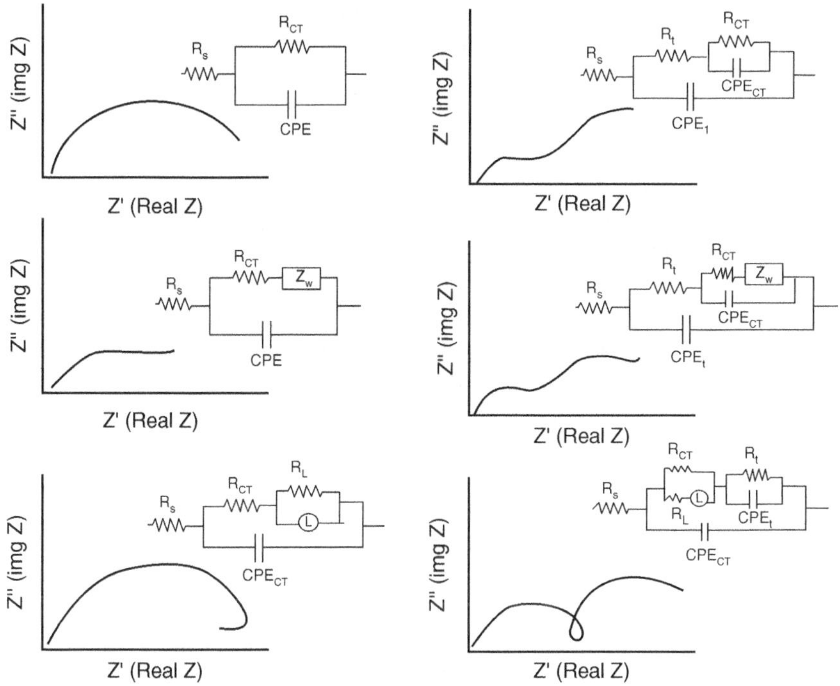

Figure 9.10 Typical Nyquist plots with a different equivalent circuit for metals in the aqueous medium with the presence of inhibitor. *Source:* Authors.

Warburg diffusion coefficient, as a result of diffusion of species through the passive layer, which is due to a less compact and protective oxide layer [51]. R_L stands for the resistance associated with the inductive element L, which is a measure of inhibitor molecule desorption from the metal surface [52]. Due to the frequency dispersion for the surface roughness, heterogeneity of the metal, active site, distribution, and grain boundaries, the constant phase element has been used instead of capacitance (double layer capacitance).

The impedance of CPE is defined as follows by its two parameters, Q and n [50]:

$$Z_{CPE} = Q^{-1}(i\omega)^{-n} \tag{9.21}$$

where n is the heterogeneity measure, Q is the CPE constant and ω is the angular frequency ($\omega = 2\pi f_{max}$).

The double-layer capacitance (C_{dl}) can be evaluated from the CPE coefficient (Q) using the following equation:

$$C_{dl} = Q^{1/n} R_s^{(1-n)/n} \tag{9.22}$$

The C_{dl} value can be associated with the thickness of the protective layer (T_{pl}) of the inhibitor molecules using the Helmholtz model [50, 53];

$$C_{dl} = \frac{\varepsilon_0 \varepsilon_r}{T_{pl}} \tag{9.23}$$

where ε_0 and ε_r symbolize the vacuum and the relative dielectric constants, respectively. The dielectric constant of water molecules is higher than that of organic inhibitor molecules, and the absorption of inhibitor molecules on the metal surface causes the decreasing dielectric constant due to the replacement of water molecules by inhibitor molecules [54]. Thereby, the C_{dl} values decrease by inhibitor adsorption on the metal surface, indicating the inhibitors' absorption behavior. In addition, the obtained R_{CT} values from curve fitting analysis are used to calculate the corrosion inhibition efficiency of the inhibitor molecules by the following equation:

$$\eta\% = \left[1 - \frac{R_{ct(0)}}{R_{ct(inh)}}\right] \times 100 \tag{9.24}$$

where $R_{CT(0)}$ and $R_{CT(inh)}$ represent the charge transfer resistance of metal absence and the presence of inhibitors, respectively. In addition to the foregoing, EIS approaches are successfully employed to study metal corrosion and inhibitor corrosion inhibition capabilities in the aqueous medium.

9.2.2.5 Electrochemical Noise Analysis (ENA)

The ENA is a non-destructive and non-intrusive technology for monitoring the corrosion reaction of metals/alloys, and it can provide statistics about the corrosion mechanism [55]. ENA is defined as a series of low-frequency, low-magnitude variations in electrochemical potential and current at the metal-solution interface over time in an aqueous medium. This ENA is achieved by simultaneous or separate measurement of current fluctuation and potential caused by spontaneous electrochemical reactions. It is one of the three-electrode systems, i.e., two of the electrodes are identical in terms of material and geometry, while the third is a typical reference electrode. A high-impedance voltmeter is used to detect the potential between the reference and the two working electrodes, whereas a zero-resistance ammeter measures the current between the two working electrodes [56, 57]. The simultaneous measurement of electrochemical potential and current noise is preferable, and can be used for instantaneous determination of the electrochemical noise resistance, R_N (with time), and impedance R_{SN} (with frequency). The prevailing corrosion process can be determined by the fluctuations' magnitude, time, and frequency. The following are some parameters that can be obtained from time and frequency domain analysis to measure corrosion using ENA [57, 58].

Mean (M_a): The mean is defined as the average of data from potential and current measurements with time, as expressed by the following equation [58]:

$$M_Z = \frac{1}{N}\sum_{t}^{N}(Z_t) \tag{9.25}$$

where M_Z is a mean value, N denotes the number of points on the metal, and Z denotes the potential (E) or current (I) noise value.

The variance, commonly known as noise power, measures the average of the AC power in the signal. The square root of the variance is called the standard deviation.

$$\sigma_Z = \sqrt{\frac{1}{N}\sum_{t=1}^{N}(Z_t - M_Z)^2} \tag{9.26}$$

where σ_Z is a standard deviation, Z is a potential (E) or current (I).

The skew of a distribution is a measure of its symmetry, and is a parameter with no dimensions. A time record made up entirely of unidirectional transients will be substantially skewed, which could be beneficial for detecting transients related to metastable pitting. Suppose the current noise is measured between two identical electrodes. In this case, the transients can be unidirectional (if only one electrode is active) or bidirectional (if both electrodes are active), and the skew will be an unpredictable characteristic [58].

$$Skew = \frac{\frac{1}{N}\sum_{t=1}^{N}(Z_t - M_Z)^3}{Z_I^3} \tag{6.27}$$

The square root of the average value of the square of potential or current is the root mean square. It is a measurement of the signal's available power.

$$Z_{r.m.s} = \sqrt{\frac{1}{N}\sum_{t=1}^{N}Z_t^2} \tag{9.28}$$

where $Z_{r.m.s}$ is a root mean square of potential (E) or current (I)

Kurtosis is likewise non-dimensional, as it measures the form of distribution for a normal distribution. There are two types of kurtosis, i.e., positive and negative kurtosis, which implies a spiked (positive) and flatter distribution (negative)

$$\frac{\frac{1}{N}\sum_{t=1}^{N}(Z_t - M_z)^4}{(\sigma_Z)^4} \tag{6.29}$$

Noise resistance (R_N) is determined as the standard deviation of potential divided by the standard deviation of current.

$$R_N = \frac{\sigma_E}{\sigma_I} \tag{6.30}$$

The localization index/pitting index (PI) parameter is the standard deviation divided by the root mean square of current

$$LI = \frac{\sigma_I}{I_{r.m.s}} \tag{6.31}$$

Under certain conditions, the noise resistance (RN) can be assumed to be equal to the polarization resistance (RP). The obtained RP value from ENS can be used to calculate the corrosion current density and corrosion rate using Equations (9.5) and (9.8). Furthermore, the calculated LI values are used to identify the type of corrosion phenomena. For example, 0.001≤PI≤0.01 indicates uniform corrosion, 0.01≤PI≤0.1 specifies mixed corrosion, and 0.1≤LI≤1 indicates pitting/localized corrosion [57–59]. The examples of the types of corrosion phenomena identified from the ENA spectrum were exhibited in Figure 9.11.

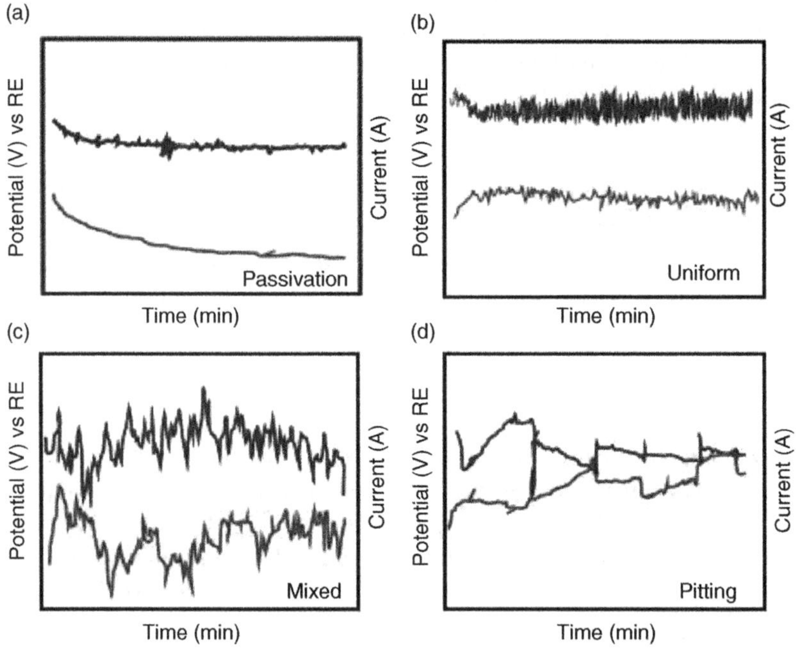

Figure 9.11 Types of corrosion phenomena from ENA spectrum, Passivation (a); uniform corrosion (b); mixed corrosion (c); and pitting/localized corrosion (d). *Source:* Authors.

9.3 Surface Characterization Methods

Weight loss and electrochemical methods provide strong quantitative and qualitative insights into the corrosion behavior of metals in electrolytes. However, these techniques cannot obtain information about the chemical state of metals' surfaces and changes that occurred before, within, or after the corrosion process. Surface analytical methods are widely used to overcome this limitation by providing strong information about the mechanism and chemistry of observed corrosion processes. Several surface characterization techniques are applied for the investigation of the surface's state of metals, such as X-ray diffraction (XRD), Auger electron spectroscopy (AES), X-ray photoelectron spectroscopy (XPS), Fourier transform infrared absorption spectroscopy (FTIR), etc. A detailed description of these and other techniques and their application in corrosion testing are discussed comprehensively in *Analytical Methods in Corrosion Science and Engineering*, edited by Philippe Marcus and Florian Mansfeld [60]. However, in line with the present book's objective, in this section, a brief description of the most used surface analysis techniques in corrosion inhibition studies is given; particularly X-ray photoelectron spectroscopy (XPS), X-ray diffraction (XRD), scanning electron microscopy (SEM), and atomic force microscopy (AFM).

9.3.1 X-ray Photoelectron Spectroscopy (XPS)

XPS is an indispensable surface-sensitive analytical technique to investigate the surface state of materials exposed to liquid and gaseous environments. Besides corrosion, XPS is used in many research and industrial fields such as materials science, catalysis, electrode kinetics, and the control and optimization of products. The principle of XPS is to detect and measure photoelectrons ejected from the material's surface after radiating the sample with X-rays. These photoelectrons are characteristic of the surface atoms and, therefore, the chemical state of the material's surface. Such information is essential for better understanding and interpreting corrosion processes and underlying mechanisms. Recently, Samaniego-Gámez et al. applied XPS techniques to investigate the corrosion behavior of AA2055 Aluminum-Lithium Alloys exposed to the sulfuric acid solution [61]. The surface chemical composition of Al-Li AA2055 alloys determined by XPS is represented in Figures 9.12 and 9.13.

In recent corrosion inhibition studies, Ko et al. studied the corrosion inhibition performance of an imidazole compound for carbon steel weldments in District Heating Water [62]. The authors used electrochemical techniques to evaluate the inhibition efficiency of investigated imidazole and surface characterization techniques to evaluate the carbon steel's surface. XPS analysis was performed to determine potential interactions between the imidazole molecule and the carbon steel

(a)

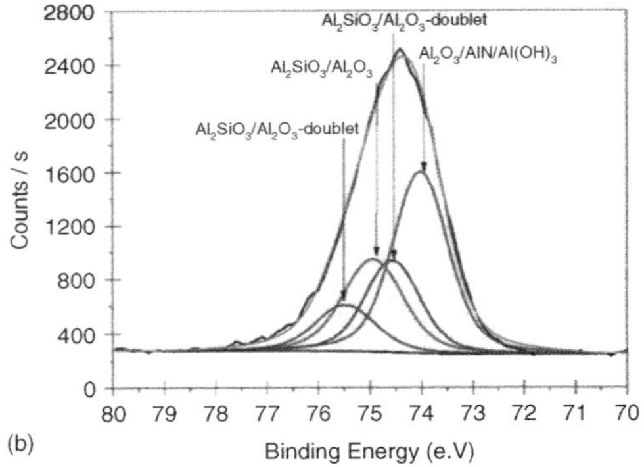

(a')

S1 (0.19 $A{\cdot}cm^{-2}/H_2O$)

Compound	Peak Binding Energy (e.V)	Area Ratio	FWHM fit param (eV)
$Al_2O_3/AlN/Al(OH)_3$	74.02	1	1.26
$Al_2O_3/AlN/Al(OH)_3$–doublet	74.56	0.5	1.26
Al_2SiO_5/Al_2O_3	74.93	0.55	1.38
Al_2iSiO_5/Al_2O_3–doublet	75.47	0.28	1.38

(b)

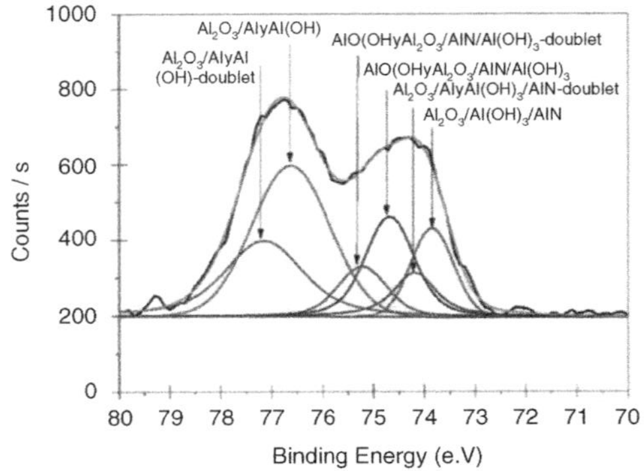

(b')

S2 (1.0 $A{\cdot}cm^{-2}/H_2O$)

Compound	Peak Binding Energy (e.V)	Area Ratio	FWHM fit param (eV)
$Al_2O_3/Al(OH)_3/AlN$	73.85	0.32	0.99
Al_2O_3/Al(OH)3/AlN-doublet	74.19	0.22	0.99
$(Al_2O_3/Al)/Al(OH)$	76.62	1	1.8
$(Al_2O_3/Al)/Al(OH)$-doublet	77.15	0.59	1.8
$AlO(OH)/Al_2O_3/AlN/Al(OH)_3$	74.68	0.46	1.13
$AlO(OH)/Al_2O_3/AlN/Al(OH)_3$-doublet	75.22	0.23	1.13

Figure 9.12 "X-ray photoelectron spectra for Al-Li AA2055 alloy anodized: (a) Al 2p/S1, (b) Al 2p/S2, (c) Al 2p/S3, (d) Al 2p/S4. Parameters obtained from the peak binding energy and (FWHM) fitting (a) S1, (b) S2, (c) S3, (d) S4" [61]. Reproduced with permission from reference 62 under Creative Commons CC BY 4.0 license. Copyright 2022 MDPI.

(c)

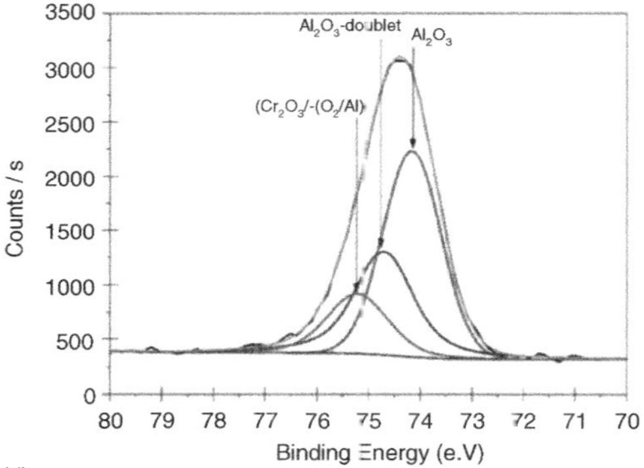

(c')

S3 (0.19 $A.cm^{-2}/Na_2Cr_2O_7$)

Compound	Peak Binding Energy (e.V)	Area Ratio	FWHM fit param (eV)
Al_2O_3	74.16	1	1.34
Al_2O_3-doublet	74.7	0.61	1.34
Cr_2O_3-(O_2/Al)	75.21	0.32	1.39

(d)

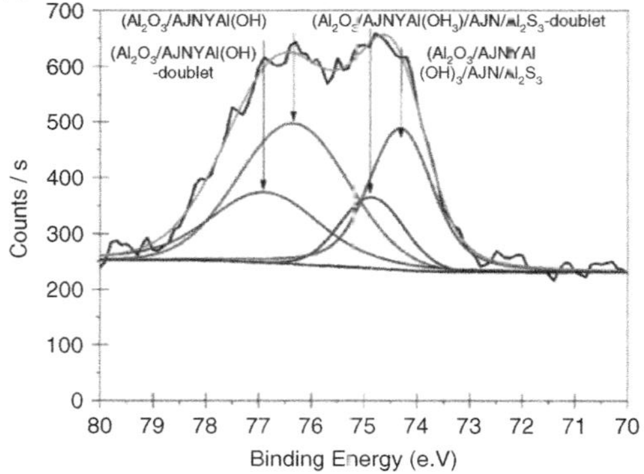

(d')

S4 (1.0 $A.cm^{-2}/Na_2Cr_2O_7$)

Compound	Peak Binding Energy (e.V)	Area Ratio	FWHM fit param (eV)
$(Al_2O_3/AlN/Al(OH)_3/$ AlN/Al_2S_3	74.31	0.66	1.42
$(Al_2O_3/AlN/AlO(OH)_3/$ AlN/Al_2S_3-doublet	74.85	0.29	1.42
$(Al_2O_3/Al)/AlO(OH)$	76.34	1	2.45
$(Al_2O_3/Al)/AlO(OH)$ -doublet	76.88	0.56	2.45

Figure 9.12 (Cont'd)

(a)

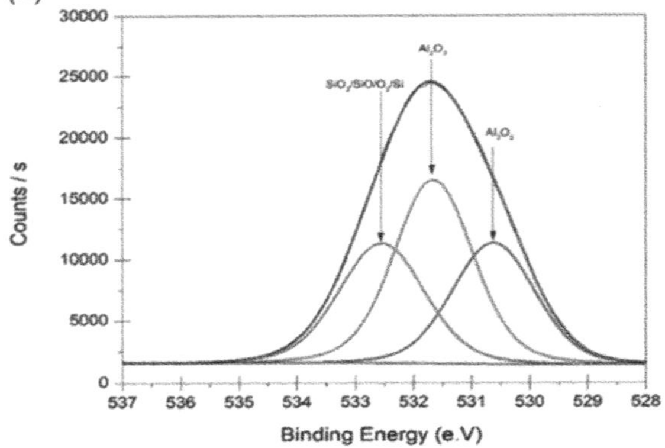

(a')

S1 (0.19 $A \cdot cm^{-2}/H_2O$)

Compound	Peak Binding Energy (e.V)	Area Ratio	FWHM fit param (eV)
Al_2O_3	531.65	1	1.59
Al_2O_3	530.62	0.68	1.66
$SiO_2/SiO/(O_2/Si)$	532.55	0.7	1.71

(b)

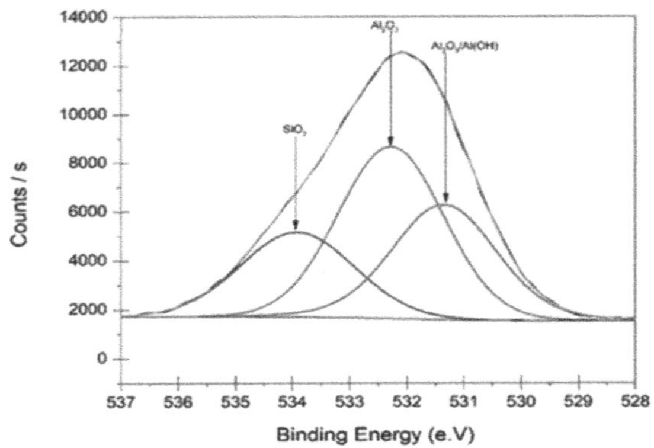

(b')

S2 (1.0 $A \cdot cm^{-2}/H_2O$)

Compound	Peak Binding Energy (e.V)	Area Ratio	FWHM fit param (eV)
Al_2O_3	532.28	1	1.59
$Al_2O_3/Al(OH)$	531.35	0.7	1.71
SiO_2	533.92	0.68	1.66

Figure 9.13 "X-ray photoelectron spectra for Al-Li AA2055 alloy anodized: (**a**) O 1s /S1, (**b**) O 1s/S2, (**c**) O 1s/S3, (**d**) O 1s/S4. Parameters obtained from the peak binding energy and (FWHM) fitting (**a**) S1, (**b**) S2, (**c**) S3, (**d**) S4" [61]. Reproduced with permission from reference 62 under Creative Commons CC BY 4.0 license. Copyright 2022 MDPI.

(c)

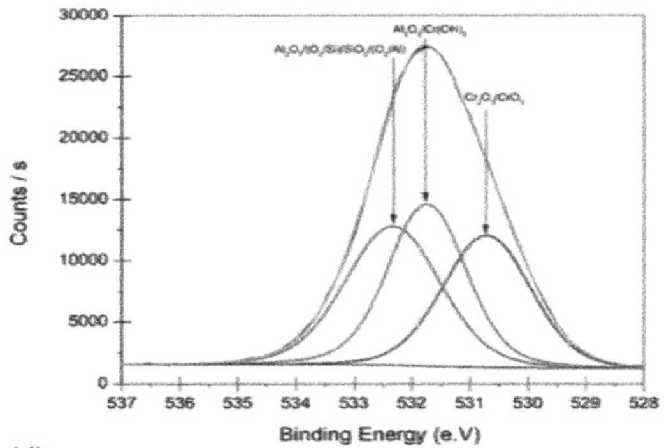

(c')

S3 (0.19 A·cm^{-2}/$Na_2Cr_2O_7$)

Compound	Peak Binding Energy (e.V)	Area Ratio	FWHM fit param (eV)
Al_2O_3/$Cr(OH)_3$	531.75	0.97	1.58
Cr_2O_3/CrO_3	530.72	0.87	1.78
Al_2O_3/(O_2/Si)/SiO_2/(O_2/Al)	532.34	1	1.92

(d)

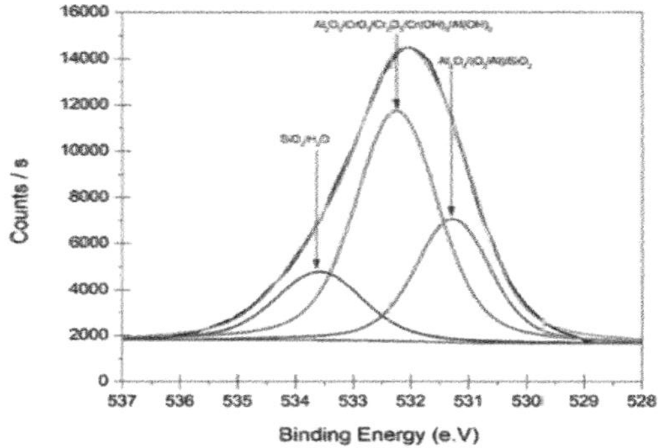

(d')

S4 (1.0 A·cm^{-2}/$Na_2Cr_2O_7$)

Compound	Peak Binding Energy (e.V)	Area Ratio	FWHM fit param (eV)
Al_2O_3/(O_2/Al)/SiO_2	531.28	0.49	1.53
SiO_2/H_2O	533.59	0.32	1.79
Al_2O_3/CrO_3/Cr_2O_3/$Cr(OH)_3$/$Al(OH)_3$	532.26	1	1.68

Figure 9.13 (Cont'd)

surface. As shown in Figure 9.14, the authors found that the imidazole interacts with the carbon steel surface mostly through -C=NC and -C-NH-C present in the imidazole's molecule. In the same context, Cui et al. investigated the potential of dopamine-produced carbon dots as carbon steel corrosion inhibitors in hydrochloric acid solutions [63]. Along with electrochemical studies, authors used XPS to examine the possibility of coordination with carbon steel (Figure 9.15). The authors found strong evidence of carbon dots adsorption and bonding on the carbon steel surface based on XPS analysis.

These works along with many others [64–74] confirm that XPS is an indispensable surface characterization technique for understanding corrosion and corrosion inhibition processes.

9.3.2 X-ray Diffraction (XRD)

The XRD is a useful technique to analyze the crystallization and structure of materials. Its principle is the Bragg equation: $2d\sin\theta = n\lambda$. A diffraction pattern is created when an incident x-ray beam passes through the crystal. This diffraction pattern is characteristic of the material and provides detailed and precise information about its microstructure. In corrosion inhibition studies, oxides and hydroxides are the most common corrosion products developed on a metal's surface after exposure to a corrosive environment. Thus, the XRD technique has successfully identified different corrosion products formed on uninhibited metal and microstructure changes after adding corrosion inhibitors. More recently, Asaad et al. reported the corrosion inhibition abilities of Gum Arabic nanoparticles (GA-NPs) for reinforced concrete exposed to a carbon dioxide environment using several electrochemical and surface characterization techniques [14]. Among these techniques, authors applied XRD to identify the microstructure of the

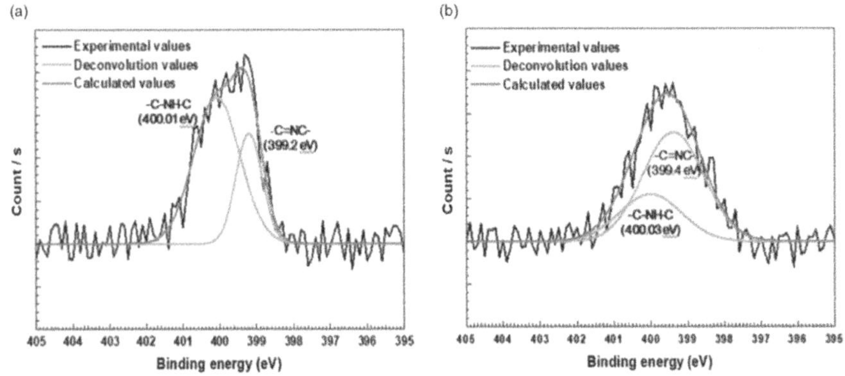

Figure 9.14 "XPS results (N1s) of carbon steel weldment after 6h immersion in (a) 500ppm and (b) 1000ppm of imidazole" [62]. Reproduced with permission from reference 63 under Creative Commons CC BY 4.0 license. Copyright 2022 MDPI.

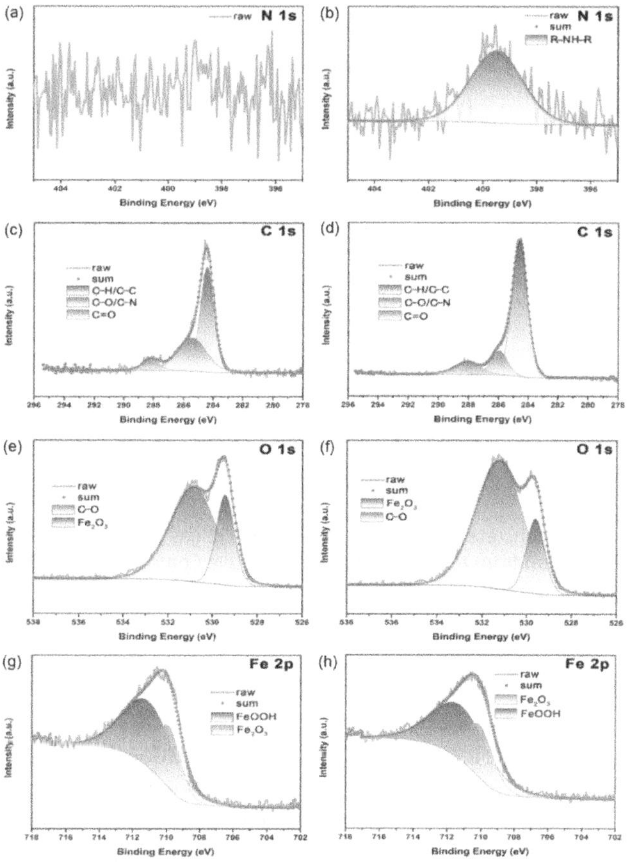

Figure 9.15 "The high-resolution XPS spectra of the steel surface after exposure to 1 M HCl solution and in the presence of 400ppm NCDs, (a,b) N 1s, (c,d) C1s, (e,f) O 1s and (g,h) Fe 2p" [63]. Reproduced with permission from reference 64 under Creative Commons CC BY 4.0 license. Copyright 2022 MDPI.

control concrete specimen and the specimens treated with green GA-NPs inhibitor. After 180 days of exposure to CO_2 gas, authors found quartz (Q), portlandite (P), and calcite (C) as the main component of the control sample; while specimens treated with inhibitor showed increased peaks for quartz and reduced peaks for portlandite and calcite (Figure 9.16).

9.3.3 Scanning Electron Microscopy (SEM)

SEM is a widely used technique to produce microscopic imaging of the surface structure of materials. Its principle is based on scanning the surface of the sample with an electron beam by an electron microscope, producing secondary electrons used to generate the surface image and magnify the size of the sample by 30,000X.

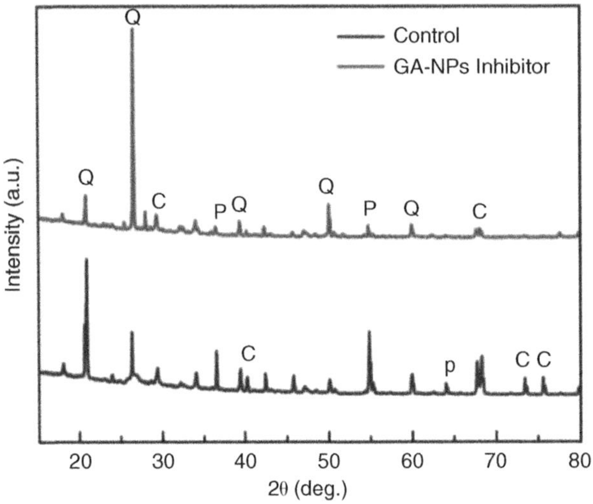

Figure 9.16 "XRD patterns of concrete specimens exposed to CO2 gas for 180 days" [14]. Reproduced with permission from reference 76 under Creative Commons CC BY 4.0 license. Copyright 2022 MDPI.

The SEM analysis is used to study metals' surface morphologies in corrosive mediums. Although this technique does not provide insights into the mechanisms of corrosion inhibition processes, it includes important information on changes that occurred on the metal's surface in corroded and inhibited environments. In addition, when used with energy dispersive x-ray spectroscopy (EDS) analysis, which is the case in most of the studies, it can provide a useful elemental identification of the element present at the characterized surface. There are many research works reporting SEM and SEM/EDS analysis of metals and alloys under a variety of corrosive mediums [75–85]. For instance, Liu et al. studied the surface morphologies of polished J55 Steel and exposed it to an acid solution in the absence and presence of dextrin and its graft copolymer [86]. As shown in Figure 9.17, the authors observed several corrosion products on the steel surface exposed to a corrosive solution compared to polished steel. After adding the corrosion inhibitor, the steel surface showed a compact and smooth morphology similar to the surface that had not been corroded, thus concluding that inhibitor addition to corrosive solution prevents the steel corrosion.

9.3.4 Atomic Force Microscopy (AFM)

The AFM is another surface morphology technique that is based on a microscopic force sensor (cantilever). This cantilever is used to measure the force between the small tip of the AFM and the sample, generating a two and three-dimensional

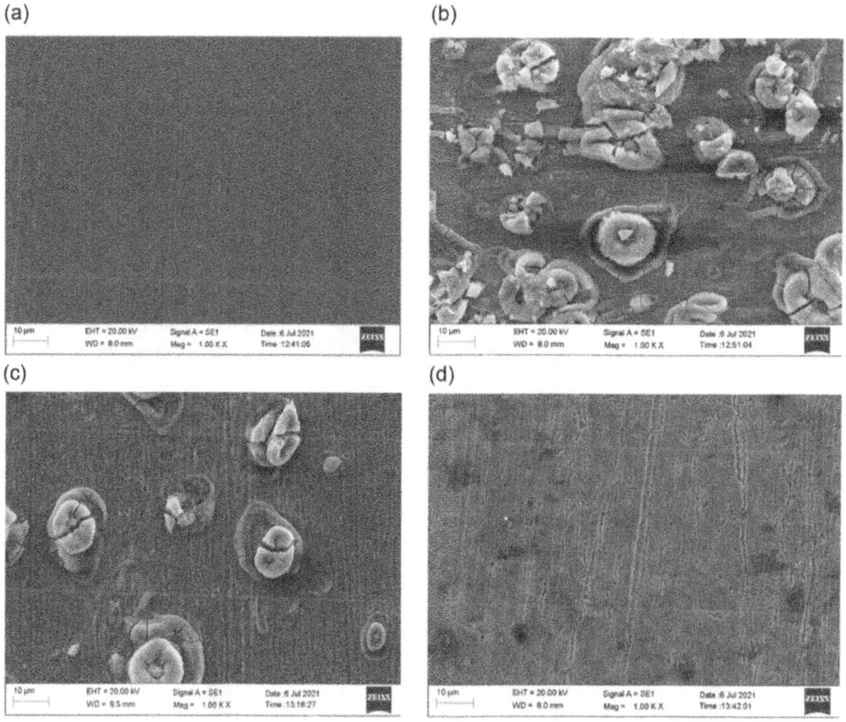

Figure 9.17 "Surface morphology of J55 steel under different conditions. (a) Not corroded; (b) 1 M HCl; (c) 300 mg/L Dxt; (d) 300 mg/LDxt-g-CPL" [86]. [86] / Reproduced with permission from Liu M et al. [2021] / MDPI / CC BY 4.0.

morphology map. This high-resolution instrument can investigate material surfaces at high magnification, at the nanoscale, and even at the atomic structure. The AFM analysis has been extensively used in corrosion inhibition studies for surface morphology characterization of different metals and alloys in various corrosive conditions [81–96]. In a recent paper, Lgaz and Lee investigated the corrosion inhibition properties of hydrazone derivatives using electrochemical, SEM, AFM, and computational techniques for carbon steel in 15wt HCl [97]. The authors used surface roughness as a critical factor in identifying changes in carbon steel surfaces exposed to uninhibited and inhibited HCl solutions. As the authors mentioned, "*A lower roughness of a steel surface indicates lower corrosion products and higher protection*" [97]. According to this condition, the authors observed that adding a corrosion inhibitor to the HCl solution significantly reduced the total roughness of the whole scan range of carbon steel from 1.31μm (blank) to 53.94 nm. In addition, as shown in Figure 9.18, the carbon steel exposed to the blank solution showed a substantial crack due to corrosion attack, while a smoother surface is observed when exposed to the inhibited solution.

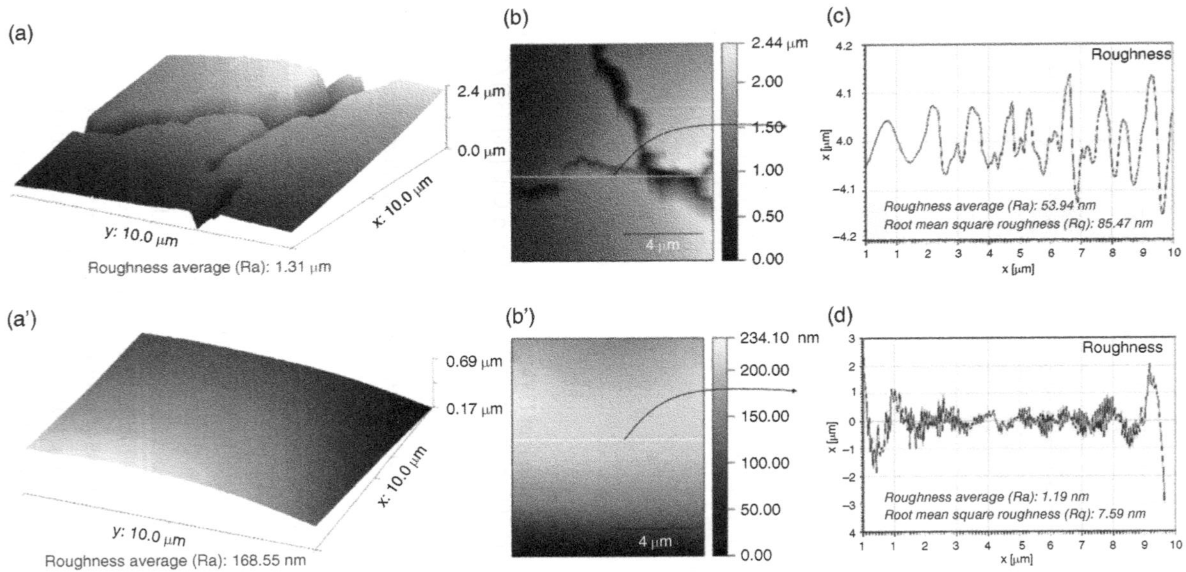

Figure 9.18 "3D and 2D AFM graphs of N80 steel after 24 h immersion in 15% HCl (a, b) without and (a, b) with 5×10−3 mol/L of FHDZ at 303 K, (c, c) the corresponding height profiles of 2D selected area of N80 steel" [97]. [101] / Reproduced with permission from Lgaz H et al. [2022] / ELSEVIER.

Compared to the techniques mentioned above, there are few reported research works on the application of Auger electron spectroscopy (AES), scanning tunneling microscopy (STM) [98, 99], glow discharge optical emission spectrometry (GD-OES) [100], secondary ion mass spectrometry (SIMS) [101, 102], and extended x-ray absorption fine structure spectroscopy (EXAFS) [103] in corrosion inhibition studies.

9.4 Computational Methods

Adsorption is the main corrosion inhibition mechanism by which a corrosion inhibitor prevents metals' corrosion. The adsorption process can be through physical, chemical, or a combination of physical and chemical interactions between the inhibitor molecule and a metal surface. Based on this fact, researchers have made great efforts to theoretically figure out the possible corrosion inhibition mechanisms and active sites of inhibitors responsible for their inhibition effect. Theoretical methods such as Density Functional Theory (DFT), molecular dynamics (MD), and Monte Carlo (MC) simulations can provide insights into the most reactive adsorption sites of an inhibitor molecule and its adsorption configuration over a metal surface; several other theoretical methods have also been applied in corrosion inhibition studies. However, these methods have remarkable differences in terms of both models and outcomes. The application of these methods has been reviewed and discussed in detail in several works. The reader can refer to [104–111] for an in-depth overview of all theoretical methods applied in corrosion inhibition studies. This section will focus on the widely used theoretical methods and their advantages and limitations for corrosion inhibitor studies.

9.4.1 Density Functional Based Theoretical Methods

DFT-based methods are the most frequently used theoretical tools for modeling the corrosion inhibition process. This theoretical approach can be divided into calculations: Quantum-chemical descriptors (QCDs) and first-principles calculations.

Quantum-chemical descriptors: QCDs refer to molecular electronic descriptors derived from DFT calculations. These calculations have been used to find a direct relationship between the inhibitor's electronic parameters and its corrosion inhibition performance. It is based on the assumption that an inhibitor's performance is a function of its donor-acceptor abilities and reactive parameters derived from its highest-occupied (HO) and lowest-unoccupied (LU) molecular orbital (MO) energies. In this regard, HOMO and LUMO are considered significant factors related to the donor and acceptor tendency of an inhibitor molecule, respectively, and thus they have been used extensively to theoretically classify the performance of corrosion inhibitors. Based on Koopmans' theorem, negative values of HOMO and LUMO energies are equal to ionization energy (IP) and electron affinity (EA), respectively. Based on this concept, other quantum chemical parameters such as chemical potential (μ), absolute electronegativity (χ), global hardness (η), and global softness (σ) are calculated using the following equations [105, 111]:

$$\mu = -\chi = \left[\frac{\partial E}{\partial N}\right]_{\nu(r)} = -\left(\frac{IP + EA}{2}\right) \tag{9.32}$$

$$\eta = \frac{1}{2}\left[\frac{\partial^2 E}{\partial N^2}\right]_{\nu(r)} = \frac{IP - EA}{2} \tag{9.33}$$

$$\sigma = 1/2\eta \tag{9.34}$$

The energy gap, which is defined as follows, is one of the most used theoretical descriptors to discuss the inhibitor's performance.

$$\Delta E = E_{LUMO} - E_{HOMO} \tag{9.35}$$

Inhibitor molecules with a low energy gap are less stable and are assumed to have a high reactivity when interacting with a metal surface. In addition, using the same DFT concept, the fraction of the transferred electron (ΔN) can also be predicted using the following equation [105, 111]:

$$\Delta N = \frac{\phi_{metal} - \chi_{inh}}{2(\eta_{metal} + \eta_{inh})} \tag{9.36}$$

where ∅ is the work function calculated for the metal surface, and χ_{inh} & η_{inh} represent the electronegativity of the inhibitor and the global hardness of metal and the inhibitor, respectively.

In the case of ΔN, a positive value indicates the possibility of charge transfer between an inhibitor molecule and a metal surface, and vice versa if the ΔN value is negative [111].

Quantum-chemical descriptors give information about the inhibitor's reactivity. Particularly, inhibitors' reactive sites and geometries can be successfully identified using these descriptors. Other parameters such as Fukui functions, atomic charge, and molecular electrostatic potential have also been used to determine the most reactive sites of inhibitor molecules [112–114].

Lgaz et al. have reported a theoretical study using quantum-chemical descriptors and molecular dynamics simulation to model mild steel corrosion inhibition using four hydrazone compounds [115]. A graphical representation of the optimized molecular structure and frontier molecular orbitals of investigated hydrazone derivatives calculated at DFT/6–31+G(d,p) is given in Figure 9.19, while

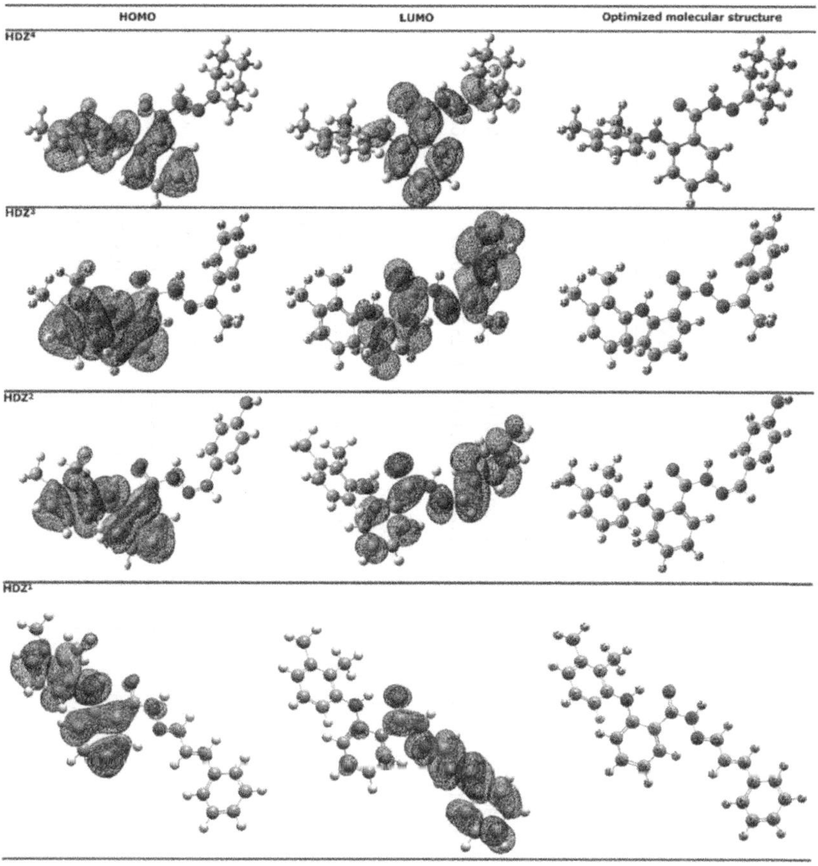

Figure 9.19 "The optimized molecular structure and frontier molecular orbitals distribution of hydrazone derivatives calculated at DFT/6−31+G(d,p)" [115]. Reproduced with permission from reference 120. Copyright 2022 Elsevier.

quantum-chemical descriptors are listed in Table 9.2. The authors used the generated graphical and numerical parameters to draw the following conclusions:

- HOMO densities are concentrated on the mefenamic acid (MA) moiety of compounds, suggesting higher electron-donating properties.
- The electron-donating ability of compounds represented by their HOMO energy values does not predict the experimental order of inhibition performance.
- The order of LUMO energy values agrees with the experimental inhibition efficiency.
- The energy gap values predict a reactivity order in line with experimental results, i.e., the most effective compounds have a smaller energy gap value.

Table 9.2 "The computed quantum chemical for hydrazone derivatives obtained at different theoretical levels" [115].

	E_{HUMO}(eV)	E_{LUMO}(eV)	ΔE(eV)	MV(cm3/mol)	ε(eV)	η(eV)	χ(eV)	ω(eV)	σ(eV)	ΔN(eV)	μ(eV)
Semi-empirical/PM3											
HDZ^4	−8.457	0.037	8.494	564.2	0.027	4.247	4.210	37.639	0.235	0.072	2.03
HDZ^3	−8.437	−0.082	8.355	571.3	0.026	4.177	4.259	37.890	0.239	0.067	1.94
HDZ^2	−8.237	−0.475	7.762	574.3	0.027	3.881	4.356	36.816	0.258	0.060	3.64
HDZ^1	−8.502	−0.732	7.769	583.1	0.024	3.885	4.617	41.404	0.257	0.026	1.66
HF/6–31+G(d.p)											
HDZ^4	−7.654	1.660	9.314	242.5	0.048	4.657	2.997	20.918	0.215	0.196	3.44
HDZ^3	−7.669	1.631	9.300	321.0	0.047	4.650	3.019	21.196	0.215	0.194	3.10
HDZ^2	−7.683	1.525	9.209	353.2	0.046	4.604	3.079	21.825	0.217	0.189	3.63
HDZ^1	−7.731	1.619	9.350	355.1	0.046	4.675	3.056	21.833	0.214	0.189	2.91
DFT/B3LYP/6–31 + G(d.p)											
HDZ^4	−5.208	−1.333	3.875	231.9	0.097	1.937	3.271	10.362	0.516	0.400	4.00
HDZ^3	−5.226	−1.555	3.748	239.2	0.095	1.874	3.352	10.529	0.534	0.392	3.68
HDZ^2	−5.261	−1.663	3.580	278.9	0.093	1.790	3.471	10.783	0.559	0.377	4.32
HDZ^1	−5.309	−1.903	3.104	284.9	0.091	1.552	3.757	10.951	0.644	0.343	3.07

DFT/GGA/DNP											
HDZ^4	−3.853	−1.186	2.973	223.1	0.215	1.486	2.502	4.652	0.673	0.780	3.04
HDZ^3	−4.102	−1.555	2.547	228.5	0.196	1.274	2.829	5.096	0.785	0.782	3.54
HDZ^2	−4.001	−1.663	2.337	237.0	0.213	1.169	2.832	4.688	0.855	0.850	4.23
HDZ^1	−4.114	−1.903	2.211	265.3	0.200	1.105	3.009	5.004	0.905	0.819	3.01
DFTGGA/DNP											
HDZ^4_ O31H	−7.477	−5.636	1.841	254.2	0.051	0.920	6.557	19.781	1.087	−0.944	4.67
HDZ^3_ N32H	−7.444	−5.445	1.998	267.8	0.048	0.999	6.444	20.749	1.001	−0.813	4,81
HDZ^2_ O47H	−6.242	−5.187	1.055	283.9	0.116	0.527	5.715	8.614	1.896	−0.848	4.08
HDZ^1_ N32H	−7.800	−5.717	2.083	301.5	0.042	1.042	6.759	23.790	0.960	−0.931	4.15

- There is no correlation between the dipole moment values and the experimental inhibition efficiency.
- The corrosion inhibition performance is associated with the increased molecular volume of compounds.
- The increasing order of compounds' softness is in line with inhibition efficiency.
- The fraction of the transferred electron (ΔN) follows the opposite trend of the inhibition efficiency.

While this approach seems useful and provides essential information about the reactivity of inhibitor molecules, it has several limitations. The main limitation of this approach is that it only investigates the inhibitor's molecule without considering the metals' surface and/or corrosive environments. A simplistic relationship between quantum-chemical descriptors and the inhibitor's performance is not always a solid factor to consider for conclusions about a compounds' inhibition performance. Also, in some cases, and primarily when dealing with many compounds, such correlation cannot be found [116–119].

First-principles calculations: These DFT-based calculations use the fundamental quantum theory principles to describe the interaction and bonding of inhibitor molecules with metal surfaces. Since corrosion inhibition is an adsorption process that is mostly based on physical and/or chemical interactions, it is of paramount importance to model the potential charge transfer between molecules and surfaces. This theoretical approach can provide useful insights into corrosion inhibition mechanisms and exceeds other theoretical methods such as QCDs and MD simulations. However, there is limited research dealing with corrosion inhibition modeling using this approach. A reasonable explanation may be the high computational costs and the necessity for high-performance computing resources, which make it difficult to model inhibitor-metal interactions involving large molecules. Despite its performance, most of the works applying this approach have focused on one or two small aromatic ring compounds. Kokalj and his co-workers are a pioneer research team in this category [117–126]. For instance, Gustinčič and Kokalj used first-principles DFT calculations to study azole-based corrosion inhibitors' bonding on Cu2O copper surfaces [127]. Figure 9.20 represents different azole molecules in their neutral and deprotonated forms, while Figure 9.21 shows the most stable identified adsorption structures of azole molecules on different surface sites for non-dissociative and dissociative adsorption modes. Figure 9.22 summarizes the adsorption energy values of molecules in different adsorption modes. After a series of investigations, the authors concluded that unsaturated Cu sites bond adsorbates much stronger than saturated sites. Additionally, non-dissociative adsorption energies were found to be similar in the three molecules, while dissociated imidazole showed less adsorption strength compared to dissociated tetrazole and triazole.

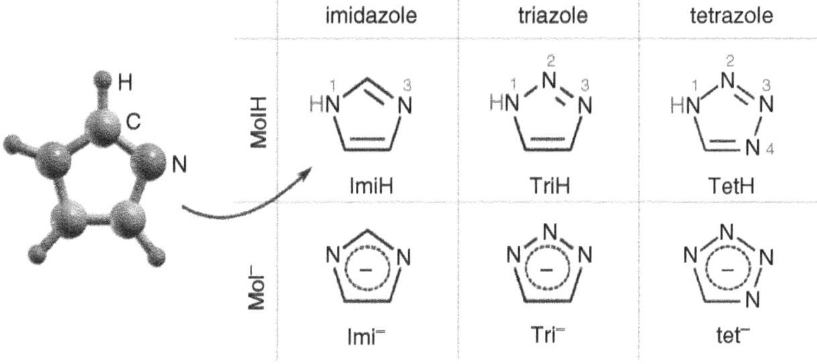

Figure 9.20 "Skeletal formulas of imidazole, triazole, and tetrazole in neutral form (MolH, **top** row) and deprotonated form (Mol−, **bottom** row)" [127]. Reproduced with permission from reference 132 under Creative Commons CC BY 4.0 license. Copyright 2022 MDPI.

Figure 9.21 "The most stable identified adsorption structures of imidazole (left), triazole (middle), and tetrazole (right) at the Cu^{CUS} site on $Cu_2O(111)w/o+1Cu^{CUS}$ surface model for non-dissociative (MolH*, top) and dissociative (Mol*, bottom) adsorption modes" [127]. Reproduced with permission from reference 132 under Creative Commons CC BY 4.0 license. Copyright 2022 MDPI.

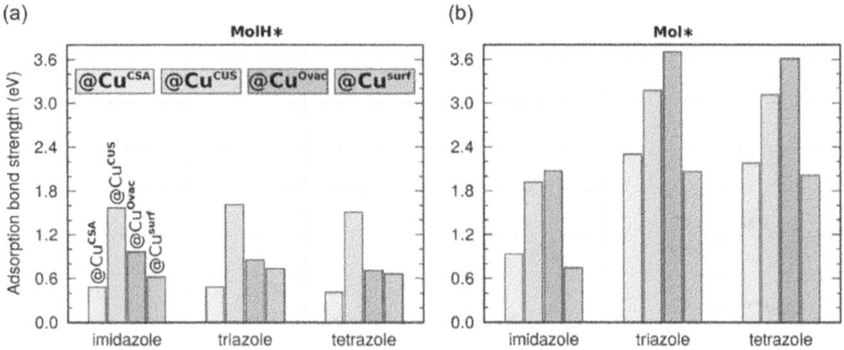

Figure 9.22 "Bond-strengths for (a) MolH* and (b) Mol* adsorption modes of imidazole, triazole, and tetrazole at considered Cu sites"[127]. Reproduced with permission from reference 132 under Creative Commons CC BY 4.0 license. Copyright 2022 MDPI.

9.4.2 Molecular Dynamics (MD) Simulation

MD is a widely used theoretical method to investigate the interaction of corrosion inhibitors with metal surfaces. Inhibitor-metal adsorption systems can be built and simulated using the MD method in the presence of some corrosive species like water, chloride ions, and hydronium. It is a simple, high speed and low-cost computational method that is based on a force field such as COMPASS and CVFF. In recent years, many research works have been conducted using this theoretical approach combined with DFT and experimental corrosion monitoring methods. A more recent review paper by Ebenso et al. reviews in detail the application of MD simulation for corrosion inhibition modeling. The main advantage of this simulation method is obtaining the adsorption configuration of inhibitor molecules on metal surfaces in the presence of a simulated solvent. For instance, the work reported by Lgaz et al. investigated the adsorption configurations of four hydrazone molecules on an Fe(110) surface in the presence of water molecules, chloride, and hydronium ions [115]. Authors have concluded that hydrazone molecules adopted a parallel adsorption mode when interacting with the Fe(110) surface (Figure 9.23). Quantitative analysis of results revealed that interaction energies of adsorbed neutral and protonated molecules were in line with experimental results, with HDZ^1 having the most substantial negative $Z'' = Z_{imag} = Z_0 sin\theta$, followed by HDZ^2, HDZ^3, and then HDZ^4. However, adsorbed protonated molecules showed a low E_{inter} value compared to the neutral form of molecules.

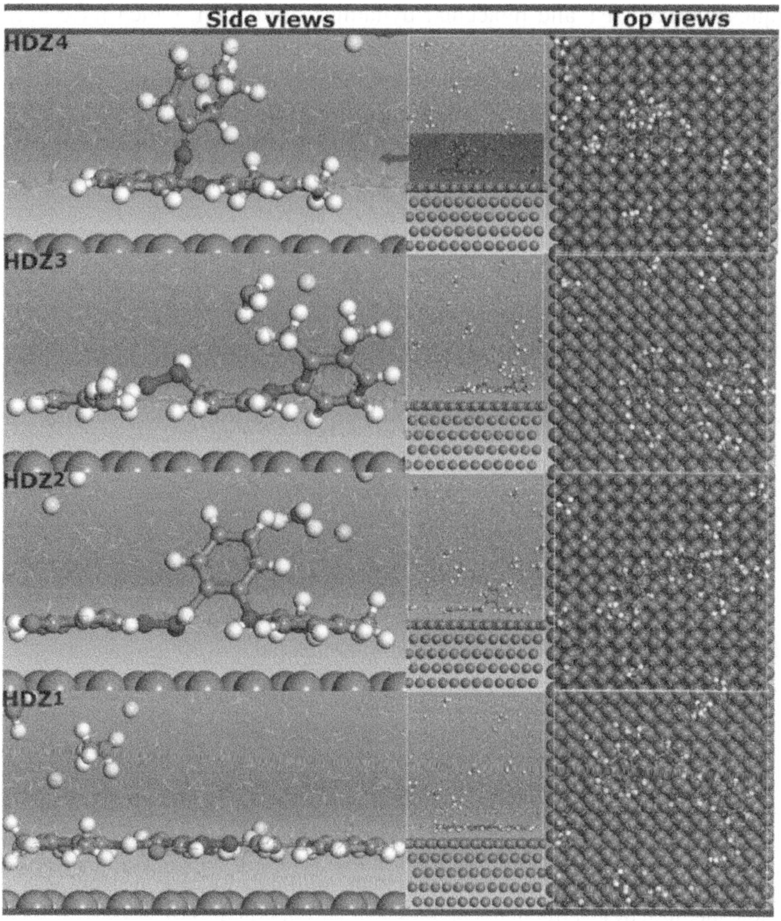

Figure 9.23 "Side and top views of the equilibrium adsorption configurations of the hydrazone derivatives on the Fe (110) surface in solution [115]". Reproduced with permission from reference 120. Copyright 2022 Elsevier.

9.5 Conclusions

In this chapter, a brief overview of corrosion monitoring techniques was reported. In line with the book's objective, the main focus was placed on corrosion monitoring techniques applied to corrosion inhibition studies. More specifically, a large part of this chapter was devoted to chemical (weight loss) and electrochemical techniques while highlighting common surface characterization techniques used to evaluate the surface state of metals under corrosion. In addition, computational techniques such as DFT and molecular dynamics simulation, which have been applied to the theoretical modeling of corrosion inhibition studies were also reviewed. The application of these techniques and methods in corrosion inhibition studies are reported on in detail in other chapters of this book.

Acknowledgment

"This work was supported by the National Research Foundation of Korea (NRF) grant funded by the Korea government(MSIT) (No. NRF-2018R1A5A1025137), and the research fund of the Grand Information & Communication Technology Research Center support program (IITP-2020–0-101741) supervised by the IITP (Institute for Information & communications Technology Planning & Evaluation)".

References

1 Metals ACG-1 on C of. (2017). Standard practice for preparing, cleaning, and evaluating corrosion test specimens. ASTM international.
2 NACE International. (2005). Standard recommended practice: preparation, installation, analysis, and interpretation of corrosion coupons in oilfield operations. NACE international.
3 ASTM G46-94. (2005). Standard guide for examination and evaluation of pitting corrosion. ASTM International.
4 Wan, S., Chen, H., Zhang, T. et al. (2021). Anti-corrosion mechanism of parsley extract and synergistic iodide as novel corrosion inhibitors for carbon steel-Q235 in acidic medium by electrochemical, XPS, and DFT methods. *Frontiers in Bioengineering and Biotechnology* [Internet]. [cited 2022 Mar 2]; 9. Available from: https://www.frontiersin.org/article/10.3389/fbioe.2021.815953.
5 Ouakki, M., Galai, M., Benzekri, Z. et al. (2021 December 15). A detailed investigation on the corrosion inhibition effect of newly synthesized pyran derivative on mild steel in 1.0 M HCl: experimental, surface morphological (SEM-EDS, DRX& AFM) and computational analysis (DFT & MD simulation). *Journal of Molecular Liquids* 344: 117777.
6 Kahkesh, H. and Zargar, B. (2021 December 15). Corrosion protection evaluation of Allium Jesdianum as a novel and green source inhibitor for mild steel in 1M HCl solution. *Journal of Molecular Liquids* 344: 117768.
7 Teng, Y., Zhang, W., Wang, M. et al. (2021 December 15). Anthocyanin as sustainable and non-toxic corrosion inhibitor for mild steel in HCl media: electrochemical, surface morphology and theoretical investigations. *Journal of Molecular Liquids* 344: 117721.

8 Hassan, R.M. and Ibrahim, S.M. (2021 December 15). Performance and efficiency of methyl-cellulose polysaccharide as a green promising inhibitor for inhibition of corrosion of magnesium in acidic solutions. *Journal of Molecular Structure* 1246: 131180.

9 Pan, L., Li, G., Wang, Z. et al. (2021 December 14). Carbon dots as environment-friendly and efficient corrosion inhibitors for Q235 steel in 1 M HCl. *Langmuir* 37 (49): 14336–14344.

10 de Sousa Rodrigues, F.A., Gonçalves, Y.M.H., Horta, B.A.C. et al. (2021 December 5). Experimental and theoretical studies of isonitrosoacetanilides derivatives as corrosion inhibitors for mild steel in 1 mol L−1 HCl. *Journal of Molecular Structure* 1245: 131256.

11 Sargolzaei, B. and Arab, A. (2021 December 1). Synergism of CTAB and NLS surfactants on the corrosion inhibition of mild steel in sodium chloride solution. *Materials Today. Communications* 29: 102809.

12 Sanni, O., Ren, J., and Jen, T.-C. (2021 December). Surface and corrosion properties of type 430 ferritic stainless steel in parsley (Petroselinum Sativum) essential oil-containing sulphuric acid solution. *Surface Topography: Metrology and Properties* 9 (4): 045050.

13 Alkadir Aziz, I.A., Annon, I.A., Abdulkareem, M.H. et al. (2021 December). Insights into corrosion inhibition behavior of a 5-mercapto-1, 2, 4-triazole derivative for mild steel in hydrochloric acid solution: experimental and DFT studies. *Lubricants* 9 (12): 122.

14 Asaad, M.A., Huseien, G.F., Baghban, M.H. et al. (2021 January). Gum Arabic nanoparticles as green corrosion inhibitor for reinforced concrete exposed to carbon dioxide environment. *Materials* 14 (24): 7867.

15 Cáceres, L., Frez, Y., Galleguillos, F. et al. (2021 December). Aqueous dried extract of Skytanthus acutus Meyen as corrosion inhibitor of carbon steel in neutral chloride solutions. *Metals* 11 (12): 1992.

16 Zhuoke, L., Jun, C., Ting, M., and Dan, N. (2021 November 3). Synthesis of bimannich base with thiazole and its corrosion inhibition effect on H2S and CO2 at high temperature. *BMC Chemistry* 15 (1): 59.

17 Fouda, A.E.-A.S., El-Askalany, A.H., Molouk, A.F.S. et al. (2021 November 4). Experimental and computational chemical studies on the corrosion inhibitive properties of carbonitrile compounds for carbon steel in aqueous solutions. *Scientific Report* 11 (1): 21672.

18 Narenkumar, J., Devanesan, S., AlSalhi M.S. et al. (2021 December 1). Biofilm formation on copper and its control by inhibitor/biocide in cooling water environment. *Saudi Journal of Biological Sciences* 28 (12): 7588–7594.

19 Attar, T., Nouali, F., Kibou, Z. et al. (2021 October 19). Corrosion inhibition, adsorption and thermodynamic properties of 2-aminopyridine derivatives on the corrosion of carbon steel in sulfuric acid solution. *Journal of Chemical Sciences* 133 (4): 109.

20 Wang, J., Liu, J., Liu, Q., and Chong, Y. (2021 December 1). The inhibition performance of heterocyclic compounds on Q235 steel in methanol/formic acid medium: experimental and theory. *Journal of Molecular Liquids* 343: 117663.

21 Mirinioui, A., El Attari, H., Fdil, R. et al. (2021 September 23). Dysphania ambrosioides essential oil: an eco-friendly inhibitor for mild steel corrosion in hydrochloric and sulfuric acid medium. *Journal of Bio- and Tribo-Corrosion* 7 (4): 150.

22 Pernaje, S., Bhat, I.J., and Maddani, M.R. (2021 December 1). Spectroscopic characterization and evaluation of tender Theobroma cacao pod extract as an ecofriendly inhibitor for mild steel in 1 M HCl. *Research on Chemical Intermediates* 47 (12): 5369–5388.

23 Abbout, S., Hsissou, R., Erramli, H. et al. (2021 December 5). Gravimetric, electrochemical, and theoretical study, and surface analysis of novel epoxy resin as corrosion inhibitor of carbon steel in 0.5 M H2SO4 solution. *Journal of Molecular Structure* 1245: 131014.

24 Loto, R.T., Loto, C.A., Leramo, R., and Oyebade, B. (2021 November). Corrosion inhibition and statistical data of low carbon steel in HCl media by admixed organic compounds. *Journal of Physics. Conference Series* 2070 (1): 012149.

25 Loto, R.T. and Ogaga, A. (2021 November). Gravimetric study and statistical data on the symbiotic effect of organic intermixtures on corrosion inhibition of P4 mold steel. *Journal of Physics. Conference Series* 2070 (1): 012150.

26 Cherrak, K., El Massaoudi, M., Outada, H. et al. (2021 November). Electrochemical and theoretical performance of new synthetized pyrazole derivatives as promising corrosion inhibitors for mild steel in the acid environment: molecular structure effect on efficiency. *Journal of Molecular Liquids* 15 (342): 117507.

27 El Faydy, M., About, H., Warad, I. et al. (2021 November 15). Insight into the corrosion inhibition of new bis-quinolin-8-ols derivatives as highly efficient inhibitors for C35E steel in 0.5 M H2SO4. *Journal of Molecular Liquids* 342: 117333.

28 Chen, T., Gan, H., Chen, Z. et al. (2021 November 15). Eco-friendly approach to corrosion inhibition of AA5083 aluminum alloy in HCl solution by the expired Vitamin B1 drugs. *Journal of Molecular Structure* 1244: 130881.

29 Singh, A. (2019). *Corrosion Inhibitors*. BoD – Books on Demand. 206 p.

30 Kissinger, P.T., Preddy, C.R., Shoup, R.E., and Heineman, W.R. (1996). Fundamental concepts of analytical electrochemistry. In: *Laboratory Techniques in Electroanalytical Chemistry*, 2e (ed. P. Kissinger and W.R. Heineman). CRC Press.

31 Ramachandran, V.S. and Beaudoin, J.J. (2000). *Handbook of Analytical Techniques in Concrete Science and Technology: Principles, Techniques and Applications*. Elsevier. 995 p.

32 Popov, B.N. (2015). *Corrosion Engineering: Principles and Solved Problems*. Elsevier. 813 p.

33 Riskin, J. and Khentov, A. (2019). *Electrocorrosion and Protection of Metals*. Elsevier. 420 p.

34 Streeter, I., Wain, A.J., Davis, J., and Compton, R.G. (2005 October 1). Cathodic reduction of bisulfite and sulfur dioxide in aqueous solutions on copper electrodes: an electrochemical ESR study. *The Journal of Physical Chemistry. B* 109 (39): 18500–18506.

35 Faisal, M., Saeed, A., Shahzad, D. et al. (2018 December 1). General properties and comparison of the corrosion inhibition efficiencies of the triazole derivatives for mild steel. *Corrosion Review* 36 (6): 507–545.

36 Aliofkhazraei, M. (2018). *Corrosion Inhibitors, Principles, and Recent Applications*. BoD – Books on Demand. 264 p.

37 El-Katori, E.E. and Hashem, N. (2021 January 1). The inhibition activity of lemongrass extract (LGE) for mild steel acid corrosion: electrochemical, surface and theoretical investigations. *Anti-Corrosion Methods and Materials* 69 (1): 71–80.

38 Charles-Granville, U.-E., Liu, C., Scully, J.R., and Kelly, R.G. (2020 January). An RDE approach to investigate the influence of chromate on the cathodic kinetics on 7XXX series Al alloys under simulated thin film electrolytes. *Journal of the Electrochemical Society* 167 (11): 111507.

39 Berke, N.S., Escalante, E., Nmai, C.K., and Whiting, D.A. (1996). *Techniques to Assess the Corrosion Activity of Steel Reinforced Concrete Structures*. ASTM International. 198 p.

40 Stern, M. and Geary, A.L. (1957 January 1). Electrochemical polarization: I. A theoretical analysis of the shape of polarization curves. *Journal of the Electrochemical Society* 104 (1): 56.

41 Andrade, C. and González, J.A. (1978). Quantitative measurements of corrosion rate of reinforcing steels embedded in concrete using polarization resistance measurements. *Materials and Corrosion* 29 (8): 515–519.

42 Ahmad, Z. (2006). *Principles of Corrosion Engineering and Corrosion Control*. Elsevier. 674 p.

43 Rahman, S.U. and Atta Ogwu, A. (2020). Chapter 13 – corrosion and Mott-Schottky probe of chromium nitride coatings exposed to the saline solution for engineering and biomedical applications. In: *Advances in Medical and Surgical Engineering* [Internet] (ed. W. Ahmed, D.A. Phoenix, M.J. Jackson, and C.P. Charalambous), 239–265. Academic Press. [cited 2022 March 9]. Available from: https://www.sciencedirect.com/science/article/pii/B9780128197127000139.

44 Tait, W.S. (2018). Chapter 5 – electrochemical corrosion basics. In: *Handbook of Environmental Degradation of Materials* [Internet], 3e (ed. M. Kutz), 97–115. William Andrew Publishing. [cited 2022 Mar 9]. Available from: https://www.sciencedirect.com/science/article/pii/B9780323524728000058.

45 Kr Saha, S., Dutta, A., Ghosh, P. et al. (2016). Novel Schiff-base molecules as efficient corrosion inhibitors for mild steel surface in 1 M HCl medium: experimental and theoretical approach. *Physical Chemistry Chemical Physics: PCCP* 18 (27): 17898–17911.

46 Yadav, D.K. and Quraishi, M.A. (2012 November 21). Application of some condensed uracils as corrosion inhibitors for mild steel: gravimetric, electrochemical, surface morphological, UV–visible, and theoretical investigations. *Industrial & Engineering Chemistry Research* 51 (46): 14966–14979.

47 Quraishi, M.A., Chauhan, D.S., and Saji, V.S. (2020). *Heterocyclic Organic Corrosion Inhibitors: Principles and Applications*. Elsevier. 300 p.

48 Lillard, R.S. and Pint, B. (2020). *Principles of Corrosion in Nuclear Systems: Theory and Analytical Methods.* Oak Ridge, TN (United States): Oak Ridge National Lab. (ORNL).

49 Berke, N.S. and Chaker, V. (1990). *Corrosion Rates of Steel in Concrete*. ASTM International. 197 p.

50 Orazem, M.E. and Tribollet, B. (2008). *Electrochemical Impedance Spectroscopy*. New Jersey. 383–389.

51 Tabarelli, A., Garcez, E.O., Nunes, F.M., et al. (2019 September 16). Investigating the effect of curing on the chloride diffusion coefficient of conventional concrete. Materials Research [Internet]. [cited 2022 Mar 9]; 22. Available from: http://www.scielo.br/j/mr/a/NJzsRJDNkRVMVR7qfqBqGQp/?lang=en&format=html.

52 Yadav, D.K., Quraishi, M.A., and Maiti, B. (2012 February 1). Inhibition effect of some benzylidenes on mild steel in 1M HCl: an experimental and theoretical correlation. *Corrosion Science* 55: 254–266.

53 Hirschorn, B., Orazem, M.E., Tribollet, B. et al. (2010). Determination of effective capacitance and film thickness from constant-phase-element parameters. *Electrochimica Acta* 55 (21): 6218–6227.

54 Subbiah, K., Lee, H.-S., Mandal, S., and Park, T. (2021 September 15). Conifer cone (Pinus resinosa) as a green corrosion inhibitor for steel rebar in chloride-contaminated synthetic concrete pore solutions. *ACS Applied Materials & Interfaces* 13 (36): 43676–43695.

55 Bertocci, U., Mullen, J.L., and Ye, Y.-X. (1983). Electrochemical noise measurements for the study of localized corrosion and passivity breakdown. In: *Passivity of Metals and Semiconductors* [Internet] (ed. M. Froment), 229–234. Elsevier. [cited 2022 Mar 9]. Available from: https://www.sciencedirect.com/science/article/pii/B9780444422521500400.

56 Mansfeld, F. (2005 August 25). The electrochemical noise technique — applications in corrosion research. *AIP Conference Proceedings* 780 (1): 625–630.

57 Obot, I.B., Onyeachu, I.B., Zeino, A., and Umoren, S.A. (2019 July 3). Electrochemical noise (EN) technique: a review of recent practical applications to corrosion electrochemistry research. *Journal of Adhesion Science and Technology* 33 (13): 1453–1496.

58 Perdomo, J.J. and Singh, P.M. (2002 August 1). Electrochemical Noise (ECN) measurements as a corrosion monitoring tool: a review. *Corrosion Review* 20 (4–5): 359–378.

59 Cottis, R.A. (2001 March 1). Interpretation of Electrochemical Noise data. *Corrosion* 57 (03).

60 Marcus, P. and Mansfeld, F.B. (ed.) (2005). *Analytical Methods in Corrosion Science and Engineering*. Boca Raton: CRC Press. 776 p.

61 Samaniego-Gámez, P.O., Almeraya-Calderon, F., Maldonado-Bandala, E. et al. (2021 November). Corrosion behavior of AA2055 aluminum-lithium alloys anodized in the presence of sulfuric acid solution. *Coatings* 11 (11): 1278.

62 Ko, S.-J., Choi, S.-R., Hong, M.-S. et al. (2021 January). Effect of imidazole as corrosion inhibitor on carbon steel weldment in district heating water. *Materials* 14 (16): 4416.

63 Cui, M., Yu, Y., and Zheng, Y. (2021 January). Effective corrosion inhibition of carbon steel in hydrochloric acid by dopamine-produced carbon dots. *Polymers* 13 (12): 1923.

64 Wang, X., Huang, A., Lin, D. et al. (2020). Imidazolium-based ionic liquid as efficient corrosion inhibitor for aa 6061 alloy in hcl solution. *Materials* 13 (20): 1–18.

65 Palumbo, G., Kollbek, K., Wirecka, R. et al. (2020). Effect of CO2 partial pressure on the corrosion inhibition of N80 carbon steel by gum arabic in a CO2-water saline environment for shale oil and gas industry. *Materials* 13 (19).

66 Lin, B., Tang, J., Wang, Y. et al. (2020). Study on synergistic corrosion inhibition effect between Calcium Lignosulfonate (CLS) and inorganic inhibitors on Q235 carbon steel in an alkaline environment with Cl−. *Molecules* 25 (18).

67 Kwolek, P. (2020). Corrosion behavior of 7075 aluminum alloy in acidic solution. *RSC Advances* 10 (44): 26078–26089.

68 Guimarães, T.A.S., da Cunha, J.N., de Oliveira, G.A. et al. (2020). Nitrogenated derivatives of furfural as green corrosion inhibitors for mild steel in HCl solution. *Journal of Materials Research and Technology* 9 (4): 7104–7122.

69 Ysiwata-Rivera, A.P., Hernández-Hernández, E., Cadenas-Pliego, G. et al. (2020). Effect of modified hexagonal boron nitride nanoparticles on the emulsion stability, viscosity, and electrochemical behavior of nanostructured acrylic coatings for the corrosion protection of AISI 304 stainless steel. *Coatings* 10 (5).

70 Kwolek, P., Koscielniak, B., and Wytrwal-Sarna, M. (2020). Pentavalent vanadium species as potential corrosion inhibitors of Al2Cu intermetallic phase in the sulfuric(VI) acid solutions. *Materials* 13 (8).

71 Petrunin, M., Maksaeva, L., Gladkikh, N. et al. (2020). Thin benzotriazole films for inhibition of carbon steel corrosion in neutral electrolytes. *Coatings* 10 (4).

72 Yuan, Q., Cheng, R., Zou, S. et al. (2020). Isatin thiosemicarbazone derivatives as inhibitors against corrosion of AA6060 aluminum alloy in acidic chloride medium: substituent effects. *Journal of Materials Research and Technology* 9 (5): 11935–11947.

73 El Faydy, M., Lakhrissi, B., Jama, C. et al. (2020). Electrochemical, surface, and computational studies on the inhibition performance of some newly synthesized 8-hydroxyquinoline derivatives containing benzimidazole moiety against the corrosion of carbon steel in a phosphoric acid environment. *Journal of Materials Research and Technology* 9 (1): 727–748.

74 Lgaz, H., Chung, I.-M., Albayati, M.R. et al. (2020). improved corrosion resistance of mild steel in acidic solution by hydrazone derivatives: an experimental and computational study. *Arabian Journal of Chemistry* 13 (1): 2934–2954.

75 Alamiery, A.A. (2022). Study of corrosion behavior of n'-(2-(2-oxomethylpyrrol-1-yl)ethyl)piperidine for mild steel in the acid environment. *Biointerface Research in Applied Chemistry* 12 (3): 3638–3646.

76 Benhiba, F., Hsissou, R., Abderrahim, K. et al. (2022). Development of new pyrimidine derivative inhibitor for mild steel corrosion in acid medium. *Journal of Bio- and Tribo-Corrosion* 8 (2).

77 Rezaeivala, M., Karimi, S., Sayin, K., and Tüzün, B. (2022). Experimental and theoretical investigation of corrosion inhibition effect of two piperazine-based ligands on carbon steel in acidic media. *Colloids and Surfaces A: Physicochemical and Engineering Aspects* 641.

78 Hameed, W.F., Rashid, K.H., and Khadom, A.A. (2022). Investigation of tetraazaadamantane as corrosion inhibitor for mild steel in oilfield-produced water under sweet corrosive environment. *Journal of Bio- and Tribo-Corrosion* 8 (1).

79 Beniken, M., Salim, R., Ech–chihbi, E. et al. (2022). Adsorption behavior and corrosion inhibition mechanism of polyacrylamide on C–steel in 0.5 M H2SO4: electrochemical assessments and molecular dynamic simulation. *Journal of Molecular Liquids* 348.

80 Kalia, V., Kumar, P., Kumar, S. et al. (2022). Synthesis, characterization and corrosion inhibition potential of oxadiazole derivatives for mild steel in 1M HCl: electrochemical and computational studies. *Journal of Molecular Liquids* 348.

81 Abdelshafi, N.S., Ibrahim, M.A., Badran, A.-S., and Halim, S.A. (2022). Experimental and theoretical evaluation of a newly synthesized quinoline derivative as corrosion inhibitor for iron in 1.0 M hydrochloric acid solution. *Journal of Molecular Structure* 1250.

82 Ouknin, M., Ponthiaux, P., Costa, J., and Majidi, L. (2022). Adsorption properties and electrochemical behavior of thymus willdenowii boiss and reut essential oil as a green inhibitor for mild steel corrosion in 1 M HCl. *Portugaliae Electrochimica Acta* 40 (1): 1–17.

83 Hashem, H.E., Farag, A.A., Mohamed, E.A., and Azmy, E.M. (2022). Experimental and theoretical assessment of benzopyran compounds as inhibitors to steel corrosion in aggressive acid solution. *Journal of Molecular Structure* 1249.

84 Hamani, H., Daoud, D., Benabid, S. et al. (2022). Investigation on corrosion inhibition and adsorption mechanism of azomethine derivatives at mild steel/0.5 M H2SO4 solution interface: gravimetric, electrochemical, SEM and EDX studies. *Journal of the Indian Chemical Society* 99 (2).

85 Kahlouche, A., Ferkous, H., Delimi, A. et al. (2022). Molecular insights through the experimental and theoretical study of the anticorrosion power of a new eco-friendly Cytisus multiflorus flowers extract in a 1 M sulfuric acid. *Journal of Molecular Liquids* 347.

86 Liu, M., Xia, D., Singh, A., and Lin, Y. (2021 September). Analysis of the anti-corrosion performance of dextrin and its graft copolymer on J55 steel in acid solution. *Processes* 9 (9): 1642.

87 Shenoy, K.V., Venugopal, P.P., Reena Kumari, P.D., and Chakraborty, D. (2022). Anti-corrosion investigation of a new nitro veratraldehyde substituted imidazopyridine derivative Schiff base on mild steel surface in hydrochloric acid medium: experimental, computational, surface morphological analysis. *Materials Chemistry and Physics* 281.

88 Qiang, Y., Zhi, H., Guo, L. et al. (2022). Experimental and molecular modeling studies of multi-active tetrazole derivative bearing sulfur linker for protecting steel from corrosion. *Journal of Molecular Liquids* 351.

89 El Aatiaoui, A., Daoudi, W., El Boutaybi, A. et al. (2022). Synthesis and anticorrosive activity of two new imidazo[1, 2-a]pyridine Schiff bases. *Journal of Molecular Liquids* 350.

90 Anadebe, V.C., Nnaji, P.C., Onukwuli, O.D. et al. (2022). Multidimensional insight into the corrosion inhibition of salbutamol drug molecule on mild steel in oilfield acidizing fluid: an experimental and computer-aided modeling approach. *Journal of Molecular Liquids* 349.

91 Sarkar, T.K., Yadav, M., and Obot, I.B. (2022). Mechanistic evaluation of adsorption and corrosion inhibition capabilities of novel indoline compounds for oil well/tubing steel in 15% HCl. *Chemical Engineering Journal* 431.

92 Luo, W., Zhang, S., Wang, X. et al. (2022). Ionic macromolecules based on non-halide counter anions for super prevention of copper corrosion. *Journal of Molecular Liquids* 349.

93 Zeng, Y., Kang, L., Wu, Y. et al. (2022). Melamine-modified carbon dots as high effective corrosion inhibitor for Q235 carbon steel in neutral 3.5 wt% NaCl solution. *Journal of Molecular Liquids* 349.

94 Bairy, M., Pais, M., Kumari, P.P., and Rao, S.A. (2022). Hydrazinecarbothioamide derivative as an effective inhibitor for corrosion control: electrochemical, surface and theoretical studies. *Journal of Bio- and Tribo-Corrosion* 8 (1).

95 El Azzouzi, M., Azzaoui, K., Warad, I. et al. (2022). Moroccan, Mauritania, and Senegalese gum Arabic variants as green corrosion inhibitors for mild steel in HCl: weight loss, electrochemical, AFM and XPS studies. *Journal of Molecular Liquids* 347.

96 Huong Pham, T., Lee, W.-H., and Kim, J.-G. (2022). Chrysanthemum coronarium leaves extract as an eco-friendly corrosion inhibitor for the aluminum anode in the aluminum-air battery. *Journal of Molecular Liquids* 347.

97 Lgaz, H. and Lee, H. (2022 February 1). Facile preparation of new hydrazone compounds and their application for long-term corrosion inhibition of N80 steel in 15% HCl: an experimental study combined with DFTB calculations. *Journal of Molecular Liquids* 347: 117952.

98 Wu, X., Wiame, F., Maurice, V., and Marcus, P. (2020). Adsorption and thermal stability of 2-mercaptobenzothiazole corrosion inhibitor on metallic and pre-oxidized Cu(1 1 1) model surfaces. *Applied Surface Science* 508.

99 Wu, X., Wiame, F., Maurice, V., and Marcus, P. (2020 October 15). 2-Mercaptobenzimidazole films formed at ultra-low pressure on copper: adsorption, thermal stability, and corrosion inhibition performance. *Applied Surface Science* 527: 146814.

100 Pilbáth, A., Bertóti, I., Sajó, I. et al. (2008). Diphosphonate thin films on zinc: preparation, structure characterization and corrosion protection effects. *Applications of Surface Science* 255 (5PART 1): 1841–1849.

101 Vaghefinazari, B., Wang, C., Mercier, D. et al. (2021). Adverse effect of 2,5PDC corrosion inhibitor on PEO coated magnesium. *Corrosion Science* 192.

102 Finšgar, M. (2021). Surface analysis by gas cluster ion beam XPS and ToF-SIMS tandem MS of 2-mercaptobenzoxazole corrosion inhibitor for brass. *Corrosion Science* 182.

103 Stumm, W. (1997). Reactivity at the mineral-water interface: dissolution and inhibition. *Colloids and Surfaces A: Physicochemical and Engineering Aspects* 120 (1–3): 143–166.

104 Ani, J.U., Obi, I.O., Akpomie, K.G. et al. (2020 November 4). Corrosion inhibition studies of metals in acid media by fibrous plant biomass extracts and density functional theory: a mini-review. *Journal of Natural Fibers* 0 (0): 1–11.

105 Obot, I.B., Macdonald, D.D., and Gasem, Z.M. (2015 October 1). Density functional theory (DFT) is a powerful tool for designing new organic corrosion inhibitors. Part 1: an overview. *Corrosion Science* 99: 1–30.

106 Ebenso, E.E., Verma, C., Olasunkanmi, L.O. et al. (2021). Molecular modeling of compounds used for corrosion inhibition studies: a review. *Physical Chemistry Chemical Physics* 23 (36): 19987–20027.

107 Ali, A. (2018). *Advanced Engineering Testing*. BoD – Books on Demand. 118 p.

108 Islam, N. and Kaya, S. (2018). *Conceptual Density Functional Theory and Its Application in the Chemical Domain*. CRC Press. 423 p.

109 Obot, I.B., Haruna, K., and Saleh, T.A. (2019 January 1). Atomistic simulation: a unique and powerful computational tool for corrosion inhibition research. *Arabian Journal for Science and Engineering* 44 (1): 1–32.

110 Verma, C., Lgaz, H., Verma, D.K., et al. (2018 June 15). Molecular dynamics and Monte Carlo simulations as powerful tools for the study of interfacial adsorption behavior of corrosion inhibitors in aqueous phase: a review. *Journal of Molecular Liquids* 260: 99–120.

111 Lgaz, H., Chaouiki, A., Lamouri, R. et al. (2021). Computational methods of corrosion inhibition assessment. In: *Sustainable Corrosion Inhibitors I: Fundamentals, Methodologies, and Industrial Applications* (ed. C.M. Hussain and C. Verma) [Internet], 87–109. American Chemical Society. [cited 2022 Mar 3]. (ACS Symposium Series; vol. 1403). doi: 10.1021/bk-2021-1403.ch006.

112 Daoud, D., Douadi, T., Hamani, H. et al. (2015 mai). Corrosion inhibition of mild steel by two new S-heterocyclic compounds in 1 M HCl: experimental and computational study. *Corrosion Science* 94: 21–37.

113 Rodríguez-Valdez, L.M., Villamisar, W., Casales, M. et al. (2006 décembre). Computational simulations of the molecular structure and corrosion properties of aminoethyl, aminoethyl, and hydroxyethyl imidazolines inhibitors. *Corrosion Science* 48 (12): 4053–4064.

114 Mishra, A., Verma, C., Srivastava, V. et al. (2018 September 1). Chemical, electrochemical and computational studies of newly synthesized novel and environmentally friendly heterocyclic compounds as corrosion inhibitors for mild steel in acidic medium. *Journal of Bio- and Tribo-Corrosion* 4 (3): 32.

115 Lgaz, H., Salghi, R., Masroor, S. et al. (2020 June 15). Assessing corrosion inhibition characteristics of hydrazone derivatives on mild steel in HCl: insights from electronic-scale DFT and atomic-scale molecular dynamics. *Journal of Molecular Liquids* 308: 112998.

116 Guo, L., Zhu, S., Zhang, S. et al. (2014 October 1). Theoretical studies of three triazole derivatives as corrosion inhibitors for mild steel in acidic medium. *Corrosion Science* 87: 366–375.

117 Kokalj, A. (2010 December 30). Is the analysis of the molecular electronic structure of corrosion inhibitors sufficient to predict the trend of their inhibition performance. *Electrochimica Acta* 56 (2): 745–755.

118 Kokalj, A., Xie, C., Milošev, I., and Crespo, D. (2021). How relevant are molecular electronic parameters for predicting corrosion inhibition efficiency: imidazoles as corrosion inhibitors of Cu/Zr materials in NaCl solution. *Corrosion Science* 193.

119 Kokalj, A. (2021). Molecular modeling of organic corrosion inhibitors: calculations, pitfalls, and conceptualization of molecule–surface bonding. *Corrosion Science* 193.

120 Kovačević, N. and Kokalj, A. (2012 November 15). Chemistry of the interaction between azole-type corrosion inhibitor molecules and metal surfaces. *Materials Chemistry and Physics* 137 (1): 331–339.

121 Kokalj, A., Behzadi, H., and Farahati, R. (2020 June 1). DFT study of aqueous-phase adsorption of cysteine and penicillamine on Fe(110): role of bond-breaking upon adsorption. *Applied Surface Science* 514: 145896.

122 Kovačević, N., Milošev, I., and Kokalj, A. (2017 August 1). How relevant is the adsorption bonding of imidazoles and triazoles for their corrosion inhibition of copper? *Corrosion Science* 124: 25–34.

123 Kokalj, A. (2015). Ab initio modeling of the bonding of benzotriazole corrosion inhibitor to reduced and oxidized copper surfaces. *Faraday Discussions* 180 (0): 415–438.

124 Kokalj, A., Peljhan, S., Finšgar, M., and Milošev, I. (2010 November 24). What determines the inhibition effectiveness of ATA, BTAH, and BTAOH corrosion inhibitors on copper. *Journal of the American Chemical Society* 132 (46): 16657–16668.

125 Kovačević, N., Milošev, I., and Kokalj, A. (2015 September 1). The roles of mercapto, benzene, and methyl groups in the corrosion inhibition of imidazoles on copper: II. Inhibitor–copper bonding. *Corrosion Science* 98: 457–470.

126 Kokalj, A., Peljhan, S., and Koller, J. (2014 January 16). The effect of surface geometry of copper on dehydrogenation of benzotriazole. Part II. *Journal of Physical Chemistry C* 118 (2): 944–954.

127 Gustinčič, D. and Kokalj, A. (2018 May). DFT study of azole corrosion inhibitors on Cu2O model of oxidized copper surfaces: I. Molecule–surface, and Cl–surface bonding. *Metals* 8 (5): 310.

10

Experimental and Computational Methods of Corrosion Assessment

Recent Updates on Concluding Remarks

Vandana Saraswat[1], *Tarun K. Sarkar*[2], *and Mahendra Yadav*[3]

[1] *Department of Chemistry, University Institute of Sciences, Chandigarh University, Mohali-140301, India*
[2] *Department of Chemistry, IFTM University, Moradabad, Uttar Pradesh-24410, India*
[3] *Indian Institute of technology (Indian School of Mines), Dhanbad-826004, India*

10.1 Introduction

Corrosion is a natural phenomenon that relates to the gradual deterioration of metallic structures due to gradual interaction with the environment. It is a surface phenomenon that occurs due to chemical and biochemical interactions and has burdened the industrial sector since metals became a centerpiece of manufacturing in the industrial revolution. Much industrialization has occurred since then and the economic loss due to corrosion has also increased impeccably. The study of metallic corrosion hence gained much importance to both industrial and academic sectors. The corrosion study of an alloy not only provides valuable insight into its performance but also generates important data about the measures adopted to prevent corrosion. This chapter presents a detailed insight into several methodologies adopted to study the corrosion of metallic samples. The chapter explains the chemical, electrochemical, surface, and quantum chemical methodologies adopted to analyze the corrosion of metal.

The chemical methodologies of corrosion study can be summarized to gravimetric methodologies, where a suitable metal coupon is exposed to the corrosive electrolyte for different periods at different temperatures. This study provides a detailed insight into the nature of metallic corrosion and the effect of the preventive measures taken to mitigate it. The apparent energy, enthalpy, and entropy of activation along with the nature of corrosion inhibitor adsorption are obtained through this study. The electrochemical means of corrosion analysis reflects the

Grafted Biopolymers as Corrosion Inhibitors: Safety, Sustainability, and Efficiency, First Edition.
Edited by Jeenat Aslam, Chandrabhan Verma, and Ruby Aslam.

variation in electrochemical properties of the corrosion mechanism. The electrochemical studies can mostly be classified into open circuit potential (OCP), electrochemical impedance spectroscopy (EIS), and potentiodynamic polarization studies. The OCP studies reflect the steady potential of a corrosion system at equilibrium. The variation of OCP with the dosage of corrosion inhibitors states the cathodic, anodic, or mixed-type attributes of the inhibitors. The EIS technique is used to analyze the transfer of charge on the metal-electrolyte interface. The thickening of the double layer at the metal solution interface due to the adsorption of the corrosion inhibitor reduces the charge transfer. The reduction of charge transfer hence reduces the overall corrosion rate. The potentiodynamic polarization studies provide detailed information about the corrosion process through careful study of the cathodic and anodic curves. The study reflects any change in the corrosion mechanism in presence of the corrosion inhibitors. The apparent corrosion current density and corrosion potential are obtained through the study [1, 2, 3].

The variation of metallic surface, before and after the corrosion process is studied through FE-scanning electron microscopy (FESEM) and atomic force microscopy (AFM). These analyses reflect the variation in substrate surface texture and morphology. The chemical composition of the substrate surface is however confirmed through energy-dispersive X-ray spectroscopy (EDX), X-ray photoelectron spectroscopy (XPS), and FT-IR studies. The EDX provides an approximate content of the elements present upon the substrate surface while XPS generates a comprehensive understanding of the chemistry of the metallic substrate. The FT-IR study usually confirms the effect of polar functionality on corrosion inhibition. A comparison between FT-IR spectra of pristine inhibitor molecules and inhibited metallic substrate provides valuable insight into the corrosion inhibition mechanism. The adsorption of corrosion inhibitors to a metallic substrate can be studied through quantum chemical analysis. These computational methods have come up as an effective tool for corrosion study. Two of the most widely used software are Gaussian and Materials Studio. The computational analysis through the Gaussian software package provides the ability to optimize the molecular structure through DFT. This analysis hence generates highly optimized molecular structure and provides valuable data regarding the reactivity of the molecule. The Materials Studio package however provides a very effective methodology to study the adsorption of inhibitors to the metallic substrate [4, 4, 5].

10.2 Chemical Methodologies

10.2.1 Gravimetric Methods

Gravimetric or weight loss method is one the easiest methods of corrosion analysis. This method of analysis requires exposure of the metal surface to the respective corrosive electrolyte solution for a fixed period. To execute the process, the objected metal sheets are often cut into small coupons of the required dimension. Then the coupons are polished thoroughly with emery papers of different grades to remove all types of imperfections upon the surface. The polished coupons are degreased with organic solvents and dried under a vacuum to remove the presence of adsorbed solvents. These polished and degreased coupons are then subjected to corrosion studies by exposure to corrosive electrolytes under different thermal conditions for a required period. The temperature is often varied within the range of 303 K–353 K while the exposure periods are reported to lie within 6 h–240 h, in accordance with the required study. The gravimetric studies are carried out for corrosive solutions with and without corrosion inhibitors. The following are the equations used to study the apparent corrosion rate (ϕ); mmy^{-1}, the fraction of the surface covered (θ), and corrosion inhibition efficiency (%E) are determined by the following equations.

$$\phi = \frac{87.6 \times \Delta W}{D \times A \times T} \tag{10.1}$$

$$\theta = \frac{\phi_0 - \phi_i}{\phi_0} \tag{10.2}$$

$$\%E = \frac{\phi_0 \phi_i}{\phi_0} \times 100 \tag{10.3}$$

Where ΔW represents the loss in a metal sample (mg), A represents the area of metal sample used in the study (cm^2), D is the density of the metal sample. The symbols ϕ_0 and ϕ_i represent the apparent corrosion rate in the absence and presence of the corrosion inhibitors.

10.2.2 Effect of Inhibitor Dosage and Temperature

The corrosion mitigation capabilities of a corrosion inhibitor can be understood by observing the effect of inhibitor dosage and temperature. It is usually observed that the corrosion rate of metallic substrates increases with temperature. Upon the addition of corrosion inhibitors to the corrosive electrolyte, the corrosion rate is observed to reduce substantially. The corrosion inhibition efficiency of a corrosion inhibitor is observed to increase with the increase in dosage. This phenomenon occurs due to the adsorption of corrosion inhibitors on the metal surface. Hence a protective thin film is generated over the metal surface. This film protects the metal underneath from further exposure and thereby protects it from corrosion. It is usually observed that the corrosion inhibition efficiency increases significantly with dosage at lower concentrations. With the gradual increase in corrosion inhibitor dosage, the inhibition efficiency is observed to reach a point beyond which no further increase is observed. This concentration of the inhibitor is recognized as the optimum dosage. This phenomenon is further explained here in Figure 10.1.

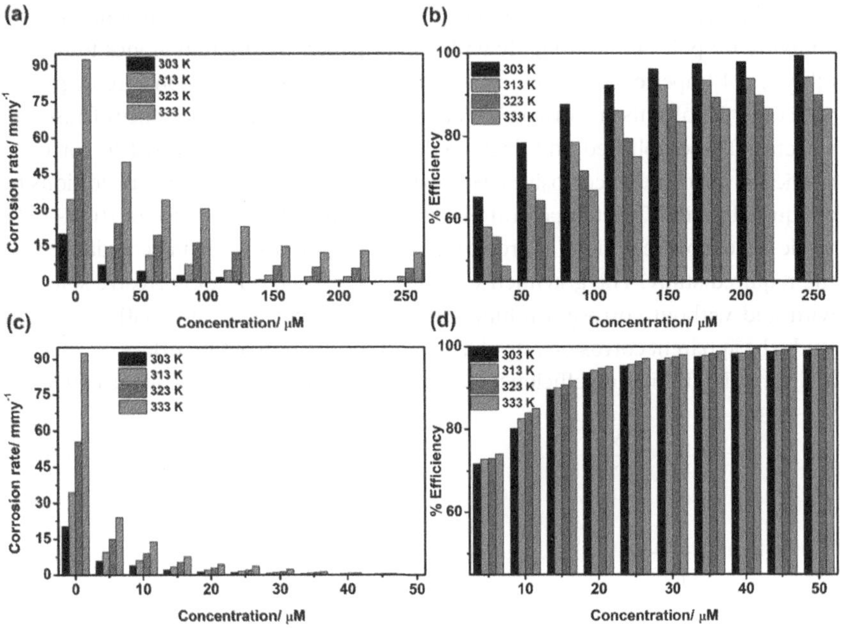

Figure 10.1 Representation of corrosion inhibition capabilities of corrosion inhibitors with respect to temperature and dosage. Images (a, b) represent the physisorption of the pyrimidine compound (MPA) and the chemisorption of the indoline compound, FTTI (c, d). The data are taken from T.K. Sarkar et al. Mater. Today Commun. 26 (2021) 101862, Chem. Eng. J. 431 (2022) 133481.

Images a and b represent the corrosion inhibition capabilities of a pyrimidine compound (MPA) [6] while c and d represent the similar capabilities of an indoline compound (FTTI) [7]. It is observed from the data of both compounds that the optimum concentration is attained at $150\,\mu Ml^{-1}$ for MPA and $25\,\mu Ml^{-1}$ of FTTI beyond which no significant increase in corrosion inhibition efficiency is observed. This occurrence is owed to the adsorption of the corrosion inhibitor molecules upon the optimum number of the corrosion active sites and further adsorption is not feasible. The data is represented in Figure 10.1. also, provide a detailed insight into the corrosion inhibition nature of the inhibitors. It is obvious from Figures 10.1a and 10.1b that the corrosion rate is comparatively higher at elevated temperatures for all the dosages of MPA, while the images of Figures 10.1c and 10.1d reflect that the corrosion rate and inhibition efficiency of the metallic substrate remains nearly constant at higher temperatures for all the dosage of FTTI. this observation relates to the physisorption of MPA and chemisorption of FTTI over the metallic substrate. It is well documented that the corrosion inhibitors that display higher efficiencies at higher temperatures effectively chemisorb upon the metal surface and generates a much more stable protective film. The protective film generated by chemisorbed inhibitors is much more stable at higher temperatures and thereby the extent of corrosion inhibition is much better. The majority of modern publications however refer to the fact that organic corrosion inhibitors display both chemisorption and physisorption capabilities. Hence the inhibitors displaying a higher extent of chemisorption display more efficiency in comparison to the others [8, 9]. The extent of adsorption of any organic corrosion inhibitors cannot be verified from this much data alone. The other required parameters are further explained within the other sections of the chapter.

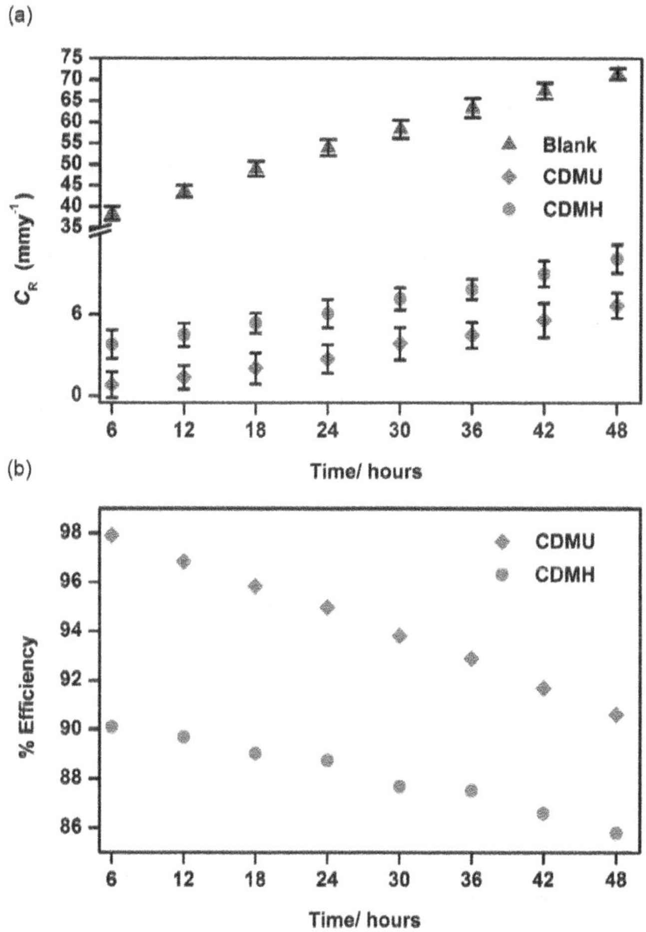

Figure 10.2 Representation of the effect of prolonged exposure of mild steel sample in HCl solution in the presence and absence of corrosion inhibitors. V. Saraswat et al. J. Phys. Chem. Solids. 160(2022) 110341. © Elsevier.

10.2.3 Effect of Exposure Periods

The effect of a prolonged exposure period is often carried as shown in Figure 10.2 out to study the stability of the protective film generated due to the adsorption of the corrosion inhibitors upon the metal substrate. The study is carried out through weight loss methodology. The temperature is kept fixed for the purpose (mostly room temperature) and the exposure period is prolonged up to 240h for the study. The corrosion inhibition efficiency of the corrosion inhibitor is observed to decrease with the exposure period. This observation is explained as the dissolution of the protective film generated due to inhibitor adsorption. The gradual dissolution of the adsorbed layer exposes the metallic substrate underneath and an increase in apparent corrosion rate is observed to increase [10, 56]. It has also been reported by several authors that prolonged exposure of the metal samples to corrosive electrolytes (>48h) leads to a near constancy in corrosion rate with respect to further exposure. This observation refers to the fact that prolonged exposure to metal coupons essentially contaminates the corrosive medium. Some of the corrosion products get dissolved in the electrolyte and reduces its corrosive capabilities.

10.2.4 The Thermodynamic & Activation Parameters

The thermodynamic and activation parameters of the corrosion process are usually obtained through gravimetric analysis. The thermodynamic parameters such as apparent activation energy (E_a) is calculated through the Arrhenius equation (Equation 10.4) while, the enthalpy of activation ($\Delta H^{\#}{}_a$) and the entropy of activation ($\Delta S^{\#}{}_a$) are calculated by transition state equation (Equation 10.5).

$$\log(\phi) = -\frac{E_a}{2.303RT} + \log\lambda \tag{10.4}$$

$$\log\left(\frac{\phi}{T}\right) = \log\left(\frac{R}{N_a h}\right) + \left(\frac{\Delta S_a^{\#}}{2.303R}\right) - \left(\frac{\Delta H_a^{\#}}{2.303RT}\right)\log\left(\frac{\phi}{T}\right)$$

$$= \log\left(\frac{R}{N_a h}\right) + \left(\frac{\Delta S_a^{\#}}{2.303R}\right) - \left(\frac{\Delta H_a^{\#}}{2.303RT}\right) \tag{10.5}$$

The parameters used in the equations are R, the universal gas constant (8.314 J $K^{-1}mol^{-1}$); T, temperature (K), λ, pre-exponential factor; h, Planck's constant; and N_a, Avagadro number.

10.2.4.1 Apparent Activation Energy

The apparent activation energy (E_a) is analyzed through Arrhenius equation mentioned in Equation (10.4). Here plots of log(ϕ) vs. 1000/T (Figure 10.3) are prepared to calculate the required values and the slope of the plot is used to calculate the value of E_a. The relation used for the purpose is $E_a = -(\text{slope}) \times 2.303 \times R$. The values of E_a play a very important part in understanding the nature of corrosion

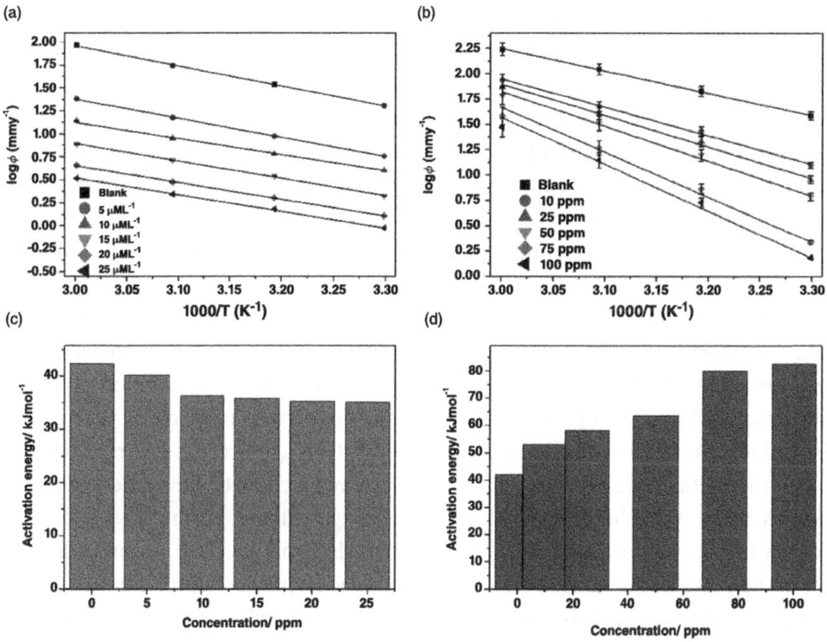

Figure 10.3 Arrhenius plots and the corresponding data obtained. The data for Figure 10.3(a, c) was obtained from T.K. Sarkar et al., 2022. 431 (2022) 133481, represents predominant chemisorption and the Figure 10.3(b, d) was obtained with permission of Elsevier 297 (2020) 111883. © Elsevier.

inhibitor adsorption. It has been reported that the decrease or near constancy of the E_a values indicates chemical interaction between the corrosion inhibitors and the substrate. Here the electron-rich sites of corrosion inhibitors directly donate the π, n electrons to the vacant d orbitals of the metallic surface. The protective film hence generated is very stable and can provide superior protection at elevated temperatures. Figure 10.3a represents the Arrhenius plots of such a molecule (FTTI). The obtained E_a values are represented in Figure 10.3b, and it can be observed that the obtained values are slightly lower than the blank solution. With the increase in the concentration of the corrosion inhibitors, the values of E_a attain a neat constancy. These observations suggest the chemisorption of the corrosion inhibitors upon the metal surface [11, 57]. If the values of E_a rise increase in corrosion inhibitor dosage, then the adsorption of the inhibitor molecules occurs through charged interaction with the metal surface, i.e., physisorption.

Here, the charged part of the inhibitor molecules interacts with the oppositely charged sites upon the metallic substrate, which leads to the formation of the protective layer [53]. This type of observation is highly reported in literature and a similar observation is presented in Figure 10.3(b, d). Here it is reported that the E_a values gradually increase with the inhibitor dosage, with no sign of constancy attainment. Hence such kind of observation reflects the predominant physisorption of corrosion inhibitor molecules.

10.2.4.2 Activation Enthalpy and Entropy

The values activation enthalpy ($\Delta H^{\#}{}_{a}$) and activation enthalpy ($\Delta S^{\#}{}_{a}$) are calculated using linear plots of log(ϕ/T) vs. 1/T (Figure 10.5) using the relations for slope = $[-\Delta H^{\#}{}_{a}/2.303R]$ and intercept = $[\log(R/N_a h)+\Delta S^{\#}{}_{a}/(2.303R)]$. The trends of $\Delta H^{\#}{}_{a}$ and $\Delta S^{\#}{}_{a}$ for different dosages of corrosion inhibitors provide valuable information about the nature of interactions at the substrate solution interface. It has been reported the positive values of $\Delta H^{\#}{}_{a}$ stand for the endothermic nature of the corrosion process. When the values of $\Delta H^{\#}{}_{a}$ increase with the corrosion inhibitor dosage, then the corrosion inhibitor molecules have physiosorbed upon the substrate surface. The opposite trend is usually observed in the case of the chemisorption of corrosion inhibitors. The values of $\Delta S^{\#}{}_{a}$ represent the randomness upon the metal surface in presence of a corrosive electrolyte. Very often values of $\Delta S^{\#}$ increase with the corrosion inhibitor dosage (Figure 10.4b and 10.4d). The increase in inhibitor dosage occurs due to the desorption of the water molecules from the metallic substrate surface during the adsorption of corrosion inhibitors. This kind of observation is usually observed during the physisorption of corrosion

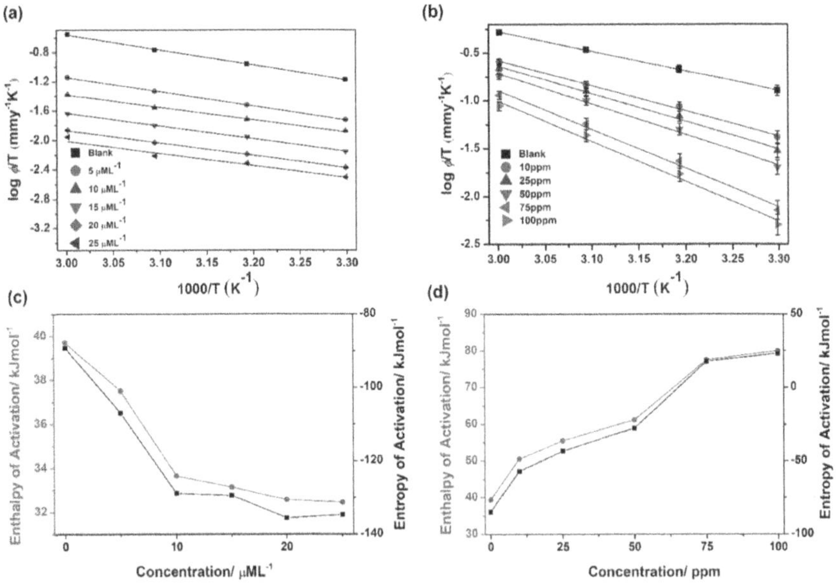

Figure 10.4 Transition state plots and the corresponding data obtained. The data for Figure 10.4(a, c) was obtained from T.K. Sarkar et al., 2022 with permission of Elsevier. 431 (2022) 133481 and represents predominant chemisorption, and the Figure 10.4(b, d) was obtained from V. Saraswat et al. J. Mol. Liq. 297 (2020) 111883. © Elsevier.

inhibitors. The reverse trend is however also reported in several cases. These cases refer to a more ordered activated state of the corrosion reactions in presence of the corrosion inhibitors [12]. Such observations are reported for corrosion inhibitors displaying chemisorption upon the metallic substrate (Figure 10.4a and 10.4c).

10.2.5 Adsorption Isotherms

The corrosion mitigation capacity of organic compounds is usually classified into two categories. If they undergo charge-based interaction with the substrate, then the interaction is classified as physisorption while the interaction through sharing electrons between electron-rich sites of corrosion inhibitors and vacant d orbitals of metals is classified as chemisorption. However, the adsorption of corrosion inhibitor molecules does not occur by pure physisorption or chemisorption methodology. The corrosion inhibitors undergo mixed-type interactions with the metallic substrate. Here both charge-based and electron-sharing mechanisms occur simultaneously. While some of the corrosion inhibitors undergo predominant physisorption with the metal substrate. the corrosion inhibition efficiency usually reduces with temperature in such cases due to the dissolution of the protective layer. When the corrosion inhibitors undergo predominant chemisorption the corrosion inhibition efficiency stays nearly constant at elevated temperatures. This occurrence represents the superior stability of the protective film generated through the chemisorption of the corrosion inhibitors [13, 14, 15, 16]. The nature of corrosion inhibitor adsorption upon the metal surface is justified by fitting the obtained data against various isotherm models. It is popular isotherms used to study the adsorption of corrosion inhibitors are Langmuir [17, Saraswat and Mahendra Yadav 2020], Freundlich [18], Temkin [19, 20], Frumkin [21], and Flory-Huggins [22] isotherm models. It is reported by several renowned authors that, if the value-free energy of adsorption lies around $-20\,\mathrm{kJ\,Mol^{-1}}$ the molecules undergo predominant physisorption. While the values lesser than $-40\,\mathrm{kJ\,Mol^{-1}}$ represent the predominant chemisorption of the inhibitor molecules, the values lying within the intermediate range signify simultaneous physisorption and chemisorption [23, 24].

10.2.5.1 Langmuir Isotherm

The isotherm model is prepared upon the assumption that the film generated over the metallic substrate is monolayered. The inhibitors get adsorbed upon a fixed number of localized equilibrium sites upon the substrate surface. The isotherm is usually represented as:

$$\frac{C_{inh}}{\theta} = \frac{1}{K_{ads}} + C_{inh} \frac{C_{inh}}{\theta} = \frac{1}{K_{ads}} + C_{inh} \tag{10.6}$$

Here C_{inh} represents the concentration of the corrosion inhibitors in the corrosive electrolyte solution, θ is the ratio of metal surface area covered by the inhibitor molecules, K_{ads} is the equilibrium constant for the adsorption process of the inhibitor molecules. A representation of Langmuir adsorption isotherm is presented in Figure 10.5.

10.2.5.2 Temkin Isotherm

The Temkin isotherm (Figure 10.6) is represented as:

$$A\theta = \log(K_{ads}) + \log(C_{inh})A\theta = \log(K_{ads}) + \log(C_{inh}) \quad (10.7)$$

Here C_{inh} represents the concentration of the corrosion inhibitors in the corrosive electrolyte solution, θ is the ratio of metal surface area covered by the inhibitor molecules, K_{ads} is the equilibrium constant for the adsorption process of the inhibitor molecules, a is the interaction parameter, A is a constant; $A = -2a/2.303$.

If a<0 an indication of repulsion between the adsorbed molecules.

If a>0 an indication between the adsorbed molecules.

If a=0 an indication of a lack of interaction between the adsorbed species.

10.2.5.3 Freundlich Isotherm

The adsorption isotherm model explains the nonideal and reversible adsorption of corrosion inhibitors. Here it is considered that the protective film generated over the substrate surface is effectively multilayered. The generation of the

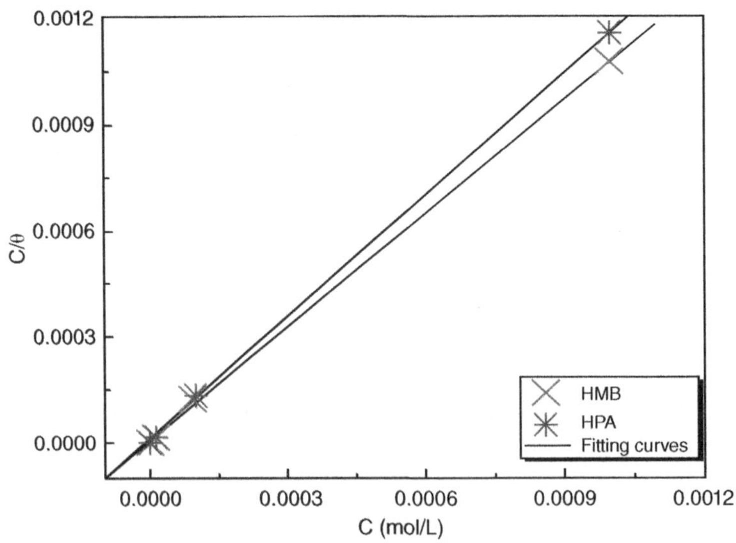

Figure 10.5 Langmuir adsorption isotherm model for quinoline derivatives; mild steel in HCl. Images were obtained from Faydy et al. J. Mol. Liq. 325 (2021) 11524. © Elsevier.

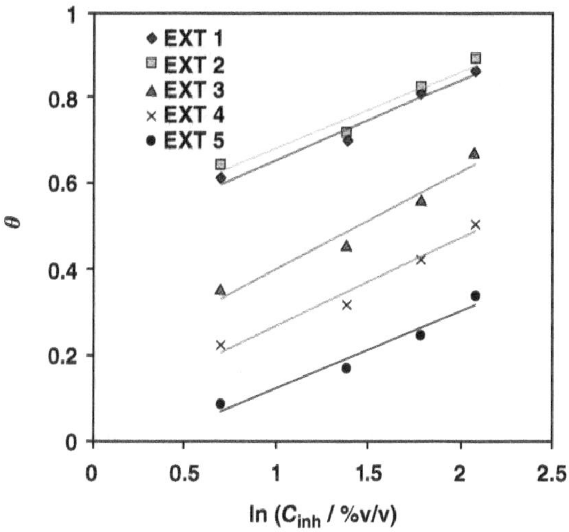

Figure 10.6 Temkin adsorption isotherm obtained from M.A. DeyabJ. Ind. Eng. Chem. 22 (2015) 384–389. © Elsevier.

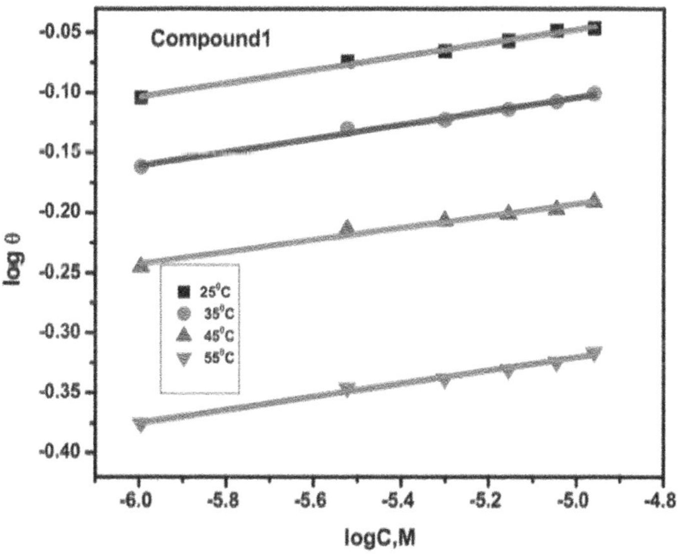

Figure 10.7 Freundlich adsorption isotherm obtained from H.S. Gadow et al., 2019 / With permission of Elsevier.

multilayered film is often associated with heterogeneity at the adsorption sites. The isotherm is represented below in Equation (10.7) and Figure 10.7.

$$\log(\theta) = \log(K_{ads}) + \log(C_{inh})\log(\theta) = \log(K_{ads}) + \log(C_{inh}) \tag{10.8}$$

10.2.5.4 Frumkin Isotherm

Frumkin isotherm is one of the more popular isotherm models used to study the adsorption mechanism of corrosion inhibitors. The isotherm model generally used for corrosion inhibitor studies is represented in Equation (10.9) and an isotherm plotted image is presented in Figure 10.8.

$$\log\left[\frac{\theta}{C_{inh}(1-\theta)}\right]=\log(K_{ads})+A\theta\log\left[\frac{\theta}{C_{inh}(1-\theta)}\right]=\log(K_{ads})+A\theta \quad (10.9)$$

If a is the interaction parameter, A is a constant; $A = -2a/2.303$.
If $a<0$ an indication of repulsion between the adsorbed molecules.
If $a>0$ an indication between the adsorbed molecules.
If $a=0$ an indication of a lack of interaction between the adsorbed species.

10.2.5.5 Flory-Huggins Isotherm

The Flory-Huggins isotherm provides a direct correlation between the simultaneous adsorption of corrosion inhibitor molecules and the desorption of corrosion inhibitor molecules [26]. The isotherm is presented in Equation (10.9) and a model plot of the isotherm id is presented in Figure 10.9. Here plots of $\log(\theta/C_{inh})$ vs. $\log(1-\theta)$ are prepared and the values of slope and intercept are used to obtain the required parameters. According to Javadian et al. [27], the isotherm parameter x increases in presence of the corrosion inhibitors, and the corresponding value obtained determined the number of water molecules desorbed by a corrosion inhibitor molecule.

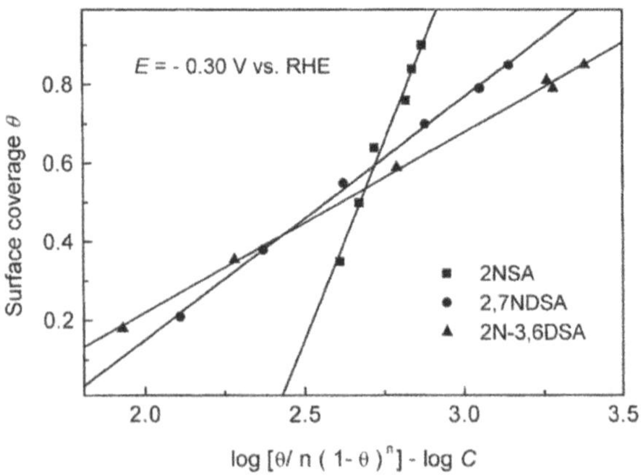

Figure 10.8 Frumkin adsorption isotherm obtained from Vracar et al. Corros. Sci. 44 (2002) 1669–1680 [25] © Elsevier.

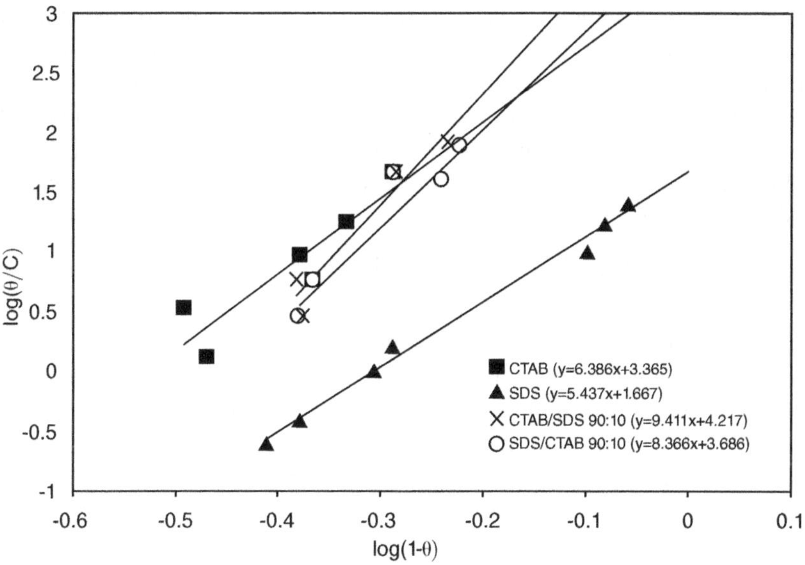

Figure 10.9 Flory-Huggins adsorption isotherm obtained from Javadian et al. Appl. Surf. Sci. 285 (2013) 674–681 © Elsevier.

$$\log\left[\frac{\theta}{C_{inh}}\right] = \log(xK_{ads}) + x\log(1-\theta)\log\left[\frac{\theta}{C_{inh}}\right] = \log(xK_{ads}) + x\log(1-\theta) \quad (10.10)$$

10.2.5.6 Free Energy of Adsorption

The standard free energy (ΔG_{ads}) of adsorption is obtained from the value of the equilibrium constant of the corrosion inhibitor adsorption process (K_{ads}). The values K_{ads} are obtained through linear fitting of the experimental data to a standard isotherm model, as mentioned above. The model which displays the best fitting of the data is considered for the calculation of ΔG_{ads} through Equation (10.11).

$$\Delta G_{ads} = -RT\ln(55.5K_{ads}) \quad (10.11)$$

Where R is the universal gas constant (8.314 J $K^{-1}mol^{-1}$), T is the temperature (K) and 55.5 is the value of the concentration of water in solution in $molL^{-1}$. The values of ΔG_{ads} provide imperative inference about the nature of corrosion inhibitor adsorption. It is observed within standard literature that if the value of ΔG_{ads}

is greater than $-20\,kJmol^{-1}$ then the corrosion inhibitors undergo predominant physisorption. However, if the value of ΔG_{ads} is lesser than $-40\,kJmol^{-1}$ then the corrosion inhibitors undergo predominant chemisorption. The corrosion inhibitors with intermediate values of ΔG_{ads} represent both the physisorption and chemisorption of corrosion inhibitor molecules upon the metallic surface [23, 24].

The values for entropy (ΔS_{ads}) and enthalpy (ΔH_{ads}) of adsorption can be calculated using Equation (10.12).

$$\Delta G_{ads} = \Delta H_{ads} - T\Delta S_{ads} \Delta G_{ads} = \Delta H_{ads} - T\Delta S_{ads} \tag{10.12}$$

Here linear plot for ΔG_{ads} vs. T is plotted first. The value of intercept is ΔH_{ads} and ΔG_{ads} is the value of the slope. The positive value of ΔH_{ads} represents the endothermic nature of the inhibitor adsorption process while the negative values stand for the exothermic process. The corrosion inhibitors that undergo predominant chemisorption display positive ΔH_{ads}. The corrosion inhibitors with negative ΔH_{ads} display predominant physisorption. Hence the inhibitors with positive values of ΔH_{ads} are observed to be more efficient at elevated temperatures [28, 29, 30].

10.3 Electrochemical Methodologies

The electrochemical analyses of corrosion are usually carried out with three-electrode systems. Here the desired metal is taken as the working electrode while Pt is used as the counter electrode and SCE or Ag/AgCl is used as a reference electrode. The working electrode is usually of the dimension of $1\,cm \times 1\,cm$ with exposed surface area and the reference electrode is attached to a Luggin capillary. Initially, the electrode system is exposed to the corrosive electrolyte to attain equilibrium the potential of the equilibrium is recorded as the OCP for the system. The OCP plays a vital role in EIS and polarization studies. The polarization study is usually carried out for a potential window ±0.5 V from the OCP at a scan rate of $0.1\,Vs^{-1}$. The electrochemical impedance studies (EIS) measurement is usually carried out for the frequency range of 100 kHz–10 MHz with an amplitude of 10 mV, peak to peak AC signal at OCP. These analyses provide valuable information about the working electrode-electrolyte interface.

10.3.1 Open Circuit Potential

Open circuit potential (OCP/ E_{OCP}) is the potential of the working electrode when it is in equilibrium with a corrosive electrolyte [31]. The values of OCP are usually obtained from pristine and corrosion inhibitors containing electrolytes before any other electrochemical experimentation. Generally, plots of OCP vs. time (as represented in Figure 10.10) are obtained from the electrochemical cell setup and the obtained data obtained in presence of corrosion inhibitors are compared against a pristine electrolyte system. It reported that the more negative values of OCP in corrosion inhibitors containing electrolytes indicate protection of the working electrode surface. However, it has also been reported that any deviation of OCP within the range of ±85 mV reflects the mixed-type nature of the corrosion inhibitors. The mixed-type corrosion inhibitors adsorb in both cathodic and anodic sites of corrosion. Hence the apparent rate of corrosion gets reduced [32, 33, 34].

10.3.2 Electrochemical Impedance Spectroscopy

The resistance of alternating current (AC) flow in a complex electrochemical system is known as impedance. This phenomenon is used to study the frequency response from an electrochemical cell including energy storage and degeneracy. The process is non-destructive and provides very accurate mechanistic data for corrosion systems. The technique applies to low-conductivity electrolytes, and it provides accurate data on solution resistance. The data obtained from the electrochemical impedance studies (EIS) is usually fitted against a suitable circuit using a nonlinear least square methodology. The data obtained through EIS studies are displayed in form of three different plots. The first type is called the Nyquist where the resistance is plotted in terms of real and imaginary axes (Z_{real} vs. $-Z_{img}$). The Bode impedance plots are prepared by plotting mod. Resistance against the log of frequency (log*f* vs. IZI) and the Bode phase plots are prepared by plotting -phase angle against the log of frequency (log*f* vs. −phase). These plots are commonly

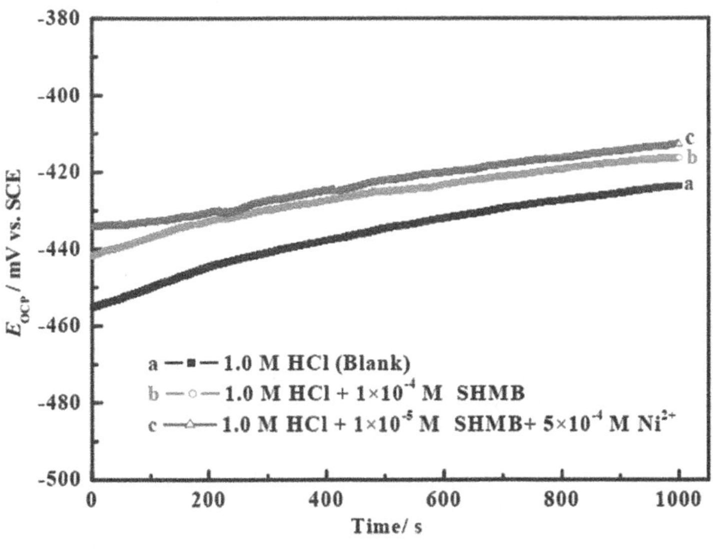

Figure 10.10 OCP Vs Time plots obtained from El-Lateef et al. Appl. Surf. Sci. 501 (2020) 144237. © Elsevier.

used to interpret the nature of electrochemical reactions happening at the working electrode/electrolyte interface. The goodness of the fitting of an EIS spectrum is usually represented as χ^2 and if the values lie in the order of 10^{-4} to 10^{-3} then the fitting is considered to be acceptable.

The equivalent circuit of a corroding system usually consists of a parallel combination of constant phase element (CPE) and resistance (R_t) representing the metal solution interface and another uncompensated resistance is usually applied in series to represent the corrosive electrolyte (Rs), represented in Figure 10.11. Generally, R_t represents the resistance at the working electrode amid the highest and lowest frequencies. For a corrosion, cell R_t represents the feasibility of charge transfer during the corrosion of the metal substrate. It is usually observed that the values of R_t are low for pristine electrolyte solution and higher for corrosion inhibitor-containing solution. This phenomenon suggests that the transfer of charge is restricted in presence of corrosion inhibitors within the electrolyte solution. The adsorption of corrosion upon the substrate surface thickens the double layer at the metal/solution interface and hence the charge transfer becomes hindered. Hence both the cathodic and anodic processes of corrosion get reduced. The extent of change in the values of R_t in presence of the corrosion inhibitors with respect to the pristine electrolyte is utilized to calculate the corrosion inhibition efficiency using Equation (10.13). The representative of Nyquist plots is presented in Figure 10.12 as depressed semi-circular plots whose center lies beneath the x-axis. The depressing nature of Nyquist plots represents the inhomogeneities at the metal surface. The double layer capacitance (C_{dl}) and is calculated using Equation (10.14). The parameters obtained through EIS studies are presented here as charge transfer resistance in presence of inhibitors (R_t), charge transfer resistance in absence of inhibitors ($R_{t(0)}$), constant phase element (CPE) parameters; constant Y_0 and exponent n ($0<n<1$). The values of corrosion inhibition efficiency ($\%E_i$) and double-layer capacitance (C_{dl}) were calculated using the expressions.

$$\%E_i = \frac{R_t - R_{t(0)}}{R_t} \times 100 \tag{10.13}$$

$$C_{dl} = \left(Y_0 R_t^{1-n}\right)^{1/n} C_{dl} = \left(Y_0 R_t^{1-n}\right)^{1/n} \tag{10.14}$$

Though the Nyquist plots are usually used to represent the EIS data, the plots have a drawback. These plots do not represent the data associated with scanning frequencies. To present the variation with frequencies the Bode plots are used (represented in Figure 10.13). It is usually observed that the slope of the body impedance (log*f* vs. IZI) is linear, and the slope is close to 0.9. The values of the impedance curve are observed to increase in all the frequencies with corrosion inhibitor. The observation is owed to, the reduction of corrosion rate due to the

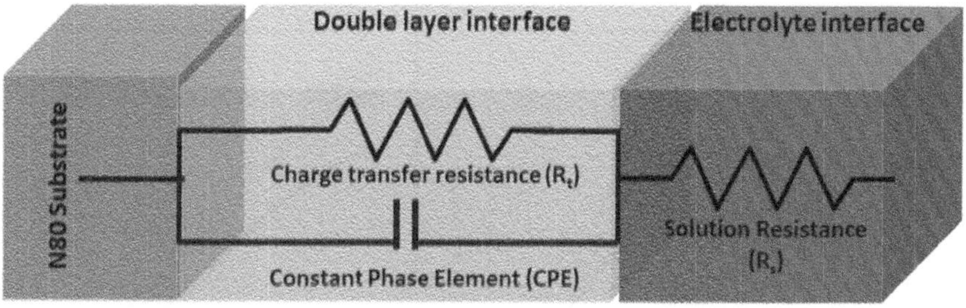

Figure 10.11 Representation of Randel's circuit, presenting the metal solution interface. Here N80 steel is considered as the substrate. The image is obtained from T.K. Sarkar et al. Chem. Eng. J. 431 (2022) 133481. © Elsevier.

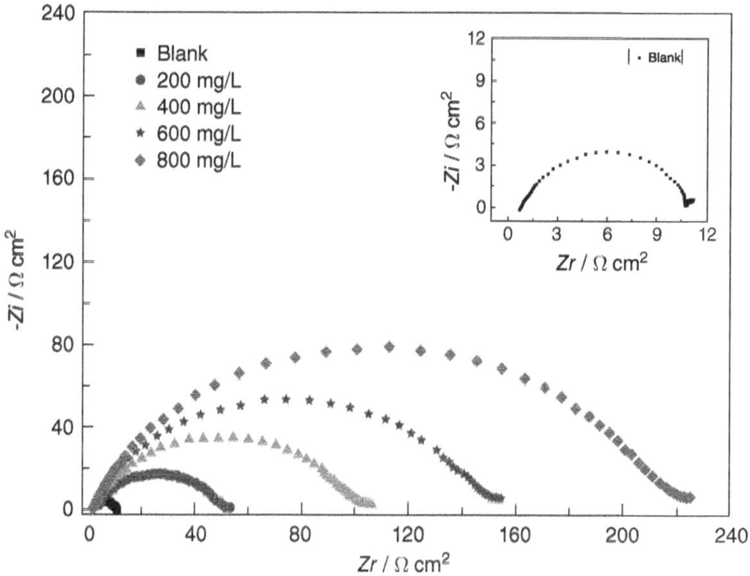

Figure 10.12 Representation of depressed semi-circular Nyquist plots for mild steel HCl in the absence and presence of inhibitor inhibitors. W. Zhanget al. J. Mol. Liq. 321 (2021) 114923. © Elsevier.

generation of a protective thin film upon the metal substrate. The Bode phase angle plots usually represent the relaxation processes at the metal solution interface. The Bode phase plots presented in Figure 10.13 signifies the presence of one relaxation process at the metal solution interface; i.e., the charge transfer process [35]. The increase in the negative value of phase angle with inhibitor dosage indicates the attainment of ideal capacitor capabilities. It is theorized that if a system attains a slope value of −1 for Bode impedance and −90 for Bode phase then behaves as an ideal capacitor. However, such observation is not obtained for real-life samples and a corrosion system never attains ideal capacitor properties.

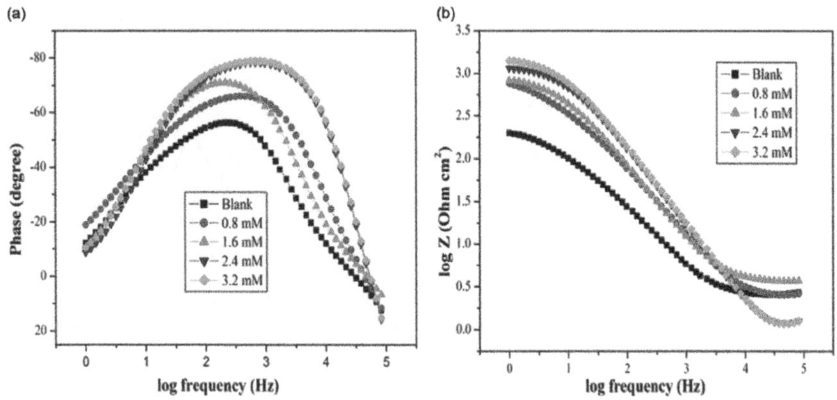

Figure 10.13 The representation of Bode impedance and phase plots for mild steel in HCl in the presence and absence of corrosion inhibitors. Chaitra et al. J. Mol. Liq. 211 (2015) 1026–1038. © Elsevier.

10.3.3 Potentiodynamic Polarization

The potentiodynamic polarization study provides detailed information about the kinetics of corrosion at the working electrode. The science of electrode kinetics has allowed us a clear understanding of corrosion and practical measurement of corrosion rates. The well-known interpretation of the corrosion process by superimposing the partial cathodic and anodic curves was developed by Wagner and Traud [36]. The kinetics of corrosion are developed by plotting the potential of the working electrode against the observed current density. These curves are known as polarization curves and the representation of the polarization curve is represented in Figure 10.14. Here the potential is varied externally with respect to OCP and if it is far enough in either cathodic or anodic direction, then the influence of the other reaction becomes negligible. The curves of potentiodynamic polarization curves are evaluated using Butler-Volmer equation:

$$i = i^0 e^{((2.303E - E_{corr})/\beta_a)} - e^{((2.303E - E_{corr})/\beta_c)} \tag{10.15}$$

Here i represents the measured current density of the working electrode, E represents the potential of the working electrode, i_{corr} is the corrosion current density, E_{corr} is the corrosion potential, β_a is anodic and β_c is the cathodic Tafel slopes.

The potentiodynamic polarization curves are analyzed through Tafel extrapolation (Figure 10.14). Here a tangent is drawn upon both the cathodic and anodic curves respectively. Two perpendicular lines are then drawn for the point where both tangents meet. Perpendicular lines are drawn from the intersection point of

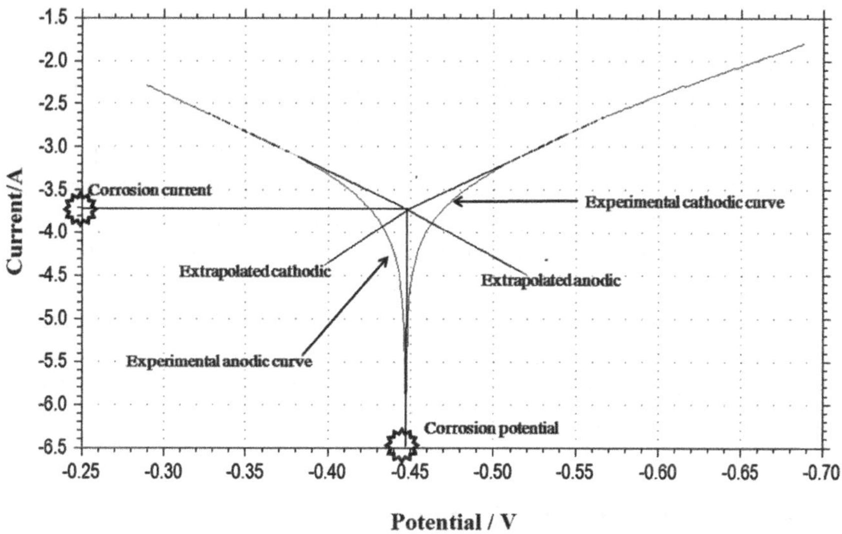

Figure 10.14 Representation of potentiodynamic polarization curve and the corrosion parameters associated with it.

the tangents towards potential and corrosion current density. The point in which the perpendicular line intersects with the potential axis represents the corrosion potential while the other point represents the corrosion current density. The corrosion of metals is usually explained by the cathodic and anodic polarization curves. However, the rate of corrosion reactions may be controlled by either cathodic or anodic reactions. In most cases the cathodic reduction controls the overall corrosion rate.

The potentiodynamic polarization curves provide an efficient insight into the corrosion process of a working electrode. The increase in corrosion current density (i_{corr}) indicates an enhancement in corrosion while the decrease represents the protection of the electrode surface. This phenomenon is often utilized to calculate the inhibition efficiencies (%Ep) were elucidated through the following equation, where $i^0{}_{corr}$ is the corrosion current density in absence of corrosion inhibitors and i_{corr} is the corrosion density in presence of the corrosion inhibitors.

$$\%E_p = \frac{i^0_{corr} - i_{corr}}{i^0_{corr}} \tag{10.16}$$

The corrosion values of corrosion potential (E_{corr}) provide information about the nature of the adsorption of corrosion inhibitors. It is usually observed that the values of E_{corr} commensurate with the values of OCP. If the values of E_{corr} lie beyond

+85 mV from the blank then the corrosion inhibitors efficiently adsorb upon the cathodic sites of corrosion, while the values beyond −85 mV represent the anodic type of protection. If the values lie within the range of ±85 mV then the inhibitors effectively adsorb upon both cathodic and anodic sites of corrosion. These types of corrosion inhibitors are often termed mixed-type corrosion inhibitors [37].

The values of β_a and β_c represent the cathodic and anodic reactions of corrosion. If these values remain unchanged in presence of the corrosion inhibitors, then it is confirmed that the corrosion inhibitors do not change the nature of the corrosion reactions. However, the higher values of β_a in presence of corrosion inhibitors indicate the retardation of cathodic reaction, and the retardation of anodic reactions is confirmed by the higher values of β_c [38, 39, 40]. The representation of potentiodynamic polarization curves is presented in Figure 10.13, while Figure 10.15 represents the curves in presence of corrosion inhibitors.

10.4 Surface Characterizations

10.4.1 Scanning Electron Microscopy

Scanning electron microscopy (SEM) is an analytical tool that provides highly magnified images of a substrate surface. The resolution of these images usually approaches to a few nm and the magnifications lie within the range of 10x to 1000000x. these high-resolution images provide a detailed visual insight into the

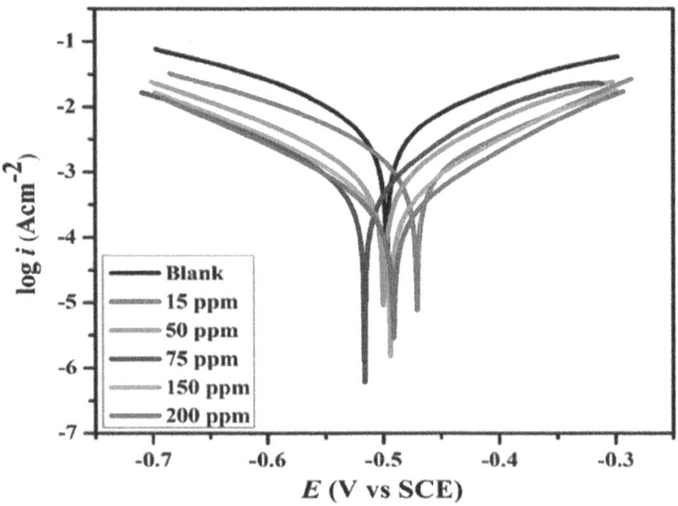

Figure 10.15 Potentiodynamic polarization curves for mild steel in 15% HCl solution with N & S doped quantum dots as corrosion inhibitors. Saraswat, V., & Yadav, M. 2021 / With permission of Elsevier.

substrate surface that incriminate a detailed perception of both corrosion and its mitigation. Prior to SEM analysis, the metal samples of the required dimensions (1 cm × 1 cm × 0.1 cm) are finely polished and exposed to the corrosive electrolyte for a period of at least 24 hours. Separate samples are exposed to pristine electrolyte solution and inhibitor-containing solutions. It is usually observed that the polished surface is free from cracks and pits. The abrasion lines generated during the polishing process are observed for the polished surface (Figure 10.16a). The samples exposed to pristine electrolytes display significant signs of corrosion. The generation of a significant number of cracks and pits is usually confirmed through these images, which indicates the dissolution of metallic substrate (Figure 10.16b). The metal samples exposed to corrosion inhibitor-containing electrolytes however display a different picture in comparison to the samples exposed to pristine electrolytes. The inhibitor-containing samples display a significant reduction in corrosion-related deformities upon the substrate surface. This observation confirms the protective capabilities of corrosion inhibitors[11].

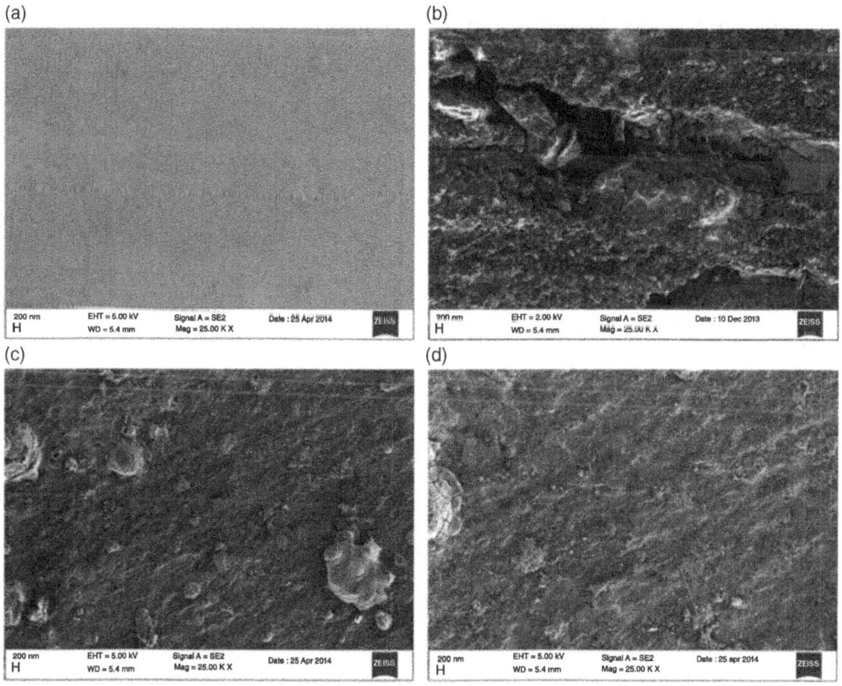

Figure 10.16 SEM images of N80 steel obtained from Yadav et al. (2015) / from ELSEVIER. Figure 10.16a represents polished N80 steel surface, Figure 10.16b represents N80 steel surface exposed to pristine 15% HCl, and Figures 10.16c and 10.16d represents the N80 steel surface exposed to corrosion inhibitor containing 15% HCl. © Elsevier.

10.4.2 Atomic Force Microscopy

The surface topography of metal samples is usually studied through atomic force microscopy (AFM). The instrumentation of AFM includes a sharp tip attached to a flexible cantilever and the scanning of the sample is carried out in a raster fashion. When the tip scans over the substrate surface, the deflections within the cantilever are recorded to obtain the required images. The samples used for the study are prepared using the same methodology as the SEM analysis. The AFM samples are exposed to pristine and corrosion inhibitor-containing solutions and after at least 24 hours, the samples are dried in a vacuum. Then the dried samples are subjected to the AFM analysis. Several parameters such as average roughness (Ra), and root mean square roughness (Rq) are used to analyze the surface topography [54, 55].

It is reported that the roughness of the polished metallic samples is the lowest. The patterns observed over the polished sample are attributed to the remaining roughness after the polishing process (Figure 10.17a). The highest roughness is usually observed in the samples exposed to the electrolyte solution. This observation relates to the corrosion of the metallic samples within the aggressive solution. Since the samples display high corrosion, it is evident that they display high values of surface roughness (Figure 10.17b). The roughness values of the metal samples

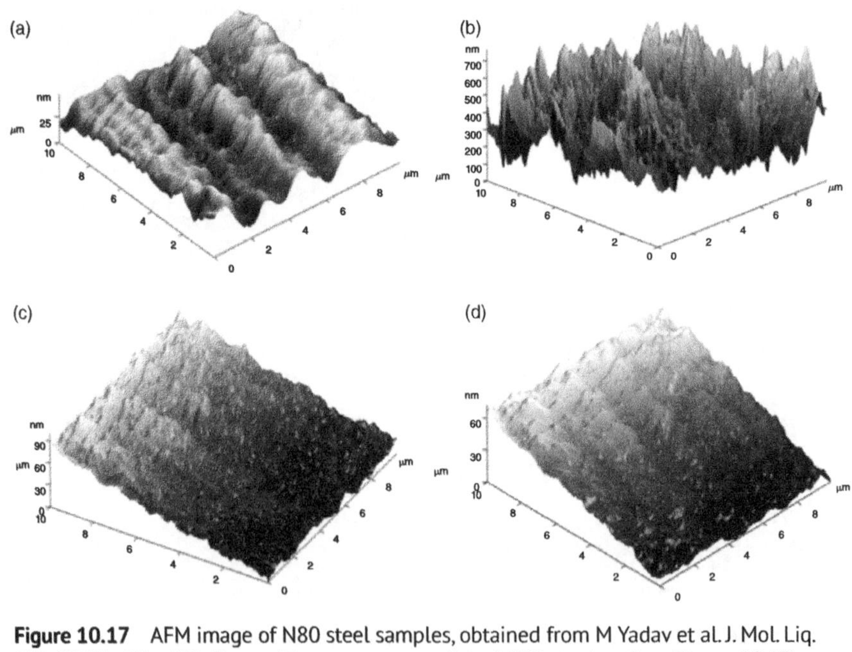

Figure 10.17 AFM image of N80 steel samples, obtained from M Yadav et al. J. Mol. Liq. 212 (2015) 451–460. Figure 17 a. represents polished N80 steel surface, Figure 10.17b. represents N80 steel surface exposed to pristine 15% HCl, Figures 10.17c and 10.7d represents the N80 steel surface exposed to corrosion inhibitor containing 15% HCl. © Elsevier.

exposed to the corrosion inhibitor-containing electrolytes are observed to lie between the previously mentioned two samples. This observation supports the decrease in corrosion of metal samples due to the corrosion inhibition capabilities of the studied corrosion inhibitors.

10.4.3 Chemical Composition Analysis of Substrate Surface

The SEM images provide a visual insight into the corrosion process of the metallic substrate but provide no idea about the chemical composition of the substrate surface. The chemical composition of the substrate surface provides valuable information about the corrosion products and the nature of the protective films generated during the corrosion inhibition process. The information about chemical composition is usually obtained through energy-dispersive X-ray spectroscopy (EDX) and X-ray photoelectron spectroscopy (XPS). The data of EDX are displayed in the form of plots of energy vs. intensity (counts) while the XPS is presented as binding energy vs. intensity plots. The EDX analysis provides qualitative analysis of the elements present upon the metallic substrate, while the XPS analysis provides a detailed insight into the chemical composition within the substrate surface.

10.4.3.1 EDX Analysis

A representative EDX image is presented in Figure 10.18. The image represents three samples of N80 steel exposed to 15% HCl in the presence and absence of corrosion inhibitors. Figure 10.18b represents the sample exposed to pristine electro lytes displaying the presence of Fe and Cl. These elements confirm the presence of the corrosion products generated. While images 18 (c,d) represent the samples exposed to corrosion inhibitor-containing electrolytes. This image confirms the presence of C, N, O, and Fe, hence the image displays the presence of corrosion products along with the adsorbed corrosion inhibitors. This observation is supportive of the phenomenon of corrosion inhibition of metallic substrate in presence of corrosion inhibitors [11].

10.4.3.2 XPS Analysis

The quantitative analysis of the chemical composition of corroded samples is carried out through XPS analysis. The analysis provides a detailed insight into the nature of the corrosion-inhibited metallic samples. The analysis is designed upon the theory that X-rays have enough energy to displace electrons of an element's core orbitals. By studying these displaced electrons, we can obtain the values of the binding energy of each element and recognize the chemical nature of the element. The X-rays used for this purpose can penetrate up to 4nm of the substrate surface and that is enough to study the nature of a thin film generated through the corrosion process. Detailed analysis of the high-resolution XPS

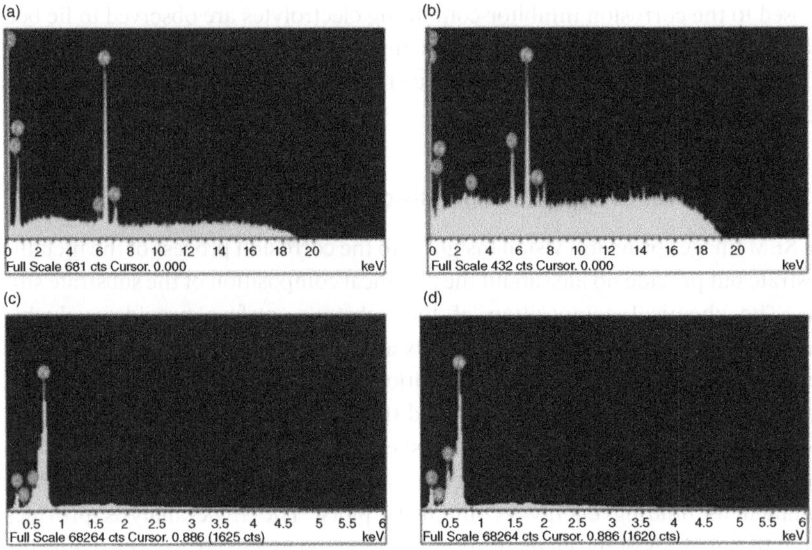

Figure 10.18 EDX image of N80 steel samples exposed to 15 % HCl, obtained from Yadav et alJ. Mol. Liq. 212 (2015) 451–460. Figure 16 a. freshly polished metal sample, Figure 16 b. polished metal sample exposed to pristine 15% HCl, Figure 16 c, d polished sample exposed to corrosion inhibitors containing 15% HCl. © Elsever.

spectra 0 through proper deconvolution provides a proper understanding of the chemical nature of the protective film generated. The presence of elements such as C, N, and S signify the presence of organic components, and the presence of transition metals, O, Cl, etc. indicates the presence of inorganic counterparts. The deconvoluted XPS spectra of N80 steel in corrosion inhibitor containing 15% HCl are presented and explained in detail for the convenience of the reader.

The XPS spectra of C1s is usually observed at 284.5 eV and it is usually deconvoluted into three peaks. The peak at ~284.5 eV represents C-C, C=C, and CH bonds within the sample, the peak at ~285.88 eV represent C– (S, N, O) functionalities and the peak at ~287.49 eV confirms the presence of C=O [41].

The XPS peak of N1s is usually observed a ~399.0 eV, which consists of two more peaks at ~398.46 eV and ~399.94 eV for C=N and C–/=N^+ [42]. The S2p peak is observed to be doublet, where the peak at ~161.0 eV stands for $2p_{3/2}$ and ~162.0 eV stands for $2p_{1/2}$ [43, 44, 45]. The peak of $2p_{3/2}$ usually consists of two more peaks at ~161.65 eV and 161.27 eV. Similar is the situation for $2p_{1/2}$ where the peaks can be deconvoluted into ~162.19 eV and ~162.70 eV. The presence of these organic functionalities within the corrosion product confirms the formation of a protective thin film with organic compounds [46].

The inorganic part of the inhibitor-protected film is confirmed by the presence of transition metals and oxygen. In the case of the example being discussed, the Fe 2p

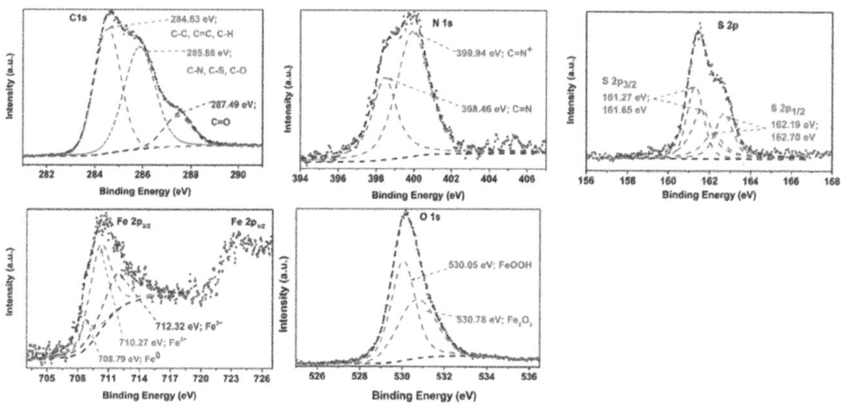

Figure 10.19 XPS spectra of C1s, N1s, S2p, $Fe2p_{3/2}$ and O1s corrosion product of N80 steel in 15% HCL. The corrosion inhibitors used for the study are indoline compounds. The image is obtained from T.K. Sarkar et al. Chem. Eng. J. 431 (2022) 133481. © Elsevier.

spectra is obtained and it is observed as doublet where the peak of $2p_{3/2}$ appears at ~710.51 eV. This peak can further be deconvoluted into three other peaks at ~708.79 eV for Fe^0, 710.27 eV for $Fe^{2+,}$ and 713.32 eV for Fe^{3+}. The O1s spectra are observed at 530.14 eV and it is usually deconvoluted into two more peaks, representing FeOOH at ~530.05 eV and Fe (+2 &+3) oxides at ~530.78 eV. The presence of these peaks confirms the inorganic presence within the protective film. A representative presentation of deconvoluted high resolution XPS spectra is presented in Figure 10.19.

10.5 Computational Methodologies

10.5.1 Density Functional Theory (DFT)

The computational methodologies are important tools for corrosion inhibitor analysis. The computational analysis is usually applied to co-relate the experimental data with the details of the molecular structure. Generally, the molecular structure of the corrosion inhibitor is optimized using density functional theory (DFT) and Gaussian 09 is a popular software package in this regard. BL3YP-631G module is very popular for this purpose. Very often the geometry optimization is carried out for the inhibitor molecules in gaseous and aqueous conditions. Through geometry optimization, various theoretical parameters such as energies of the highest occupied molecular orbital (E_H) and lowest unoccupied molecular orbital (E_L) along with dipole moment (μ). Through the values of these E_H and E_L various other parameters such as the electronegativity (χ), global hardness (γ), softness (σ), and a fraction of electrons transferred from the inhibitor molecule to the metal surface (ΔN) are obtained. The following are the equations used to obtain such parameters.

$$\Delta E = E_L - E_H \tag{10.17}$$

$$\chi = \frac{E_L + E_H}{2} \tag{10.18}$$

$$\gamma = \frac{E_L - E_H}{2} \tag{10.19}$$

$$\sigma = \frac{1}{3} \tag{10.20}$$

$$\Delta N = \frac{\chi_{Fe} - \chi_{inh}}{2\gamma_{inh}};\ldots\chi_{Fe} = 7.0\,\text{Ev} \tag{10.21}$$

The computational analysis of corrosion inhibitor molecules begins with geometry optimization and based upon that the previously mentioned parameters are obtained. The geometry optimization is carried out in the gaseous phase and the aqueous solvent phase for an appropriate comparison. The aqueous solvent is selected for this purpose because most of the corrosive electrolytes are aqueous. A representation of DFT analyzed organic corrosion inhibitors is presented in Figure 10.20 and Figure 10.21.

The molecular reactivity of an organic corrosion inhibitor molecule is studied through the frontier molecular orbitals (FMO). The energies of FMOs draw a clear picture of the interaction capabilities between the corrosion inhibitors and the metal substrate. The highest occupied molecular orbital (HOMO) is often associated with the electron donation capabilities of the inhibitor molecules and the lowest unoccupied molecular orbital (LUMO) is associated with the electron acceptation capabilities of the corrosion inhibitor molecules [47]. The difference between the HOMO and LUMO (ΔE) represents the overall interaction capabilities of the inhibitor molecules. It has been reported that the lower values of ΔE signify the greater interaction capabilities of the corrosion inhibitor molecules. The values of ΔE increase for the organic molecules under aqueous solvation conditions [21].

Values of the dipole moment (μ) represent the polarity of the corrosion inhibitor molecules. It has been reported that the corrosion inhibitor molecules with higher values of dipole moment represent higher interaction with the metal surface [48]. The fraction of electron transferred ΔN is a useful tool to obtain valuable information about the adsorption of corrosion inhibitors onto the metal substrate [49]. If the value of ΔN is less than 3.6 then the corrosion inhibitor molecules have a superb electron donation capability and thereby they behave ass prominent soft base properties [50]. The other theoretical parameters of softness (σ) and global hardness (γ) predict the corrosion inhibitor adsorption through the soft hard acid base principle (SHAB). According to SHAB soft acids interact better with a soft base. The metal surface is considered soft due to its electron acceptation capabilities. The corrosion inhibitor molecules are considered a soft base due to their

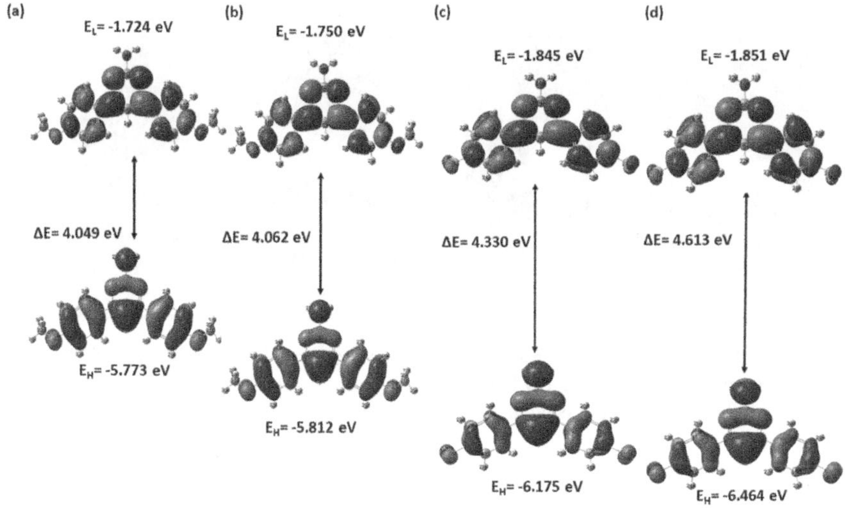

Figure 10.20 Frontier orbitals (HOMO & LUMO) of pyrimidine molecules obtained through DFT analysis using Gaussian 09, BL3YP-631G (d, p). Molecules (MPA and CPA) were studied under aqueous (a, c) and gaseous (b, d) conditions. Images obtained from T.K. Sarkar et al. Mater. Today Commun. 26 (2021) 101862, © Elsevier.

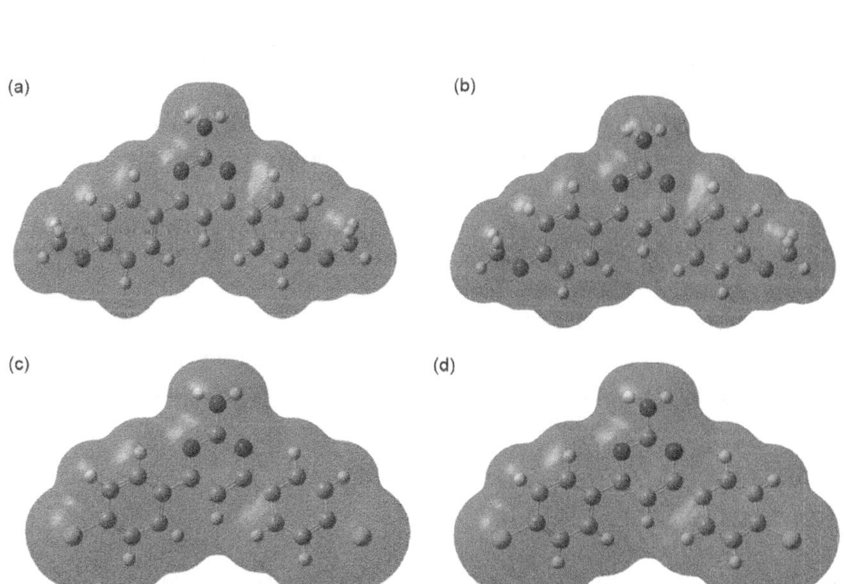

Figure 10.21 Cumulative charge distribution of pyrimidine molecules obtained through DFT analysis using Gaussian 09, BL3YP-631G (d, p). Molecules (MPA and CPA) were studied under aqueous (a, c) and gaseous (b, d) conditions. Images obtained from T.K. Sarkar et al. Mater. Today Commun. 26 (2021) 101862, © Elsevier.

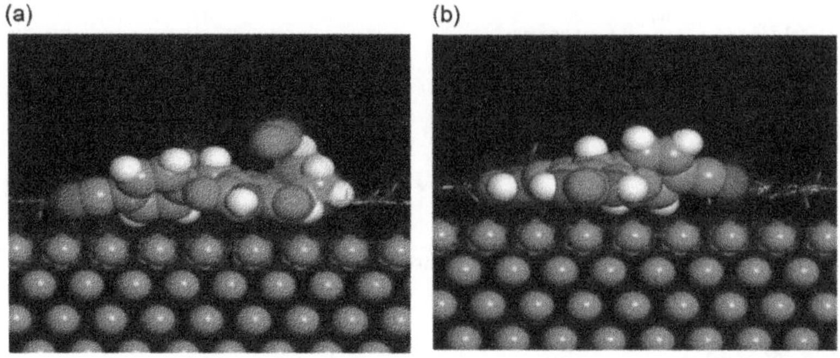

Figure 10.22 MDS simulation for adsorption of corrosion inhibitors, (a) CTDP and (b) OPDB. Images obtained from V Saraswat et al. Colloids Surfaces A Physicochem. Eng. Asp. 599 (2020) 124881. © Elsevier.

electron donation capabilities. Hence the interaction between the meal and the inhibitors is predicted through low values of γ and high values of σ [51].

10.5.2 Molecular Dynamics Simulation Studies

Molecular dynamics simulation study (MDS) is a very important tool to simulate the adsorption phenomenon of corrosion inhibitors. The software used for this purpose is Material Studio Software 7.0 from BIOVIA-Accelrys, USA. This study is carried out on the most stable facet of an alloy. It is considered that a stable facet of steel (1 1 0) is often termed Fe (1 1 0). The most stable facet is different for every transition metal-based alloy. Generally, the facet of the alloy, the corrosion inhibitor molecule, and 50–200 water molecules are used for the simulation purpose. Figure 10.22 presents one similar simulated CTDP and OPDB / Fe (110) /50 H2O system, reported by V Saraswat et al. [52]. It is evident from the image that the corrosion inhibitor CTDP (Figure 10.22a) adsorbs upon Fe (1 1 0) facet through interaction between pyrimidine, imidazole, and pyridine rings. However, the corrosion inhibitor OPDB adsorbs upon Fe (1 1 0) adsorb through the interaction of pyrimidine and pyrone ring.

It is the general trend that corrosion inhibitors with a lower value of adsorption provide better corrosion inhibition capabilities.

References

1 Ech-chihbi, E. et al. (2020). Computational, MD simulation, SEM/EDX, and experimental studies for understanding the adsorption of benzimidazole derivatives as corrosion inhibitors in 1.0 M HCl solution. *Journal of Alloys and Compounds* 844: 155842. doi: 10.1016/j.jallcom.2020.155842.

2 Hou, B.S. et al. (2021). In-depth insight into the inhibition mechanism of pyrimidine derivatives on the corrosion of carbon steel in a CO2-containing

environment based on experiments and theoretical calculations. *Corrosion Science* 181 (November 2020): doi: 10.1016/j.corsci.2021.109236.

3 Wint, N. et al. (2021). The ability of Mg2Ge crystals to behave as "smart release" inhibitors of the aqueous corrosion of Zn-Al-Mg alloys. *Corrosion Science* 179 (February 2020): doi: 10.1016/j.corsci.2020.109091.

4 Yadav, M., Sarkar, T.K., and Purkait, T. (2015c). Amino acid compounds as eco-friendly corrosion inhibitor for N80 steel in HCl solution: electrochemical and theoretical approaches. *Journal of Molecular Liquids* 212: 731–738. doi: 10.1016/j.molliq.2015.10.021.

5 Yadav, M., Sarkar, T.K., and Obot, I.B. (2016). Carbohydrate compounds as green corrosion inhibitors: electrochemical, XPS, DFT and molecular dynamics simulation studies. *RSC Advances* 6 (111): 110053–110069. doi: 10.1039/c6ra24026g.

6 Sarkar, T.K. et al. (2021). Mitigation of corrosion in petroleum oil well/tubing steel using pyrimidines as efficient corrosion inhibitor: an experimental and theoretical investigation. *Materials Today Communications* 26: 101862. doi: 10.1016/j.mtcomm.2020.101862.

7 Sarkar, T.K., Yadav, M., and Obot, I.B. (2022). Mechanistic evaluation of adsorption and corrosion inhibition capabilities of novel indoline compounds for oil well/tubing steel in 15% HCl. *Chemical Engineering Journal* 431 (P4): 133481. doi: 10.1016/j.cej.2021.133481.

8 El Faydy, M. et al. (2021). Synthesis and anti-corrosion characteristics of new 8-quinolinol analogs with amide-substituted on C35E steel in acidic medium: experimental and computational ways. *Journal of Molecular Liquids* 325: 115224. doi: 10.1016/j.molliq.2020.115224.

9 Zhang, W. et al. (2021). Stevioside–Zn2+ system as an eco-friendly corrosion inhibitor for C1020 carbon steel in hydrochloric acid solution. *Colloids and Surfaces A: Physicochemical and Engineering Aspects* 612 (September 2020): 126010. doi: 10.1016/j.colsurfa.2020.126010.

10 Saraswat, V., Kumari, R., and Yadav, M. (2022). Novel carbon dots as efficient green corrosion inhibitor for mild steel in HCl solution: electrochemical, gravimetric and XPS studies. *Journal of Physics and Chemistry of Solids* 160 (July 2021): 110341. doi: 10.1016/j.jpcs.2021.110341.

11 Yadav, M., Gope, L., and Sarkar, T.K. (2015a). Synthesized amino acid compounds as eco-friendly corrosion inhibitors for mild steel in hydrochloric acid solution: electrochemical and quantum studies. *Research on Chemical Intermediates* 42 (3): 2641–2660. doi: 10.1007/s11164-015-2172-5.

12 Sukul, D. et al. (2018). Newly synthesized quercetin derivatives as corrosion inhibitors for mild steel in 1 M HCl: combined experimental and theoretical investigation. *Physical Chemistry Chemical Physics* 20 (9): 6562–6574. doi: 10.1039/c7cp06848d.

13 Amin, M.A. et al. (2011). Monitoring corrosion and corrosion control of iron in HCl by non-ionic surfactants of the TRITON-X series –Part II. Temperature

effect, activation energies, and thermodynamics of adsorption. *Corrosion Science* 53 (2): 540–548. doi: 10.1016/j.corsci.2010.09.019.

14 Bahrami, M.J., Hosseini, S.M.A., and Pilvar, P. (2010). Experimental and theoretical investigation of organic compounds as inhibitors for mild steel corrosion in sulfuric acid medium. *Corrosion Science* 52: 2793–2803. doi: 10.1016/j.corsci.2010.04.024.

15 Heydari, H. et al. (2018). Comparison of two Schiff bases containing O-methyl and nitro substitutes for corrosion inhibiting of mild steel in 1 M HCl solution. *Journal of Molecular Liquids* 254: 177–187. doi: 10.1016/j.molliq.2018.01.112.

16 Olasunkanmi, L.O. et al. (2015). Some quinoxalin-6-yl derivatives as corrosion inhibitors for mild steel in hydrochloric acid: experimental and theoretical studies. *The Journal of Physical Chemistry C* 119 (28): 16004–16019. doi: 10.1021/acs.jpcc.5b03285.

17 Cheng, S. et al. (2007). Carboxymenthylchitosan as an eco-friendly inhibitor for mild steel in 1M HCl. *Materials Letters* 61 (14–15): 3276–3280. doi: 10.1016/j.matlet.2006.11.102.

18 Gadow, H.S., Farghaly, T.A., and Eldesoky, A.M. (2019). Experimental and theoretical investigations for some spiropyrazoles derivatives as corrosion inhibitors for copper in 2 M HNO 3 solutions. *Journal of Molecular Liquids* 294: 111614. doi: 10.1016/j.molliq.2019.111614.

19 Deyab, M.A. (2015). Egyptian licorice extract as a green corrosion inhibitor for copper in hydrochloric acid solution. *Journal of Industrial and Engineering Chemistry* 22: 384–389. doi: 10.1016/j.jiec.2014.07.036.

20 Quraishi, M.A. and Jamal, D. (2001). Corrosion inhibition by fatty acid oxadiazoles for oil well steel (N-80) and mild steel. *Materials Chemistry and Physics* 71 (2): 202–205. doi: 10.1016/S0254-0584(00)00378-3.

21 Mi, H., Xiao, G., and Chen, X. (2015). Theoretical evaluation of corrosion inhibition performance of three antipyrine compounds. *Computational and Theoretical Chemistry* 1072: 7–14. doi: 10.1016/j.comptc.2015.08.023.

22 Youse, A., Javadian, S., and Neshati, J. (2014). A new approach to study the synergistic inhibition effect of cationic and anionic surfactants on the corrosion of mild steel in HCl solution. *Industrial & Engineering Chemistry Research* 53 (13): 5475–5489. doi: 10.1021/ie402547m.

23 Abdallah, Y.M., Shalabi, K., and Bayoumy, N.M. (2018). Eco-friendly synthesis, biological activity, and evaluation of some new pyridopyrimidinone derivatives as corrosion inhibitors for API 5L X52 carbon steel in 5% sulfamic acid medium. *Journal of Molecular Structure* 1171: 658–671. doi: 10.1016/j.molstruc.2018.06.045.

24 Xia, G. et al. (2015). Synergic effect of methyl acrylate and N-cetylpyridinium bromide in N-cetyl-3- (2-methoxycarbonylvinyl) pyridinium bromide molecule for X70 steel protection. *Corrosion Science* 94: 224–236. doi: 10.1016/j.corsci.2015.02.005.

25 Vračar, L.M. and Draži, D.M. (2002). Adsorption and corrosion inhibitive properties of some organic molecules on an iron electrode in sulfuric acid. *Corrosion Science* 44 (8): 1669–1680. doi: 10.1016/S0010-938X(01)00166-4.

26 Madkour, L.H. et al. (2018). Quantum chemical calculations, molecular dynamic (MD) simulations, and experimental studies of using some azo dyes as corrosion inhibitors for iron. Part 2: bis–azo dye derivatives. *Journal of Molecular Structure* 1163: 397–417. doi: 10.1016/j.molstruc.2018.03.013.

27 Javadian, S., Yousefi, A., and Neshati, J. (2013). Synergistic effect of mixed cationic and anionic surfactants on the corrosion inhibitor behavior of mild steel in 3.5% NaCl. *Applied Surface Science* 285 (PARTB): 674–681. doi: 10.1016/j.apsusc.2013.08.109.

28 Guo, L. et al. (2020). Multidimensional insights into the corrosion inhibition of 3,3-dithiodipropionic acid on Q235 steel in H2SO4 medium: a combined experimental and in silico investigation. *Journal of Colloid and Interface Science* 570: 116–124. doi: 10.1016/j.jcis.2020.03.001.

29 Tan, B. et al. (2018). A combined experimental and theoretical study of the inhibition effect of three disulfide-based flavoring agents for copper corrosion in 0.5 M sulfuric acid. *Journal of Colloid and Interface Science* 526: 268–280. doi: 10.1016/j.jcis.2018.04.092.

30 Tao, Z. et al. (2009). Corrosion inhibition of mild steel in acidic solution by some oxo-triazole derivatives, 51: 2588–2595. doi: 10.1016/j.corsci.2009.06.042.

31 El-lateef, H.M.A. (2020). Applied surface science corrosion inhibition characteristics of a novel salycilidene isatin hydrazine sodium sulfonate on carbon steel in HCl and a synergistic nickel ions additive: a combined experimental and theoretical perspective. *Applied Surface Science* 501 (October 2019),: 144237. doi: 10.1016/j.apsusc.2019.144237.

32 Belakhdar, A. et al. (2020). Computational and experimental studies on the efficiency of Rosmarinus officinalis polyphenols as green corrosion inhibitors for XC48 steel in an acidic medium. *Colloids and Surfaces A: Physicochemical and Engineering Aspects* 606 July: 125458. doi: 10.1016/j.colsurfa.2020.125458.

33 El-Tabei, A.S. et al. (2021). Experimental and theoretical (DFT&MC) studies for newly synthesized cationic amphiphilic substance based on a naphthol moiety as a corrosion inhibitor for carbon steel during the pickling process. *Journal of Molecular Liquids* 331: 115692. doi: 10.1016/j.molliq.2021.115692.

34 Nadi, I. et al. (2021). Insights into the inhibition mechanism of 2,5-bis(4-pyridyl)-1,3,4-oxadiazole for carbon steel corrosion in hydrochloric acid pickling via experimental and computational approaches. *Journal of Molecular Liquids* 116958. doi: 10.1016/j.molliq.2021.116958.

35 El Aoufir, Y. et al. (2020). The effect of the alkyl chain length on corrosion inhibition performances of 1,2,4-triazole-based compounds for mild steel in 1.0 M HCl: insights from experimental and theoretical studies. *Journal of Molecular Liquids* 303: 112631. doi: 10.1016/j.molliq.2020.112631.

36 Wagner, C. and Traud, W. (1938). Über die Deutung von Korrosionsvorgängen durch Überlagerung von elektrochemischen Teilvorgängen und über die Potentialbildung an Mischelektroden. *Zeitschrift für Elektrochemie und angewandte physikalische Chemie* 44 (7): 391–402. doi: 10.1002/bbpc.19380440702.

37 Saraswat, V., Yadav, M., and Obot, I.B. (2020a). Investigations on eco-friendly corrosion inhibitors for mild steel in the acid environment: electrochemical, DFT and Monte Carlo simulation approach. *Colloids and Surfaces A: Physicochemical and Engineering Aspects* 599: 124881. doi: 10.1016/j.colsurfa.2020.124881.

38 Abd El-Lateef, H.M., Sayed, A.R., and Shalabi, K. (2021). Synthesis and theoretical studies of novel conjugated polyazomethines and their application as efficient inhibitors for C1018 steel pickling corrosion behavior. *Surfaces and Interfaces* 23: 101037. doi: 10.1016/j.surfin.2021.101037.

39 Yang, Z. et al. (2015). Structure of a novel Benzyl Quinolinium chloride derivative and its effective corrosion inhibition in 15wt.% hydrochloric acid. *Corrosion Science* 99: 281–294. doi: 10.1016/j.corsci.2015.07.023.

40 Zuo, X. et al. (2021). Research of lilium brownii leaves extract as a commendable and green inhibitor for X70 steel corrosion in hydrochloric acid. *Journal of Molecular Liquids* 321: 114914. doi: 10.1016/j.molliq.2020.114914.

41 Cao, S. et al. (2019). Task-specific ionic liquids as corrosion inhibitors on carbon steel in 0.5 M HCl solution: an experimental and theoretical study. *Corrosion Science* 153 (May 2018): 301–313. doi: 10.1016/j.corsci.2019.03.035.

42 Tourabi, M. et al. (2013). Electrochemical and XPS studies of the corrosion inhibition of carbon steel in hydrochloric acid pickling solutions by 3,5-bis(2-thienylmethyl)-4-amino-1,2,4-triazole. *Corrosion Science* 75: 123–133. doi: 10.1016/j.corsci.2013.05.023.

43 Chouhan, A. et al. (2020). Synergistic lubrication performance by incommensurately stacked ZnO-decorated reduced graphene oxide/MoS2 heterostructure. *Journal of Colloid and Interface Science* 580: 730–739. doi: 10.1016/j.jcis.2020.07.033.

44 Feng, L. et al. (2019). Synergistic corrosion inhibition effect of thiazolyl-based ionic liquids between anions and cations for copper in HCl solution. *Applied Surface Science* 483 (March): 901–911. doi: 10.1016/j.apsusc.2019.03.299.

45 Luo, W. et al. (2021). A new pyridazine derivative synthesized as an efficient corrosion inhibitor for copper in sulfuric acid medium: experimental and theoretical calculation studies. *Journal of Molecular Liquids* 341: 117370. doi: 10.1016/j.molliq.2021.117370.

46 Zarrok, H. et al. (2012). Corrosion control of carbon steel in phosphoric acid by purpald – Weight loss, electrochemical and XPS studies. *Corrosion Science* 64: 243–252. doi: 10.1016/j.corsci.2012.07.018.

47 El Adnani, Z. et al. (2013). DFT theoretical study of 7-R-3methylquinoxalin-2(1H)-thiones (RH; CH3; Cl) as corrosion inhibitors in hydrochloric acid. *Corrosion Science* 68: 223–230. doi: 10.1016/j.corsci.2012.11.020.

48 Obot, I.B. and Gasem, Z.M. (2014). Theoretical evaluation of corrosion inhibition performance of some pyrazine derivatives. *Corrosion Science* 83: 359–366. doi: 10.1016/j.corsci.2014.03.008.

49 Cao, Z. et al. (2014). Novel benzimidazole derivatives as corrosion inhibitors of mild steel in the acidic media. Part II: theoretical studies. *Corrosion Science* 83: 292–298. doi: 10.1016/j.corsci.2014.02.025.

50 Lukovits, I., Kálmán, E., and Zucchi, F. (2001). Corrosion inhibitors—correlation between electronic structure and efficiency. *Corrosion* 57 (1): 3–8. doi: 10.5006/1.3290328.

51 Guo, L. et al. (2017). Theoretical insight into an empirical rule about organic corrosion inhibitors containing nitrogen, oxygen, and sulfur atoms. *Applied Surface Science* 406: 301–306. doi: 10.1016/j.apsusc.2017.02.134.

52 Saraswat, V., Yadav, M., and Obot, I.B. (2020b). Investigations on eco-friendly corrosion inhibitors for mild steel in the acid environment: electrochemical, DFT and Monte Carlo simulation approach. *Colloids and Surfaces A: Physicochemical and Engineering Aspects* 599 (April): 124881. doi: 10.1016/j.colsurfa.2020.124881.

53 Saraswat, V. and Yadav, M. (2020). Computational and electrochemical analysis on quinoxalines as corrosion inhibitors for mild steel in acidic medium. *Journal of Molecular Liquids* 297: 111883. doi: 10.1016/j.molliq.2019.111883.

54 Yadav, M. et al. (2015). Corrosion inhibition effect of pyrazole derivatives on mild steel in hydrochloric acid solution. *Journal of Adhesion Science and Technology* 29 (16): 1690–1713. doi: 10.1080/01694243.2015.1040979.

55 Yadav, M. et al. (2015). Corrosion inhibition effect of spiropyrimidinethiones on mild steel in 15% HCl solution: insight from electrochemical and quantum studies. *RSC Advances* 5 (87): 70832–70848. doi: 10.1039/c5ra14406j.

56 Saraswat, V. and Yadav, M. (2020). Carbon dots as green corrosion inhibitor for mild steel in HCl solution. *Chemistry Select* 5 (25): 7347–7357. doi: 10.1002/slct.202000625.

57 Yadav, M., Sarkar, T.K., and Purkait, T. (2015b). Studies on adsorption and corrosion inhibitive properties of indoline compounds on n80 steel in hydrochloric acid. *Journal of Materials Engineering and Performance* 24 (12): 4975–4984. doi: 10.1007/s11665-015-1765-x.

11

Grafted Natural Gums Used as Sustainable Corrosion Inhibitors

Brahim El Ibrahimi[1,*], *Elyor Berdimurodov*[2], *Walid Daoudi*[3], *and Lei Guo*[4]

[1] *Department of Applied Chemistry. Faculty of Applied Sciences, Ibn Zohr University, 86153 Aït Melloul, Morocco*
[2] *Faculty of Chemistry, National University of Uzbekistan, Tashkent 100034, Uzbekistan*
[3] *Laboratory of Molecular Chemistry, Materials and Environment (LCM2E), Department of Chemistry, Multidisciplinary Faculty of Nador, University Mohamed I, 60700 Nador, Morocco*
[4] *School of Material and Chemical Engineering, Tongren University, Tongren 554300, PR China*
[*] *Corresponding author*

11.1 Introduction

Corrosion-induced destruction of metallic-based structures is one of the most critical problems for a varied range of industries because it leads to a severe decline in investment and loss in production, as well as can cause environmental contamination and human loss. The use of corrosion inhibitors is an extraordinary approach to suppressing or limiting metallic degradation under environmental and service conditions (i.e., aggressive media) [1, 2].

A wide range of organic and inorganic compounds are applied as anti-corrosion agents against metallic dissolution in corrosive media. This relates to their intrinsic molecular structure that contains either heteroatoms (e.g., O, N, and S) or conjugated systems (including multiple bonds and aromatic rings). It is well known that these regions of molecules are the suitable and active sites for electron donation toward the protected metal surface, which leads to the adsorption of inhibitor molecules resulting in the formation of a protective film upon metal. Nevertheless, the major disadvantage associated with the application of these compounds (i.e., traditional inhibitors) is their dangerous impact on health and ecology. This has compelled the search for an alternative and a non-toxic class of inhibitors, which could offer supreme protection and cause the least effect on the environment and human health [3, 4].

Several reports have come across the literature describing the application of inhibitors based on environmentally benign heterocyclic compounds, biological

Grafted Biopolymers as Corrosion Inhibitors: Safety, Sustainability, and Efficiency, First Edition.
Edited by Jeenat Aslam, Chandrabhan Verma, and Ruby Aslam.

polymers, amino acids, ionic liquids, and their derivatives, as well as natural extracts [5–9]. Moreover, surfactants and pharmaceutical compounds are applied as sustainable anti-corrosion agents for several metallic materials in different corrosive environments [10, 11].

11.2 Gum-based Corrosion Inhibitors

Recently, the use of natural polymers (i.e., biopolymers) as alternative and green inhibitors has been largely reported [12, 13]. In this regard, owing to their eco-friendliness, sustainability, and non-toxicity, natural gums are gaining more importance. In this subject, the application of gums (e.g., gum *Arabic*, locust bean gum, *Albizia* gum, Ficus tricopoda gum, Khaya ivorensis gum, etc.) as environmentally safe corrosion inhibitor for metallic-based materials has in recent times received tremendous attention by several researchers. Table 11.1 collects some applied natural gums as green corrosion inhibitors for different metal/solution systems.

Table 11.1 Examples of gums used as corrosion inhibitors for various metal/solution systems.

Gum	Metal	Corr. Med.	IE (%)	References
Gum arabic	Aluminium Mild steel	0.1 M H_2SO_4	80% at 0.5 g L^{-1} (at 303 K) 38% at 0.5 g L^{-1} (at 333 K)	[14]
Guar gum	Carbon steel	1 M H_2SO_4	93% at 1.5 g L^{-1} (at 298 K)	[15]
Ficus glumosa gum	Mild steel	H_2SO_4	75% at 0.5 g L^{-1} (at 303 K)	[16]
Ficus Benjamina gum	Aluminium	0.1 M HCl	41% at 0.5 g L^{-1} (at 303 K)	[17]
Anogessius leocarpus gum	Mild steel	0.1 M HCl	51% at 0.5 g L^{-1} (at 303 K)	[18]
Commiphora pendunculata gum	Al. alloy	0.1 M HCl	64% at 0.5 g L^{-1} (at 303 K)	[19]
Ficus platyphylla gum	Mild steel	0.1 M HCl	71% at 0.5 g L^{-1} (at 303 K)	[20]
Ficus tricopoda gum	Aluminium	0.1 M H_2SO_4	85% at 0.5 g L^{-1} (at 303 K)	[21]
Khaya ivorensis gum	Mild steel	0.1 M HCl	84% at 0.5 g L^{-1} (at 303 K)	[22]
Gloriosa superba gum	Aluminium	0.1 M HCl	86% at 0.5 g L^{-1} (at 303 K)	[23]

These substances consist of a mixture of long-chain polysaccharides in their original form generally used in the field of food processing, cosmetics, and pharmaceutical applications. Naturally occurring gums are of various types such as seed gums, exudate gums, microbial gums, mucilage gums, seaweed gums, etc. They are grouped depending on their sources, either neutral (e.g., guar gum extracted from seeds of guar beans) or ionic (e.g., agar extracted from red seaweeds) gums [24].

11.3 Application of Grafted Gums as Corrosion Inhibitors

Recent advances and modifications of these natural gums are reported to be beneficial in terms of their sustainability, as well as their good corrosion inhibition effect for metallic material under various environments. Such modifications can also resolve or minimize some of the known drawbacks allied with the application of gums, namely thermal instability, uncontrolled hydration rates, microbial contamination, and pH-dependent solubility.

Conducting chemical modifications like grafting, crosslinking, esterification, etherification, and nanocomposites on the existing gums is the current approach in the application of gum-based inhibitors. This is due to the availability of polar functional groups (e.g., hydroxyl groups) and the feasibility of the modification for introducing conjugate pi-electrons and additional heteroatoms on the backbone of these biopolymers. Moreover, it can be combined with various additives such as surfactants or salts of halides (KI and KBr) to increase their inhibitive effect as well. Recently, a polymeric blend of Arabic and cashew gums was produced and is applied as an effective anti-corrosive substance for mild steel in acidic media [25].

Exudate gums are obtained as the natural exudates of different tree species and exhibit unique properties in a wide variety of applications. Among the most industrially relevant exudate gums, there are gum Arabic, gum karya, and gum tragacanth. The unique compositions of natural exudate gums make them suitable for corrosion inhibitor applications. Gum Arabic (GA), which is also known as acacia gum, Senegal gum, and gum Sudani is a plant exudate, consisting of glycoproteins and polysaccharides mainly consisting of galactose and arabinose. Commercially, the gum is harvested mainly from wild trees, mostly in Sudan and all through the Sahel. Gum Arabic is the oldest and most popular in all-natural gums. Because of its non-toxic behavior, natural origin, commercial availability, various adsorption sites, polymeric nature, and eco-friendly character, gum Arabic is widely used as anticorrosive material, especially in aqueous electrolytes. This type of gum contains various electron-rich polar sites through which they easily get adsorbed on the metallic surface and behaves as effective anti-corrosive material.

The well acidizing procedure is one of the most important stages in the gas and oil industry, which is a stimulation technique for rock reservoirs. Acidic solutions are employed during this process, and the hydrochloric acid at a higher concentration (at 15%) is a commonly used stimulation fluid for this purpose [26, 27]. In this regard, the inhibiting effect of Gum Acacia grafted with polyacrylamide (GA-g-PAM) at different monomer concentrations against mild steel corrosion in a 15% HCl solution has been outlined [28]. The scheme of the graft co-polymerization process of GA-g-PAM is presented in Figure 11.1. It was noted that the corrosion inhibition efficiency varies with the amount of grafted polyacrylamide and leads to a significant increase in the inhibitive behavior at a lower concentration than the unmodified Gum Acacia. Among different grafted co-polymers, gum acacia (1 g) and acrylamide (3 g) resulted in maximum corrosion inhibition efficiency of 94% at $0.3\,gL^{-1}$ concentration (at 298 K). The inhibitive effect of the developed grafted gum was related to the formation of a preventive film as confirmed using energy-dispersive X-ray spectroscopy, atomic force microscopy, and also FE-SEM. Nevertheless, at a high grafting level of GA and after long immersion of mild steel in the acid medium, the observed inhibitive effect was gradually decreased.

Guar gum (GG), a plant-side gum, is an endospermic gum of *Cyamopsis tetragonolobus* and is a non-ionized, water-dissolvable gum with an average ratio of galactose to mannose equal to 1:2. Chemically or biochemically modified derivatives of GG polymer are highly stable and show great application potentials in wastewater treatment, biomedical, hydraulic fracturing, coatings, and in corrosion inhibition as well [24]. In this regard, the successful use of polyacrylamide grafted

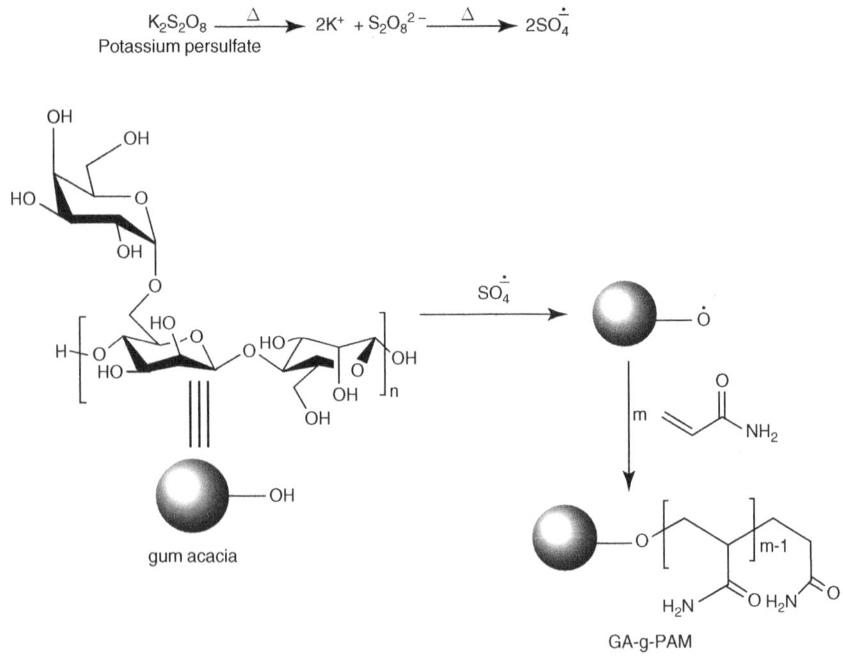

Figure 11.1 Reaction scheme of synthesized GA-g-PAM inhibitor. [28] / Elsevier.

GG (GG-g-PAM) with different grafting percentages as a corrosion inhibitor for mild steel in an acidic solution has been outlined [29]. The scheme of the graft copolymerization process of GG-g-PAM is presented in Figure 11.2. The supreme inhibition effectiveness of GG-g-PAM recorded after five hours of immersion was around 93%. The observed protective effect is related to the adsorption of grafted product on the metal via their endo- and exo-cyclic oxygen atoms, which are demonstrated an intramolecular synergism effect. On the other hand, a better anti-corrosion outcome (≈ 95%) is noted in the case of GG grafted with 2-acrylamide-2-methyl propane sulfonic acid (GG-g-AMPS, Figure 11.3) for carbon steel and copper metals under saline circumstances [30, 31]. In the case of copper immersed in NaCl solution, the prevention efficacy of GG-g-AMPSA varied proportionally with the immersion time. Besides, the protective action of GG-grafted methyl methacrylate (GG-g-MMA, Figure 11.3) in the presence of iodide ions was confirmed for steel in the NaCl solution saturated with CO_2. The recorded prevention percentage of the "GG-g-MMA (300 ppm) +KI (5 mM)" inhibitor formulation was about 97% at 323 K [32]. Recently, a similar functionalized GG, this time with ethyl acetoacetate (GG-g-EEA, Figure 11.3), also demonstrated an interesting anti-corrosion effect. For P110 steel, which is largely used to construct tubing and casing pipes in the oil and gas industry, the inhibition performance increased from 78% (at 500 ppm) for unmodified GG to 92% for the newly synthesized GG-g-EEA in 15% HCl solution. According to a multi-surface analysis (including SEM, EDX, AFM, and XPS

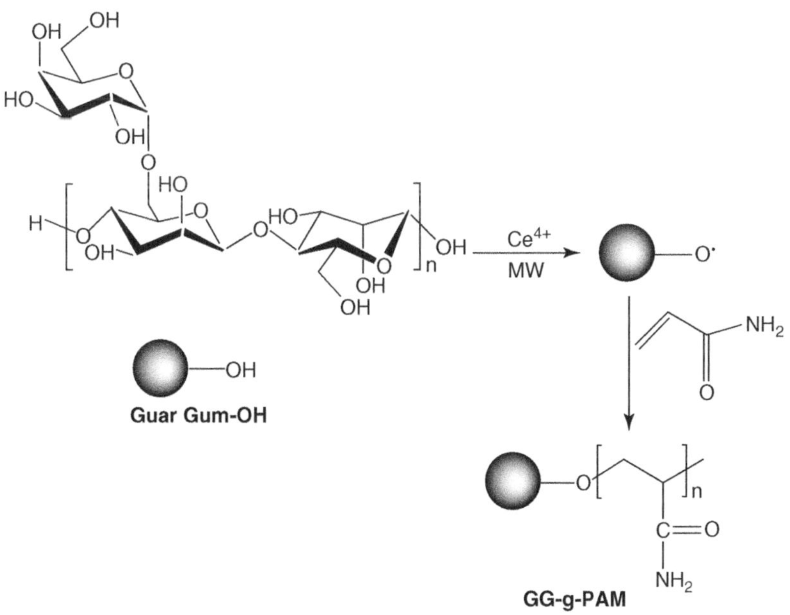

Figure 11.2 Mechanistic pathway for synthesis of GG-g-PAM inhibitor. [29] / Elsevier.

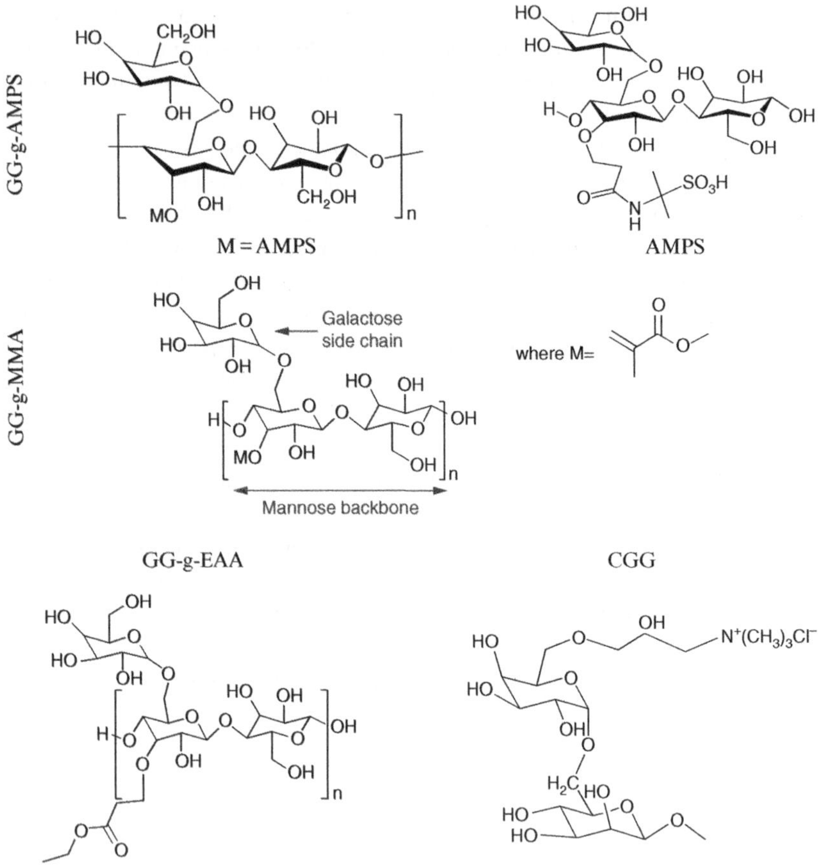

Figure 11.3 Molecular structure of GG-g-AMPS, GG-g-MMA, GG-g-EAAand CGG modified gums.

techniques) and computational studies, the observed retardation of steel corrosion in the presence of GG-g-EEA was attributed to its effective adsorption onto the metal surface [33]. The application of chemically modified GGs as green inhibitors against metallic corrosion can be extended to be applicable under extreme corrosive conditions. For instance, the cationic GG (noted as CGG, Figure 11.3) is used for the inhibition of mild steel corrosion in a 15% HCl solution. Under such circumstances, the modified GG shows better inhibition efficiency (95% at $0.4\,g\,L^{-1}$) compared with raw GG (86% at $0.4\,g\,L^{-1}$) [34]. In recent years, by employing chemical computational tools, corrosion scientists attempt to understand in more detail the action mechanism of corrosion inhibitors both under atomic and electronic scales [35–38]. In this regard and based on quantum chemical calculations, the observed improvement in the inhibitive effect of CGG was attributed to its particular electronic structure in comparison with the unmodified GG.

Xanthan gum (XG), a microbial gum, is an extracellular polysaccharide secreted by *Xanthomonas campestris* bacteria and is largely applied in food industries, as well as in non-food industries such as in the manufacture of suspensions, foams, and emulsions [39]. The inhibitive effect of natural XG has been outlined for mild steel in an acidic corrosive medium, which can suppress 74% of corrosion at 1 g L^{-1} (at 303 K) [40]. To improve its anti-corrosion effect, polyaniline grafted XG (XG-g-PANI, Figure 11.4 (a)) under microwave irradiation demonstrated an interesting prevention capability against the corrosion of aluminium in molar hydrochloric acid [41]. The conducted modification upgraded the inhibition efficiency from 63% for XG to 91% for XG-g-PANI. Figure 11.4 (b) shows the effect of the addition of XG and XG-g-PANI substances on the polarization behavior and surface morphology of metal. Such a better corrosion hindrance of the grafted product is related to the reinforcing of the inhibitor/metal interaction through conjugate

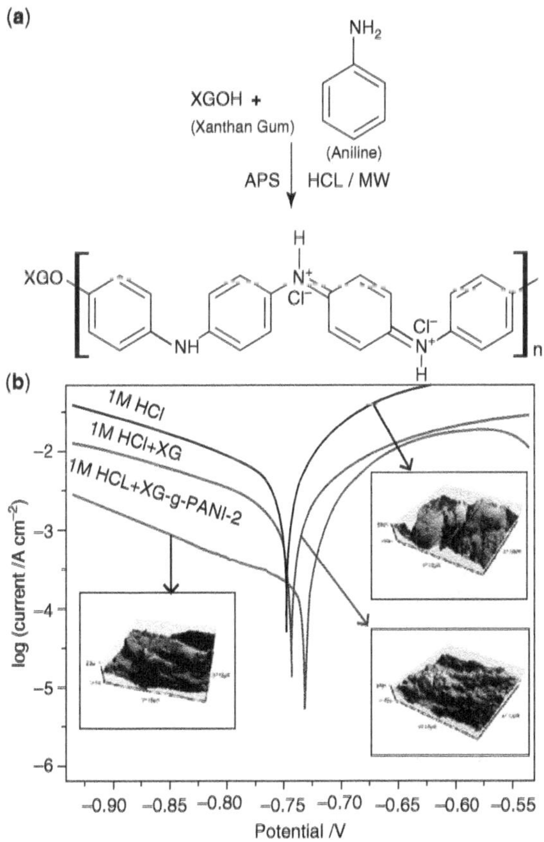

Figure 11.4 (**a**) Synthesis rout and molecular structure of XG-g-PANI. (**b**) Tafel plots for Aluminium in 1M HCl, 1M HCl+ XG, and 1M HCl+XG-g-PANI, as well as AFM photographs samples. [41] / With permission of Elsevier.

π-electrons and lone pair of electrons of nitrogen atoms delivered by PANI, as well as the free hydroxyl groups of XG. Considering the good complex ability of phosphate groups with metal ions, an XG derivative has been synthesized by introducing these groups on the gum backbone (i.e., phosphorylated XG, Fig. 11.5) and then is applied as an environment-friendly and efficient inhibitor for mild steel in NaCl solution [42]. An inhibition percentage of 93% was achieved in the presence of a developed derivative at a lower concentration of 150 ppm.

Recently, a novel green-grafted XG polymer with β-cyclodextrin (XG-g-β-CD, Figure 11.6) was synthesized according to the schema in Figure 11.6 and then

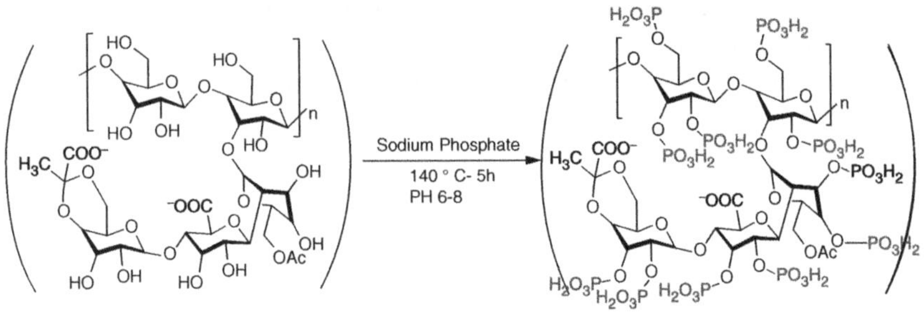

Figure 11.5 Synthesis rout and molecular structure of phosphorylated XG. [42] / Elsevier.

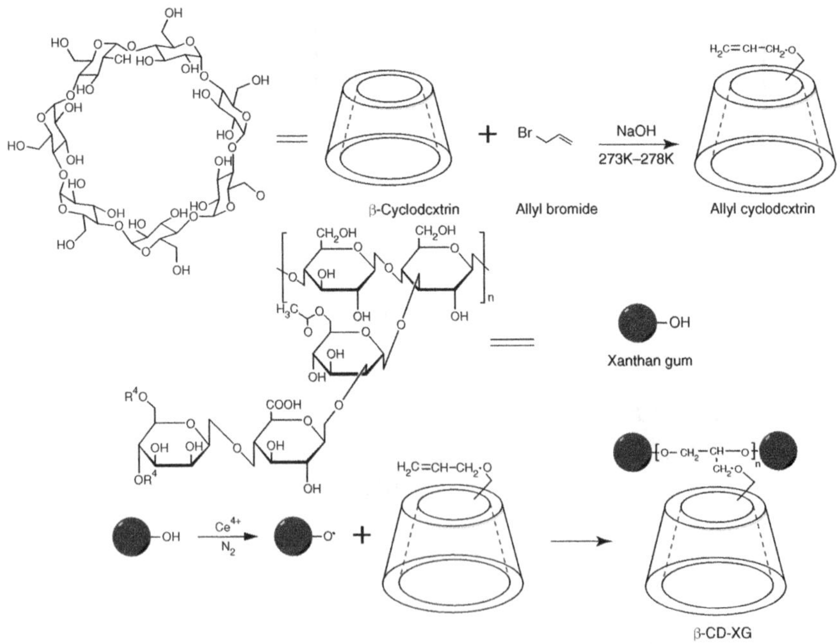

Figure 11.6 Synthetic route of XG-g-β-CD modified gum. [43] / With permission of Elsevier.

used to suppress the corrosion of X80 and L80 steels in molar sulfuric and hydrochloric acids, respectively [43, 44]. As compared to natural XG, the modified form has shown superior corrosion inhibition, reaching around 95% at 200 ppm (at 239K). The observed anti-corrosion effect was related to the chemical adsorption of XG-g-β-CD on the metal surface (leading to the film formation) through –OH, $-COO^-$, $-CH_2-O-$, and $-CH_2-O-CH_2-$ groups. This encourages its application for corrosion prevention of steel-based pipelines in the petroleum industry.

Likewise grafted GGs, the functionalized XGs also exhibited good abilities to slow the degradation of metallic materials even in extremely aggressive environments, like those in oil and gas industries in which acidic solutions at higher concentrations are commonly applied such as 15% HCl solution. Under this particular situation, a novel developed xanthan gum-graft-polyacrylamide (XG-g-PAM) according to the synthetic route in Figure 11.7 is used as a good inhibitor (IE = 93% at $0.4\,g\,L^{-1}$ and 298 K) against the corrosion of mild steel in a 15% HCl medium. The benefic effect of XG-g-PAM addition for metal protection can be easily revealed from the SEM images presented in Figure 11.8. As can be seen, the implementation of XG-g-PAM was considerably declined the corrosion attack of the aggressive environment. The electronic-scale analysis, via quantum chemical calculations, of the grafted XG molecule outlined its outstanding reactivity This can engender its ease of adsorption onto metal surface leading to the formation of a protective layer [45].

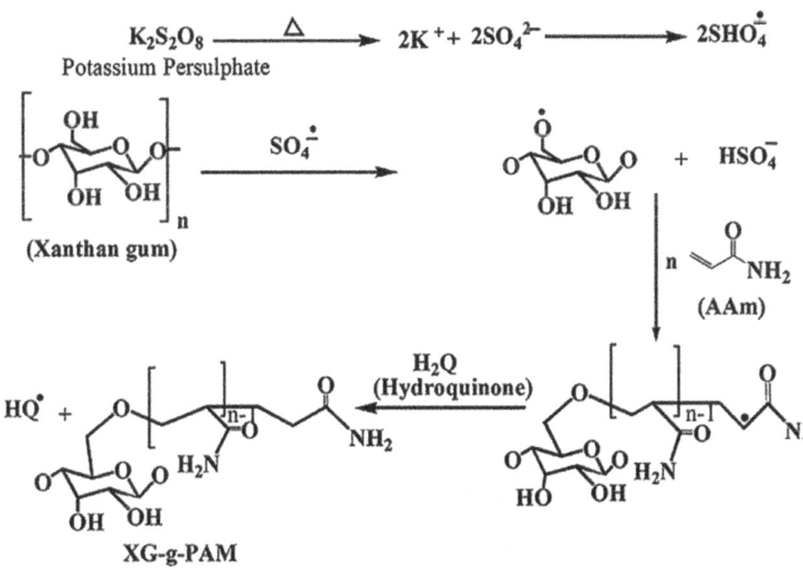

Figure 11.7 Synthetic route for XG-g-PAM inhibitor. [45] / Elsevier.

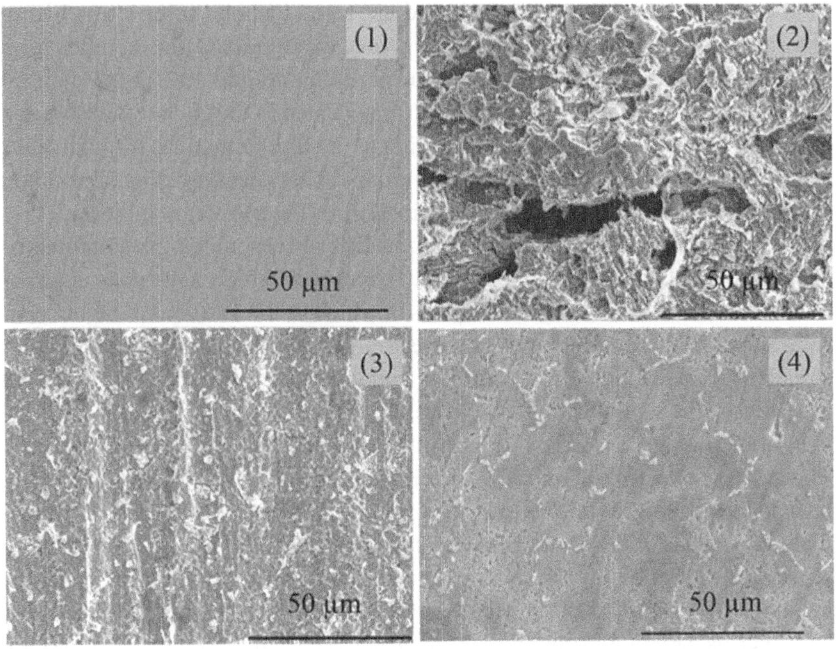

Figure 11.8 SEM images of mild steel surface: (1) prior immersion in acid solution, (2) in the absence of inhibitors; (3) in the presence of XG; (4) in the presence of XG-g-PAM (at 0.4 g L^{-1} and 298 K)for 6 h. [45] / from ELSEVIER.

Cashew tree gum grafted with poly(acrylamide) (CTG-g-PAM) at different percentages has been also reported to act as a green inhibitor for mild steel in an acidic solution. The maximum inhibition efficiency noted for the grafted product (≈ 57%) was found to be greater than that of the non-grafted one (≈ 11%). Furthermore, the increase of the incorporated PAM within the CTG-g-PAM graft copolymer has exhibited a benefic effect on the inhibition efficiency [46]. On the other hand, the implementation of a non-ionic surfactant within the gum backbone also demonstrated a benefic effect on its inhibitive action. For instance, two bio-based surfactants on gum rosin have been applied as inhibitors for line-pipe steel. These compounds are prepared by reacting gum rosin with linoleic acid via the Diels-Alder reaction. According to the electrochemical testing, both anti-corrosive agents act as mixed-type inhibitors, with excellent inhibition efficiencies of over 80% at 100 ppm in molar HCl solution [47].

11.4 Conclusion and Perspectives

The application limitations allied with the toxicity of traditional organic/inorganic substances as corrosion inhibitors have been motivated and requested the use of naturally occurring substances as anti-corrosion agents. In this regard, natural gum polymers were shown the capability and potentiality to be applied as environmental and sustainable anti-corrosion materials. In order to improve their inhibitive effect and to overcome some drawbacks associated with their application, the graft copolymerization of natural gums was largely applied. As a result, the grafted gums copolymers showed better corrosion hindrance and enhanced properties compared with unmodified ones.

The chemical modifications of natural gums are commonly performed on Arabic, Guar, and Xanthan gums, whereas the grafting on other gums is not yet conducted. Hence, more attention is required to this subject in upcoming studies in this field. On the other hand, the use of grafted gums as corrosion inhibitors was generally limited to iron-based alloys in acidic conditions. Therefore, their inhibitive action for other metal/medium systems remains insufficiently studied or unknown, which implies its exploration in future research works. Up to now, the implementation of these grafted gums copolymers within corrosion inhibitor formulations, which include surfactants, intensifiers, solvents, and co-solvent, is still unexplored for a wide range of applications (e.g., in the petroleum industry). This can be listed as a future promising perspective in the effective application of gums co-polymers as anti-corrosion materials.

Useful Links

www.nace.org
www.imoa.info
www.corrosion.org

References

1 El Ibrahimi, B., Baddouh, A., Oukhrib, R. et al. (2021). Electrochemical and in silico investigations into the corrosion inhibition of cyclic amino acids on tin metal in the saline environment. *Surfaces and Interfaces* 23: 100966.

2 El Ibrahimi, B., Bazzi, L., and EL Issam, S. (2020). The role of pH in corrosion inhibition of tin using the proline amino acid: theoretical and experimental investigations. *RSC Advances* 10: 29696.

3 Patni, N., Agarwal, S., and Shah, P. (2013). Greener approach towards corrosion inhibition. *Chinese Journal of Engineering* 2013.

4 Popoola, L.T. (2019). Progress on pharmaceutical drugs, plant extracts, and ionic liquids as corrosion inhibitors. *Heliyon* 5: e01143.

5 Obot, I.B. and Edouk, U.M. (2017). Benzimidazole: a small planar molecule with diverse anti-corrosion potentials. *Journal of Molecular Liquids* 246: 66–90.

6 Verma, C., Ebenso, E.E., and Quraishi, M.A. (2017). Ionic liquids as green and sustainable corrosion inhibitors for metals and alloys: an overview. *Journal of Molecular Liquids* 233: 403–414.

7 Ansari, K.R., Chauhan, D.S., Quraishi, M.A., Mazumder, M.A.J. and Singh, A. (2020). Chitosan Schiff base: an environmentally benign biological macromolecule as a new corrosion inhibitor for oil & gas industries. *International Journal of Biological Macromolecules* 144: 305–315.

8 El Ibrahimi, B., Jmiai, A., Bazzi, L., and El Issami, S. (2020). Amino acids and their derivatives as corrosion inhibitors for metals and alloys. *Arabian Journal of Chemistry* 13: 740–771.

9 Verma, C., Ebenso, E.E., Quraishi, M.A., and Hussain, C.M. (2021). Recent developments in sustainable corrosion inhibitors: design, performance and industrial scale applications. *Materials Advances* 2: 3806–3850.

10 Zhu, Y., Free, M.L., Woollam, R., and Durnie, W. (2017). A review of surfactants as corrosion inhibitors and associated modeling. *Progress in Materials Science* 90: 159–223.

11 Tanwer, S. and Shukla, S.K. (2022). Recent advances in the applicability of drugs as corrosion inhibitor on a metal surface: a review. *Current Research in Green and Sustainable Chemistry* 5: 100227.

12 Umoren, S.A. and Eduok, U.M. (2016). Application of carbohydrate polymers as corrosion inhibitors for metal substrates in different media: a review. *Carbohydrate Polymers* 140: 314–341.

13 Sabirneeza, A.A.F., Geethanjali, R., and Subhashini, S. (2015). Polymeric corrosion inhibitors for iron and its alloys: a review. *Chemical Engineering Communications* 202: 232–244.

14 Umoren, S.A. (2008). Inhibition of aluminum and mild steel corrosion in acidic medium using Gum Arabic. *Cellulose* 15: 751–761.

15 Abdallah, M. (2004). Guar gum as corrosion inhibitor for carbon steel in sulphuric acid solutions. *Portugaliae Electrochimica Acta* 22: 161–175.

16 Ameh, P.O., Magaji, L., and Salihu, T. (2012). Corrosion inhibition and adsorption behavior for mild steel by Ficus glumosagum in H_2SO_4 solution. *African Journal of Pure and Applied Chemistry* 6: 100–106.

17 Umoren, S.A., Obot, I.B., and Ebenso, E.E. (2008). Corrosion inhibition of aluminum using exudate gum from pachylobus edulis in the presence of halide ions in HCl. *E-Journal of Chemistry* 5: 138407.

18 Ameh, P.O., Odiongenyi, A.O., and Eddy, N.O. (2012). Joint effect of Anogessius Leocarpus Gum (AL Gum) exudate and halide ions on the corrosion of mild steel in 0.1 M HCl. *Portugaliae Electrochimica Acta* 30: 235–245.

19 Ameh, P.O. and Eddy, N.O. (2014). Commiphora pedunculata gum as a green inhibitor for the corrosion of aluminum alloy in 0.1 M HCl. *Research on Chemical Intermediates* 40: 2641–2649.

20 Eddy, N.O., Ameh, P., Gimba, C.E., and Ebenso, E.E. (2012). Chemical information from GCMS of ficus Platyphylla Gum and its corrosion inhibition potential for mild steel in 0.1 M HCl. *International Journal of Electrochemical Science* 7: 5677–5691.

21 Eddy, N.O., Ameh, P.O., Gwarzo, Y.M., Okop, J.I., and Dodo, N.S. (2013). Physicochemical study and corrosion inhibition potential of Ficus tricopoda for aluminum in acidic medium. *Portugaliae Electrochimica Acta* 3: 79–93.

22 Ameh, P.O. (2012). Adsorption and inhibitive properties of khaya ivorensis gum for the corrosion of mild steel in HCl. *International Journal of Modern Chemistry* 2: 28–40.

23 Eddy, N.O., Ibok, U.J., Ameh, P.O., Alobi, N.O., and Sambo, M.M. (2014). Adsorption and quantum chemical studies on the inhibition of the corrosion of aluminum in HCl by gloriosa superba (GS) gum. *Chemical Engineering Communications* 201: 1360–1383.

24 Peter, A., Obot, I.B., and Sharma, S.K. (2015). Use of natural gums as green corrosion inhibitors: an overview. *International Journal of Industrial Chemistry* 6: 153–164.

25 Arthur, D.E. (2020). Computational and experimental study on corrosion inhibition potential of the synergistic 1:1 combination of Arabic and cashew gums on mild steel. *Petroleum Research* 5: 170–180.

26 Migahed, M.A., Zaki, E.G., and Shaban, M.M. (2016). Corrosion control in oil wells tubing steel during matrix acidizing operations. *RSC Advances* 6: 71384–71396.

27 Obot, I.B., Onyeachu, I.B., Umoren, S.A., et al. (2020). High-temperature sweet corrosion and inhibition in the oil and gas industry: progress, challenges and future perspectives. *Journal of Petroleum Science and Engineering* 185: 106469.

28 Biswas, A., Mourya, P., Mondal, D., Pal, S., and Udayabhanu, G. (2018). Grafting effect of gum acacia on mild steel corrosion in acidic medium: gravimetric and electrochemical study. *Journal of Molecular Liquids* 251: 470–479.

29 Roy, P., Karfa, P., Adhikari, U., and Sukul, D. (2014). Corrosion inhibition of mild steel in acidic medium by polyacrylamide grafted Guar gum with various grafting percentage: effect of intramolecular synergism. *Corrosion Science* 88: 246–253.

30 Singh, A., Samih Mohamed, H., Singh, S., Yu, H. and Lin, Y. (2020). Corrosion inhibition using guar gum grafted 2-acrylamido-2-methyl propane sulfonic acid (GG-AMPS) in tubular steel joints. *Construction and Building Materials* 258: 119728.

31 Singh, A., Liu, M., Ituen, E., and Lin, Y. (2020). Anti-corrosive properties of an effective guar gum grafted 2-Acrylamido-2-Methylpropanesulfonic acid (GG-AMPS) coating on copper in a 3.5% NaCl solution. *Coatings* 10: 241.

32 Singh, A., Ansari, K.R., and Quraishi, M.A. (2020). Inhibition effect of natural polysaccharide composite on hydrogen evolution and P110 steel corrosion in 3.5 wt% NaCl solution saturated with CO2: a combination of experimental and surface analysis. *International Journal of Hydrogen Energy* 45: 25398–25408.

33 Singh, A., Ansari, K.R., Quraishi, M.A., Kaya, S., and Erkan, S. (2021). Chemically modified guar gum and ethyl acrylate composite as a new corrosion inhibitor for a reduction in hydrogen evolution and tubular steel corrosion protection in an acidic environment, International. *Journal Of Hydrogen Energy* 46: 9452–9465.

34 Biswas, A., Pal, S., and Udayabhanu, G. (2017). Effect of chemical modification of a natural polysaccharide on its inhibitory action on mild steel in 15% HCl solution. *Journal of Adhesion Science and Technology* 31: 2468–2489.

35 Obot, I.B., Haruna, K., and Saleh, T.A. (2019). Atomistic simulation: a unique and powerful computational tool for corrosion inhibition research. *Arabian Journal for Science and Engineering* 44: 1–32.

36 Obot, I.B., Macdonald, D.D., and Gasem, Z.M. (2015). Density functional theory (DFT) is a powerful tool for designing new organic corrosion inhibitors. Part 1: an overview. *Corrosion Science* 99: 1–30.

37 Bourzi, H., Oukhrib, R., El Ibrahimi, B. et al. (2020). Furfural analogs as sustainable corrosion Inhibitors – predictive efficiency using DFT and Monte Carlo Simulations on the Cu(111), Fe(110), Al(111), and Sn(111) surfaces in acid media. *Sustainability* 12: 3304.

38 Bourzi, H., Oukhrib, R., El Ibrahimi, B. et al. (2020). Understanding of anti-corrosive behavior of some tetrazole derivatives in acidic medium: adsorption on Cu (111) surface using quantum chemical calculations and monte Carlosimulations. *Surface Science* 702: 121692.

39 Patel, J., Maji, B., Moorthy, N.S.H.N., and Maiti, S. (2020). Xanthan gum derivatives: a review of synthesis, properties, and diverse applications. *RSC Advances* 10: 27103–27136.

40 Mobin, M. and Rizvi, M. (2016). Inhibitory effect of xanthan gum and synergistic surfactant additives for mild steel corrosion in 1M HCl. *Carbohydrate Polymers* 136: 384–393.

41 Babaladimath, G., Vishalakshi, B., and Nandibewoor, S.T. (2018). Electrical conducting Xanthan Gum-graft-polyaniline as corrosion inhibitor for aluminum in a hydrochloric acid environment. *Materials Chemistry and Physics* 205: 171–179.

42 Kumar, K.V. and Rao, B.V.A. (2019). Phosphorylated Xanthan Gum, an environment-friendly, efficient inhibitor for mild steel corrosion in aqueous 200 ppm NaCl. *Materials Today: Proceedings* 15: 155–165.

43 Cao, Y., Zou, C., Wang, C. et al. (2021). Green corrosion inhibitor of β-cyclodextrin modified xanthan gum for X80 steel in 1 M H_2SO_4 at different temperatures. *Journal of Molecular Liquids* 341: 117391.

44 Cao, Y., Zou, C., Wang, C. et al. (2021). β-cyclodextrin modified xanthan gum as an eco-friendly corrosion inhibitor for L80 steel in 1 M HCl. *Cellulose* 28: 11133–11152.

45 Biswas, A., Pal, S., and Udayabhanu, G. (2015). Experimental and theoretical studies of xanthan gum and its graft copolymer as corrosion inhibitor for mild steel in 15% HCl. *Applied Surface Science* 353: 173–183.

46 Ramesan, M.T. and Surya, K. (2016). Studies on electrical, thermal, and corrosion behavior of cashew tree gum grafted Poly(Acrylamide). *Polymers from Renewable Resources* 7: 81–99.

47 Atta, A.M., El-Mahdy, G.A., Elsaeed, A.M., and Al-Lohedan, H.A. (2014). Synthesis of bio-based corrosion inhibitors based on Rosin for line-pipe steel. *Digest Journal of Nanomaterials and Biostructures* 9: 1047–1058.

12

Grafted Pectin as Sustainable Corrosion Inhibitors

Dan-Yang Wang, Hui-Jing Li, and Yan-Chao Wu

Weihai Marine Organism & Medical Technology Research Institute, Harbin Institute of Technology, Weihai 264209, P. R. China

12.1 Introduction

Pectin is a natural heteropolysaccharide obtained from the cell walls of terrestrial higher plants, with the characteristics of biocompatibility, environmental, and non-toxicity [1, 2]. The sources of pectin generally are the leaves, skins, stems, and fruits of plants such as citrus, apple, grapefruit, beet, fig, orange, potato, peanut, and sunflower. At present, pectin is widely applied in the food industry as a gelling agent and potential functions to make full use of pectin are being explored [3]. In nature, the structure and function of pectin rely on plant species and subordinate parts. Pectin present in the fruit is reduced by pectinase to pectinesterase during its ripening process [4]. It is the second most abundant polysaccharide in plant cells besides cellulose. Pectin is a ubiquitous component of the cell wall, which reduces the porosity of the cell wall and promotes adhesion between adjacent cells in the middle lamella [5]. It can release signaling oligosaccharides to regulate biological processes such as plant growth and development and prevent the invasion of pathogenic bacteria [6]. The relative content of pectin depends on the plant and cell type, for example, the relative content of pectin in the primary and secondary cell walls of grasses is less than 10%, while it is about 35% in other dicot and monocot plants [7].

The complex chemical pectin structure shown in Figure 12.1 consists of varied structural units, namely the homogalacturonan (HG) unit, rhamnogalacturonan I (RG-I) units, and rhamnogalacturonan II (RG-II) units. Specifically, the HG unit is composed of α -(1,4) linked d-galacturonic acid residues, and the RG-I and RG-II units are alternately formed by galacturonic acid and rhamnose residues. Diverse patterns have been proposed according to various linkages between the aforementioned structural elements. The common characteristic in different

Grafted Biopolymers as Corrosion Inhibitors: Safety, Sustainability, and Efficiency, First Edition.
Edited by Jeenat Aslam, Chandrabhan Verma, and Ruby Aslam.

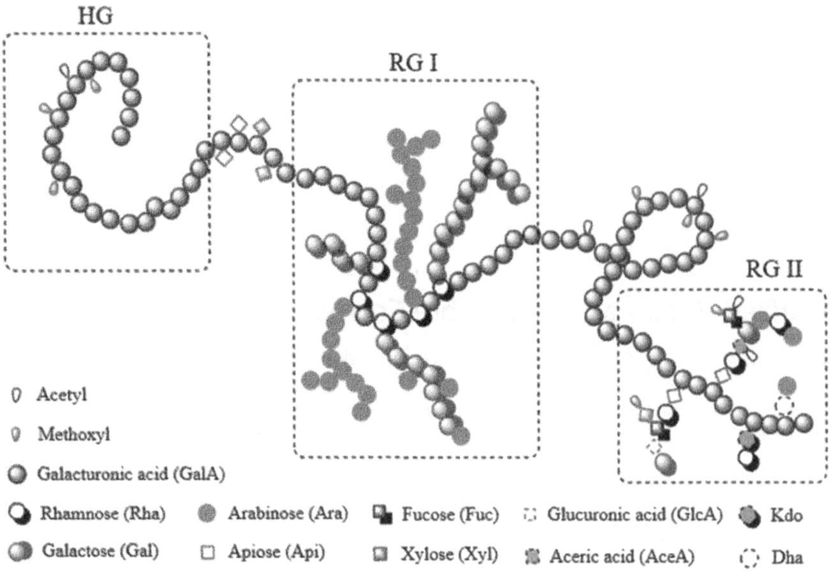

Figure 12.1 Diagrammatic representation of pectin molecule. *Source:* D.Y. Wang.

models is that the backbone of galacturonic acid is cut off by rhamnogalacturonan units. The D-GalA residues of HG and RG units can not only be methylated on the C-6 position but also acetylated on the O-2 or O-3 position. Moreover, amino acid and phenolic groups can be also adhered to the backbone of the pectin compound.

Generally, a common method of forming a passivation layer to retard metal corrosion reactions is a corrosion inhibitor, which is an electrochemical damage process, that has a great impact on the metal itself and even the industrial economy. In most industrial fields, different precautions are applied to retard metal corrosion by using additives, such as corrosion inhibitors. The usual requirements and characteristics for corrosion inhibitors are as follows: (1) have enough special and functional structure (atoms with lone pairs of electrons: N, O, S, P, etc.; conjugated π structures; empty d-orbitals), (2) be able to form a firm and uniform film on the metal surface, (3) the ability to make a passive protective film covering the metallic surface, (4) the solubility in the corrosive medium, (5) cost-effective, (6) non-toxic and green. In short, they are produced by renewable and environmental precursors and consume less energy during the preparation process. At present, the research work of corrosion inhibitors is developing in the direction of economy, eco-friendliness, biodegradability, and non-toxicity.

Surprisingly, pectins have abundant functional groups, oxygen-containing groups containing lone pairs of electrons on its carbohydrate backbone, such as carboxylic acid, carboxymethyl, and hydroxyl, to satisfy the needs of candidate molecules for corrosion in different media (Figure 12.2). Another characteristic of

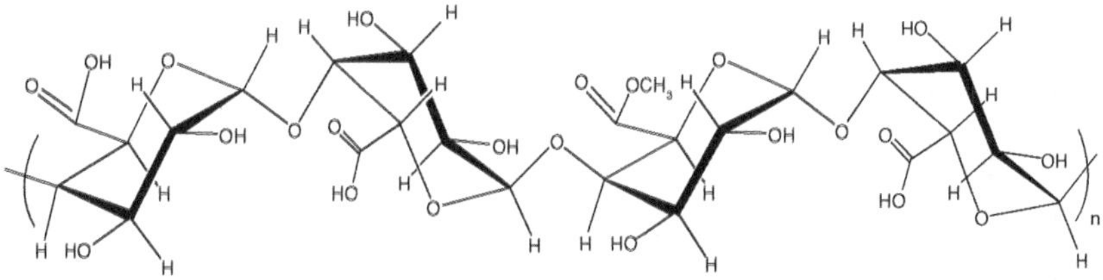

Figure 12.2 Chemical structure of pectin from citrus peel. *Source:* Fares.

pectin is that the glycosidic bonds of the polygalacturonic acid backbone chains are more resistant to the hydrolysis of acids, such as HCl and H_2SO_4, under mild conditions. What's more interesting is that due to the different residues of the extracted pectin, the pectin composition is also different, including the ratio of sugar, the length of the chain, and the content of protein and phenol, which will greatly affect the inhibitive effectiveness of pectin. Most pectin exists in the form of original pectin with large molecular weight, poor solubility, and gelability. Modification of pectin is beneficial to enhance its physiological function, and such modification has become a research hotspot in recent years. Smart designing and preparing of broad-spectrum derivatives in pectin molecules which have a mass of hydroxyl (–OH) and carboxyl (–COOH) groups scattered on the backbone, and neutral sugars served as branch chains make it easier to form complexes with metals and then build up a protective barrier for more efficient corrosion inhibition. The grafted groups are anionic, cationic, nonionic, or ampholytic due to the specialty and selectivity of pectin structure and chemical composition.

12.2 Pectins as Corrosion Inhibitors

The corrosion inhibiting effect of pectin/ascorbic acid on the tin in a neutral sodium chloride solution has been carried out. At room temperature, 200 ppm of ascorbic acid was combined with pectin, and greater corrosion inhibition was observed for the binary system. The inhibition is due to the formation of protective complexes at the tin/solution interface. The results of the potentiodynamic polarization technique showed that this inhibition only affects the cathodic process, which indicates that the pectin/ascorbic acid binary system is a cathodic inhibitor. Pectin can be engineered into many applications by carefully altering its functional chemistry, like other carbohydrate biopolymers, as shown in Figure 12.3. For some anti-corrosion polymers, the overall performance of the material is improved by coupling one or more adsorption sites capable of binding to the metal surface resulting in structural changes and physical displacement of corrosive molecules such as water at the metal/solution interface.

No.	Inhibitors	Metal	Medium	Major findings	References
1	0.2 g / dm^3 pectin	cadmium	0.5 M HCl	The IE was 67.5%. Pectin was mainly an anode inhibitor and had weak adsorption capacity on the cathode.	[8] Khairou and El-Sayed, 2003
2	2 g / L citrus peel pectin	aluminum	2 M HCl	The IE was 91% at 283 K. Pectin replaces water-adsorbed molecules with a pectin skeleton, which forms a protective layer on the aluminum surface and may prevent aluminum dissolution from acidic corrosive media.	[9] Fares et al., 2012
3	250 ppm. pectin+50 ppm PPA+20 ppm Zn^{2+}	carbon steel	neutral aqueous	The IE was 94%. A possible corrosion inhibition mechanism is that the [Zn(II)-PPA-pectin] diffuses onto the metal surface and reacts with the Fe2+/Fe3+ions at the anode to produce a [Fe(II)/Fe(III)/Zn(II)-PPA-pectin] complex that covers the anodic zone and controls the reaction in this zone.	[10] Prabakaran et al., 2015
4	2 g / L citrus peel pectin	mild steel	1 M HCl	The IE was 94.2% at 318 K. Pectin is a mixed-type inhibitor, which allows both anodic and cathodic reactions to be affected simultaneously, caused by the geometric blocking effect of the inhibitor species adsorbed on the metal surface.	[11] Fiori-Bimbi, 2015
5	1000 ppm apple pectin	X60 steel	0.5 M HCl	The IE was 78.7%. HOMO orbital is located mainly on the d-galacturonic acid unit whereas the LUMO is located on the d-galacturonic methyl ester unit. These places can be directly adsorbed onto the steel surface based on donor-acceptor interactions between the π-electrons of the carbonyl groups of these units, several O atoms with excess negative charges, and the vacant d-orbitals of iron atoms.	[12] Umoren et al., 2015

6	1 g / L *opuntia ficus-indica* pectin	mild steel	1 M HCl	The IE was 96% at 308 K. Potentiodynamic polarization study showed that the inhibitor acts as a cathodic-type inhibitor.	[13] Saidi et al., 2015
7	4 g / L tomato peel pectin	tin	2% NaCl 1% acetic acid 0.5% citric acid	The application of cheap sources can implement the viable cyclical economy principle in solving the main problem of waste disposal.	[14] Grassino et al., 2016
8	500 ppm apple pectin+5 mM CeO_2	X60 steel	0.5 M HCl	The IE was 91.3% at 333 K. Pectin and CeO_2 act as moderate corrosion inhibitors for X60 steel in a 0.5 M HCl environment. The addition of CeO_2 to pectin exerted both antagonistic and synergistic.	[15] Umoren and Madhankumar, 2016
9	pectin	AZ31 Mg alloy	3.5 wt.% NaCl	The corrosion acceleration effect was linked to chelates formation between the natural polymers and Mg^{2+} ions in the electrolyte rather than on the alloy surface.	[16] Umoren et al., 2019
10	2 g / L sunflower head pectin	mild steel	1 M HCl	Sunflower head pectin (SFHP), obtained by enzyme-assisted extraction, exhibited an IE_{max} of 92.05%. Surface tests confirmed the formation of inhibitor films.	[17] Ma et al., 2019
11	10% vol. peels crude extract (PPE)	mild steel	1 M HCl	The inhibition efficiency (96%) is related to the formation of a hydrophobic organic film, probably made of various antioxidant polyphenol molecules, acting in synergy with other filming macromolecules such as pectin or fatty acids.	[18] Aghzzaf et al., 2019
12	pectin from tomato peel waste	tin	2% NaCl and 1% acetic acid	The similar values gained for CAP (75.9%) and TPP (73.9%), and fine agreement between applied methods confirmed that pectin isolated from tomato peel waste is capable to prevent the corrosion of tin in an aggressive, acid food-simulated solution.	[19] Halambek et al., 2020

(Continued)

(Continued)

No.	Inhibitors	Metal	Medium	Major findings	References
13	1 g/L pectin	6061Al–15%$_{(V)}$ $SiC_{(P)}$ composite (AlMMCs)	0.025 M HCl	The adsorption of the pectin onto the AlMMCs can be attributed either to the sharing up of lone pair of electrons from the heteroatom oxygen or to the transfer of p electrons to the vacant 3p-orbitals of the aluminum leading to the formation of a coordinate bond between the inhibitor molecule and metal surface. The inhibitor molecule replaces the already adsorbed water molecules and gets adsorbed onto the surface of the metal through the coordinate bond formation.	[20] Charitha and Rao, 2020
14	800 ppm citrus peel pectin	Mild steel	1 M H_3PO_4	The IE was 70%. Pectin acted as a mixed-type inhibitor by bringing down both cathodic and anodic reactions. Kinetic and thermodynamic parameters revealed the chemical adsorption of pectin onto the mild steel surface.	[21] Sushmitha and P. Rao, 2020
15	0.8 g/L pectin	6061 aluminum alloy	0.025 N HCl	The IE was 79.79% at 323 K. Adsorption of PECT was reiterated by quantum chemical calculations using density functional theory (DFT). DFT calculations contributed a deeper insight into the systematic understanding of the mechanism of the corrosion inhibition process.	[22] Fathima et al., 2021
16	1 g/L enzyme- and acid-extracted sugar beet pectin	mild steel	1 M HCl	The results showed the dependence of corrosion efficiency on the method of pectin extraction. Enzymatic-extracted pectin (93.9%) showed moderately higher inhibition efficiency than acid-extracted pectin (91.5%).	[23] Abou-Elseoud et al., 2021
17	5 g/L pectin	dual-phase AISI 1040 steel	0.5 M H_2SO_4	The better inhibition efficiency (83.36%) is achieved in the case of ferrite bainite when compared to the ferrite martensite structure	[24] Prabhu et al., 2021

18	5 g/L pectin	EN8 dual-phase steel	0.5 M H_2SO_4	The highest inhibition efficiency of 76.43% was achieved at 1 h of exposure at 40 °C. Physical adsorption can occur due to electrostatic interaction among protonated pectin molecules with the sulfate ion-covered metal surface. Interaction between electrons of oxygen of OH group with the metallic surface may also take place	[25] Prabhu et al., 2021
19	10 g/L pectin nanoparticles	mild steel	1 M H_2SO_4	The IE was 92% at 25 °C. Fourier-transform infrared spectroscopy (FTIR) results indicate that these nanoparticles contain various chemical bonds (C-C, CH_2, C-O-C, and cellulose) with metal surfaces, leading to producing a barrier layer that protects the surface.	[26] Abdulhessein and Ali, 2021

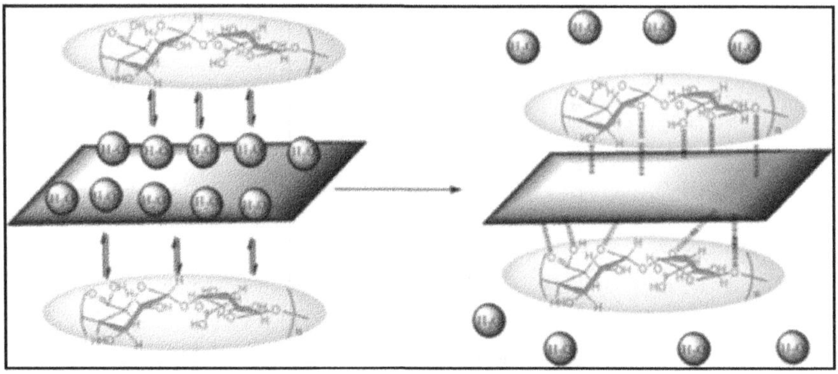

Figure 12.3 Schematic representation for the replacement of water molecules with pectin macromolecules and protective layer formation. *Source:* Fares.

12.3 Pectins Grafted Derivatives as Corrosion Inhibitors

Pectin has a large number of functional groups distributed onto its structural skeleton, as well as a certain number of neutral sugar side chains, which is the basis for the preparation of a wide range of derivatives of pectin [27]. Pectin could be modified by changing the functional groups to obtain modified compounds, including substitution (alkylation [28–31] (Fig.12.4), amidation [32–34] (Fig.12.5), quaternization [35, 36] (Fig.12.6), thiolation [37–39], sulfation [40–42], oxidation [43]), chain extension (cross bonding [44–46] and graft polymerization [47]), and degradation (chemical [48, 49], physical [50–52], and enzymatic reaction [53, 54]), as is the case with most carbohydrate polymers. Structural modifications of the anticorrosive polymers result in the coupling of single or multiple adsorption sites to the metalline surface. Corrosive molecules are mechanically displaced through the metal/solution interface, thereby improving the overall performance of the material.

Geethanjali et al. [55] synthesized pectin-grafted polyacrylamide (Pec-g-PAAm) and pectin-grafted polyacrylic acid (Pec-g-PAA) and carried out the corrosion inhibition of compounds' ability on carbon steel in 3.5% NaCl by electrochemical techniques (Figure 12.7). The testing results showed that the grafted polymers, in which pectin was employed as precursors, were able to inhibit the corrosion of steel in neutral media. The inhibitive efficiency of the pectin polymer grafted by polyacrylamide was slightly higher than that of the pectin polymer grafted by polyacrylate. In general, the inhibitive efficiency is about 85% for carbon steel in 3.5% NaCl.

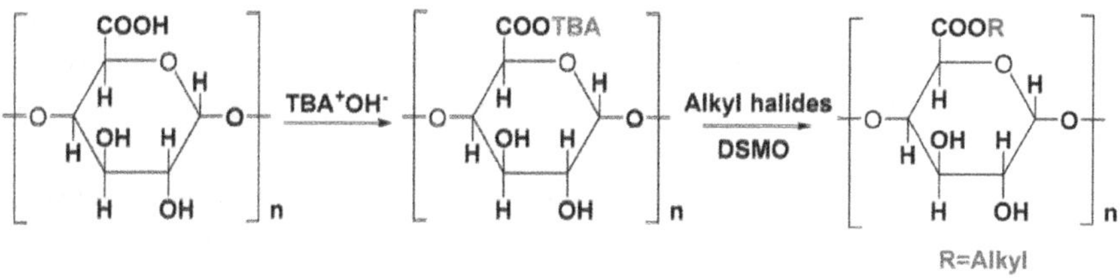

Figure 12.4 Alkylation of pectin with alkyl halide. *Source:* Chen.

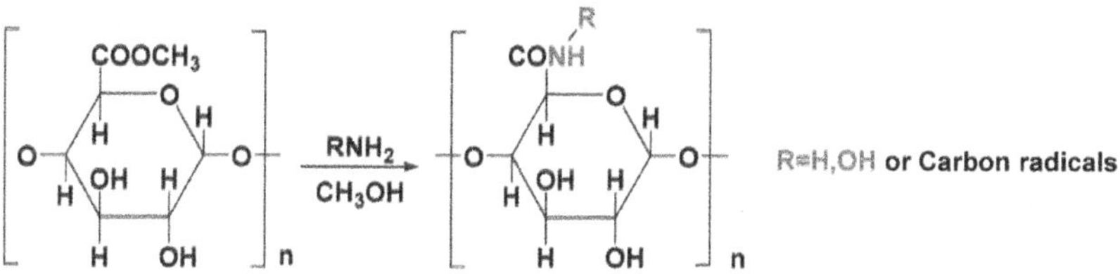

Figure 12.5 Amidation of pectin. *Source:* Chen.

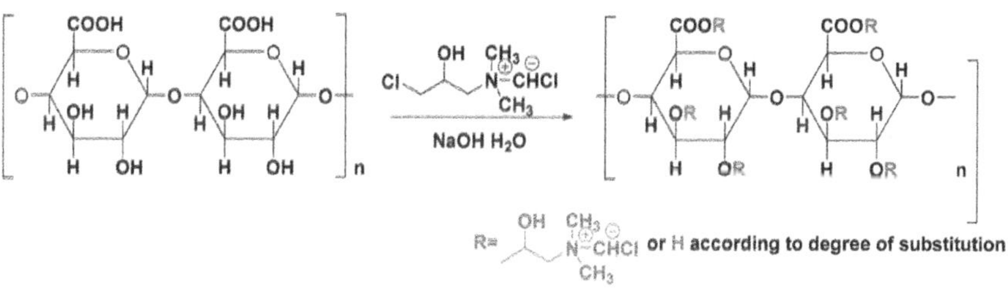

Figure 12.6 Reaction scheme for quaternization of pectin with 3-chloro-2-hydroxypropyltrimethylammonium chloride. *Source:* Fan.

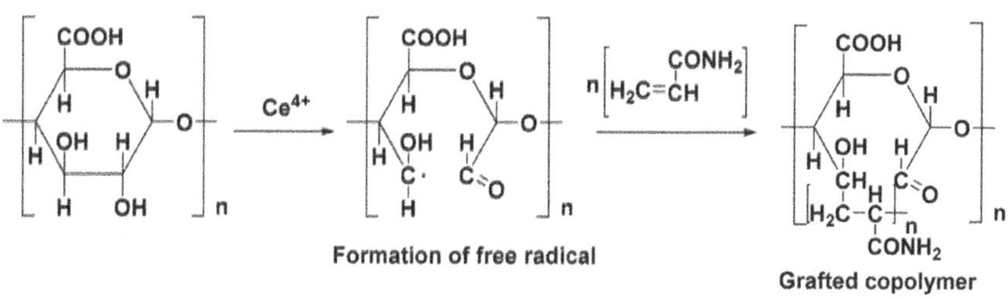

Figure 12.7 Mechanism of graft copolymerization of PAM on pectin. *Source:* Geethanjali.

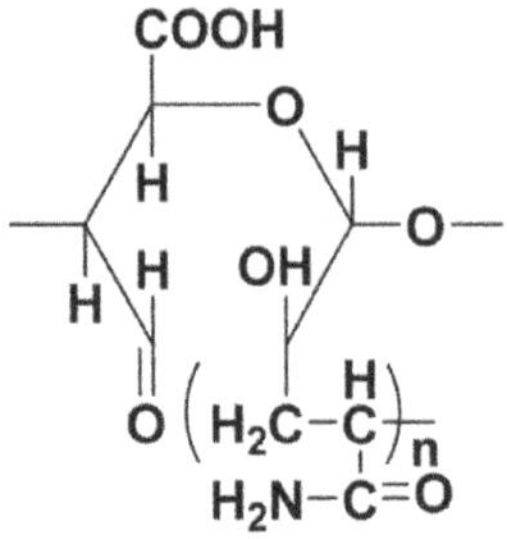

Figure 12.8 Structure of O-g-PAM. *Source:* Banerjee.

Pectin-modified products have also been reported for serving as antiscalants in water treatment showed in Figure 12.8 [56]. This process is only used to reduce the molecular weight of the pectin by acid hydrolysis before grafting, resulting in a graft copolymer with better scale protection for carbonates, sulfates, and phosphates. Scaling repair experiments with the ultimate hydrolyzed material showed better precipitation of carbonates, but the reaction was independent of temperature and pH. The crystal morphology of the precipitated phases was carried out by using SEM and XRD techniques.

12.4 Prospects and Challenges

Since the discovery of pectin in 1825, many advances have been made in the extraction of raw material sources, extraction processes, structural modifications, and applications of pectin polysaccharides. Pectin is ubiquitous in plants, and due to its wide range of applications far beyond the common usage as a food additive, pectin has found a variety of applications at the forefront of science, such as prebiotic effects, hypoglycemic effects, hypocholesterolemic effects, and anticancer effects. Research on pectins is increasing; their different applications in the food and pharmaceutical industries, along with biomedical applications are becoming more and more important and widespread. However, some questions of stability, ideal molecular weight, and interactions with other compounds remain to be investigated. Although many studies have been conducted to investigate the application of pectin derivatives in various fields, few studies have investigated their use in corrosion inhibitors. Therefore, this can also be considered a challenge to be overcome to improve the application of this class of compounds in a wider range of corrosion areas.

It is well known that the chemical constituents in some carbohydrates can form a passivation layer to block the metal/solution interface by blocking the passage of corrosive ions or molecules. Pectin is readily available in nature, is benign, renewable, and eco-friendly, and is an alternative to other potentially toxic organic inhibitors. In terms of corrosion inhibitive performance, pectin is a complex of structurally stable, naturally degradable, green macromolecules with high metal surface protection inhibition strength by forming an effective protective layer. Structural variations of pectins can yield different functions and greater bioactivity. Due to the diversity of their functional groups, they are an ideal candidate for modification. Therefore, physicochemical reactions are generally performed on hydroxyl, carboxyl, methoxy, and amide groups. In addition, the source, isolation method, and modification of pectin affect the molecular weight, viscosity, and different applications.

In summary, pectins can be tailored to new applications. This paper reviews the prospects in the application of pectin and its derivatives in corrosion inhibitors to obtain pectin polysaccharides with new bioactive structures and functions for a variety of applications. With the further development of modification technology, the research on modification and the application of modified pectins will certainly be further developed. Currently, modification of pectins is mainly to obtain to obtain some specific properties, but the modified pectins can also obtain other auxiliary properties, which can be further developed and utilized. Moreover, the application of pectin derivatives should be strongly promoted in commercial, industrial, and many other fields. In the future, the overall development of the industry is expected, and future research directions will be based on biopolymers, biotechnology, and renewable resources. This review will contribute to the research of novel corrosion inhibitors and provide new possibilities for the design of practical pectin-based corrosion inhibitors for metal corrosion solutions.

Acknowledgments

This work was supported by the National Natural Science Foundation of China (21672046, 21372054).

Declaration of Competing Interest

On behalf of all authors, the corresponding author states that there is no conflict of interest.

References

1 Rodsamran, P. and Sothornvit, R. (2019). Lime peel pectin integrated with coconut water and lime peel extract as a new bioactive film sachet to retard soybean oil oxidation. *Food Hydrocolloids* 97: 105173.
2 Sheldon, R.A. (2016). Green chemistry, catalysis, and valorization of waste biomass. *Journal of Molecular Catalysis A: Chemical* 422: 3–12.
3 Yang, J.S., Mu, T.H., and Ma, M.M. (2018). Extraction, structure, and emulsifying properties of pectin from potato pulp. *Food Chemistry* 244: 197–205.
4 Keppler, F., Hamilton, J.T.G., Bra, M., and Röckmann, T. (2006). Methane emissions from terrestrial plants under aerobic conditions. *Nature* 439: 187–191.

5 Kumar, A.M., Adesina, A.Y., Hussein, M.A., Umoren, S.A., and Saravanan, S. (2020). Preparation and characterization of Pectin/Polypyrrole based multifunctional coatings on TiNbZr alloy for orthopedic applications. *Carbohydrate Polymers* 242: 116285.

6 Haas, K.T., Wightman, R., Meyerowitz, E.M., and Peaucelle, A. (2020). Pectin homogalacturonan nanofilament expansion drives morphogenesis in plant epidermal cells. *Science* 367: 1003–1007.

7 Mohnen, D. (2008). Pectin structure and biosynthesis. *Current Opinion in Plant Biology* 11: 266–277.

8 Khairou, K.S., and El-Sayed, A. (2003). Inhibition effect of some polymers on the corrosion of cadmium in a hydrochloric acid solution. *Journal of Applied Polymer Science* 88: 866–871.

9 Fares, M.M., Maayta, A.K., and Al-Qudah, M.M. (2012). Pectin is a promising green corrosion inhibitor of aluminum in hydrochloric acid solution. *Corrosion Science* 60: 112–117.

10 Prabakaran, M., Ramesh, S., Periasamy, V., and Sreedhar, B. (2015). The corrosion inhibition performance of pectin with propyl phosphonic acid and Zn^{2+} for corrosion control of carbon steel in an aqueous solution. *Research on Chemical Intermediates* 41: 1–23.

11 Fiori-Bimbi, M.V., Alvarez, P.E., Vaca, H., and Gervasi, C.A. (2015). Corrosion inhibition of mild steel in HCL solution by pectin. *Corrosion Science* 92: 192–199.

12 Umoren, S.A., Obot, I.B., Madhankumar, A., and Gasem, Z.M. (2015). Performance evaluation of pectin as ecofriendly corrosion inhibitor for X60 pipeline steel in acid medium: experimental and theoretical approaches. *Carbohydrate Polymers* 124: 280–291.

13 Saidi, N., Elmsellem, H., Ramdani, M., Chetouani, A., and Hammouti, B. (2015). Using pectin extract as an eco-friendly inhibitor for steel corrosion in 1M HCl media. *Der Pharma Chemica* 7: 87–94.

14 Grassino, A.N., Halambek, J., Djakovic, S. et al. (2016). Utilization of tomato peel waste from a canning factory as a potential source for pectin production and application as a tin corrosion inhibitor. *Food Hydrocolloids* 52: 265–274.

15 Umoren, S.A. and Madhankumar, A. (2016). Effect of addition of CeO2 nanoparticles to pectin as an inhibitor of X60 steel corrosion in HCl medium. *Journal of Molecular Liquids* 224: 72–82.

16 Umoren, S.A., Solomon, M., Madhankumar, A., and Obot, I.B. (2019). Exploration of natural polymers for use as green corrosion inhibitors for AZ31 magnesium alloy in a saline environment. *Carbohydrate Polymers* 230: 115466.

17 Ma, X.M., Wang, J.B., Xu, J.J. et al. (2019). Sunflower head pectin with different molecular weights is a promising green corrosion inhibitor of carbon steel in a hydrochloric acid solution. *Acs Omega* 4: 21148–21160.

18 Aghzzaf, A.A., Veys In enaux, D., and Rocca, E. (2019). Pomegranate peels crude extract as a corrosion inhibitor of mild steel in HCl medium: passivation and hydrophobic effect. *Materials and Corrosion* 71: 148–154.

19 Halambek, J., Cindrić, I., and Grassino, A.N. (2020). Evaluation of pectin isolated from tomato peel waste as a natural tin corrosion inhibitor in sodium chloride/acetic acid solution. *Carbohydrate Polymers* 234: 115940.

20 Charitha, B.P., and Rao, P. (2020). Pectin as a potential green inhibitor for corrosion control of 6061Al-15%((V))SiC((P))Composite in acid medium: electrochemical and surface studies. *Journal of Failure Analysis and Prevention* 20: 1684–1696.

21 Sushmitha, Y. and Rao, P. (2020). Electrochemical Investigation on the acid corrosion control of mild steel using biopolymer as an Inhibitor. *Portugaliae Electrochimica Acta* 38: 149–163.

22 Fathima, H., Pais, M., and Rao, P. (2021). Anticorrosion performance of biopolymer pectin on 6061 aluminum alloy: electrochemical, spectral and theoretical approach. *Journal of Molecular Structure* 1243.

23 Abou-Elseoud, W.S., Abdel-Karim, A.M., Hassan, E.A., and Hassan, M.L. (2021). Enzyme- and acid-extracted sugar beet pectin as green corrosion inhibitors for mild steel in hydrochloric acid solution. *Carbohydrate Polymer Technologies and Applications* 2: 100072.

24 Prabhu, D., Sharma, S., Prabhu, P.R. Jomy, J. Sadanand, R.V. (2021). Analysis of the inhibiting action of pectin on corrosion of AISI1040 dual-phase steel with ferrite-martensite and ferrite-bainite structure: a comparison in 0.5 M sulphuric acid. *Journal of the Iranian Chemical Society* 19: 1109–1128.

25 Prabhu, P.R., Hiremath, P., Prabhu, D., Gowrishankar, M.C., and Gurumurthy, B.M. (2021). Chemical, electrochemical, thermodynamic, and adsorption study of EN8 dual-phase steel with ferrite-martensite structure in 0.5 M H2SO4 using pectin as an inhibitor. *Chemical Papers* 75: 6083–6099.

26 Abdulhessein, B.A. and Ali, A.M. (2021). Green corrosion inhibitor of low carbon steel in acidic media by pectin nanoparticles using potentiostatic and weight losses method. *Materials Science Forum* 1021: 241–250.

27 Chen, J., Liu, W., Liu, C.M. et al. (2015). *Pectin Modifications: A Review, Critical Reviews in Food Science & Nutrition* 55: 1684–1698.

28 Fischer, A., Houzelle, M.C., Hubert, P. et al. (1998). Detection of intramolecular associations in hydrophobically modified pectin derivatives using fluorescent probes. *Langmuir* 14: 4482–4488.

29 Miralleshouzelle, M.C., Hubert, P., and Dellacherie, E. (2001). Hydrophobic Alkyl Chains–Pectin conjugates. Comparative study of some physicochemical properties in relation to covalent coupling vs ionic association. *Langmuir* 17: 1384–1391.

30 Fraeye, I., Roeck, A.D., Duvetter, T. et al. (2007). Influence of pectin properties and processing conditions on thermal pectin degradation. *Food Chemistry* 105: 555–563.

31 Kim, Y., Yoo, Y.H., Kim, K.O. et al. (2008). Textural properties of the gelling system of low-methoxy pectins produced by the demethylating reaction of pectin methyl esterase. *Journal of Food Science* 73: C367–C372.

32 Musabayane, C., Munjeri, O., Bwititi, P., and Osim, E. (2000). Orally administered, insulin-loaded amidated pectin hydrogel beads sustain plasma concentrations of insulin in streptozotocin-diabetic rats. *Journal of Endocrinology* 164: 1–6.

33 Einhorn-Stoll, U., Salazar, T., Jaafar, B., and Kunzek, H. (2001). Thermodynamic compatibility of sodium caseinate with different pectins. Influence of the milieu conditions and pectin modifications. *Die Nahrung* 45: 332–337.

34 Rha, H.J., Bae, I.Y., Lee, S. et al. (2011). Enhancement of anti-radical activity of pectin from apple pomace by hydroxamation. *Food Hydrocolloids* 25: 545–548.

35 Fan, L., Gao, S., Wang, L. et al. (2012). Synthesis and anticoagulant activity of pectin sulfates. *Journal of Applied Polymer Science* 124: 2171–2178.

36 Simkovic, I., Uhliariková, I., Yadav, M.P., and Mendichi, R. (2010). Branched arabinan obtained from sugar beet pulp by quaternization under acidic conditions. *Carbohydrate Polymers* 82: 815–821.

37 Smart, J.D. (2005). The basics and underlying mechanisms of mucoadhesion. *Advanced Drug Delivery Reviews* 57: 1556–1568.

38 Sharma, R. and Ahuja, M. (2011). Thiolated pectin: synthesis, characterization, and evaluation as a mucoadhesive polymer. *Carbohydrate Polymers* 85: 658–663.

39 Majzoob, S., Atyabi, F., Dorkoosh, F. et al. (2006). Pectin-cysteine conjugate: synthesis and in-vitro evaluation of its potential for drug delivery. *Journal of Pharmacy and Pharmacology* 58: 1610–1610.

40 Martinichen-Herrero, J.C., Carbonero, E.R., Gorin, P., and Iacomini, M. (2005). Anticoagulant and antithrombotic activity of a sulfate obtained from a glucan component of the lichen Parmotrema mantiqueirense Hale. *Carbohydrate Polymers* 60: 7–13.

41 Maas, N.C., Gracher, A., Sassaki, G.L. et al. (2012). Sulfation pattern of citrus pectin and its carboxy-reduced derivatives: influence on anticoagulant and antithrombotic effects. *Carbohydrate Polymers* 89: 1081–1087.

42 Bae, I.Y., Yu, N.J., Rha, H.J. et al. (2009). Effect of sulfation on the physicochemical and biological properties of citrus pectins. *Food Hydrocolloids* 23: 1980–1983.

43 Takei, T., Sato, M., Ijima, H., and Kawakami, K. (2010). In situ gellable oxidized citrus pectin for localized delivery of anticancer drugs and prevention of homotypic cancer cell aggregation. *Biomacromolecules* 11: 3525–3530.

44 Willats, W., Knox, J.P., and Mikkelsen, J.D. (2006). Pectin: new insights into an old polymer are starting to gel. *Trends in Food Science & Technology* 17: 97–104.

45 Vincken, J.-P. (2003). If homogalacturonan were a side chain of Rhamnogalacturonan I. Implications for cell wall architecture. *Plant Physiology* 132: 1781–1789.

46 Semdé, R., Moës, A.J., Devleeschouwer, M.J., and Amighi, K. (2003). Synthesis and enzymatic degradation of epichlorohydrin cross-linked pectins. *Drug Development & Industrial Pharmacy* 29: 203–213.

47 Fares, M.M., Assaf, S.M., and Abul-Haija, Y.M. (2010). Pectin grafted poly(N-vinylpyrrolidone): optimization and in vitrocontrollable theophylline drug release. *Journal of Applied Polymer Science* 117: 1945–1954.

48 Round, A.N., Rigby, N.M., Macdougall, A.J., and Morris, V.J. (2010). A new view of pectin structure was revealed by acid hydrolysis and atomic force microscopy. *Carbohydrate Research* 345: 487–497.

49 Elboutachfaiti, R., Delattre, C., Michaud, P., Courtois, B., and Courtois, J. (2008). Oligogalacturonans production by free radical depolymerization of polygalacturonan. *International Journal of Biological Macromolecules* 43: 257–261.

50 Roeck, A.D., Duvetter, T., Fraeye, I. et al. (2009). Effect of high-pressure/high-temperature processing on chemical pectin conversions in relation to fruit and vegetable texture. *Food Chemistry* 115: 207–213.

51 Chen, J., Liang, R.H., Wei, L. et al. (2012). Degradation of high-methoxyl pectin by dynamic high-pressure microfluidization and its mechanism – ScienceDirect. *Food Hydrocolloids* 28: 121–129.

52 Makuuchi, K. (2010). A critical review of radiation processing of hydrogel and polysaccharide. *Radiation Physics & Chemistry* 79: 267–271.

53 Yadav, S., Yadav, P.K., Yadav, D., and Yadav, K.D.S. (2009). Pectin lyase: a review. *Process Biochemistry* 44: 1–10.

54 Bonnin, E., Dolo, E., Goff, A.L., and Thibault, J.F. (2002). Characterization of pectin subunits released by an zoptimized combination of enzymes. *Carbohydrate Research* 337: 1687–1696.

55 Geethanjali, R., Fathima, S., and Subhashini, S. (2014). Water-soluble and biodegradable pectin-grafted polyacrylamide and pectin-grafted polyacrylic acid: electrochemical investigation of corrosion-inhibition behavior on mild steel in 3.5% NaCl media. *Indian Journal of Materials Science* 2014: 1–9.

56 Chauhan, K., Kumar, R., Kumar, M., Sharma, P., and Chuahan, S.G. (2012). Modified pectin-based polymers as green antiscalants for calcium sulfate scale inhibition. *Desalination* 305: 31–37.

13

Grafted Chitosan as Sustainable Corrosion Inhibitors

Elyor Berdimurodov[1,], Abduvali Kholikov[1], Khamdam Akbarov[1], Khasan Berdimuradov[2], Nilufar Tursunova[1], Omar Dagdag[3], Rajesh Haldhar[4], Mohamed Rbaa[5], Brahim El Ibrahimi[6], and Dakeshwar Kumar Verma[7]*

[1] *Faculty of Chemistry, National University of Uzbekistan, Tashkent, 100034, Uzbekistan*
[2] *Faculty of industrial Viticulture and Food Production Technology, Shahrisabz branch of Tashkent Institute of Chemical Technology, Shahrisabz, 181306, Uzbekistan*
[3] *Institute of Nanotechnology and Water Sustainability, College of Science, Engineering and Technology, the University of South Africa, Johannesburg, South Africa*
[4] *School of Chemical Engineering, Yeungnam University, Gyeongsan, 712749, South Korea*
[5] *Laboratory of Organic Chemistry, Catalysis and Environment, Faculty of Sciences, Ibn Tofail University, PO Box 133, 14000, Kenitra, Morocco*
[6] *Department of Applied Chemistry, Faculty of Applied Sciences, Ibn Zohr University, 86153, Morocco*
[7] *Department of Chemistry, Government Digvijay Autonomous Postgraduate College, Rajnandgaon, Chhattisgarh 491441, India*
* *Corresponding author*

13.1 Introduction

13.1.1 Importance of Grafted Chitosan as Sustainable Corrosion Inhibitors

Metallic materials comprised of steel, copper, and aluminium are widely employed in the petroleum, oil-gas, electronic, and automobile industries [1–3]. These metal materials are typically exposed to the environment when they come in contact with the atmosphere, soil, petroleum fuels or gases, organic solvents, organic acids, natural gases, water, and moisture [4]. As a result, the metal surface becomes corroded and is destroyed. This is not just an economic issue of supply chain scarcity for the many industries that rely on these raw materials—it is an environmental problem as well [5, 6]. To address these concerns, corrosion inhibitors are widely used to protect metal from corrosion processes [7].

The grafted chitosan has intriguing applications as a corrosion inhibitor in acidic, saline, and alkaline mediums. The grafted chitosan boasts natural corrosion inhibitors that are less toxic, environmentally friendly, good water-soluble, biodegradable,

Grafted Biopolymers as Corrosion Inhibitors: Safety, Sustainability, and Efficiency, First Edition.
Edited by Jeenat Aslam, Chandrabhan Verma, and Ruby Aslam.

non-volatile, non-flammable, low cost, and don't require dangerous technology or syntheses methodology. Currently, the chitosan has undergone adaptations to create new chitosan modifications [8, 9]. In these compounds, the phosphor, sulphur, oxygen, and nitrogen atoms transform the grafted chitosan into a better inhibitor that is more polar [10]. The chitosan was modified with more electron-rich functional groups, aromatic rings, and heteroatoms [11]. For example, the chitosan was grafted with the following functional groups: >C=O–N=O, –N=N– [12], >C=N–>C=O, >C=C<, amide (–$CONH_2$), thioether (–S–), thiol (–SH), nitrile (–CN), acid halide (–COX), nitro (–NO_2), ester (–COOR), ether (–O–), Alkoxy (–OR), trialkyl amine (–NR_2), dialkyl amine (–NHR), amino/monoalkyl amine (–NH_2) and hydroxyl (–OH) [13, 14]. Several review papers [4, 15, 16] discussed the main properties of chitosan and its derivatives as corrosion inhibitors for various metals in acidic, neutral, alkaline, and saline corrosion medium [17]. It is indicated in [4, 15, 16] that chitosan and its derives are high potential corrosion inhibitors for aluminium, steel, and copper in the acidic, neutral, alkaline, and saline corrosion mediums. The inhibition efficiency was increased by the amino-functional groups, which is more convenient for functionalization [18]. Some research works suggest that chitosan is the interface and mixed type with the anodic dominance. This suggestion was confirmed by the results of Langmuir adsorption isotherms [19]. Naturally, chitosan is not soluble and has no good corrosion inhibition efficiency. The corrosion of metal samples in the aqueous phase is not favourably blocked [20, 21]. However, the various compounds of chitosan such as Schiff bases, carbon dots, ethers, nitrogen, and hydroxyl-contained modifications maintain effectiveness when water-soluble. In some research works [22], the corrosion protection degree of water-soluble chitosan compounds has been thoroughly investigated. The results indicate that these compounds effectively blocked both the metal oxidation and hydrogen formation by the high degree of adsorption performance [22]. Other research works [23, 24] show that the inhibition protection of chitosan depends on various factors from the metal type, the nature of the corrosion solution, temperature and pH of the environment, inhibitor concentration, as well as immersion time. For example, the inhibition rose with the increase in immersion time. The reason is that the adsorption degree rose with the increased immersion time.

13.2 Main Part

13.2.1 Grafted Chitosan-Based Compounds as Sustainable Anti-corrosion Agents for Steel-based Materials: Theoretical and Investigational Insights

Currently, the grafted chitosan derivatives are broadly performed as anti-corrosion agents for steel-based materials in various corrosion mediums [25, 26]. Table 13.1 indicates the main characteristics of grafted chitosan-based compounds as sustainable anti-corrosion agents for steel-based materials. For example, Hasanin and Kiey

Table 13.1 Main characteristics of grafted chitosan-based compounds as sustainable protective agents for steel-based materials.

No.	Molecular structure and name	Metal type	Analysed Solution	Protection degree/inhibitor type/adsorption isotherm	References
1.	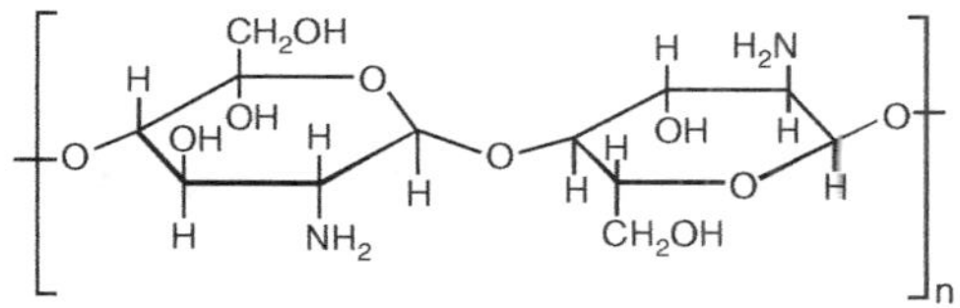 Chitosan	Mild steel	0.1 M HCl	96% at 60 °C and 4 μM/ mixed type/Langmuir	[53]
2.	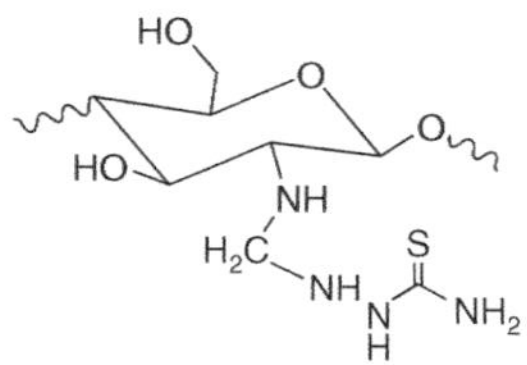 Thiosemicarbazide modified chitosan	304 steel	2% acetic acid	55% at 30 mg/L and 30 2 °C/ mixed-type/-	[37]
3.	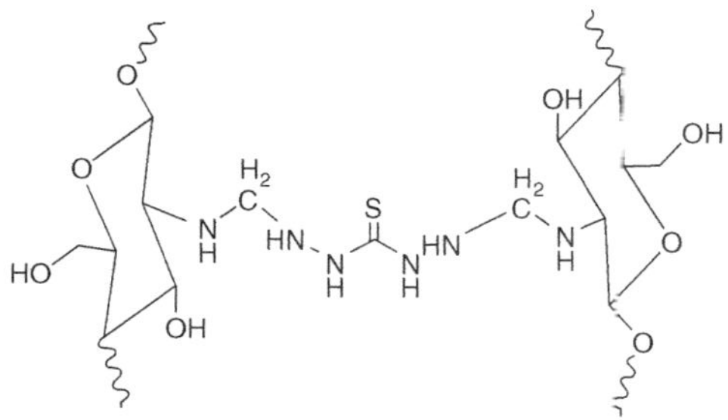 Thiocarbohydrazide modified chitosan			92% at 30 mg/L and 30±2 °C/mixed-type/-	

(*Continued*)

Table 13.1 (Continued)

No.	Molecular structure and name	Metal type	Analysed Solution	Protection degree/inhibitor type/adsorption isotherm	References
4.	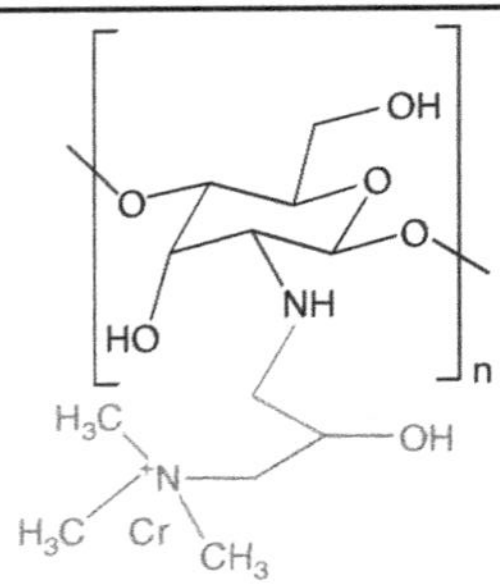 N-(2-hydroxy-3-trimethyl ammonium) propyl chitosan chloride	Mild steel	1 M HCl	98.9% at 500 ppm/mixed-type/Langmuir	[54]
5.	Zinc oxide nanoparticles ZnO NPs/Chitosan	Mild steel	1 M HCl	93.95% at 100 ppm/mixed type/Langmuir	[27]
6.	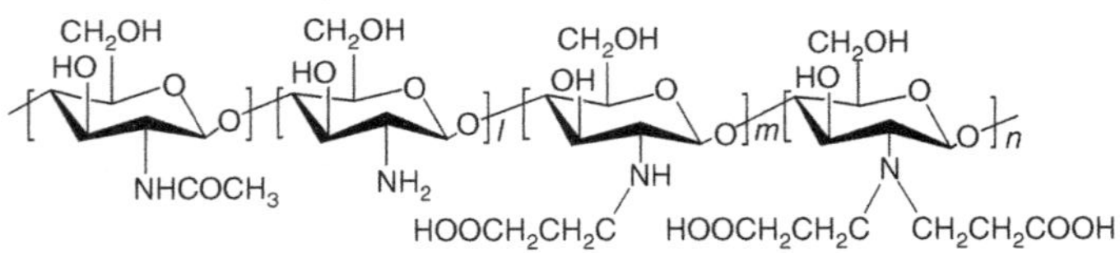 N-carboxylated chitosan with glacial acetic acid	Carbon steel	1 M HCl	Corrosion rate is 0.0254 mm/y at 40 °C/mixed-type/-	[55]

7.	Grafted chitosan with graphene oxide and allyl sulfonate	X70 pipeline steel	CO_2 saturated NaCl electrolyte at 60 °C	89.7% at 25 ppm/-	[32]
8.	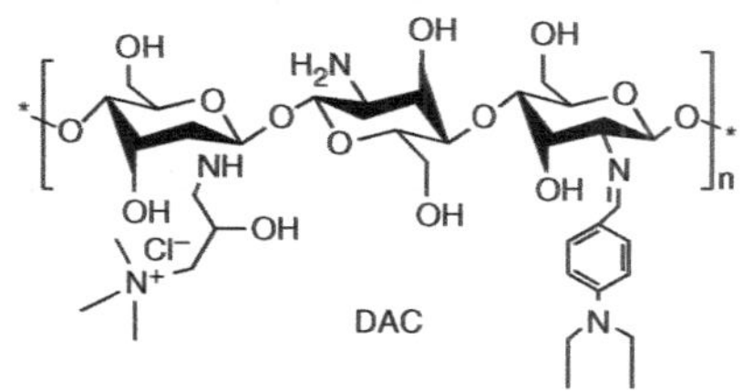 N-4-(diethylamino)benzyl quaternary ammonium chitosan	X80 pipeline steel	1 M HCl	93.83% at 6 mg/L/mixed inhibitors dominated by the anodic reaction/ Langmuir	[56]
9.	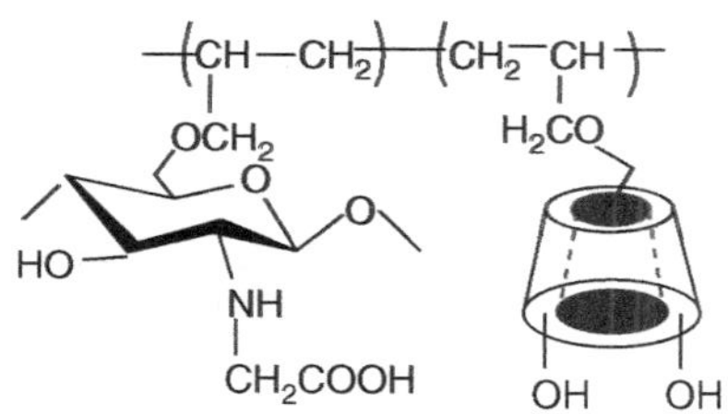 Grafted chitosan with β-cyclodextrin	Carbon steel	1 M HCl	96.02% at 230 ppm and 298 K/mixed type/ Langmuir	[57]

(Continued)

Table 13.1 (Continued)

No.	Molecular structure and name	Metal type	Analysed Solution	Protection degree/inhibitor type/adsorption isotherm	References
10.	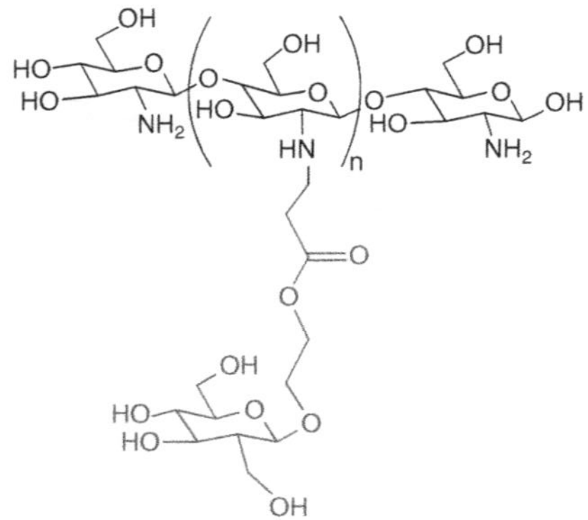 Glucosyl oxyethyl acrylate graft chitosan derivative	5 L X70 pipeline steel	1 M HCl	90.4% at 500 ppm/-	[58]
11. 12.	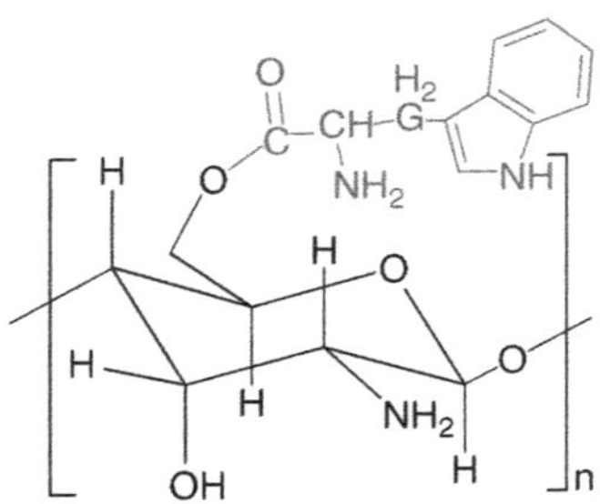 Grafted chitosan (MCO) with sodium silicate (SS)	Carbon steel	3.5 wt% NaCl	82.73% for MCO and 98.25% for MCO+SS at 350 ppm/mixed type/ Langmuir	[59]

13.	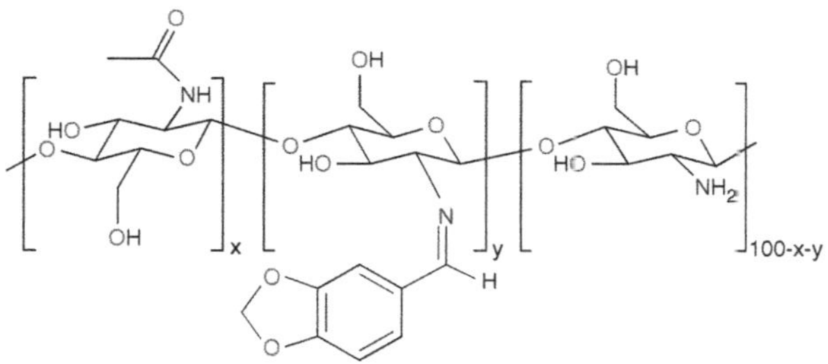 Piperonal-chitosan Schiff base (y=number of replacements and x=number of acetylation)	Carbon steel	15% HCl	85.16% at 600 mg/L/ (91.15% at a low dosage of 10 mm of KI)/mixed-type with cathodic dominate/-	[50]
14.	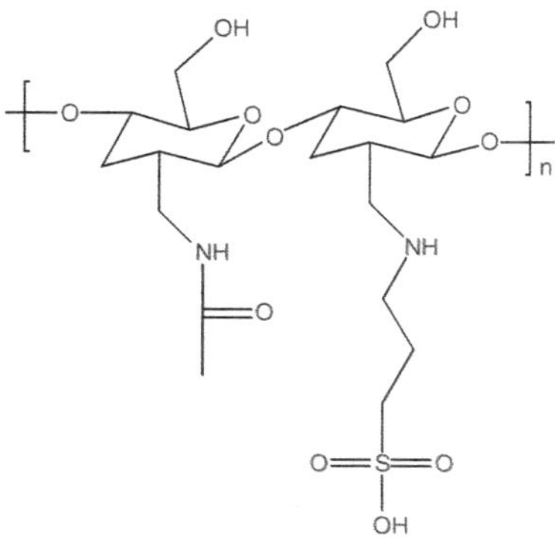 Sulfonated chitosan	Carbon steel	2 M HCl	95.6±0.1 at 5000 ppm/ mixed type/Langmuir	[39]

(Continued)

Table 13.1 (Continued)

No.	Molecular structure and name	Metal type	Analysed Solution	Protection degree/inhibitor type/adsorption isotherm	References
15.	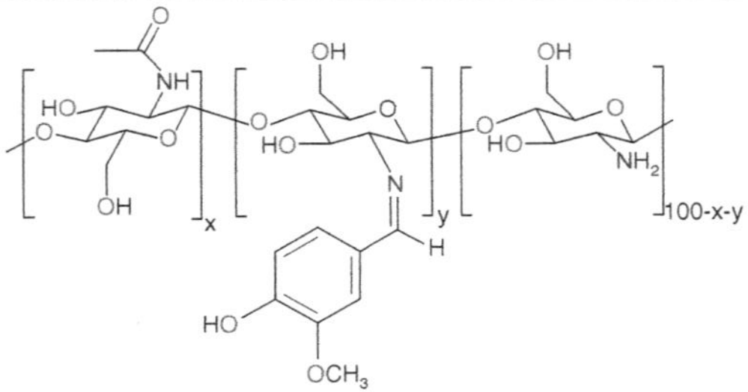 Vanillin-modified chitosan (y=number of replacements and x=number of acetylation)	Carbon steel	15% HCl	98% at 500 mg/L/ mixed-type behaviour with cathodic pre-dominance/Langmuir	[60]
16.	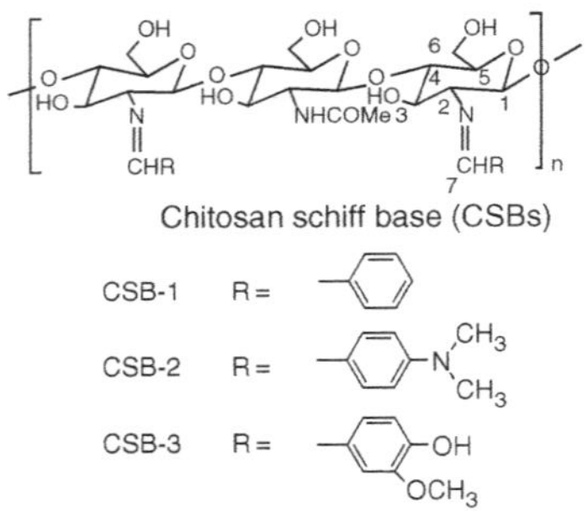 4-Hydroxy-3-methoxybenzaldehyde (CSB-3), 4-(dimethylamino)-benzaldehyde (CSB-2), benzaldehyde (CSB-1).	Mild steel	1 M HCl	90.65% at 50 ppm/mixed-type/-	[49]

[27] introduced the nanocomposite-based chitosan (CH), which was doped with zinc oxide nanoparticles ZnO NPs. The structural characteristics and inhibition performance for mild steel in 1 M HCl of this grafted chitosan were investigated in this research work. This grafted chitosan was prepared using environmentally friendly and cost-effective methods. The spectroscopic methods were used to characterize the prepared nanocomposite-based chitosan (CH), which was doped with zinc oxide nanoparticles ZnO NPs. The corrosion inhibition protection of this selected compound was analysed at different concentrations. It is indicated that the defender's ability rose with the increase in concentration due to the nature of corrosion inhibitors covering a large area of metal surface. As a result, the inhibition efficiency rose and the contact regions of metal surface with electrolytes decreased. The results of the DLS analysis confirmed that this CH-ZnO NPs corrosion inhibitor is highly stable. Its corrosion inhibition efficiency is more favourable for mild steel in aggressive hydrochloric acid solutions. The size of the nanomaterial is also the next important factor to increase inhibition efficiency. The surface morphology of mild steel after corrosion and inhibition with the selected nanomaterial was investigated by applying the EDX and SEM methods. The maximum corrosion protection degree of studied chitosan-based nanomaterials was 93.95%, meaning the corrosion processes on the anode and cathode regions were maximally blocked by the formation of a protective film. The corrosion inhibitor effectively controlled the charge transfer resistance. The kinetic information at the metal/solution contact was characterized by various electrochemical methods. The electron transformation during the corrosion and inhibition processes was described by the phase angle and Body plots. The obtained results confirmed that the rate of electron transfer was very high in corrosion processes, which means that the metal electrode was destroying data at a high rate. In comparison, the rate of electron transfer is very low in inhibited solutions, indicating that the corrosion inhibitor effectively blocked the metal surface. The data also revealed that the Nyquist curves for corrosive and inhibited solutions were a single depressed semicircle, representing that an anodic steel dissolution and cathodic hydrogen evolution are controlled by the charge transfer mechanism [28]. The size of obtained Nyquist plots rose with the presence of selected chitosan-based corrosion inhibitors, confirming that the polarization resistance is increased by the formation of a protective film on the metal surface [29]. It was also determined that the obtained Body plots were one-time constant, showing that the corrosion and inhibition processes are controlled by the charge transfer resistance. The size of Body plots rose with the increase in the inhibitor concentration, confirming that a dense film was formed on the metal surface by the strong adsorption of CH-ZnO NPs inhibitor. The Z and phase angle shifted towards the negative region with the inhibitor addition, revealing that the surface coverage expanded with the inhibitor introduction [30, 31]. As can be indicated in Fig. 1D, the constant phase element, polarization resistance, and solution resistance shows the capacitance of the electrical double layer and adsorbed inhibitor film [27].

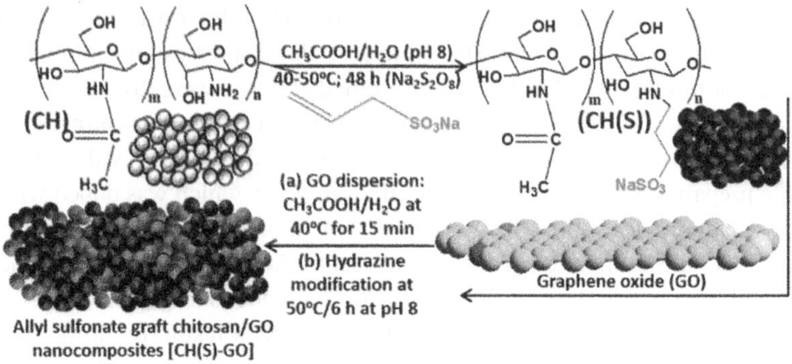

Figure 13.1 Schematic illustrations of preparation of chitosan nano compound for corrosion inhibitor. [32] / Royal Society of Chemistry / Licensed under CC BY 3.0.

Eduok et al. [32] introduced the allyl sulfonate graft chitosan/graphene oxide nanocomposite material as an anticorrosive material for X70 pipeline steel substrate in a CO_2-saturated NaCl electrolyte at 60 °C. This material is green in corrosion protection since chitosan is a natural product. This focal chitosan-based nanocomposite in the study was synthesized through the reaction between the amine-bound glucosamine molecular units and reactive chemical groups according to the syntheses schema in Figure 13.1, which indicates the schematic illustrations of the preparation of chitosan nano compounds for the corrosion inhibitor [32]. In this preparation, the allyl sulfonate reacted with the chitosan in the acidic catalytic environment at alkaline pH. Then, the formed allyl sulfonate graft chitosan CH(S) was modified to graphene oxide (GO) to form the CH(S)–GO corrosion inhibitor. The inhibition characteristics of this material were investigated for X70 pipeline steel substrate in a CO_2-saturated NaCl electrolyte at 60 °C by the Fourier transform infrared spectroscopy, X-ray diffraction spectrometry, thermo-gravimetric analysis, scanning electron microscope, energy-dispersive X-ray spectroscopy, and transmission electron microscopy methods. The final results suggest that this grafted chitosan formed the defender of the polymeric chitosan-GO hybrid nanocomposite films on the metallic surface. As a result, the CO_2 dissolution and NaCl salt deposits on the metal surface were considerably blocked. The results of the surface analysis confirm that the surface protection of this composite is superior. The hydrophobicity of the protective film is increased by the reduced GO nanosheet contents. The graphene oxide is responsible for insulating the metal surface from the aqueous corrosion solution [33, 34]. The hydroxyl and amino functional groups of nanocomposites are mainly responsible for corrosion inhibition. As can be observed in the obtained result, the allyl sulfonate graft chitosan CH(S) effectively insulates the metal surface from contact exposure in an aqueous environment. The inhibitor forms a protective film on the metal surface that is more thermodynamically stable in the acidic corrosion medium. The coordination bonds between the metal surface and grafted chitosan are also more stable [35, 36].

Li et al. [37] investigated the inhibition performance of two newly grafted chitosan—namely thiocarbohydrazide (TCFCS) and thiosemicarbazide (TSFCS)—for

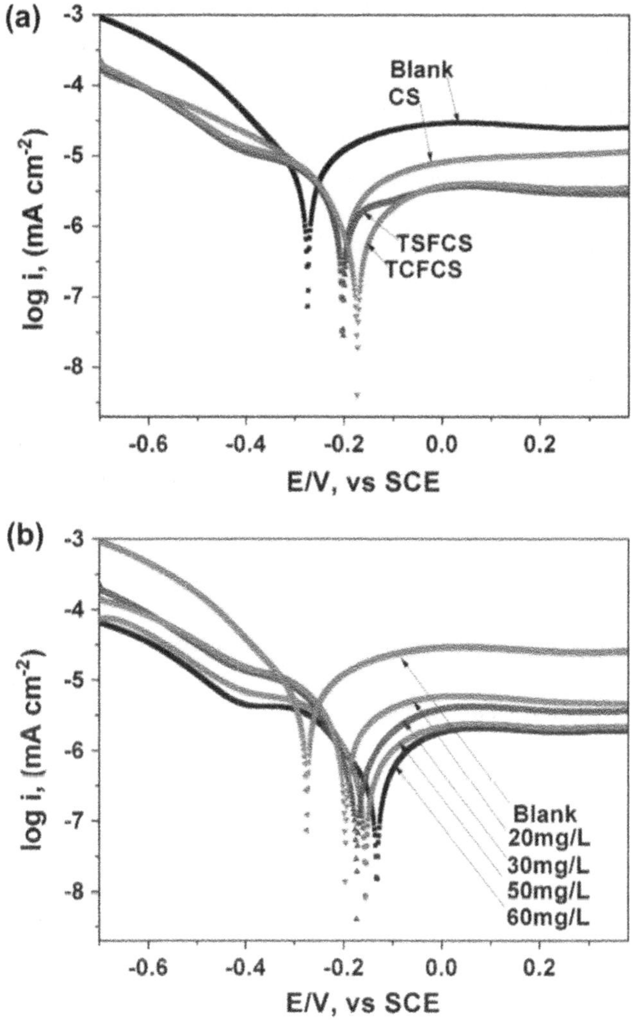

Figure 13.2 Tafel curves for blank and TCFCS corrosion inhibitors at various concentrations. [37] / With permission of Elsevier.

304 steel in a 2% acetic acid corrosive solution. The main properties of synthesized TCFCS and TSFCS are studied by differential scanning calorimetry, thermal gravity analysis, elemental analysis, and Fourier-transform infrared spectroscopy methods. The surface morphology was investigated through scanning electron microscopy. It was found that (i) the TCFCS and TSFCS are mixed-type corrosion inhibitors; (ii) the maximum inhibition efficiency was over 92% at 60 mg/L; they are good adsorbents for heavy ions.

Figure 13.2 indicates the Tafel curves for blank and TCFCS corrosion inhibitors at various concentrations [37]. It should be stressed that the polarization curves shifted to more positive regions with the presence of TCFCS, suggesting that the hydrogen evolution and steel dissolution were decreased with the formation of a protective film. The Tafel curves depend on the change in concentration, which means that when the concentration rose, the Tafel curves shift to lower corrosion current density regions. This is due to the corrosion inhibitor increasing the polarization resistance [38]. On the other hand, the high polarization resistance depends on the molecular structure of the studied grafted chitosan. The amino-functional groups are chemically linked with the metal surface for rigid covalent bonds. The main electrochemical kinetics parameters such as the corrosion current density and corrosion potential were estimated using the obtained polarization curves.

Farhadian et al. [39] synthesized the sulphonated chitosan (SCS) and studied its inhibition performance for steel in a 2 M HCl solution. The obtained results show that the maximum protection degree of SCS was 95.6±0.1% at 5000 ppm. The corrosion current density was reduced and polarization resistance increased with the presence of SCS related to the results of electrochemical experiments. It is also indicated that the surface roughness of carbon steel was reduced with the inhibitor adsorption according to the results in the profilometer and SEM-EDX surface analysis. As can be observed in the quantum chemical calculations, the covalent bonds were formed between the chitosan modification and carbon steel surface through the sulphide functional groups. The stability of inhibitor-formed protective film is estimated by the open circuit potential analysis. Figure 13.3 shows the obtained OCP for blank and SCS corrosion inhibitors at various

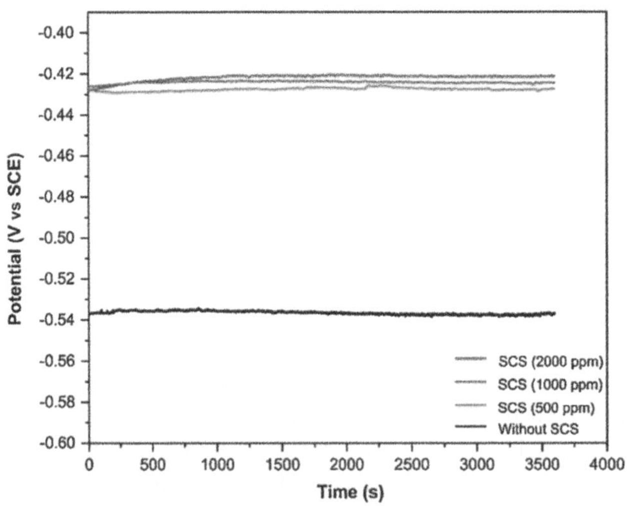

Figure 13.3 Obtained OCP for blank and SCS corrosion inhibitors at various concentrations. [39] / With permission of Elsevier.

concentrations. The results in OCP confirmed that the defensive thin layer was formed on the metal surface and this film is more stable. The OCP was maintained around more positive potential, showing more stability of the formed defensive layer. It is also revealed that the open circuit potential is nearly stable during this period, confirming that the inhibitor-formed protective film is more stable in the acidic corrosion solution.

The adsorption performances of grafted chitosan were described by the adsorption isotherms. Currently, the Frumkin, Temkin, and Langmuir isotherms are widely used to identify the adsorption characters on the metal surface. The thermodynamic effect of the corrosion inhibitor/metal surface interactions is also detailed by the Frumkin [40, 41], Temkin [42], and Langmuir [43, 44] isotherms. The main adsorption thermodynamic parameters such as the entropy, enthalpy, Gibbs free energy, and constant for adsorption-desorption were calculated.
It was found that the negative values of enthalpy are responsible for the adsorption of grafted chitosan-based corrosion inhibitors are exothermic. This means that the heat was released through the inhibitor adsorption on the metal surface. The values in entropy indicate the reaction order on the metal surface. It is also revealed that the values in Gibbs free energy show the type of adsorption. When the value in Gibbs free energy is under −20 kJ/mol, then the inhibitor is adsorbed on the metal surface by the physical sorption, in which the positively charged nitrogen atoms in the molecular structure of grafted chitosan interact with the negatively charged metal surface. On the other hand, the value of Gibbs free energy over −40 kJ/mol is responsible for the chemisorption [45, 46]. It is indicated that the heteroatoms of grafted chitosan interact chemically to form the covalent bonds. In this chemisorption, the electron-rich heteroatoms are attributed to an increase in inhibition performance [47, 48].

The quantum chemical calculation of grafted chitosan was performed in several research works. For example, the quantum chemical calculation of select grafted chitosan were reviewed in the following cited sources [49, 50]. These collective works of research studied the quantum chemical properties of 4-hydroxy-3-methoxy benzaldehyde (CSB-3), 4-(dimethylamino) benzaldehyde (CSB-2), and benzaldehyde (CSB-1). These compounds are grafted with chitosan. Figure 13.4 shows the optimized molecular structure and LUMOs as well as the HOMOs of protonated CSB molecules in the aqueous phase [49]. The HOMO regions are mainly cited around the nitrogen and hydroxyl functional groups. In HOMO, the electrons are localized and easily transferred to the metal surface to form coordination bonds. On the other hand, the LUMO regions are localized around hetero rings [51]. These results confirmed the high inhibition efficiency of grafted chitosan. Some important theoretical parameters such as the frontier molecular orbital energies, electro-negativity, inverse of the hardness, and global softness were estimated by the HOMO and LUMO energies [52].

Figure 13.4 Optimized molecular structure and LUMOs, HOMOs of protonated CSB molecule in the aqueous phase. [49] / American Chemical Society.

13.2.2 Grafted Chitosan Based-Compounds as Sustainable Anti-corrosion Agents for Copper-Based Materials: Investigational and Theoretical Insights

Copper materials are widely used in the electronic industry because they have good electrical, mechanical, and thermal properties [10, 61]. The accumulation of corrosion products and the deposition are large problems for copper metal-based materials [62]. This affects the mechanical and electrical properties of copper materials [63, 64]. Applying a de-scaling treatment and periodic acid cleaning can be used to remove the accumulated corrosion products and the deposition from the metal surface. These cleaning processes use concentrated mineral acids such as hydrochloric acid or sulphuric acid, just to name a few. The acidic solution can corrode the copper surface during the cleaning process. Then, corrosion inhibitors are added to protect the copper surface from the acidic solution [65, 66]. Grafted chitosan is appealing to use as a remarkably green and efficient corrosion inhibitor. Table13.2 shows the main characteristics of grafted chitosan derivatives as sustainable corrosion inhibitors for copper-based materials. For example, Mouaden et al. [65] introduced the new chitosan (CS) cross-linked using cinnamaldehyde (Cinn) as a biological biopolymer for copper in 1 M HCl. This inhibitor was synthesized by the microwave irradiation processes. It was found that the maximum inhibition efficiency of Cinn-CS was over 89% at 1000 mg/L. This inhibitor is the mixed type with cathodic dominance. The hydrogen evolution was significantly blocked and the steel dissolution was also stopped by the presence of

Table 13.2 Main characteristics of grafted-based compounds as sustainable protective agents for copper-based materials.

No.	Molecular structure	Metal type	Analysed solution	Protection degree/inhibitor type/adsorption isotherm	References
17.	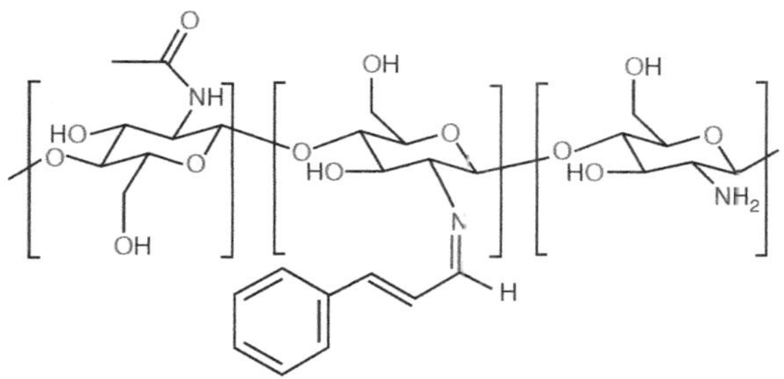 Chitosan (CS) was cross-linked using cinnamaldehyde	Copper	1 M HCl	Over 89% at a dose of 1000 mgL^{-1}/ mixed-type/-	[65]
18.	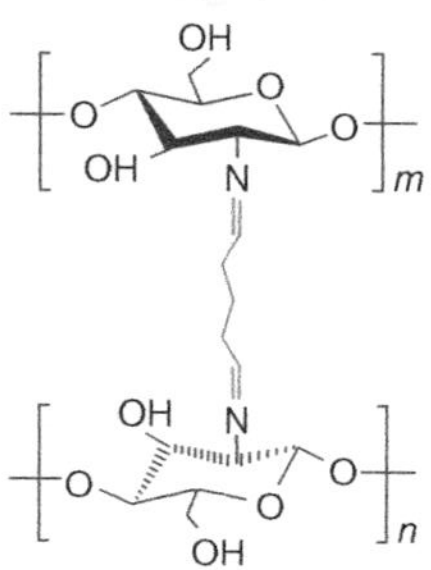 Chitosan with benzotriazole and glutaraldehyde	Copper	0.1 M Na_2SO_4 (pH = 2)	-	[62]

(Continued)

Table 13.2 (Continued)

No.	Molecular structure	Metal type	Analysed solution	Protection degree/inhibitor type/adsorption isotherm	References
19.	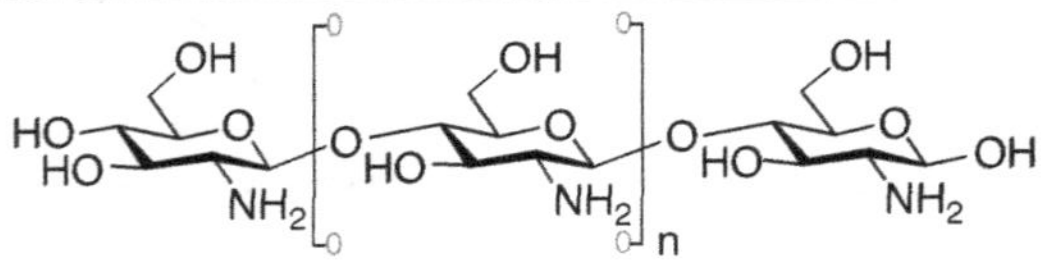	Copper	3% NaCl	86% at 1000 ppm/Langmuir	[64]
20.		Copper	0.5 M HCl	87% at 10^{-1} mg L^{-1}/ Langmuir	[61]
21.	Chitosan biopolymer	Copper	0.5 M HCl	93% at 8 μM/mixed type/ Langmuir	[10]
22.		Copper	Sulphide-contaminated synthetic seawater	89% at 800 ppm/	[63]
23.		Copper	3.5 wt% NaCl	90.32% at 1000 ppm/mixed type/Langmuir	[66]

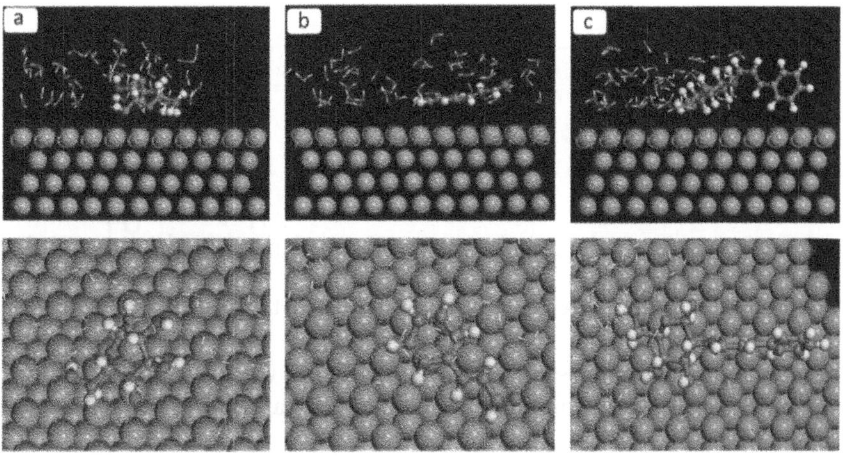

Figure 13.5 MD simulation results for (c) Cinn-CS, (b) cinnamaldehyde, and (a) chitosan on Cu (111)/ 50 H_2O. [65] / With permission of Elsevier.

a corrosion inhibitor. Monte Carlo (MC) simulations for Cinn-CS were carried out in this research initiative [65]. Universal Force Field (UFF) and Forcite module were performed to optimise the molecular structure of Cinn-CS inhibitor. The adsorption Locator module was used to estimate the adsorption energy of the selected inhibitor on the Cu (111) surface. The simulation box size is 5 × 10 × 10 Å supercell. The MC simulation was carried out in an aqueous phase with 50 water molecules in Materials studio 7.0 software because the experimental analysis was carried out in an aqueous solution. Figure 13.5 indicated the MD simulation results for (c) Cinn-CS, (b) cinnamaldehyde, and (a) chitosan on Cu (111)/ 50 H_2O [15]. It is revealed that the chitosan, cinnamaldehyde, and Cinn-CS adsorbed on the metal surface by the vertical adsorption positions; as a result, the metal surface was maximally covered and protected with the protective and density film. Deformation energy, $dE_{ad}/dNi_{Inhibitor}$, dE_{ad}/dNi_{H2O}, rigid adsorption energy, adsorption energy, and total energy for chitosan, cinnamaldehyde, and Cinn-CS on Cu (111) surface were estimated. The obtained data revealed that (i) a more stable film was formed on the metal surface; (ii) the studied corrosion inhibitors are considerably exchanged with the pre-adsorbed water molecules; (iii) the selected corrosion inhibitor has large energy for adsorption. The adsorption energy of Cinn-CS was −70.57 kJ/mol, confirming that the Cinn-CS strongly adsorbed on the metal surface by the formation of coordination bonds.

Wang et al. [62] introduced new corrosion based on grafted chitosan with benzotriazole (BTA). This inhibitor is a low-pH-sensitive green material and was synthesized through green technology. Scheme 13.1 shows the reaction procedure of grafted chitosan with benzotriazole (BTA). The corrosion protection ability of this inhibitor was analysed by the PDP and EIS. The obtained results revealed that the

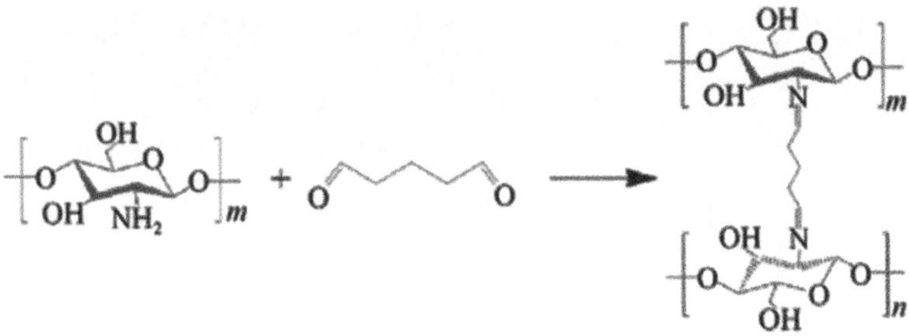

Scheme 13.1 Reaction procedure of grafted chitosan with the benzotriazole (BTA).

polarization resistance was maximally increased with the presence of anti-corrosion agents, proving that this molecule is an excellent anti-corrosion agent for the copper surface.

13.2.3 Grafted Chitosan Derivatives as Sustainable Corrosion Inhibitors for Aluminium-Based Materials: Experimental and Theoretical Insights

Aluminium-based metallic materials are widely employed in electronics, automobile, and other industries because they are low-cost and are compatible with mechanical-dependent materials. However, these materials are susceptible to corrosion when exposed to the atmosphere, saline, and acidic corrosion environments [67]. Corrosion of aluminium-based materials is a serious problem in the above industries. Currently, the chitosan and modified chitosan compounds are analysed in corrosion inhibition research with interesting resutls [68]. Table13.3 shows the main characteristics of grafted chitosan derivatives as sustainable corrosion inhibitors for aluminium-based materials. For example, Lai et al. [69] introduced a new and green corrosion inhibitor based on grafted chitosan. This inhibitor (CP) synthesized a Schiff Base reaction with chitosan and 4-pyr-idine-carboxaldehyde. Next, the TiO_2 was dispersed in CP to prepare CPT nanocomposite. The inhibition performance of CPT nanocomposite for C3003 aluminium alloy in 3.5 wt% NaCl solutions was analysed by surface morphology, electrochemical, and spectroscopic methods. The obtained results confirmed that the salt accumulations (NaCl) on the metal surface were seriously depleted by the adsorption selected corrosion inhibitor. This property is stable in aggressive saline environments. The contact angle analysis indicated that the CPT nanocomposite was a hydrophobic substance with the 120° contact angle. These results show that the selected nano compound rose the corrosion inhibition of chitosan. Figure 13.6 shows results in surface morphology analysis for (g, h) CPT (EDS), (e, f) CPT (SEM), (c, d) CP (SEM), and (a, b) TiO_2 (SEM) [69]. TiO_2 was the typical particle size of nanometres and spherical-like morphological material. The metal surface was rougher in the inhibitor presence of CP. TiO_2 importantly enhanced the

Table 13.3 Main characteristics of grafted chitosan derivatives as sustainable corrosion inhibitors for aluminium-based materials.

No.	Molecular structure	Metal type	Analysed solution	Protection degree/inhibitor type/adsorption isotherm	References
24.	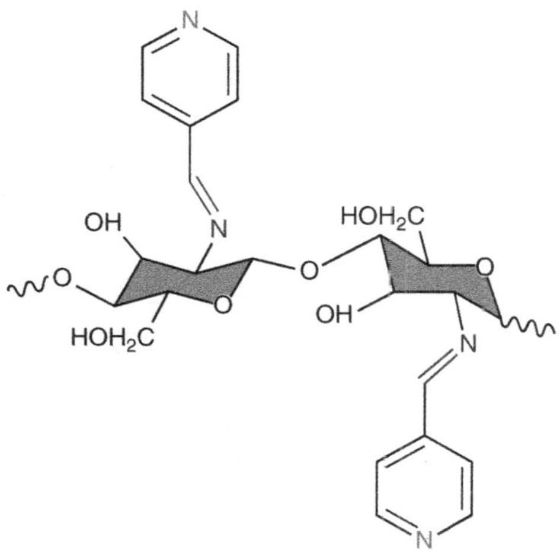 Chitosan modification with 4-pyr- idinecarboxaldehyce	C3003 aluminium	3.5 wt% NaCl solution	94.5% at 200 ppm/mixed type/Langmuir	[69]
25.	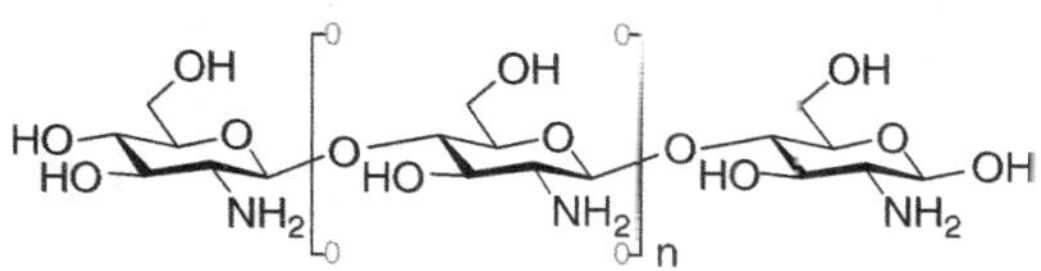 Chitosan biopolymer	Aluminium	0.1 M HCl	90% at 0.028 g/L/-	[68]
26.		Aluminium	3.65% NaCl	71.45% at 1.2 g/L/mixed type/Langmuir	[67]

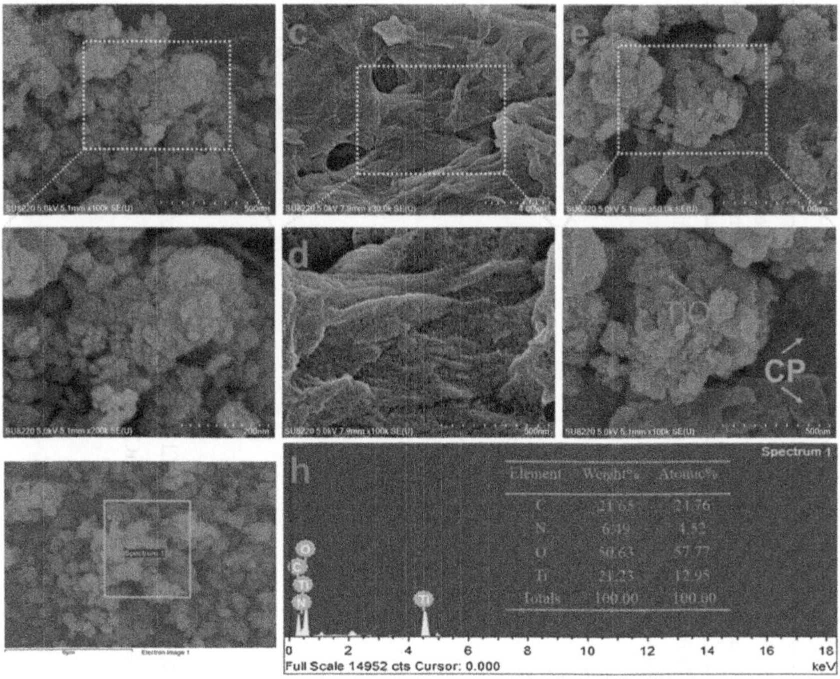

Figure 13.6 Results in surface morphology analysis for (g, h) CPT (EDS), (e, f) CPT (SEM), (c, d) CP (SEM), (a, b) TiO_2 (SEM). [69] / With permission of Elsevier.

corrosion inhibition performance of CPT. The elemental analysis of the CPT/aluminium surface was identified by EDX. As observed, titan, oxygen, nitrogen, and carbon spectra appeared in obtained spectra (Figure 13.6), confirming the high adsorption of the CPT corrosion inhibitor.

13.3 Conclusion and Future Perspective

This chapter ultimately reviewed the type of grafted chitosan corrosion inhibitors and their main inhibition properties for steel, aluminium, and copper metallic materials in acidic, saline, and other corrosive solutions. It was found that grafted chitosan is widely performed as an environmentally friendly anti-corrosion agent for various metal materials in acidic, saline, and other corrosion mediums. The grafted chitosan natural corrosion inhibitors are less toxic, environmentally friendly, effectively water-soluble, biodegradable, non-volatile, non-flammable, low-cost, and don't require dangerous technology or syntheses methodology. They are mainly mixed type for copper, steel, and aluminium surface. The adsorption characteristics were mainly described using Langmuir adsorption isotherms.

The theoretical examination established that grafted chitosan is a more polar and proficient corrosion inhibitor. The molecular dynamic simulation results suggest that the grafted chitosan adsorbed on the metal surface by the vertical adsorption positions which indicates that the metal surface was maximally covered and protected with the protective and density film. Deformation energy, $dE_{ad}/dNi_{Inhibitor}$, dE_{ad}/dNi_{H2O}, rigid adsorption energy, adsorption energy, and total energy for grafted chitosan on Fe (110), Al (111), and Cu (111) surfaces were estimated. The obtained data revealed that the adsorption energy of the grafted chitosan on the selected metal surface is very high.

The chitosan-based inhibitors will be replaced traditional compounds, because, of their greenness and high inhibition efficiency. The chitosan-based compounds would be a good source to synthesize the carbon dots and nanomaterials which will be environmentally friendly and have high inhibition efficiency. These factors indicate that these types of corrosion inhibitors would have highly proficient anti-corrosion applications for steel, copper, aluminium, magnesium, and other types of metals.

References

1 Umoren, S.A., AlAhmary, A.A., Gasem, Z.M., & Solomon, M.M. (2018). Evaluation of chitosan and carboxymethyl cellulose as ecofriendly corrosion inhibitors for steel. *International journal of biological macromolecules* 117: 1017–1028. Availabe from https://doi.org/10.1016/j.ijbiomac.2018.06.014.

2 Zhu, M., Guo, L., He, Z. et al. (2022). Insights into the newly synthesized N-doped carbon dots for Q235 steel corrosion retardation in acidizing media: a detailed multidimensional study. *Journal of Colloid and Interface Science* [Internet] 608: 2039–2049. Available from: https://www.sciencedirect.com/science/article/pii/S0021979721018373.

3 Berdimurodov, E., Kholikov, A., Akbarov, K. et al. (2022 March 20) Novel gossypol–indole modification as a green corrosion inhibitor for low–carbon steel in aggressive alkaline–saline solution. *Colloids and Surfaces A: Physicochemical and Engineering Aspects* [Internet] [cited 2022 Jan 20] 637: 128207. Available from: https://linkinghub.elsevier.com/retrieve/pii/S0927775721020768.

4 Verma, C. and Quraishi, M.A. (2021). Chelation capability of chitosan and chitosan derivatives: recent developments in sustainable corrosion inhibition and metal decontamination applications. *Current Research in Green and Sustainable Chemistry* 4: 100184. Availabe from: https://doi.org/10.1016/j.crgsc.2021.100184.

5 Bahgat Radwan, A., Mannah, C.A., Sliem, M.H. et al. (2021). Electrospun highly corrosion-resistant polystyrene–nickel oxide superhydrophobic nanocomposite coating. *Journal of Applied Electrochemistry* 51 (11): 1605–1618. Available from: https://doi.org/10.1007/s10800-021-01603-8.

6 Berdimurodov, E., Kholikov, A., Akbarov, K. et al. (2021). A gossypol derivative as an efficient corrosion inhibitor for St2 steel in 1 M HCl + 1 M KCl: an experimental and theoretical investigation. *Journal of Molecular Liquids* [Internet] 328: 115475. Available from: https://www.sciencedirect.com/science/article/pii/S0167732221002014.

7 El-Mahdy, G.A., Atta, A.M., Al-Lohedan, H.A., and Ezzat, A.O. (2015). Influence of green corrosion inhibitor based on chitosan ionic liquid on the steel corrodibility in chloride solution. *International Journal of Electrochemical Science* 10 (7): 5812–5826.

8 Badr, E.a., Hefni, H.H.H., Shafek, S.H., and Shaban, S.M. (2020). Synthesis of anionic chitosan surfactant and application in silver nanoparticles preparation and corrosion inhibition of steel. *International Journal of Biological Macromolecules* [Internet] 157: 187–201. Available from: https://doi.org/10.1016/j.ijbiomac.2020.04.184.

9 Gouda, M., Khalaf, M.M., Shalabi, K., Al-Omair, M.A., and El-Lateef, H.M.A. (2022). Synthesis and characterization of Zn – Organic frameworks containing Chitosan as a low-cost inhibitor. *Polymers (Basel)* 14 (2): 228.

10 El-Haddad, M.N. (2013). Chitosan is a green inhibitor for copper corrosion in an acidic medium. *International Journal of Biological Macromolecules* [Internet] 55: 142–149. Available from: https://doi.org/10.1016/j.ijbiomac.2012.12.044.

11 Harmami, H., Ulfin, I., Sakinah, A.H., and Ni'mah, Y.L. (2019). Water-soluble chitosan from shrimp and mussel shells as corrosion inhibitor on tinplate in 2% NaCl. *Malaysian Journal of Fundamental and Applied Sciences* 15 (2): 212–217.

12 Nemr A El Elhebshi, A., El-Deab, M.S., Ashour, I., and Ragab, S. (2022). Synergistic effect of Chitosan Biguanidine hydrochloride salt as a green inhibitor for stainless steel alloy corrosion in a 0.5 M H2SO4 solution. *Egyptian Journal of Chemistry* 65 (2): 389–398.

13 Shahmoradi, A.R., Ranjbarghanei, M., Javidparvar, A.A. et al. (2021). Theoretical and surface/electrochemical investigations of walnut fruit green husk extract as an effective inhibitor for mild-steel corrosion in 1M HCl electrolyte. *Journal of Molecular Liquids* [Internet] 338: 116550. Available from: https://www.sciencedirect.com/science/article/pii/S0167732221012745.

14 Berdimurodov, E., Verma, D.K., Kholikov, A. et al. (2021). The recent development of carbon dots as powerful green corrosion inhibitors: a prospective review. *Journal of Molecular Liquids* [Internet] 118124. Available from: https://www.sciencedirect.com/science/article/pii/S016773222102849X.

15 Ashassi-Sorkhabi, H. and Kazempour, A. (2020). Chitosan, its derivatives, and composites with superior potentials for the corrosion protection of steel alloys: a comprehensive review. *Carbohydrate Polymers* [Internet] 237 (February): 116110. Available from: https://doi.org/10.1016/j.carbpol.2020.116110.

16 Verma, C., Quraishi, M.A., Alfantazi, A., and Rhee, K.Y. (2021). Corrosion inhibition potential of chitosan-based Schiff bases: design, performance, and applications. *International Journal of Biological Macromolecules* [Internet] 184 (March): 135–143. Available from: https://doi.org/10.1016/j.ijbiomac.2021.06.049.

17 Li, H., Li, H., Liu, Y., and Huang, X. (2015). Synthesis of polyamine grafted chitosan copolymer and evaluation of its corrosion inhibition performance. *Journal of the Korean Chemical Society* 59 (2): 142–147.

18 Yang, S.F., Wen, Y., Yi, P., Xiao, K., and Fang, D.C. (2017). Effects of chitosan inhibitor on the electrochemical corrosion behavior of 2205 duplex stainless steel. *International Journal of Minerals, Metallurgy and Materials* 24 (11): 1260–1266.

19 Solomon, M.M., Gerengi, H., Kaya, T., and Umoren, S.A. (2017). Performance evaluation of a Chitosan/Silver nanoparticles composite on St37 steel corrosion in a 15% HCl solution. *ACS Sustainable Chemistry & Engineering* 5 (1): 809–820.

20 Dagdag, O., Hsissou, R., Safi, Z. et al. (2021). Rheological and simulation for macromolecular matrix epoxy bi-functional aromatic amines. *Polymer Bulletin* [Internet] Available from: https://doi.org/10.1007/s00289-021-03862-z.

21 Berdimurodov, E., Kholikov, A., Akbarov, K., Obot, I.B., and Guo, L. (2021). Thioglycoluril derivative as a new and effective corrosion inhibitor for low carbon steel in a 1 M HCl medium: an experimental and theoretical investigation. *Journal of Molecular Structure* [Internet] 1234: 130165. Available from: https://www.sciencedirect.com/science/article/pii/S0022286021002969.

22 Aghzzaf, A.A., Rhouta, B., Steinmetz, J. et al. (2012). Corrosion inhibitors based on chitosan-heptanoate-modified beidellite. *Applied Clay Science* [Internet] 65–66: 173–178. Available from: https://doi.org/10.1016/j.clay.2012.04.025.

23 Keerthana, A.K. and Ashraf, P.M. (2020). Carbon nanodots synthesized from chitosan and its application as a corrosion inhibitor in boat building carbon steel BIS2062. *Applied Nanoscience* [Internet] 10 (4): 1061–1071. Available from https://doi.org/10.1007/s13204-019-01177-0.

24 Hefni, H.H.H., Azzam, E.M., Badr, E.A. Hussein, M., and Tawfik, S.M. (2016). Synthesis, characterization, and anticorrosion potentials of chitosan-g-PEG assembled on silver nanoparticles. *International Journal of Biological Macromolecules* [Internet] 83: 297–305. Available from: https://doi.org/10.1016/j.ijbiomac.2015.11.073.

25 Zhang, Q.H., Hou, B.S., Li, Y.Y. et al. (2020). Two novel chitosan derivatives as highly efficient eco-friendly inhibitors for the corrosion of mild steel in acidic solution. *Corrosion Science* 164.

26 Abdallah, M., Fawzy, A., and Hawsawi, H. (2020). Maltodextrin and chitosan polymers as inhibitors for the corrosion of carbon steel in 1.0 M hydrochloric acid. *International Journal of Electrochemical Science* 15: 5650–5663.

27 Hasanin, M. and Al Kiey, S.A. (2021). Development of eco-friendly high-performance anti-corrosive chitosan nanocomposite material for mild steel corrosion in acid medium. *Biomass Convers Biorefinery* [Internet] (0123456789). Available from: https://doi.org/10.1007/s13399-021-02059-8.

28 Berdimurodov, E., Kholikov, A., Akbarov, K. et al. (2020). New anti-corrosion inhibitor (3ar,6ar)-3a,6a-di-p-tolyltetrahydroimidazo[4,5-d]imidazole-2,5(1 h,3h)-dithione for carbon steel in 1 M HCl medium: gravimetric, electrochemical, surface and quantum chemical analyses. *Arabian Journal of Chemistry* [Internet] 13 (10): 7504–7523. Available from: https://www.sciencedirect.com/science/article/pii/S1878535220303154.

29 Berdimurodov, E., Kholikov, A., Akbarov, K., and Guo, L. (2021). Experimental and theoretical assessment of new and eco–friendly thioglycoluril derivative as an effective corrosion inhibitor of St2 steel in the aggressive hydrochloric acid with sulfate ions. *Journal of Molecular Liquids* [Internet] 335: 116168. Available from: https://www.sciencedirect.com/science/article/pii/S0167732221008953.

30 Jessima, S.J.H.M., Berisha, A., Srikandan, S.S., and Subhashini, S. (2020). Preparation, characterization, and evaluation of corrosion inhibition efficiency of sodium lauryl sulfate modified chitosan for mild steel in the acid pickling process. *Journal of Molecular Liquids* [Internet] 320: 114382. Available from: https://doi.org/10.1016/j.molliq.2020.114382.

31 Fayyad, E.M., Sadasivuni, K.K., Ponnamma, D., and Al-Maadeed, M.A.A. (2016). Oleic acid-grafted chitosan/graphene oxide composite coating for corrosion protection of carbon steel. *Carbohydrate Polymers* [Internet] 151: 871–878. Available from: https://doi.org/10.1016/j.carbpol.2016.06.001.

32 Eduok, U., Ohaeri, E., and Szpunar, J. (2021). Anticorrosion allyl sulfonate graft chitosan/graphene oxide nanocomposite material. *Mater Advances* 2 (5): 1621–1634.

33 Yarkulov, A., Umarov, B., Rakhmatkarieva, F. et al. (2022). Diacetate cellulose-silicon bionanocomposite adsorbent for recovery of heavy metal ions and benzene vapors: an experimental and theoretical investigation. *Biointerface Research in Applied Chemistry* 12 (3): 2862–2880.

34 Verma, D.K., Kazi, M., Alqahtani, M.S. et al. (2021 October 5). N–hydroxybenzothioamide derivatives as green and efficient corrosion inhibitors for mild steel: experimental, DFT and MC simulation approach. *Journal of Molecular Structure* [Internet] [cited 2021 May 23] 1241: 130648. Available from). https://linkinghub.elsevier.com/retrieve/pii/S002228602100781X.

35 Srivastava, M., Srivastava, S.K., Nikhil, J.G., and Prakash, R. (2019). Chitosan-based new nanocomposites for corrosion protection of mild steel in aggressive chloride media. *International Journal of Biological Macromolecules* [Internet] 140: 177–187. Available from: https://doi.org/10.1016/j.ijbiomac.2019.08.073.

36 Rbaa, M., Fardioui, M., Verma, C. et al. (2020). International Journal of Biological Macromolecules 8-Hydroxyquinoline based chitosan derived carbohydrate polymer as biodegradable and sustainable acid corrosion inhibitor for mild steel: experimental and computational analyses. *International Journal of Biological Macromolecules* [Internet] 155: 645–655. Available from: https://doi.org/10.1016/j.ijbiomac.2020.03.200.

37 Li, M., Xu, J., Li, R. et al. (2014). Simple preparation of aminothiourea-modified chitosan as a corrosion inhibitor and heavy metal ion adsorbent. *Journal of Colloid and Interface Science* [Internet] 417: 131–136. Available from: https://doi.org/10.1016/j.jcis.2013.11.053.

38 Berdimurodov, E., Kholikov, A., Akbarov, K., and Guo, L. (2021). Inhibition properties of 4,5-dihydroxy-4,5-di-p-tolylimidazolidine-2-thione for use on carbon steel in an aggressive alkaline medium with chloride ions: thermodynamic, electrochemical, surface and theoretical analyses. *Journal of Molecular Liquids* [Internet] 327: 114813. Available from: https://www.sciencedirect.com/science/article/pii/S0167732220370550.

39 Farhadian, A., Varfolomeev, M.A., Shaabani, A., and Nasiri, S. (2020). Sulfonated chitosan as green and high cloud point kinetic methane hydrate and corrosion inhibitor: experimental and theoretical studies. *Carbohydrate Polymers* [Internet] 236 (April 2019): 116035. Available from: https://doi.org/10.1016/j.carbpol.2020.116035.

40 Berdimurodov, E., Guo, L., Kholikov, A. Akbarov, K., and Zhu, M. (2021). MOFs-Based corrosion inhibitors. In: *Supramolecular Chemistry in Corrosion and Biofouling Protection*, 287–305. CRC Press.

41 Dagdag, O., Berisha, A., Mehmeti, V. et al. (2021). Epoxy coating as effective anti-corrosive polymeric material for aluminum alloys: formulation, electrochemical and computational approaches. *Journal of Molecular Liquids* [Internet] 117886. Available from: https://www.sciencedirect.com/science/article/pii/S0167732221026118.

42 Berdimurodov, E., Kholikov, A., Akbarov, K. et al. (2022). Novel cucurbit[6]uril-based [3]rotaxane supramolecular ionic liquid is a green and excellent corrosion inhibitor for the chemical industry. *Colloids and Surfaces A: Physicochemical and Engineering Aspects* [Internet] 633: 127837. Available from: https://www.sciencedirect.com/science/article/pii/S0927775721017064.

43 Haldhar, R., Kim, S.-C., Berdimurodov, E., Verma, D.K., and Hussain, C.M. (2021). Corrosion inhibitors: industrial applications and commercialization. In: *Sustainable Corrosion Inhibitors II: Synthesis, Design, and Practical Applications [Internet]*, 10–219. American Chemical Society. (ACS Symposium Series; vol. 1404). Available from https://doi.org/10.1021/bk-2021-1404.ch010.

44 Berdimurodov, E., Kholikov, A., Akbarov, K. et al. (2021). Novel bromide–cucurbit[7]uril supramolecular ionic liquid as a green corrosion inhibitor for the oil and gas industry. *Journal of Electroanalytical Chemistry* [Internet] 901: 115794. Available from: https://www.sciencedirect.com/science/article/pii/S1572665721008201.

45 Chen, T., Zeng, D., and Zhou, S. (2018). Study of polyaspartic acid and chitosan complex corrosion inhibition and mechanisms. *Polish Journal of Environmental Studies* 27 (4): 1441–1448.

46 Berdimurodov, E., Kholikov, A., Akbarov, K. et al. (2022 January 18). Novel glycoluril pharmaceutically active compound as a green corrosion inhibitor for the oil and gas industry. *Journal of Electroanalytical Chemistry* [cited 2022 Jan 20] 907: 116055. Available from: https://linkinghub.elsevier.com/retrieve/pii/S1572665722000479.

47 Cui, G., Guo, J., Zhang, Y. et al. (2019). Chitosan oligosaccharide derivatives as green corrosion inhibitors for P110 steel in a carbon-dioxide-saturated chloride solution. *Carbohydrate Polymer* [Internet] 203: 386–395. Available from: https://doi.org/10.1016/j.carbpol.2018.09.038.

48 Jessima, S.J.H.M., Subhashini, S., Berisha, A. Oral, A., and Srikandan, S.S. (2021). Corrosion mitigation performance of disodium EDTA functionalized chitosan biomacromolecule – Experimental and theoretical approach. *International Journal of Biological Macromolecules* [Internet] 178: 477–491. Available from: https://doi.org/10.1016/j.ijbiomac.2021.02.166.

49 Haque, J., Srivastava, V., Chauhan, D.S. Lgaz, H., and Quraishi, M.A. (2018). Microwave-induced synthesis of chitosan schiff bases and their application as novel and green corrosion inhibitors: experimental and theoretical approach. *ACS Omega* 3 (5): 5654–5668.

50 Chauhan, D.S., Mazumder, M.A.J., Quraishi, M.A., Ansari, K.R., and Suleiman, R.K. (2020). Microwave-assisted synthesis of a new Piperonal-Chitosan Schiff base as a bio-inspired corrosion inhibitor for oil-well acidizing. *International Journal of Biological Macromolecules* [Internet] 158: 231–243. Available from: https://doi.org/10.1016/j.ijbiomac.2020.04.195.

51 Rbaa, M., Benhiba, F., Hssisou, R. et al. (2020). Green synthesis of novel carbohydrate polymer chitosan oligosaccharide grafted on d-glucose derivative as a bio-based corrosion inhibitor. *Journal of Molecular Liquids* [Internet] 114549. Available from: https://doi.org/10.1016/j.molliq.2020.114549.

52 Fekry, A.M. and Mohamed, R.R. (2010). Acetyl thiourea chitosan is an eco-friendly inhibitor for mild steel in a sulphuric acid medium. *Electrochimica Acta* 55 (6): 1933–1939.

53 Umoren, S.A., Banera, M.J., Alonso-Garcia, T., Gervasi, C.A., and Mirífico M.V. (2013). Inhibition of mild steel corrosion in HCl solution using chitosan. *Cellulose* 20 (5): 2529–2545.

54 Sangeetha, Y., Meenakshi, S., and SairamSundaram, C. (2015). Corrosion mitigation of N-(2-hydroxy-3-trimethyl ammonium)propyl chitosan chloride as an inhibitor on mild steel. *International Journal of Biological Macromolecules* [Internet] 72: 1244–1249. Available from: https://doi.org/10.1016/j.ijbiomac.2014.10.044.

55 Zheng, Y., Gao, Y., Li, H. et al. (2021). Corrosion inhibition performance of composite based on chitosan derivative. *Journal of Molecular Liquids* [Internet] 324 (xxxx): 114679. Available from: https://doi.org/10.1016/j.molliq.2020.114679.

56 Li, P., He, L., Li, X., Liu, X., and Sun, M. (2021). Corrosion inhibition effect of N-(4-diethylaminobenzyl) quaternary ammonium chitosan for X80 pipeline steel in hydrochloric acid solution. *International Journal of Electrochemical Science* 16 (1): 1–21.

57 Liu, Y., Zou, C., Yan, X. et al. (2015). B-Cyclodextrin modified natural chitosan as a green inhibitor for carbon steel in acid solutions. *Industrial & Engineering Chemistry Research* 54 (21): 5664–5672.

58 Eduok, U., Ohaeri, E., Szpunar, J., and Akpan, I. (2020). Synthesis, characterization, and application of glucosyloxyethyl acrylate graft chitosan against pipeline steel corrosion. *Journal of Molecular Liquids* [Internet] 315: 113772. Available from: https://doi.org/10.1016/j.molliq.2020.113772.

59 Wang, C., Chen, J., Hu, B. et al. (2019). Modified chitosan-oligosaccharide and sodium silicate as an efficient sustainable inhibitor for carbon steel against chloride-induced corrosion. *Journal of Cleaner Production* [Internet] 238: 117823. Available from: https://doi.org/10.1016/j.jclepro.2019.117823.

60 Quraishi, M.A., Ansari, K.R., Chauhan, D.S., Umoren, S.A., and Mazumder, M.A.J. (2020). Vanillin-modified chitosan is a new bio-inspired corrosion inhibitor for carbon steel in oil-well acidizing relevant to the petroleum industry. *Cellulose* [Internet] 0123456789. Available from: https://doi.org/10.1007/s10570-020-03239-x.

61 Jmiai, A., El Ibrahimi, B., Tara, A. et al. (2017). Chitosan as an eco-friendly inhibitor for copper corrosion in acidic medium: protocol and characterization. *Cellulose* 24 (9): 3843–3867.

62 Wang, Y.N., Dong, C.F., Zhang, D.W. et al. (2015). Preparation and characterization of a chitosan-based low-pH-sensitive intelligent corrosion inhibitor. *International Journal of Minerals, Metallurgy and Materials* 22 (9): 998–1004.

63 El Mouaden, K., El Ibrahimi, B., Oukhrib, R. et al. (2018). Chitosan polymer as a green corrosion inhibitor for copper in sulfide-containing synthetic seawater. *International Journal of Biological Macromolecules* [Internet] 119: 1311–1323. Available from: https://doi.org/10.1016/j.ijbiomac.2018.07.182.

64 Oukhrib, R., El Ibrahimi, B., Bourzi, H. et al. (2017). Quantum chemical calculations and corrosion inhibition efficiency of biopolymer "chitosan" on the copper surface in 3%NaCl. *Journal of Materials and Environmental Science* 8 (1): 195–208.

65 El, K., Singh, D., Quraishi, M.A., Bazzi, L., Hilali, M. (2020). Cinnamaldehyde-modi fi ed chitosan as a bio-derived corrosion inhibitor for acid pickling of copper: microwave synthesis, experimental and computational study. *International Journal of Biological Macromolecules* [Internet] 164: 3709–3717. Available from: https://doi.org/10.1016/j.ijbiomac.2020.08.137.

66 Brou, Y.S., Coulibaly, N.H., Diki, N.G.Y.S. Creus, J., and Trokourey, A. (2020). Chitosan biopolymer effect on copper corrosion in 3.5 wt% lNaCl solution: electrochemical and quantum chemical studies. *International Journal of Biological Macromolecules* 9 (1): 182–200.

67 Fayomi, O.S.I., Akande, I.G., and Popoola, A.P.I. (2018). Corrosion protection effect of chitosan on the performance characteristics of A6063 alloy. *J Bio- Tribo-Corrosion* [Internet]. 4 (4): 1–6. Available from: https://doi.org/10.1007/s40735-018-0192-6.

68 Abd El Nabey, B.A., Goher, Y.M., Fetouh, H.A., and Karam, M.S. (2015). Anticorrosive properties of chitosan for the acid corrosion of aluminum. *Port Electrochim Acta* 33 (4): 231–239.

69 Lai, X., Hu, J., Ruan, T., Zhou, J., and Qu, J. (2021). Chitosan derivative corrosion inhibitor for aluminum alloy in sodium chloride solution: a green organic/inorganic hybrid. *Carbohydrate Polymer* [Internet] 265 (April): 118074. Available from: https://doi.org/10.1016/j.carbpol.2021.118074.

14

Grafted Starch Used as Sustainable Corrosion Inhibitors

Taiwo W. Quadri[1,], Lukman O. Olasunkanmi[2,3], Omolola E. Fayemi[4], and Eno E. Ebenso[1]*

[1] *Centre for Material Science, College of Science, Engineering and Technology, University of South Africa, Johannesburg, 1710, South Africa*
[2] *Department of Chemistry, Faculty of Science, Obafemi Awolowo University, Ile Ife, 220005, Nigeria*
[3] *Department of Chemical Science, University of Johannesburg, P.O. Box 17011, Doornfontein Campus, Johannesburg, 2028, South Africa*
[4] *Department of Chemistry, School of Chemical and Physical Sciences and Material Science Innovation & Modelling (MaSIM) Research Focus Area, Faculty of Natural and Agricultural Sciences, North-West University, Private Bag X2046, Mmabatho, 2735, South Africa*
[*] *Corresponding author*

14.1 Introduction

14.1.1 Inhibition of Metallic Corrosion via Chemical Compounds

The use of chemical compounds commonly called corrosion inhibitors has been widely adopted as the first line of defense against metal and alloy dissolution. It has been reported to be reliable, efficient, and cost-effective for industrial applications [1, 2]. Over the years, numerous chemical compounds have been assessed as potential inhibitors of metallic deterioration [3–8]. Among these are inorganic compounds such as nitrites, silicates, and chromates which have been found to successfully inhibit metallic corrosion by adsorption. They are sources of environmental toxicity, however, and thus their usage has been discouraged. On the other hand, organic compounds possess attributes that place them as worthy alternatives to inorganic inhibitors. These attributes include their possession of polar functional groups, solubility in aqueous electrolytes, ease of adsorption on metallic surfaces, and stability of the protective layer [6, 9]. Nevertheless, they do not meet all the requirements of an eco-friendly inhibitor as the principles of green chemistry are sometimes violated in the synthetic procedures, coupled with the exorbitant cost involved in their synthesis. Besides,

Grafted Biopolymers as Corrosion Inhibitors: Safety, Sustainability, and Efficiency, First Edition.
Edited by Jeenat Aslam, Chandrabhan Verma, and Ruby Aslam.

organic compounds used as chemical inhibitors are known to undergo desorption and rearrangement of inhibitor molecules at high temperatures, leading to low inhibition abilities and physical adsorption of chemical inhibitors [10].

The utilization of polymers as anticorrosive agents has recently become popular as it offers a better approach to mitigating corrosion than the traditional inorganic and organic inhibitors. The literature study revealed that natural polymers [11–13] and synthetic polymers [14, 15] have been employed as chemical additives to control the corrosion rate of different metals and alloys in various electrolytic media. In particular, natural biopolymers such as Gum Arabic [16], xanthan gum [17], dextrin [18], pectin [19], alginate [20], tannin [21], gluten [22], chitosan [23], glycogen [24], etc. have been preferred for their anticorrosive properties in recent years because of their easy availability, eco-friendliness, low cost, large molecular size, inhibitive effectiveness, and stability in corrosive electrolytes [25–28]. It must, however, be mentioned that natural polymers are often subjected to modification to improve their solubility in the medium of application and hence their inhibitive effectiveness.

Several test methods have been reported in investigating the inhibition property of chemical inhibitors in corrosive media. Gravimetric or weight loss (WL) analysis is a traditional, inexpensive technique that has been widely utilized in measuring the inhibition performance of chemical compounds. It provides information on the impact of inhibitor concentration, acid concentration, exposure time, and temperature. Insights into the thermodynamic and kinetic aspects of corrosion are also gained from gravimetric studies. Modern electrochemical methods including open circuit potential (OCP), linear polarization technique (LPR), electrochemical impedance spectroscopy (EIS), and potentiodynamic polarization (PDP) have been implemented to monitor corrosion rate in aqueous electrolytes. Surface analyses of metal substrates immersed in inhibited and uninhibited solutions are often examined using analytical techniques such as scanning electron microscopy (SEM), Fourier-transform infrared spectroscopy (FTIR), water contact angle (WCA) measurement, energy dispersive X-ray (EDX) analysis, X-ray photoelectron spectroscopy (XPS), and atomic force microscopy (AFM). To establish the inhibitor effect of studied chemical inhibitors and further provide insight into their inhibition mechanism, corrosion experts often explore modern advancements in computer systems. Computational modelings are performed using density functional theory (DFT) calculations and molecular dynamic simulations (MDS) [29, 30].

14.2 Starch Structure, Composition and Modification

Starch is one of the most abundant carbohydrate biopolymers which has attracted attention in academia and industries [31, 32]. It is commonly stored in all plant parts such as roots, leaves, tubers, seeds, and stems as a source of energy and food. The major sources of commercial starch include corn, rice, cassava, potato, wheat, etc. Generally, starch comprises two major components in different percentages, namely, predominant linear amylose (20–30%) with α1–4 glycosidic linkage and the more complex, branched amylopectin (70–80%) with α1–4 and α1–6 glycosidic linkages. The ratio of the two polymeric components present in the starch structure determines its granular size, crystallinity, and chemical properties [33–35]. Figure 14.1 displays a molecular representation of starch, amylose, and amylopectin.

Starch is a large carbohydrate organic molecule renowned for its special properties such as low cost, biodegradability, renewability, and non-bioaccumulation. It is extracted from cereals, roots, and tubers of plants where it is stored and can be used in both native (pure) and modified forms. While starch is known to be commercially applicable in food-related industries, it has also found applications in several non-food-related industries such as the pharmaceutical, wastewater

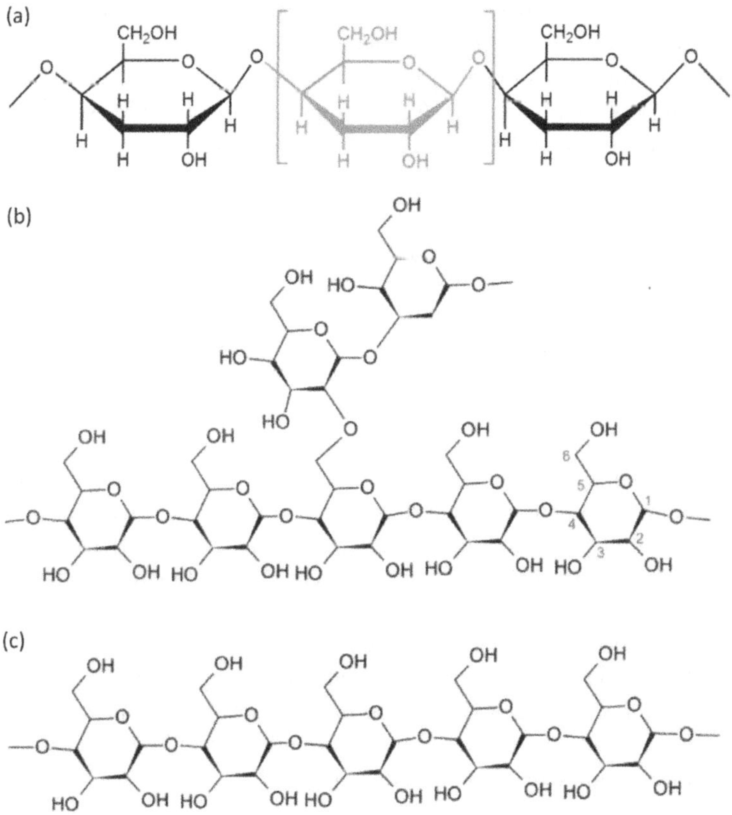

Figure 14.1 Chemical structures of starch, amylopectin, and amylose. Adapted from [34, 35].

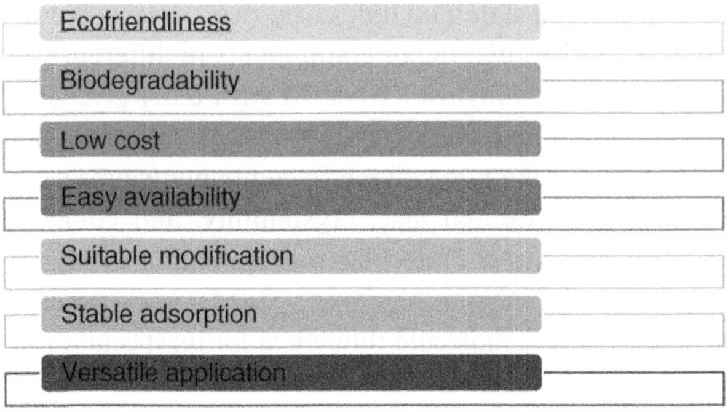

Figure 14.2 Factors responsible for the choice of starch as a chemical inhibitor.

treatment, packaging, biomedical, and textile industries [32, 36]. One of the recent applications of starch is the corrosion protection of metals and alloys. Figure 14.2 presents the factors responsible for the recent investigation and utilization of starch and its derivatives as metallic inhibitors in corrosive media.

Despite having numerous unique properties, native starch is not extensively utilized as a metallic inhibitor in corrosive solutions because of its poor solubility in aqueous electrolytes, high viscosity, poor processing tolerance, and reduced surface adhesion strength [13]. To improve its functional properties, researchers have adopted different techniques of modification. These techniques are broadly classified as chemical, physical, enzymatic, and genetic modification. Physical and chemical modification of starch are popularly implemented methods to intensify desirable functional qualities, eliminate undesirable properties, or add new characteristics [37]. The chemical modification involves the substitution of hydroxyl groups present in starch by alternative functional groups which could be carried out by oxidation, hydrolysis, esterification, crosslinking, etherification, and graft copolymerization. Graft copolymerization is a chemical modification technique that implants a synthetic monomer/polymer onto a natural polymeric material to enhance its functional properties. The addition of the sidechains does not essentially alter the polymeric matrix while enhancing properties such as solubility, thermal stability, viscoelasticity, adhesion, etc. On the other hand, physical modification entails the adjustment of particle size, gelation ability, moisture level, surface properties, and viscosity achieved through heat moisture treatment, extrusion method, annealing, high pressure, and dry heating [33, 35, 38, 39]. By modifying the starch structure, it is expected that changes will occur leading to process tolerance, solubility in water, film formation, stability in seawater medium, and interaction with other substances [37].

14.3 Application of Starch and Modified Starch as Corrosion Inhibitors

The first reported study on the inhibitive behavior of starch was conducted by Abd El Haleem and co-workers in 1986 [40]. The paper reported improved resistance of plain carbon steel in NaOH+NaCl solution to pitting corrosion due to shifting of the pitting potential in the noble region after introducing soluble starch. Ten years after this pioneering report on starch as a corrosion inhibitor, polyorganosiloxane-grafted potato starch coatings were fabricated by Sugama and Du Vall to provide surface coverage for the 6061-T6 aluminum substrate [41]. The synthesized inhibitor was confirmed using several characterization techniques and a corrosion study was carried out via impedance measurement and salt spray test supported with surface imaging and elemental analyses. The coating films were reported to improve the corrosion resistance of the aluminum substrate. Sugama, in another study conducted in 1997, modified potato starch using cerium (IV) ammonium nitrate and applied it to inhibit 6061-T6 aluminum degradation in a neutral medium using EIS, XPS, FTIR, and salt spray test [42].

Charitha and Rao investigated the protective capacity of starch for the dissolution of 6061 aluminum in 0.1 M HCl [43] and 0.25 M HCl [44] at varying concentrations (100–800 ppm) and temperature ranges (20–50 °C) using EIS and PDP. Increased dosage and temperature were recorded to increase the protective capacity of starch. With a maximum 800 ppm concentration and highest 50 °C, maximum IE of 63.44% and 58.93% from PDP measurement was obtained for aluminum disintegration in 0.1 M and 0.25 M HCl respectively. Starch was reported to retard cathodic and anodic processes and adsorb following the Langmuir adsorption model. The authors explained that the starch chemically adsorbs on the aluminum metal as a result of the donation of lone pairs and pi-electrons of the OH group. SEM and EDX analyses were conducted to establish results obtained from electrochemical assessments. Raw starch from *Dioscorea hispida (gadong* tubers) was assessed as a potential sustainable chemical additive for the protection of SAE 1045 carbon steel in 0.6 M NaCl solution [45]. The insoluble starch powder was treated with 90% dimethyl sulfoxide (DMSO) to solubilize the starch and was tested using WL and PDP. The inhibitor suppressed metal disintegration at the anodic branch and the evolution of hydrogen at the cathode but acted primarily as an anodic inhibitor. The authors found that 1500 ppm of the *gadong* tuber starch yielded 86.13% IE. The study was further supported with SEM and EDX analysis. The authors opined that the presence of different organic compounds including the hydroxyl group and anhydroglucose unit in the starch could account for its adherence properties on the steel substrate.

Tapioca starch was explored as an eco-friendly corrosion inhibitor for the dissolution of aluminum in seawater by Rosliza and Wan Nik [46]. Weight loss, electrochemical, and surface morphological analyses were performed to establish the inhibition behavior of tapioca starch for AA6061 aluminum. Electrochemical studies showed a significant decrease in double-layer capacitance (C_{dl}), corrosion rates, corrosion current densities (i_{corr}), and a corresponding rise in the values of polarization resistance (R_p) and charge transfer resistance (R_{ct}). Protection capacity increased with higher amounts of concentration and a maximum IE of 93.65% was obtained at 1000 ppm. The formation of precipitates with 1000 ppm of tapioca starch on the surface of AA606 aluminum after 2 months of immersion in the saline solution is displayed by the SEM and EDX images shown in Figure 14.3.

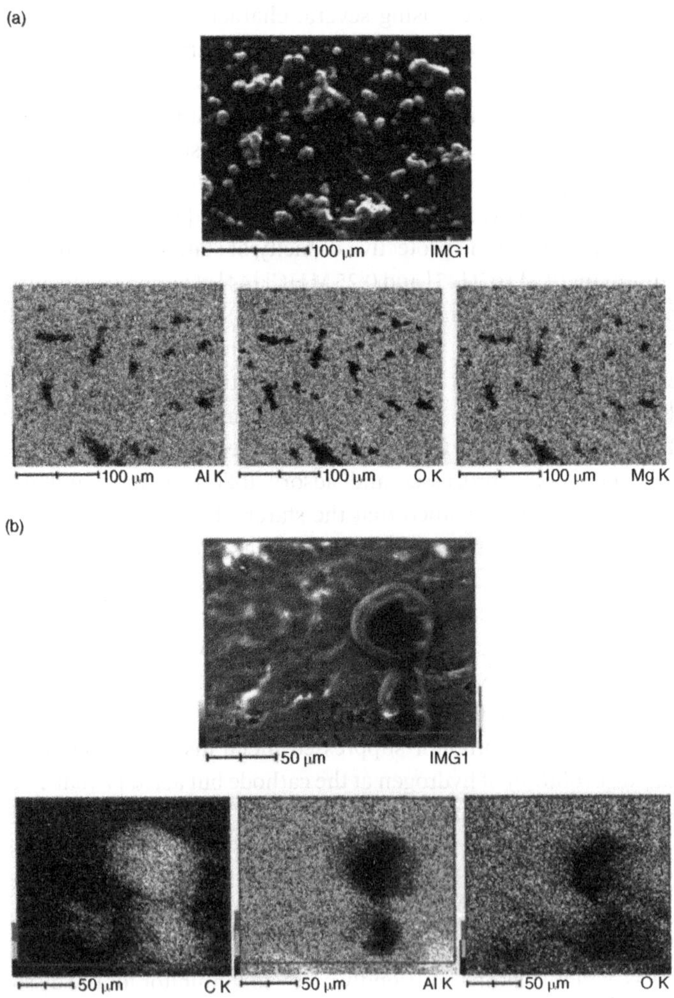

Figure 14.3 SEM and EDX analyses of aluminum immersed in (a) 3.5% NaCl (b) 1000 ppm of tapioca starch after 60 days. [46] / from ELSEVIER.

The influence of wheat starch (WS) on the disintegration of mild steel in 0.5 M H_2SO_4 was evaluated by Nwanonenyi and co-workers [47]. Traditional gravimetric analysis and the PDP method were adopted to determine the effectiveness of WS as a sustainable metallic inhibitor in acidic electrolytes. From the PDP curves, it was clear that WS displayed a mixed-type effect but primarily inhibited cathodic reactions. The impact of variation in temperature on the effectiveness of WS showed that the IE increased with increasing temperature (30 °C to 60 °C). This suggests strong interactions between the mild steel and the adsorbate which is characteristic of the chemisorption mechanism. The mode of adsorption at the metal/electrolyte interface was best explained using the Langmuir adsorption model. The presented PDP data showed that the study recorded a minimum IE of 84.32% with 0.2 g/L and a maximum IE of 92.15% with 0.8 g/L WS.

To produce a protective coating for cold rolled steel (CRS) in a corrosive saline medium, two thermoplastic coatings were selected – namely, corn starch and polystyrene. The corn starch film was prepared by introducing glycerol as a plasticizer into a suspension of corn starch. Dip coating was done by immersing ASTM 1080 steel in the dispersion of the starch-glycerol, and corrosion studies were performed using WCA measurement, optical microscopy, SEM, EDX, and AFM [48]. The polystyrene film was discovered to effectively offer surface protection to ASTM 1080 steel in a saline medium for 6 months, while the starch coating protected the metal specimen in saline solution for 14 days. In another study, computational modeling using DFT calculations and MDS were performed for eco-friendly corrosion and scale inhibitors including oxidized starch (OS), polyepoxysuccinic acid (PESA), carboxymethyl cellulose (CMC), and polyaspartic acid (PASP). The inhibitors' interactions with different surfaces including Fe (110) were probed in the gas and aqueous phases. The binding energy of the studied sustainable corrosion and scale inhibitors onto the metal surface follows the order: PESA<CMC<OS<PASP. The calculated quantum indices were used to delineate the inhibition mechanism of the selected chemical additives [49]. The geometry optimization obtained for each studied inhibitor is displayed in Figure 14.4.

Ochoa and co-workers have explored the adsorption potential and corrosion inhibition of modified starch in corrosive media for different metals. The group conducted an empirical and computational investigation on two species of modified starch for corrosion mitigation of carbon steel in 200 mg/L NaCl [50]. Cassava starch was physically and chemically modified in the fabrication of activated starch (AS) and carboxymethylated starch (CMS) with two varying degrees of substitution. The synthesized starch species were analyzed using nuclear magnetic resonance (^{13}C NMR), and experimentally evaluated using PDP and EIS techniques. Impedance spectra showed semi-circular plots that displayed one capacitive loop whose diameter size increased with the addition of the modified starch. According to the PDP results, AS offered a stronger corrosion resistance than the two different degrees of substitution of CMS ($CMS_{0.13}$ and $CMS_{0.24}$) in the saline

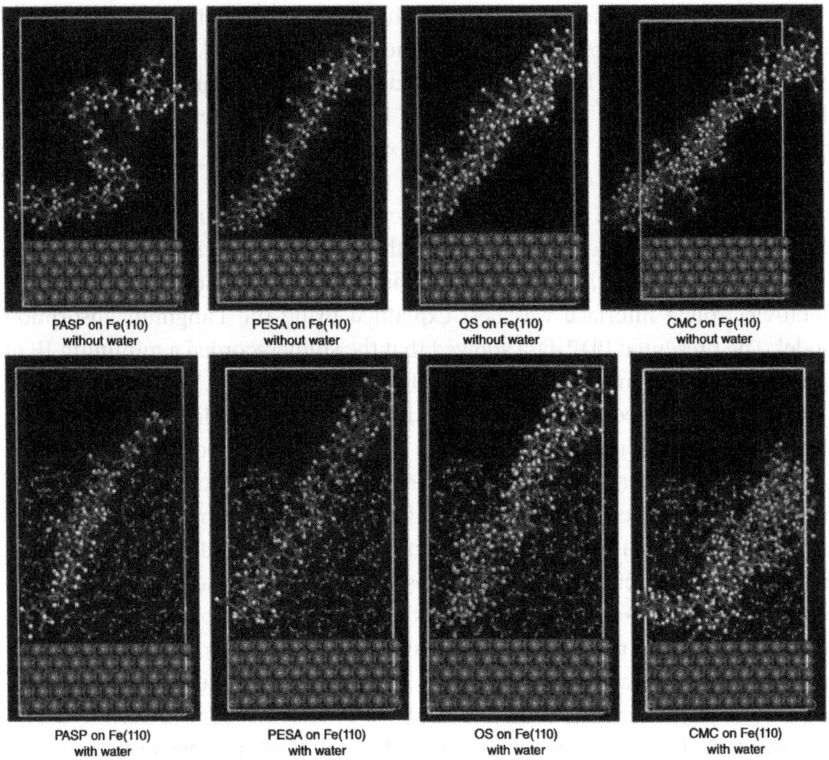

Figure 14.4 Optimized structures of inhibiting compound-Fe (110) interaction systems in gas and aqueous phase. [49] / With permission of Elsevier.

medium. The intensity of surface protection was attributed to the type and quantity of the active group present in the modified starch. The presence of alkoxy groups ($-CO^-$) in AS resulted in higher inhibition performance than CMS having alkoxy ($-CO^-$) and carboxylate ($-COO^-$) groups. Electrostatic potential mappings were performed to further gain insight into the inhibition mechanism of CMS and AS and the authors found that AS will interact strongly with the carbon steel better than CMS because it possessed more negative electrostatic potential. This was further corroborated by AFM analysis. Similar studies using modified cassava starch have been reported by Ochoa and co-workers [51–53]. To impede the rate of galvanized steel degradation in a 1 M HCl medium, native starch derived from sweet potato tubers was physically modified by the extrusion method. Anyiam and co-workers characterized the modified starch using FTIR and tested it as a chemical inhibitor using WL and PDP methods [54]. The impact of time, temperature, and concentration on the surface coverage of the inhibitor was discussed. The authors reported a maximum IE of 64.26% at 0.7 g/L.

Synergism is the most popularly adopted strategy to reduce cost while seeking to improve the effectiveness of an inhibitor at low concentrations [23, 55]. Generally, an increase in the concentration of a studied inhibitor will yield increasing inhibition performance. To circumvent the use of high concentrations, however, a different chemical substance such as a halide, surfactant, or organic compound can be combined with the main inhibitor. By doing so, the overall inhibitor effectiveness would significantly improve. Such synergistic studies involving starch have been reported by a few authors. To adequately protect the steel surface in NaCl solution, an inhibitor mixture comprising propolis (100 ppm), sodium benzoate (2000 ppm), tannin (2000 ppm), PEG400 (200 ppm), and corn starch (400 ppm) was formulated [56]. The authors hoped to exploit the various inhibition mechanisms of the components to achieve a persistent inhibitor-modified rust layer. Thus, they carefully selected each constituent of the inhibitor mixture based on their previously known inhibition action and eco-friendliness. Electrochemical evaluation of the eco-friendly synergistic inhibitor mixture was carried out via LPR and EIS. Obtained results indicated that the inhibitor mixture formed a barrier for the evolution of oxygen at the cathodic site in the electrolyte. The formulated inhibitor mixture protected steel from deterioration to the tune of 83.5%. The study was supported with FTIR, SEM, and EDX measurements. In another report, Brinda and co-workers added 0.2 mM of 2,6-diphenyl-3-methyl piperidine-4-one (DPMP) to improve the inhibition performance of 0.4 mM starch for the dissolution control of mild steel in 1 M HCl at different immersion time and temperature. At 30 °C, the IE of the starch and DPMP (2:1) yielded 72.4%, which showed better IE than the single DPMP (30.0%) and starch (46.1%). The mixture is reported to be stable at higher temperatures with a sustained protective capacity after 24 h. FTIR and DFT studies were used to further analyze the adherence of the mixture on the aluminum substrate [57]. Other reported synergistic studies in literature involved surface protection of mild steel in sulfuric acid using starch and surfactants [58], modified sweet potato starch and KI [59], and millet starch and KI [60, 61].

14.4 Utilization of Grafted Starch as Effective and Sustainable Corrosion Inhibitors

14.4.1 Grafted Cassava Starch as Eco-Friendly Corrosion Inhibitors

Over the years, the cassava plant has played vital roles including being a source of income for farmers and a food source in tropical and subtropical regions of the world [62]. It is remarkable as a major source of carbohydrates with world production reported to be 278 million tonnes in 2018 [63]. A research group in China led by Xianghong Li has extensively exploited the inhibition properties of cassava starch and its grafted forms for steel and aluminum corrosion in different corrosive media [64–71]. They have proven grafted starch to be an excellent inhibitor than its counterparts in various tested media.

Using an oxidation-reduction reaction, acrylamide was grafted onto starch to prepare cassava starch graft acrylamide copolymer (SGA) by Hui and co-workers [64]. The inhibitory potential of the prepared SGA was investigated for the deterioration of CRS in molar HCl via electrochemical (PDP and EIS) and gravimetric (WL) methods. The presented PDP data indicated that the non-inhibited solution had an i_{corr} value of 126.6 μAcm^{-2}, but the inclusion of different concentrations (10, 50, and 100 mg/L) of SGA drastically reduced the i_{corr} value to 36.4, 20.7, and 14.2 μAcm^{-2} respectively. Similarly, the prevented Nyquist diagrams indicated that the impedance spectra demonstrated one capacitive loop and the R_{ct} value increased from 48.72 Ωcm^2 (blank solution) to 454.1 Ωcm^2 (100 mg/L) due to the development of a compact and dense defensive layer on the metal substrate. The excellent inhibitory performance of SGA recorded to be 89.3% at the optimum dosage of 100 mg/L was further established using SEM. In another study, the dissolution behavior of CRS was studied in glycolic acid in the presence of cassava starch graft copolymer (CSGC) fabricated by the grafting of acrylamide onto cassava starch [65]. Using WL and electrochemical methods, the authors found out that 92% IE was attainable with 100 mg/L CSGC at 20 °C and even with a dipping time of 72 h, the inhibition efficiency was up to 89.1%. CSGC was reported to impede cathodic and anodic reactions and conformed to the Langmuir adsorption isotherm. In a similar study, the research group studied the mitigation of CRS dissolution in 1 M H_2SO_4 and obtained a higher IE of 95.7% with 100 mg/L at 20 °C. The study reported that CSGC behaved with a mixed-type effect at 20 °C while it had a predominant effect on the cathodic reaction at 50 °C [66].

Furthermore, in another published study, the research group explored the impact of cassava starch-acrylamide graft copolymer (CS-AAGC) on the degradation of 1060 aluminum in H_3PO_4. The synthetic route and the SEM images of the CS and CS-AAGC are displayed in Figure 14.5. Corrosion tests were carried out using WL and electrochemical techniques. Further corrosion analysis was conducted using XPS, AFM, SEM, and WCA analyses. At 20 °C, the maximum obtained IE was 91.9% with 1.0 g/L CS-AAGC. The Bode modulus confirmed improved adherence of CS-AAGC on the aluminum substrate at increased concentrations and revealed three time constants. Tafel plots revealed that CS-AAGC repressed both anodic and cathodic site reactions with a pronounced anodic effect. The mode of adsorption conformed to the Langmuir isotherm model. XPS analysis revealed that C, P, O, N, and Al existed in the inhibited aluminum surface. Further analysis revealed that the fitted functional groups substantiated the evidence of adherence of the inhibitor onto the metal substrate. The corrosion inhibition mechanism was related to the probable donation of electrons of N and O atoms from CS-AAGC to the vacant p-orbitals of the metal atoms to create coordinate bonds

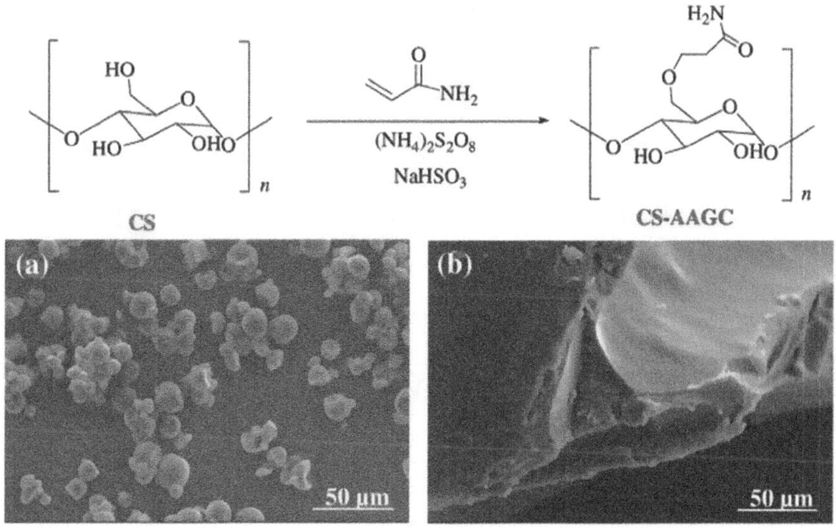

Figure 14.5 Synthetic procedure and SEM images of (a) CS (b) CS-AAGC. [67] / from Elsevier.

while accepting electrons from the metal substrate to create a back-donating bond; the presence of numerous C=O and OH groups that can coordinate with the Al^{3+} ion on the meta/electrolyte interface to create coordination bond and the interaction of CS-AAGC ion with the acid anions leading to physical adsorption on the metal surface [67].

In a recent study, the research group examined the inhibition potential of CSGC on the deterioration of aluminum in 1 M HNO_3 using conventional gravimetric, modern electrochemical, computational, and surface analytical methods [68]. The study probed the effects of temperature, immersion period, inhibitor concentration, and acid concentration on the protective capacity of CSGC. The authors chose glucose as a model for consideration in performing a computational study since starch is made up of many glucose (Glc) units. Three series of Glc-AA structures were modeled using DFT at GGA/BLYP/DND/COSMO level as portrayed in Figure 14.6. The authors explained that an increase in the number of units (AA) was responsible for stronger physisorption and the order of physisorption can be determined by the increasing values of dipole moment, μ (Debye); Glc-AA_3 (11.9741)>Glc-AA_2 (6.9104)>Glc-AA (6.1561)>Glc (5.4502). It is important to mention that the authors reasonably deduced that CSGC inhibits corrosion better than Gl-AA_3. The MDS images showed that adsorbed inhibitor molecules had a nearly flat orientation on the aluminum surface.

The synthesis of cassava starch ternary graft copolymer (CS-SAS-AAGC) was done in situ by polymerizing monomers of acrylamide (AA) and sodium allyl

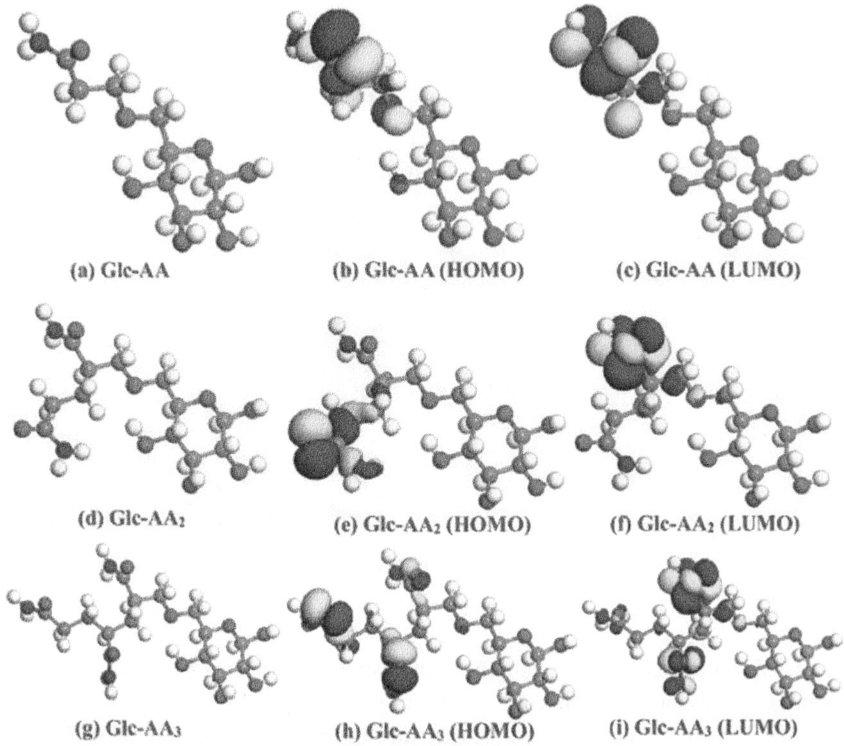

Figure 14.6 Optimized molecular structures and electron density distributions for Glc, Glc-AA, Glc-AA_2, and Glc-AA_3. [68] / With permission of Elsevier.

sulfonate (SAS) in the presence of cassava starch (CS) [69]. FTIR characterization of CS and CS-SAS-AAGC are presented in Figure 14.7. The absorption bands at 3199 cm^{-1} and 1186 cm^{-1} were assigned to N–H and the stretching vibration of S=O in sulfonate in the CS-SAS-AAGC spectrum suggest successful grafting of AA and SAS. The prepared grafted starch was utilized to retard corrosion processes in aluminum in the presence of HCl using WL, EIS, and PDP. Gravimetric analysis data showing the influence of inhibitor dosage, exposure time, HCl concentration, and operating temperatures on the surface protection of aluminum were discussed. Stable OCP measurements were attained at about 4000 s at 20 °C before electrochemical measurements were done. The OCP plots indicated that CS-SAS-AAGC predominantly impeded cathodic reaction which was further proven by the Tafel plots (Figure 14.8). With as little as 100 mg/L, the authors reported an IE of 98.2% at 20 °C from Tafel measurement, thus establishing the effectiveness of CS-SAS-AAGC. Impedance spectra indicated that aluminum degradation is characterized by a charge transfer phenomenon and the obtained imperfect semicircles

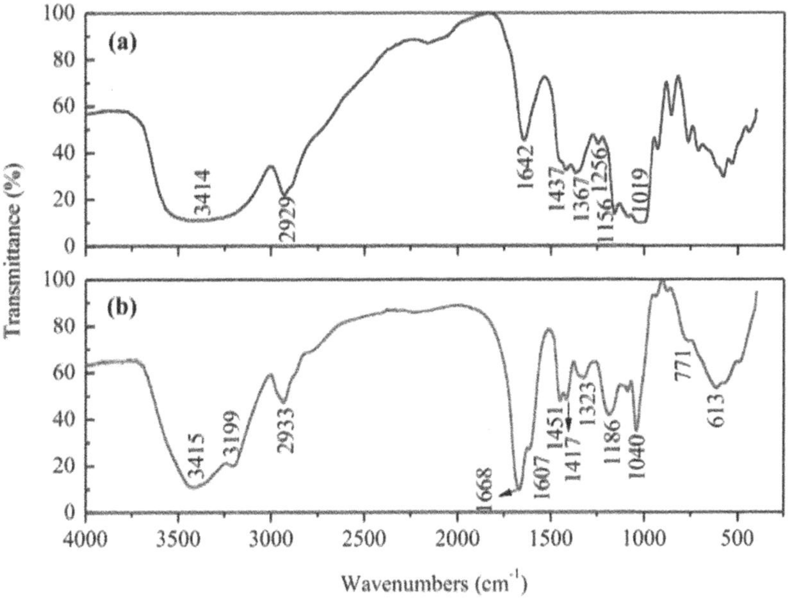

Figure 14.7 FTIR plots for (a) CS (b) CS-SAS-AAGC. [69] / With permission of Elsevier.

were attributed to the uneven nature of the metal electrode surface. SEM and WCA analyses were further applied to establish the surface protection of aluminum.

Using a different metal (CRS) and acid (1 M H_2SO_4), the authors tested the inhibition property of synthesized CS-SAS-AAGC on metallic dissolution. The study was conducted using WL, PDP, EIS, SEM, and WCA. With 50 mg/L dosages, CS-SAS-AAGC (IE_{EIS} = 95.5%) performed as an excellent inhibitor than single CS, AA, or SAS at 20 °C. However, it was detected that a rise in inhibitor dosage and temperature led to decreased IE. The mode of inhibition action of CS-SAS-AAGC was reported as mixed but with a predominant cathode effect [70]. Using another acidic medium (1 M HCl), the authors reported the inhibitive property of CS-SAS-AAGC on CRS using gravimetric, electrochemical, and surface analysis techniques [71]. An IE_{EIS} of 93.3% was obtained at an optimum concentration of 50 mg/L which showed that CS-SAS-AAGC inhibited CRS better in H_2SO_4 than HCl, probably because HCl is a more corrosive medium. The impedance diagram for the disintegration of CRS in molar HCl is presented in Figure 14.9. SEM, AFM, and WCA measurements were performed, and the obtained images are displayed in Figure 14.10. SEM and AFM images revealed a smooth metal surface for steel immersed in CS-SAS-AAGC when compared to steel immersed in HCl for 6 h. WCA analysis is used to carry out wettability studies and, in this case, the value of the WCA was amplified from 43.1–45.3° (blank solution) to 102.7–103.8° (inhibitor-containing solution) which suggests that

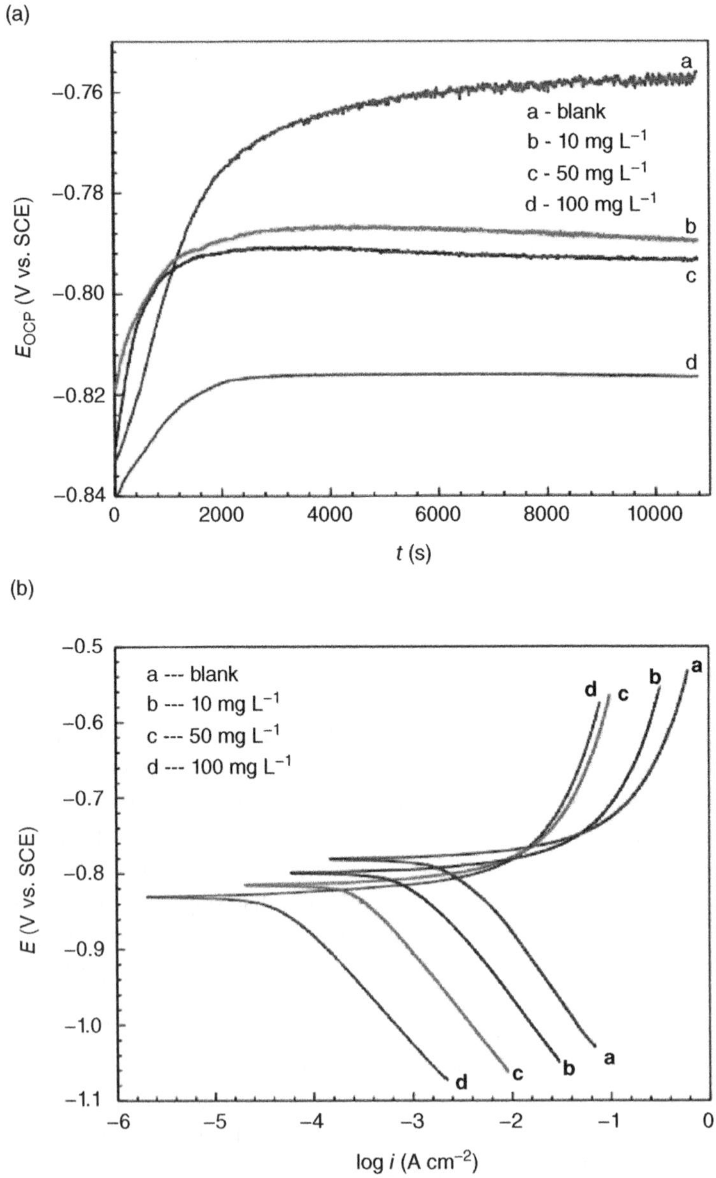

Figure 14.8 (a) OCP-time plots (b) Tafel curves for aluminum in molar HCl in the presence and absence of 100 mg/L CS-SAS-AAGC. [69] / With permission of Elsevier.

the protected metal surface is hydrophobic. Given that starch is primarily composed of amylopectin and amylose, the authors supported the study by examining the adsorption ability and inhibition mechanism of amylopectin disaccharide (APDS), amylose disaccharide (ALDS), and their grafted forms using DFT and MDS.

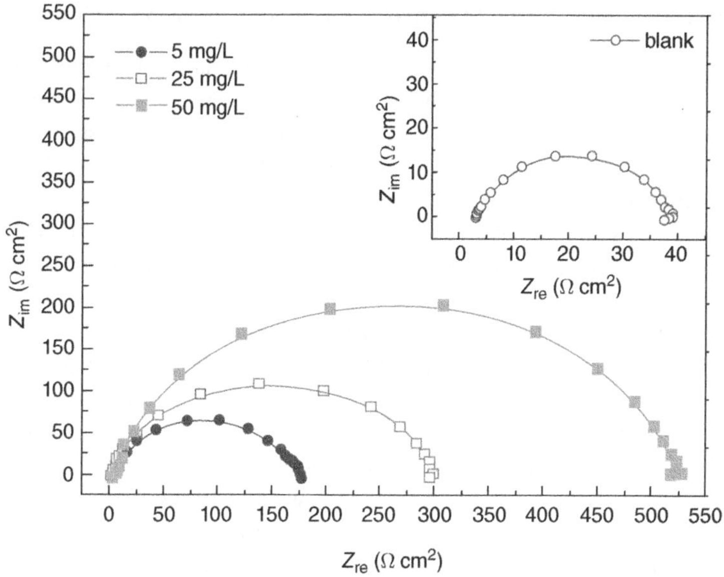

Figure 14.9 Impedance spectra for the degradation of CRS in1 M HCl in the presence and absence of various inhibitor concentrations of CS-SAS-AAGC at 293 K. [71] / With permission of Elsevier.

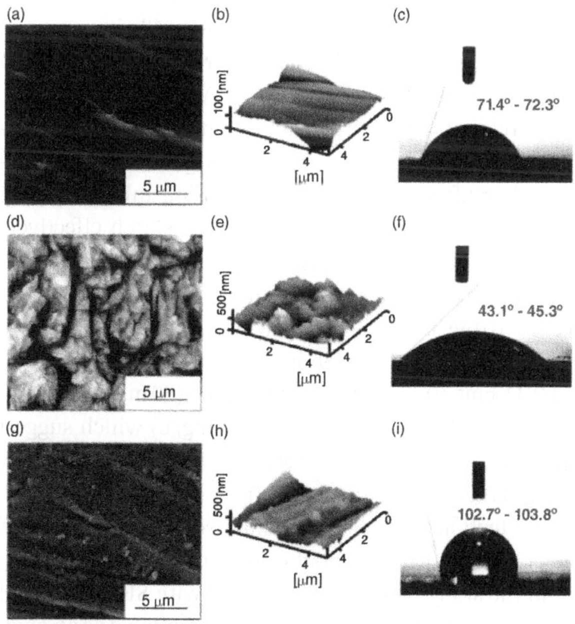

Figure 14.10 SEM (a, d, g), AFM (b, e, h) and water contact angle diagrams for CRS:(a-c) polished surface; (d-f) after 6 h exposure in 1 M HCl; (g-i) after 6 h exposure in 50mg/L CS-SAS-AAGC. [71] / With permission of Elsevier.

14.4.2 Miscellaneous Grafted Starch as Eco-Friendly Corrosion Inhibitors

From the literature study, other grafted forms of starch that have been reported as anticorrosive agents in aqueous electrolytes include corn and potato-grafted starch.

Corn starch was modified, and glycerin was grafted onto the corn starch for assessment as an environmentally benign chemical additive for C-Mn steel dissolution in molar HCl using OCP, PDP, EIS, and WL [72]. The successful synthesis of the glycerin-grafted starch designated as St63Gly37 was confirmed using ^{1}H NMR and FTIR. The ^{1}H NMR spectrum revealed that the signals of glycerin and starch were found together, and the copolymer comprised 37% glycerin and 63% starch establishing a successful grafting process. With 300 mg/L, St63Gly37 inhibited C-Mn dissolution to the maximum tune of 94.00% at 298K as determined from mass loss studies. WL studies also showed an increase in the IE at the increasing temperature on introducing St63Gly37 which suggests the adsorption mechanism might be better explained in terms of chemical adsorption. The study obtained $\Delta G°_{ads}$ values ranging from −7.65 to −6.15 kJmol^{-1} which indicates the stability of the protective layer on the steel and the spontaneity of the adsorption phenomenon. A positive value of $\Delta H°_{ads}$ (9.22 kJ mol^{-1}) and $\Delta S°_{ads}$ (52.2 Jmol^{-1}K^{-1}) suggests the corrosion inhibition process favors endothermic reaction and the adsorption process is followed by an increase in entropy respectively. Tafel curves revealed that St63Gly37 behaved predominantly as a cathodic inhibitor while impedance studies indicated that the reduced C_{dl} suggest the adsorption of St63Gly37 on the C-Mn steel surface resulting in surface protection from corrosive Cl^{-} ions.

A published study by Yating and co-workers in 2021 reported the inhibition of Zn dissolution in 1 M HCl by starch-grafted copolymer (St-g-PAM) [73]. Acrylamide monomer was grafted onto starch in the presence of an initiator (ammonium persulfate sodium bisulfite) to form St-g-PAM. The fabricated St-g-PAM was characterized using FTIR and was thereafter examined as a corrosion inhibitor using WL, PDP, EIS, and SEM. PDP results showed that the grafted starch effectively retarded the corrosion process both at the cathodic and anodic branches. Presented results showed that the corrosion current densities obtained from Tafel extrapolations dropped from 12 μA cm^{-2} (0 mg/L) to 0.992 μAcm^{-2} (50 mg/L). In addition, presented impedance parameters for Zn corrosion in molar HCl showed that R_{ct} values increased from 65.09 Ω cm^{2} (0 mg/L) to 834.5 Ωcm^{2} (50 mg/L) and C_{dl} dropped from 142.56 μFcm^{-2} (0 mg/L) to 55.61 μFcm^{-2} (50 mg/L) which suggest the formation of protective film on Zn by St-g-PAM. The inhibition efficiency increased steadily as the St-g-PAM dosage increased from 10 mg/L to 50 mg/L where it attained maximum protection of 92.20%. Langmuir isotherm was reported to best fit the inhibition data obtained. Kinetic and thermodynamic data were also presented and discussed, and the study was further enriched with SEM and AFM analyses. A summary of reported grafted starch that has been exploited as an environmentally benign corrosion inhibitor for different metals in various aggressive environments is presented in Table 14.1.

Table 14.1 A summary of grafted starch tested as effective and eco-friendly corrosion inhibitors.

S/N	Grafted starch	Metal/electrolye	IE %	Conc	Adsorption type	Nature of inhibitor	References
1.	Cassava starch graft acrylamide copolymer (SGA)	Cold rolled steel / 1 M HCl	89.3	100 mg/L	Langmuir	Mixed-type	[64]
2.	Cassava starch grafted copolymer–acrylamide onto cassava starch (CSGC)	Cold rolled steel /0.3 mol/L $C_2H_4O_3$	92.1	100 mg/L	Langmuir	Mixed-type	[65]
3.	Cassava starch graft copolymer acrylamide onto cassava starch (CSGC)	Steel /1 M H_2SO_4	95.7	100 mg/L	Langmuir	Mixed-type	[66]
4.	Cassava starch-acryl amide graft copolymer (CS-AAGC)	Aluminum / 1 mol/L H_3PO_4	90.6	1 g/L	Langmuir	Mixed-type/Anodic inhibitor	[67]
5.	Cassava starch graft copolymer acrylamide onto cassava starch (CSGC)	Aluminum / 1 M HNO_3	97.8	1 g/L	Langmuir	Mixed-type/Anodic inhibitor	[68]
6.	Cassava starch-sodium allyl sulfonate-acrylamide graft copolymer (CS-SAS-AAGC)	Aluminum/ 1 M HCl	98.2	100 mg/L	Langmuir	Mixed-type/Cathodic inhibitor	[69]
7.	Cassava starch ternary graft copolymer (CS-SAS-AAGC)	Cold rolled steel / 1 M H_2SO_4	89.6	50 mg/L	Langmuir	Mixed-type/Cathodic inhibitor	[70]
8.	Cassava starch ternary graft copolymer (CS-SAS-AAGC)	Steel / 1 M HCl	97.2	50 mg/L	Langmuir	Mixed-type/Anodic inhibitor	[71]
9.	Glycerin-grafted starch-bio copolymer (St63Gly37)	C-Mn steel / 1 M HCl	94.0	300 mg/L	Langmuir	Mixed-type/Cathodic inhibitor	[72]
10.	Starch grafted copolymer (St-g-PAM)	Zinc / 1 M HCl	92.2	50 mg/L	Langmuir	Mixed-type	[73]

14.5 Conclusions

From the foregoing discussion, it is clear that the retardation of metallic corrosion in corrosive environments can be successfully achieved using starch and its derivatives. Based on the reviewed studies in this chapter, it can be concluded that starch stands a chance of being a commendable alternative to synthetic organic inhibitors because of the numerous advantages it offers. Besides being a biodegradable and eco-friendly inhibitor in all the tested media, it was also found to favorably and effectively impede metallic degradation. Comparatively, grafted starch was reported to retard corrosion in aggressive media better than many other studied natural biopolymers. The reported inhibition efficiency of cassava grafted starch ranged from 89 to 98% which attests to the excellent inhibitive effectiveness of grafted starch. It is expected that more studies will be conducted to investigate the potential inhibitive properties of other sources of starch using modern electrochemical, surface morphological, and computational approaches. In addition, more fact-finding studies are required on the inhibition mechanism of starch and its derivatives in various corrosive solutions.

References

1 Verma, C. and Quraishi, M. (2021). Gum Arabic as an environmentally sustainable polymeric anticorrosive material: recent progress and future opportunities. *International Journal of Biological Macromolecules* 184.

2 Verma, C., Olasunkanmi, L.O., Quadri, T.W., Sherif, E.-S.M., and Ebenso, E.E. (2018). Gravimetric, electrochemical, surface morphology, DFT, and Monte Carlo simulation studies on three N-substituted 2-aminopyridine derivatives as corrosion inhibitors of mild steel in an acidic medium. *Journal of Physical Chemistry C* 122: 11870–11882.

3 Jiang, L., Qiang, Y., Lei, Z. et al. (2018). Excellent corrosion inhibition performance of novel quinoline derivatives on mild steel in HCl media: experimental and computational investigations. *Journal of Molecular Liquids* 255: 53–63.

4 Olasunkanmi, L.O., Kabanda, M.M., and Ebenso, E.E. (2016). Quinoxaline derivatives as corrosion inhibitors for mild steel in hydrochloric acid medium: electrochemical and quantum chemical studies, Physica E Low Dimens. *Systems and Nanostructures* 76: 109–126.

5 Niamien, P.M., Essy, F.K., Trokourey, A. et al. (2012). Correlation between the molecular structure and the inhibiting effect of some benzimidazole derivatives. *Materials Chemistry and Physics* 136: 59–65.

6 Quadri, T.W., Olasunkanmi, L.O., Akpan, E.D. et al. (2021). Chromeno-carbonitriles as corrosion inhibitors for mild steel in acidic solution: electrochemical, surface and computational studies. *RSC Advances* 11: 2462–2475.

7 Bastos, A., Ferreira, M., and Simões, A. (2006). Corrosion inhibition by chromate and phosphate extracts for iron substrates studied by EIS and SVET. *Corrosion Science* 48: 1500–1512.

8 Devi, M. and Kannan, K. (2013). Evaluation of corrosion inhibition performance of zinc oxide and sodium nitrite in quarry dust concrete. *Asian Journal of Chemistry* 25: 8690.

9 Umoren, S. and Solomon, M. (2014). Recent developments on the use of polymers as corrosion inhibitors-a review. *The Open Materials Science Journal* 8.

10 Verma, C., Ebenso, E.E., Quraishi, M., and Hussain, C.M. (2021). Recent developments in sustainable corrosion inhibitors: design, performance and industrial scale applications. *Materials Advanced* 2.

11 Umoren, S.A., Solomon, M.M., Madhankumar, A., and Obot, I.B. (2020). Exploration of natural polymers for use as green corrosion inhibitors for AZ31 magnesium alloy in a saline environment. *Carbohydrate Polymer* 230: 115466.

12 Rizvi, M., Natural polymers as corrosion inhibitors, organic corrosion inhibitors: synthesis, characterization, mechanism, and applications, (2021) 411–434.

13 Umoren, S.A. and Eduok, U.M. (2016). Application of carbohydrate polymers as corrosion inhibitors for metal substrates in different media: a review. *Carbohydrate Polymers* 140: 314–341.

14 Umoren, S., Ebenso, E., Okafor, P., and Ogbobe, O. (2006). Water-soluble polymers as corrosion inhibitors. *Pigment & Resin Technology* 35.

15 Basik, M. and Mobin, M. (2021). *Synthetic Polymers as Corrosion Inhibitors, Organic Corrosion Inhibitors: Synthesis, Characterization, Mechanism, and Applications*. 435–449.

16 Umoren, S. (2008). Inhibition of aluminum and mild steel corrosion in acidic medium using Gum Arabic. *Cellulose* 15: 751–761.

17 Biswas, A., Pal, S., and Udayabhanu, G. (2015). Experimental and theoretical studies of xanthan gum and its graft copolymer as corrosion inhibitor for mild steel in 15% HCl. *Applications of Surface Science* 353: 173–183.

18 Biswas, A., Das, D., Lgaz, H., Pal, S., and Nair, U.G. (2019). Biopolymer dextrin and poly (vinyl acetate) based graft copolymer as an efficient corrosion inhibitor for mild steel in hydrochloric acid: electrochemical, surface morphological and theoretical studies. *Journal of Molecular Liquids* 275: 867–878.

19 Charitha, B. and Rao, P. (2020). Pectin as a potential green inhibitor for corrosion control of 6061al–15%(v) sic (p) composite in acid medium: electrochemical and surface studies. *Journal of Failure Analysis and Prevention* 20: 1684–1696.

20 Obot, I., Onyeachu, I.B., and Kumar, A.M. (2017). Sodium alginate: a promising biopolymer for corrosion protection of API X60 high-strength carbon steel in the saline medium. *Carbohydrate Polymers* 178: 200–208.

21 Nardeli, J.V., Fugivara, C.S., Taryba, M. et al. (2019). Tannin: a natural corrosion inhibitor for aluminum alloys. *Progress in Organic Coatings* 135: 368–381.

22 Zhang, Z., Ba, H., and Wu, Z. (2019). Sustainable corrosion inhibitor for steel in simulated concrete pore solution by maize gluten meal extract: electrochemical and adsorption behavior studies. *Construction and Building Materials* 227: 117080.

23 Gupta, N.K., Joshi, P., Srivastava, V., and Quraishi, M. (2018). Chitosan: a macromolecule as a green corrosion inhibitor for mild steel in sulfamic acid useful for the sugar industry. *International Journal of Biological Macromolecules* 106: 704–711.

24 Pais, M. and Rao, P. (2021). Electrochemical, spectroscopic and theoretical studies for acid corrosion of zinc using glycogen. *Chemical Papers* 75: 1387–1399.

25 Verma, C., Quadri, T.W., Ebenso, E.E., and Quraishi, M. (2021). *Polymer Nanocomposites as Industrially Useful Corrosion Inhibitors: Recent Developments, Handbook of Polymer Nanocomposites for Industrial Applications*. 419–435.

26 Umoren, S.A. and Solomon, M.M. (2019). Protective polymeric films for industrial substrates: a critical review on past and recent applications with conducting polymers and polymer composites/nanocomposites. *Progress in Materials Science* 104: 380–450.

27 Arthur, D.E., Jonathan, A., Ameh, P.O., and Anya, C. (2013). A review on the assessment of polymeric materials used as corrosion inhibitors of metals and alloys. *International Journal of Industrial Chemistry* 4: 1–9.

28 Peter, A., Obot, I., and Sharma, S.K. (2015). Use of natural gums as green corrosion inhibitors: an overview. *International Journal of Industrial Chemistry* 6: 153–164.

29 Quadri, T.W., Akpan, E.D., Olasunkanmi, L.O., Fayemi, O.E., and Ebenso, E.E. (2022). Fundamentals of corrosion chemistry. In: *Environmentally Sustainable Corrosion Inhibitors* (ed. C. Hussain, C. Verma, and J. Aslam), 25–45. Elsevier.

30 Quadri, T.W., Olasunkanmi, L.O., Fayemi, O.E. et al. (2021). Quantitative structure-activity relationship and artificial neural network as vital tools in predicting coordination capabilities of organic compounds with the metal surface: a review. *Coordination Chemistry Reviews* 446: 214101.

31 Din, Z.u., Chen, L., Xiong, H. et al. (2020). Starch: an undisputed potential candidate and sustainable resource for the development of wood adhesive. *Starch-Stärke* 72: 1900276.

32 Jyothi, A. (2010). Starch graft copolymers: novel applications in industry. *Composite Interfaces* 17: 165–174.

33 Nawaz, H., Waheed, R., Nawaz, M., and Shahwar, D. (2020). Physical and chemical modifications in starch structure and reactivity. *Chemical Properties of Starch* 13.

34 Sharmin, E., Ahmad, S., and Zafar, F. (2012). Renewable resources in corrosion resistance. In: *Corrosion Resistance* (ed. H. Shih), 449–472. IntechOpen.

35 Cummings, S., Zhang, Y., Smeets, N., Cunningham, M., and Dubé, M.A. (2019). On the use of starch in emulsion polymerizations. *Processes* 7: 140.

36 Li, H., Qi, Y., Zhao, Y., Chi, J. and Cheng, S. (2019). Starch and its derivatives for paper coatings: a review. *Progress in Organic Coatings* 135: 213–227.

37 Bertolini, A. (2009). *Starches: Characterization, Properties, and Applications*. CRC Press.

38 Amaraweera, S.M., Gunathilake, C., Gunawardene, O.H. et al. (2021). Development of starch-based materials using current modification techniques and their applications: a review. *Molecules* 26: 6880.

39 Chen, Q., Yu, H., Wang, L. et al. (2015). Recent progress in chemical modification of starch and its applications. *RSC Advances* 5: 67459–67474.

40 Abd El Haleem, S., Abd El Rehim, S., and Shalaby, M.S. (1986). Anodic behavior and pitting corrosion of plain carbon steel in NaOH solutions containing Cl-ions. *Surface and Coatings Technology* 27: 167–173.

41 Sugama, T. and DuVall, J. (1996). Polyorganosiloxane-grafted potato starch coatings for protecting the aluminum from corrosion. *Thin Solid Films* 289: 39–48.

42 Sugama, T. (1997). Oxidized potato-starch films as primer coatings of aluminum. *Journal of Materials Science* 32: 3995–4003.

43 Charitha, B. and Rao, P. (2017). Starch as an eco-friendly green inhibitor for corrosion control of 6061-Al alloy. *Journal of Materials and Environmental Science* 8: 78–89.

44 Charitha, B. and Rao, P. (2015). Ecofriendly biopolymer as green inhibitor for corrosion control of 6061-aluminum alloy in hydrochloric acid medium. *International Journal of ChemTech Research* 8: 330–342.

45 Othman, N.K., Salleh, E.M., Dasuki, Z., and Lazim, A.M. (2018). Dimethyl Sulfoxide-treated starch of dioescorea hispida as a green corrosion inhibitor for low carbon steel in sodium chloride medium. In: *Corrosion Inhibitors, Principles and Recent Applications* (ed. M. Aliofkhazraei), 181–199. IntechOpen.

46 Rosliza, R. and Nik, W.W. (2010). Improvement of corrosion resistance of AA6061 alloy by tapioca starch in seawater. *Current Applied Physics* 10: 221–229.

47 Nwanonenyi, S., Ogbobe, O., and Oguzie, E. (2017). Protection of mild steel corrosion in a sulphfuric acid environment using wheat starch. *International Journal of Engineering and Technology* 10: 11–21.

48 Souza, C.d., Teixeira, R.L.P., Lacerda, J.C.d. et al. (2018). Polystyrene and cornstarch anti-corrosive coatings on steel. *Polímeros* 28: 226–230.

49 Chen, X., Chen, Y., Cui, J. et al. (2021). Molecular dynamics simulation and DFT calculation of "green" scale and corrosion inhibitor. *Computational Materials Science* 188: 110229.

50 Bello, M., Ochoa, N., Balsamo, V. et al. (2010). Modified cassava starches as corrosion inhibitors of carbon steel: an electrochemical and morphological approach. *Carbohydrate Polymers* 82: 561–568.

51 Bello, M., Sancristóbal, J., Ochoa, N., and Balsamo, V. (2010). Effect of the degree of substitution of carboxymethylated cassava starch tested as green corrosion inhibitor of carbon steel. In: *Proceedings of the 68th Annual Technical Conference & Exhibition* 115.

52 Bello, M., Ochoa, N., and Balsamo, V. (2011). Effect of the environmental pH on the corrosion bioinhibitive properties of modified cassava starches. In: *Proceedings of the 69th Annual Technical Conference & Exhibition* 266.

53 Ochoa, N., Bello, M., Sancristóbal, J. et al. (2013). Modified cassava starches as potential corrosion inhibitors for sustainable development. *Materials Research* 16: 1209–1219.

54 Anyiam, C., Ogbobe, O., Oguzie, E. et al. (2020). Corrosion inhibition of galvanized steel in the hydrochloric acid medium by a physically modified starch. *SN Applied Sciences* 2: 1–11.

55 Umoren, S.A. and Solomon, M.M. (2017). Synergistic corrosion inhibition effect of metal cations and mixtures of organic compounds: a Review. *Journal of Environmental Chemical Engineering* 5: 246–273.

56 Božović, S., Martinez, S., and Grudić, V. (2019). A novel environmentally friendly synergistic mixture for steel corrosion inhibition in 0.51 M NaCl. *Acta Chimica Slovenica* 66: 112–122.

57 Brinda, T., Mallika, J., and Sathyanarayana Moorthy, V. (2015). Synergistic effect between starch and substituted piperidine-4-one on the corrosion inhibition of mild steel in acidic medium. *Journal of Materials and Environmental Science* 6: 191–120.

58 Mobin, M., Khan, M., and Parveen, M. (2011). Inhibition of mild steel corrosion in an acidic medium using starch and surfactant additives. *Journal of Applied Polymer Science* 121: 1558–1565.

59 Anyiam, C., Ogbobe, O., Oguzie, E., and Madufor, I. (2020). Synergistic study of modified sweet potato starch and KI for corrosion protection of mild steel in acidic media. *Journal of Bio-and Tribo-Corrosion* 6: 1–10.

60 Nwanonenyi, S., Madufor, I., Uzoma, P., and Chukwujike, I. (2016). Corrosion inhibition of mild steel in a sulfguric acid environment using millet starch and potassium iodide. *International Research Journal of Pure and Applied Chemistry* 12: 1–15.

61 Nwanonenyi, S., Madufor, I., Chukwujike, I., and Arinze, V. (2016). Experimental and theoretical studies of inhibitive behavior of millet starch on the corrosion of aluminum in a sulfuric acid environment. *International Journal of Engineering and Technology* 8: 1–13.

62 Kolawole, F.O., Kolawole, S.K., Olugbemi, O.M., and Hassan, S.B. (2019). Green corrosion inhibitory potentials of the cassava plant (Manihot esculenta Crantz) Extract nanoparticles (CPENPs) in coatings for oil and gas pipelines. In: *Corrosion Inhibitors* (ed. A. Singh). IntechOpen.

63 Tafesse, A., Mena, B., Belay, A. et al. (2021). Cassava production efficiency in southern Ethiopia: the parametric model analysis. *Frontiers in Sustainable Food Systems* 5: 426.

64 Hui, F., Xianghong, L., Yunxian, L., and Jianxiang, L. (2011). Corrosion inhibition of cassava starch graft acryl amide copolymer for cold rolled steel in hydrochloric acid. *Journal of Chinese Society for Corrosion and Protection* 31: 265–269.

65 Li, X., Deng, S., Li, Y., and Liu, J. (2017). Corrosion inhibition of steel in glycolic acid solution by cassava starch grafted copolymer. *Chemistry and Industry of Forest Products* 37: 101–108.

66 Li, X. and Deng, S. (2015). Cassava starch graft copolymer as an eco-friendly corrosion inhibitor for steel in H2SO4 solution. *Korean Journal of Chemical Engineering* 32: 2347–2354.

67 Deng, S., Li, X., and Du, G. (2021). An efficient corrosion inhibitor of cassava starch graft copolymer for aluminum in phosphoric acid. *Chinese Journal of Chemical Engineering* 37: 222–231.

68 Li, X., Deng, S., Lin, T., and Xie, X. (2019). Cassava starch graft copolymer as a novel inhibitor for the corrosion of aluminum in HNO3 solution. *Journal of Molecular Liquids* 282: 499–514.

69 Li, X., Deng, S., Lin, T., Xie, X., and Du, G. (2018). Cassava starch-sodium allyl sulfonate-acryl amide graft copolymer as an effective inhibitor of aluminum corrosion in HCl solution. *Journal of the Taiwan Institute of Chemical Engineers* 86 252–269.

70 Xianghong, L., Shuduan, D., and Xin, X. (2020). Inhibition action of cassava starch ternary graft copolymer on steel in H2SO4 solution. *Journal of Chinese Society for Corrosion and Protection* 40: 105–114.

71 Li, X., Deng, S., Lin, T., Xie, X., and Du, G. (2020). Cassava starch ternary graft copolymer as a corrosion inhibitor for steel in HCl solution. *Journal of Materials Research and Technology* 9: 2196–2207.

72 Lahrour, S., Benmoussat, A., Bouras, B. et al. (2019). Glycerin-grafted starch as corrosion inhibitor of C-Mn steel in 1 M HCl solution. *Applied Sciences* 9: 4684.

73 Yating, W., Kexu, W., Pengxiang, G. et al. (2021). Inhibition for Zn corrosion by starch grafted copolymer. *Journal of Chinese Society for Corrosion and Protection* 41: 131–138.

15

Grafted Cellulose as Sustainable Corrosion Inhibitors

Ali Asghar Javidparvar[1], *Abdolreza Farhadian*[2,3,*], *and Ali Reza Shahmoradi*[4]

[1] *School of Metallurgy and Materials Engineering, College of Engineering, University of Tehran, P.O. Box 11155-4563, Tehran, Iran*
[2] *Department of Polymer & Materials Chemistry, Faculty of Chemistry and Petroleum Science, Shahid Beheshti University, GC, 1983969411, Tehran, Iran*
[3] *Department of Petroleum Engineering, Kazan Federal University, Kremlevskaya str. 18, 420008 Kazan, Russian Federation*
[4] *Department of Chemical Engineering, Shahreza Branch, Islamic Azad University, Shahreza, Iran*
[*] *Corresponding author*

15.1 Introduction

Biopolymers have recently attracted attention for application as organic corrosion inhibitors (OCIs) due to their large molecular size and multiple adsorption sites [1, 2]. The presence of different functional groups in the chemical structure of these OCIs facilitates their adsorption on the immersed metal surface in corrosive electrolytes and blocks a large surface area of metals [3]. The inhibition efficiency of bio-polymers as OCIs in corrosive electrolytes depends on their molecular size, the number of aromatic rings, and abundance of heteroatoms (such as O, N, S, and P), and activity centres for adsorption on the immersed electrode surface [4].

Cellulose is a well-known biopolymer introduced by Payne in 1838 with the experimental formula ($C_6H_{10}O_5$) [5]. It is widely found in plants that are used as raw materials to produce modified cellulose for practical and industrial applications [6, 7]. Due to the abundance of this polymer in nature, the production of modified cellulose is expanding rapidly [8]. In recent years, researchers have focused on improving known products as well as ways of producing new cellulose derivatives for different applications [9, 10]. Following a brief review of the structural and compositional features of cellulose and its derivatives, this chapter will also break down how the application of these substances as eco-friendly corrosion inhibitors is studied.

Once the chemical structure and some applications of cellulose and its derivatives are introduced, the case studies on the application of these bio-polymers as green corrosion inhibitors will also be examined.

Grafted Biopolymers as Corrosion Inhibitors: Safety, Sustainability, and Efficiency, First Edition.
Edited by Jeenat Aslam, Chandrabhan Verma, and Ruby Aslam.

15.1.1 Cellulose

The chemical structure of cellulose is known as the major compound in the cell walls of plants. It is a water-insoluble homopolymer that exists in linear form and with high molecular weight in nature. The crystalline regions of cellulose are connected by a large number of hydrogen bonds. It is classified as an insoluble fibre and contains at least 3,000 units of glucose ($C_6H_{12}O_6$) [11]. It is a long-chain molecular substance that has a flexible structure and is composed of β-4, 1-D units of glucopyranose, which are connected by a β-4, 1 glycosidic bond between carbon 1 and 4. For each of these units, there are three groups of hydroxyl agents—excluding units at both ends of the cellulose chain—where the attached hydroxyl group to the carbon 6 is in type I and the hydroxyl group attached to the carbons 2 and 3 areas in type II [12]. A schematic of the chemical structure of cellulose is shown in Figure 15.1.

Due to the presence of three hydroxyl groups (as shown in Figure 15.1), molecular cellulose is prone to different chemical reactions such as esterification, etherification, and oxidation. Cellulose derivatives are generally obtained by esterification or etherification of the hydroxyl groups of the cellulose chains [12]. Ether is a compound in which an oxygen atom is attached to two organic groups. Carbohydrate ethers are not as common as esters in nature. These days, the etherification of cellulose is a common process to improve the properties and increase the usability of this natural biopolymer. Methylcellulose (Cellulose-O-CH3), hydroxyl methylcellulose, and carboxymethyl cellulose (Cellulose-O-CH_2-$CO_2^-Na^+$) is cellulose ether products that have wide applications in different industries [13, 14].

The cellulose chain is linear and has amorphous and crystalline parts. In the crystalline part, the chains are quite close to each other and many hydrogen bonds are formed between their hydroxyl groups. The high degree of crystallinity leads to the high elasticity of the structure, increasing its resistance to tearing. The heating of cellulose breaks down its hydrogen bonds, decreases its crystalline

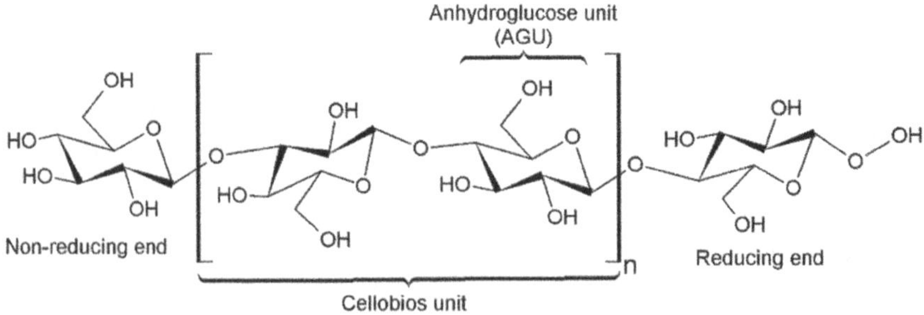

Figure 15.1 A schematic of the chemical structure of cellulose. (Zennifer, A. *et al.* (2021) 'Key advances of carboxymethyl cellulose in tissue engineering & 3D bioprinting applications', *Carbohydrate Polymers*. Elsevier, 256, p. 117561).

component while increasing its amorphous quality and improves its water absorption [15, 16]. In addition, the drying process turns the amorphous aspect into a crystalline state. The amorphous part reacts easily with chemicals and has an irregular structure so the chains are far apart and the absorption of water is limited to its amorphous part. Cellulose is insoluble in both hot and cold water which makes the most suitable solvent for it an ammonia solution of copper sulphate. The low solubility of cellulose in water is due to its strong intermolecular hydrogen bonds of cellulose chains [17].

Cellulose and lignin are two main components available in the stem of many plants and the shell of many nuts such as walnut and pistachio. Shahmoradi et al. [18] investigated the inhibition performance of cellulose and lignin extracted from the hard shell of pistachio for mild steel (MS) immersed in 1 M HCl solution. In this study, the results of electrochemical impedance spectroscopy (EIS) showed that 800 ppm of the extract containing cellulose achieved about 92% inhibition efficiency via a mixed-type inhibition mechanism. A schematic of the adsorption mechanisms of the extracted compounds on the MS surface is shown in Figure 15.2.

According to Figure 15.2, it is clear that the compounds in the extract of the pistachio hard shell can be adsorbed on the surface of MS immersed in the acid

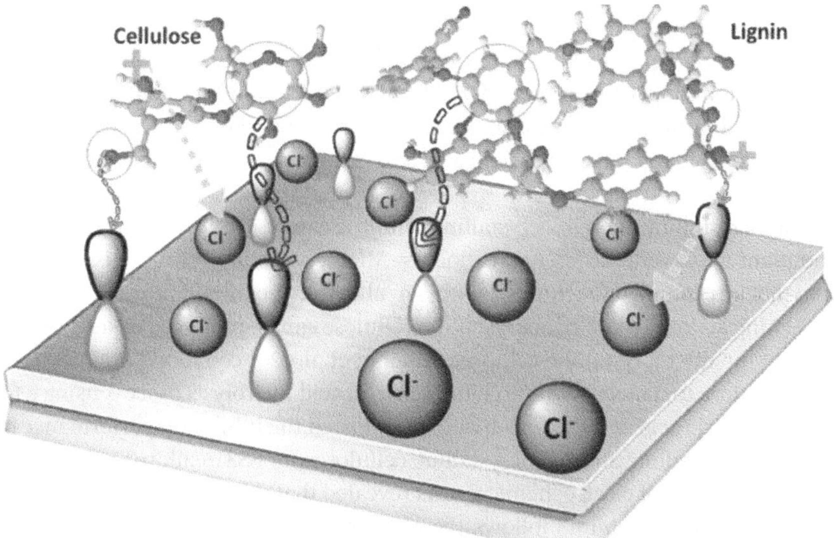

Figure 15.2 A schematic of the adsorption mechanisms of cellulose and lignin extracted from pistachio hard shell on the surface of MS immersed in 1M HCl solution. (Shahmoradi, A. R. et al., (2020) / With permission of Elsevier).

electrolyte through two mechanisms—physisorption and chemisorption—to prevent corrosion by creating an ultra-thin protective coating on the surface. For the physisorption mechanism, the extracted chemical structures can be initially protonated and positively charged in the acidic solution. Alternatively, by placing the MS electrode in the HCl electrolyte, the potential of the metal surface becomes positive causing the adsorption of anions in the solution—specifically chlorine anion—on the metal surface. Thus, the chlorine ions present on the surface of the immersed MS adsorb the protonated compounds (positively charged) in the electrolyte via electrostatic interaction wherein a physical bond is formed between the metal surface and the compounds in the extracts. On the other hand, the pairs of unsaturated electrons in the oxygen heteroatoms in these extracts are shared with the empty iron orbitals on the electrode surface leading to the creation of a covalent bond between these compounds and the steel surface. Therefore, by creating a dense layer of the physically/chemically adsorbed compounds on the surface of the MS immersed in 1 M hydrochloric acid electrolyte, this coating prevents direct contact of the electrode surface with the corrosive electrolyte and inhibits corrosion reactions [18].

Cellulose, hemicellulose, and lignin extracted from mature areca nut husk (MAH) were used as OCIs for MS and copper in both acid and alkali electrolytes by Raghavendra et al. [19]. Their results showed that the maximum inhibition efficiency of the extract for MS in 0.5 M HCl and 0.5 M NaOH was about 94% and 92%, respectively, and for copper in the same media was about 94% and 90%, respectively as well. In addition, it was found that the OCI acted as an anodic-type inhibitor for copper and a mixed-type inhibitor for MS in the mentioned conditions. Furthermore, it was proven that the adsorption of the chemical compounds on both copper and MS electrode in acid and alkali obeyed Langmuir's adsorption isotherm.

Microcrystalline and nanocrystalline cellulose can be artificially extracted from the plant via several different treatments. Microcrystalline cellulose is obtained via the interaction of cellulose with an alkaline medium followed by some bleaching actions, while nanocrystalline cellulose can be produced by acid hydrolysis of the microcrystalline cellulose. Toghan et al. [20] converted microcrystalline cellulose to nanocrystalline cellulose by an acid hydrolysis process using a 65 wt% sulphuric acid at 50 °C for 1 h and investigated the inhibition performance of the microcrystalline and nanocrystalline celluloses for SS316 alloy dipped in 2 M HCl. The results of electrochemical tests revealed that the inhibition efficiency of the investigated inhibitors follows the order nanocrystalline cellulose (96.3%)> microcrystalline cellulose (93.1%) at 200 ppm inhibitors concentration. The higher inhibition efficiency of the nanocrystalline cellulose may be due to the high aspect ratio of the nanosized inhibitor and the high steric hindrance of the microcrystalline cellulose molecule for adsorption on the electrode surface.

15.1.2 Carboxymethyl cellulose and its derivatives

Carboxymethylcellulose (CMC) sometimes known as sodium carboxymethylcellulose, is the most important water-soluble derivative of cellulose used in various industries such as food, cosmetics, pharmaceuticals, etc. CMC is a modified artificial cellulose that has a long, linear chain with an anionic polysaccharide formed by the reaction of monochloroacetic acid with alkaline cellulose. CMC was first produced and patented in 1918, and numerous efforts have since been made to optimize its properties. Depending on the percentage of purity, this material is classified into 3 types: industrial (67%), semi-pure or hygienic (98%), and pure (99.5%). Its pure form, which is used as a food additive, is also known by other names such as cellulose gum [21].

CMC has a higher solubility than cellulose due to its high carboxyl group and negatively charged electrostatic desorption. CMCs with chemical formulas Cellulose-O-CH2-COONa and Cellulose-O-CH2-COOH can be considered alkaline and acidic states, respectively. As mentioned previously, CMC is an anionic (negatively charged) polyelectrolyte compound with an acidic constant of about 4. This substance is usually prepared and synthesized as a salt and its acidic state has little solubility in water. Pure CMC has a creamy white colour and is tasteless and odourless [22, 23]. A schematic of the molecular structure of CMC is shown in Figure 15.3.

CMCs are available in several categories according to their diffusion, viscosity, particle size, and purity. The degree of substitution (DS) is one of the most important properties of CMC, which not only affects its solubility but also the properties of the solution. The DS is the average number of carboxymethyl groups per unit of glucose anhydride. The higher DS value makes the higher solubility of CMC in water, which results in the high viscosity of the solutions. The maximum theoretical DS is 3, but for CMC, the DS ranges between 0.5 and 1.5 [24]. Although commercial CMC is typically produced with a DS of 0.8–0.4, most commercial samples have a DS of about 0.7. In addition, the degree of polymerization of cellulose for commercial samples varies from 200 to 1000. Samples with a degree of substitution of 0.8–0.6 are well dissolved in water. In general, the higher degree of polymerization results in a greater chance of replacing the hydroxide group in the cellulose chain, leading to a higher DS for the final product and higher solubility. CMC has a variable acidity because of DS due to its sodium carboxylate group, and therefore its solution properties and viscosity change according to pH [25, 21].

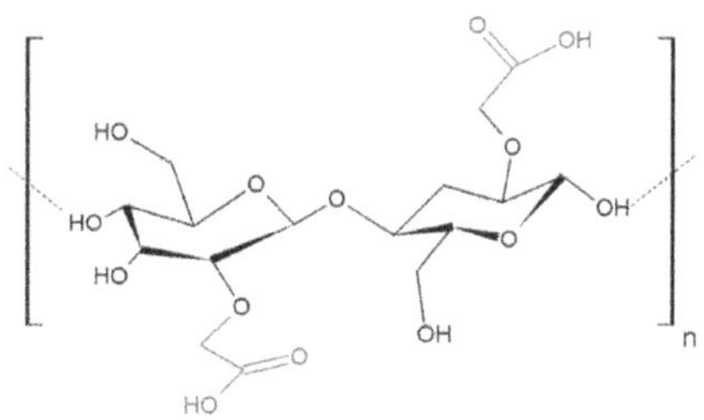

Figure 15.3 A schematic of molecular structure of CMC.

CMC is produced through the reaction of cellulose with monochloroacetic acid (MCA) in presence of NaOH. Its production includes two important stages [26, 27]:

i) Alkalinisation: Cellulose is treated with caustic soda in presence of an inert solution. Cellulose-sodium activation reaction or alkaline cellulose is often called mercerization, which usually occurs at room temperature. The temperature conditions used for the mercerization reaction have a significant effect on the quality characteristics of the final product, i.e., CMC. Cellulose swells in concentrated soda but cannot be dissolved in it.
ii) Etherification: Monochloroacetic acid ($ClCH_2COOH$) is added to the alkaline cellulose in the second stage. The reaction between alkaline cellulose and the tethering agent usually takes place at 70 °C in an aqueous system. Sodium, on the other hand, reacts with monochloroacetic acid to form sodium glycolate ($HOCH_2COONa$) and sodium chloride (NaCl). Sodium carboxymethyl cellulose is the product of the chemical reaction of alkaline cellulose with monochloroacetic acid in an alcoholic medium such as isopropyl alcohol (IPA), methanol (MeOH), or ethanol (EtOH), which must be purified in several steps. To obtain industrial carboxymethyl cellulose, the resulting material must be neutralized and dried quickly. The reactions that take place during the production of carboxymethyl cellulose are shown in Figure 15.4.

The substitution of the OH group for CMC occurs at Carbon 6. The temperature during alkaline treatment and the concentration of monochloroacetic acid have a significant effect on water solubility, viscosity, and carboxymethyl group content in the final product [28].

The purity of CMC reaches 80% after washing several times with ethanol and removing some by-products such as NaCl and sodium glycolate. Nowadays, CMC is made of thick pre-hydrolysed bleached papers, sulphite acid scraps, and linters. These scraps usually have a high degree of polymerization. The production of CMC using gas or sugarcane pulp (a by-product in the sugar industry) has also been studied. The emulsifying properties of pure sodium carboxymethyl cellulose using olive oil and coconut oil have also been studied [29, 28].

The presence of COOH and OH polar functional groups as well as several aromatic rings in the chemical structure of CMC allows it to form strong adsorption on the surface of different metals and alloys immersed in corrosive media. These properties in addition to other features such as low cost, nontoxicity, biodegradability, and water solubility introduced the CMC as a useful OCI [30].

Solomon et al. [31] investigated the effect of CMC concentration (100–500 ppm) in different electrolyte temperatures (30–60 °C) on the corrosion inhibition of MS exposed to 2 M H_2SO_4. The obtained results confirmed that the inhibition efficiency had a direct and inverse relationship, respectively, with the OCI concentration and the temperature, and the maximum inhibition efficiency was about 65%. In

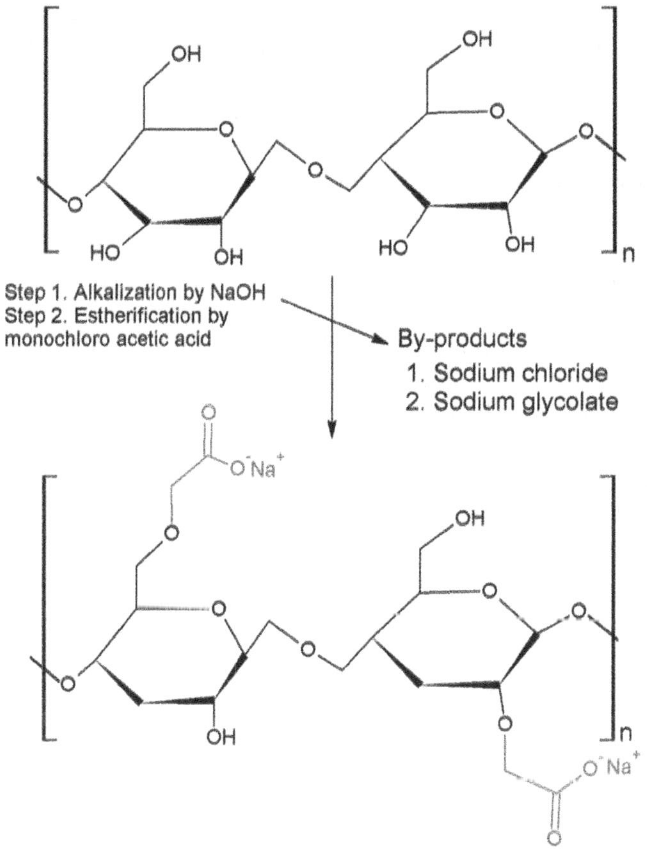

Figure 15.4 The reactions that take place during the production of CMC.

addition, it was found that the physical OCI adsorption on the MS surface followed Langmuir and Dubinin–Radushkevich adsorption isotherm models. Bayol et al. [32] evaluated the corrosion inhibition performance of sodium carboxymethyl cellulose (Na-CMC) for immersed MS in 1 M HCl solution. Their study revealed that the used OCI acted as a mixed-type inhibitor in these conditions and the adsorption on the MS surface followed Langmuir adsorption isotherm. However, the maximum inhibition efficiency obtained in presence of 0.04%wt. of the OCI was 78% indicating Na-CMC is not a potent inhibitor for MS in an HCl solution.

Li et al. [33] studied the inhibition properties of Na-CMC as an OCI for copper immersed in simulated cooling water and found that the inhibition efficiency of the inhibitor increased with increasing the OCI concentration up to 81.72% at

5 ppm. In addition, the polarization analysis suggested that the OCI acted as a mixed-type (cathodic/anodic) inhibitor and the adsorption process on the Cu substrate obeyed Langmuir's adsorption isotherm. Egbuhuzor et al. [34] and Abdallah et al. [35] used CMC as the OCI for aluminium immersed in 0.5 M HCl and both studies suggested that the used compound was physically adsorbed on the electrode surface. However, the adsorption isotherm model obtained in these two studies was different (Freundlich model in the first study and Temkin model in the second). Abdallah et al. found that the maximum inhibition efficiency at 1000 ppm of the OCI was about 81% (at 30 °C), and Egbuhuzor et al. observed that the corrosion rate (CR) can be considered as a function of time (t), temperature (T), and the OCI concentration (C) through a mathematical model: $CR = 3.8^{-5} \times T^{1.1617} - 0.00052t^{0.6176} - 0.0013C^{0.8012}$ with $R^2 = 0.8658$.

Umoren et al. [36] employed CMC for corrosion inhibition of API 5L X60 pipeline steel dipped in CO_2-saturated 3.5% NaCl solution at 25–60 °C. It was found that in presence of 100 ppm CMC, the highest inhibition efficiency was about 62% after 12 h immersion in the corrosive electrolyte. The inhibition performance was decreased by increasing the temperature, confirming the physisorption of the inhibitor on the immersed electrode. Langmuir isotherm model possessed the best fit with the experimental data indicating monolayer adsorption.

15.1.3 Cellulose Acetate and its Derivatives

Among all cellulose derivatives, cellulose acetate as an organic cellulose ester is one of the important biopolymers for industrial and commercial applications. It can dissolve in different solvents and different products can be prepared depending on the degree of substitution. Cellulose acetylation is commonly performed in a heterogeneous phase. Therefore, the reaction rate is controlled by the rate of penetration and diffusion of reactants within the cellulose fibre structure [37, 38]. Except for the fully acetylated product (cellulose triacetate), cellulose acetate is usually prepared by a solution process. In this process, the fibre pulp is first treated with acetic acid in presence of a catalyst (usually sulfuric acid) to swell the fibres. Acetylation is then performed by adding acetic anhydride and sulfuric acid. After the reaction, triacetate is formed and dissolved in the reaction medium [39]. A monomeric structure of cellulose acetate is shown in Figure 15.5.

Production of cellulose acetate from agricultural wastes such as jute, fibres attached to cotton seeds, rice straw, wheat husk, and corn fibres have been reported in the literature. One of the most important uses of cellulose acetate is its application in separation science and technology. The types of dialysis tubes that are used in research laboratories of chemistry, biochemistry, microbiology, biotechnology, and pharmacy to separate small molecular or ionic species from large molecular components, are made of cellulose acetate with different molecular pores [40, 39]. The heart of diabetic blood dialysis systems is also made from a set of very thin

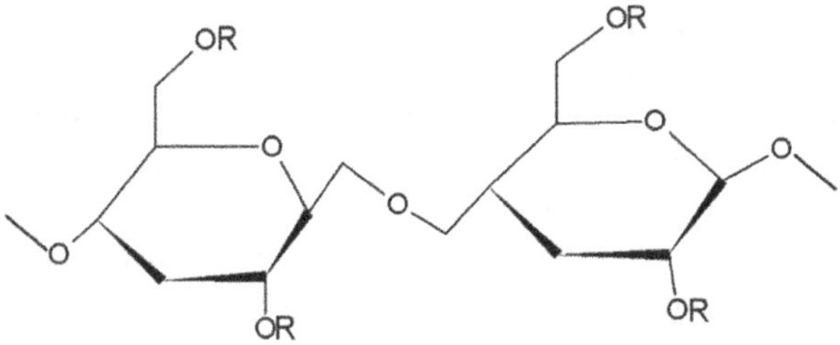

Figure 15.5 A monomeric structure of cellulose acetate.

tubes (as thin as hair) of cellulose acetate. It is noteworthy that no substance has yet been able to replace this valuable cellulose derivative for this purpose. These cellulosic derivatives are also widely used in the production of plastic films, varnishes, photographic films, thermoplastic mouldings, transparent sheets, camera accessories, magnetic tape, combs, telephones, and electrical components [41].

Cellulose acetate can be considered OCI due to its unique features such as easy production, cost-effectiveness, and eco-friendly nature. Andarany et al. [42] synthesized cellulose acetate by dissolving cellulose powder in 99% acetic acid and evaluated the corrosion inhibition of the OCI for aluminium in HCl solution with different concentrations (0.1, 0.5, 1, 2, and 3 molars). The weight loss test results showed that the employment of cellulose acetate as an OCI decreased the corrosion rate of the Al electrode in the HCl solutions with an efficiency of 50%. Tamborim et al. [39] applied an amoxicillin doped cellulose acetate film on the AA2024-T3 aluminium alloy and evaluated the corrosion inhibition performance of the film in a 0.05 M NaCl solution. The EIS analysis results showed that by doping the cellulose acetate chemical structure with 2000 ppm amoxicillin, the impedance at the lowest frequency increased from 9 kOhm.cm^2 to 29 kOhm.cm^2 and the film capacitance decreased from 15 mF/cm^2 to 1.2 nF/cm^2. The amoxicillin-metal ions complexes—Cu and Al—can form a protective layer on the substrate alloy and inhibit the corrosion reactions at anodic and cathodic sites.

15.1.4 Methylcellulose, ethyl cellulose, propyl cellulose and their derivatives

In this cellulose derivative, the hydroxyl (OH) groups in cellulose have been replaced by ethoxy groups (C_2H_{50}). Methylcellulose—substitution of one of the hydroxyls with ethoxy—contains 23.68% ethoxy. Ethyl cellulose, with a melting point of about 150 °C, contains 41.28% ethoxy, whereas the ethoxy content in propyl cellulose is 54.88% with a melting point between 240 °C to 245 °C. The completely extracted product is a white granular solid. Ethyl cellulose is a thermoplastic biopolymer and is distinguished for its easy moulding, light weight, high strength, and flexibility in a wide range of temperatures from −57 °C to 66 °C. It is one of the strongest and lightest cellulosic plastics that offers the lowest water absorption properties. Ethyl cellulose is practically colourless and tasteless, but is able to retain these properties in a wide range of applications [43, 26, 44].

Eid et al. [45] studied the possibility of using methyl cellulose as a green OCI for aluminium and aluminium silicon alloys in 0.1 M NaOH solution. Their results demonstrated that the OCI adsorption on the surface of aluminium and Al-Si substrates followed the Freundlich isotherm model and acted as a mixed-type inhibitor. In addition, it was found that the affinity of the OCI to adsorb on Si is lower than that on Al and, therefore, a higher Si content in the substrate negatively affects the adsorption strength of the OCI molecules, thus lowering the inhibition efficiency.

Hydroxyethyl cellulose is one of the ethyl cellulose derivatives which is widely used as OCI for different metals in various corrosive media. Arukalam et al. [46] investigated the corrosion inhibition performance of hydroxyethyl cellulose and hydroxypropyl methylcellulose on the oxidation of aluminium electrodes in a 1 M HCl electrolyte. The value of inhibition efficiency at different concentrations and various immersion times is presented in Figure 15.6.

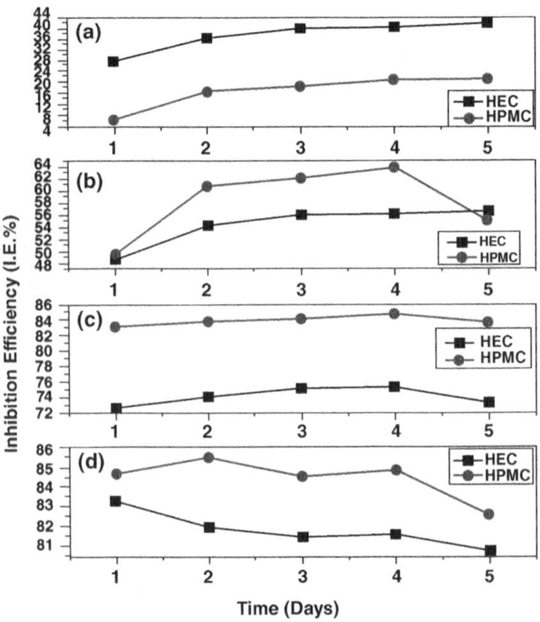

Figure 15.6 The inhibition efficiency of hydroxyethyl cellulose and hydroxypropyl methylcellulose for aluminium immersed in 1 M HCl against time at different concentrations: (a)=500 ppm; (b)=1000 ppm; (c)=1500 ppm; (d)=2000 ppm (Arukalam, I. O. et al., (2014) / Hindawi Publishing Corporation / Licensed under CC BY 3.0).

As is shown in this figure, except for Figure 15.6(a), for almost all other times and concentrations the inhibition efficiency of hydroxypropyl methylcellulose is more than that obtained for hydroxyethyl cellulose. This may be attributed to the higher molecular size and weight of hydroxypropyl methylcellulose compared to hydroxyethyl cellulose. In other words, more oxygen atoms and glucosidic rings in the chemical structure of the inhibitor can lead to higher adsorption stability and, therefore, more surface coverage of the electrode surface is covered by inhibitors. For Figure 15.6(a), the concentration of hydroxypropyl methylcellulose was not sufficient to create a compact protective layer on the surface of the immersed electrode.

Nwanonenyi et al. [47] investigated the inhibition properties of hydroxypropyl cellulose for aluminium in 0.5 M HCl and 2 M H_2SO_4 at 30–65 °C. The potentiodynamic polarization (PP) curves of the immersed electrode in hydrochloric acid and sulfuric acid solutions with or without the OCI can be observed in Figure 15.7.

According to the PP curves, hydroxypropyl cellulose act as a mixed-type inhibitor in both HCl and H_2SO_4 electrolytes with a more dominant anodic effect. In addition, by increasing the OCI concentration, both anodic and cathodic branches shift to lower current density values, confirming the inhibition of anodic and cathodic corrosion reactions of aluminium in both aggressive solutions. However, thermodynamic results suggest that the inhibition performance of the OCI decreases with increasing the temperature from 30 °C to 65 °C due to the physical adsorption of the chemical compound on the electrode surface. The maximum

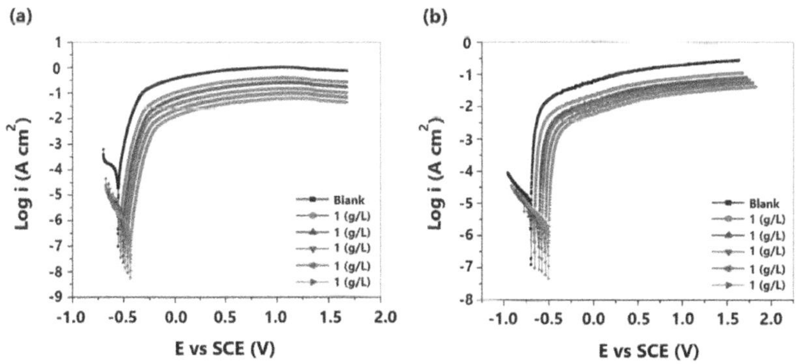

Figure 15.7 The PP curves of aluminium immersed in (a) 0.5M HCl and (b) 2M H_2SO_4 in the absence and presence of different hydroxypropyl cellulose concentrations (Nwanonenyi, S. C. et al., 2019 / With permission of Springer Nature).

inhibition efficiency values in the presence of 5 g/L OCI in the HCl and H_2SO_4 solutions were about 80% and 92%, respectively.

Sobhi et al. [48] introduced methyl hydroxyethyl cellulose as a powerful OCI for copper immersed in 1 M HCl at ambient temperature. The experimental data demonstrated that the adsorption of methyl hydroxyethyl cellulose on the electrode surface obeyed the Langmuir isotherm model and it is mainly governed by the physisorption process. The maximum inhibition efficiency was 90.29% at 30 °C.

Sangeetha et al. [49] evaluated the inhibition performance of aminated hydroxyethyl cellulose (AHEC) as an OCI for MS in 1 M HCl. A schematic of the OCI synthesis is shown in Figure 15.8.

The electrochemical analyses proved that the inhibition efficiency of the synthesized OCI increased by increasing the inhibitor concentration up to 93% for 900 ppm at room temperature. The presence of nitrogen heteroatom besides oxygen functional groups in the molecular structure of the inhibitor can help to form a stronger chemisorption between the OCI and MS surface. In addition, it was found that the synthesized inhibitor acts as a mixed-type OCI and the adsorption process obeyed the Frumkin isotherm model.

Farhadian et al. [50] chemically modified hydroxyethyl cellulose molecular structure with polyurethane (see Figure 15.9) and evaluated the synthesized OCI (named CHEC) for MS exposed to a 15% HCl solution at elevated temperatures.

Their results confirmed that the addition of only 1% of polyurethane to the chemical structure of CHEC significantly improved its inhibition performance in the corrosive electrolyte, even at high temperatures (maximum inhibition efficiency of 93% at 80 °C). It was found that the diphenyl and amine moieties acted as the main chemical adsorption sites on the synthesized inhibitor molecular structure. These results indicate that the synthesized OCI can be used as a potential corrosion inhibitor for employment at high temperatures.

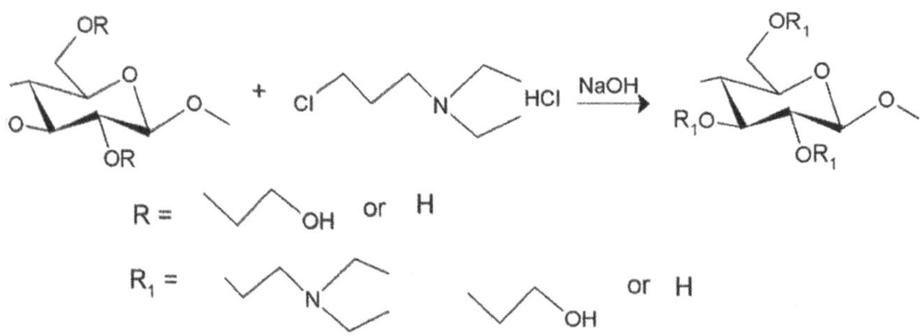

Figure 15.8 Synthesis of AHEC.

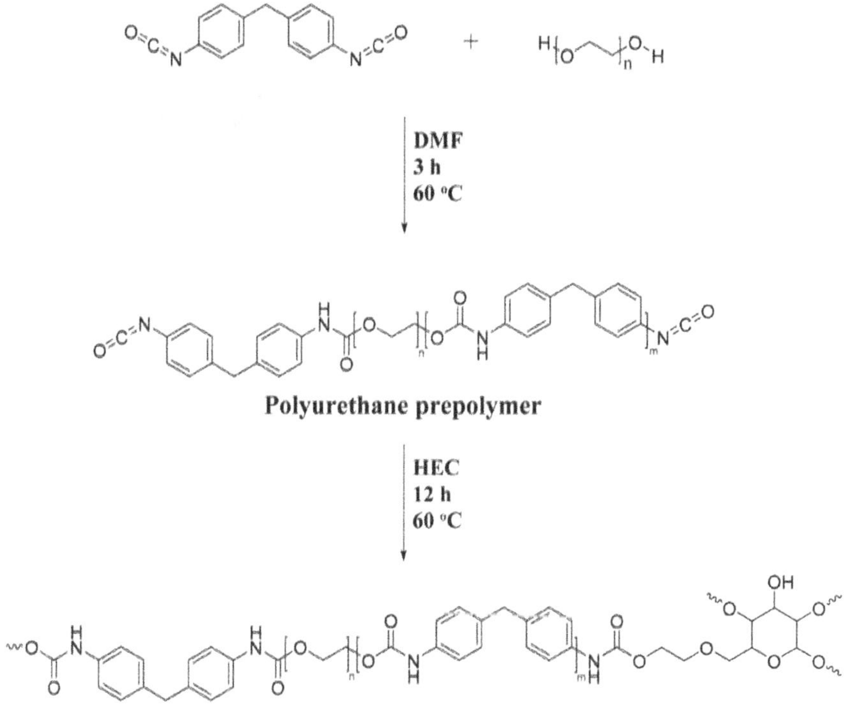

Figure 15.9 A schematic of CHEC synthesis procedure (Farhadian, A. *et al.* (2021) "Modified hydroxyethyl cellulose as a highly efficient eco-friendly inhibitor for suppression of mild steel corrosion in a 15% HCl solution at elevated temperatures, *Journal of Molecular Liquids*. Elsevier, 338, p. 116607).

15.1.5 Other cellulose derivatives

Borated aminated cellulose citrate (BACC) is another cellulose-based OCI introduced by Gan et al. [51] for A3 steel in simulated cooling water (SCW). In this OCI, oxygen and nitrogen heteroatoms can act as reaction centres for the adsorption of BACC on the immersed electrode surface. The polarization and Nyquist diagrams of the electrode immersed in the corrosive electrolyte with and without the OCI are shown in Figure 15.10.

The variations of corrosion potential in the absence and presence of different concentrations of BACC revealed that the OCI acted as a mixed-type inhibitor for the electrode in SCW and predominantly controlled the anodic reactions. What's more, the corrosion current density obtained from the polarization curves as well as the semi-circle Nyquist diameters for each sample indicated that the polarization and charge transfer resistances increased by increasing the BACC concentration in the electrolyte. The highest inhibition efficiency was 88% in

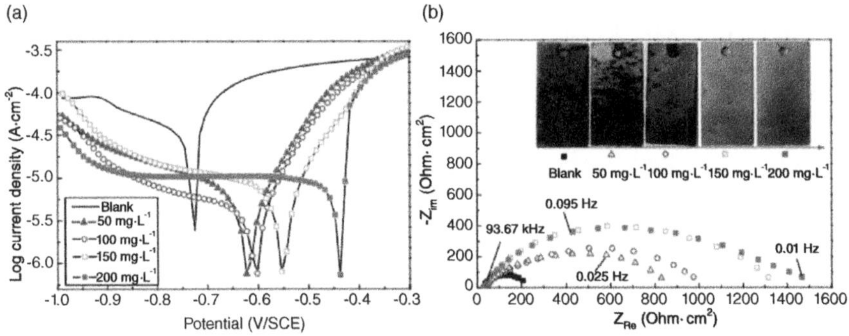

Figure 15.10 (a) Polarization and (b) Nyquist plots of the electrode immersed in SCW with different concentrations of BACC (Gan, T. et al., (2018) / With permission of American Chemical Society).

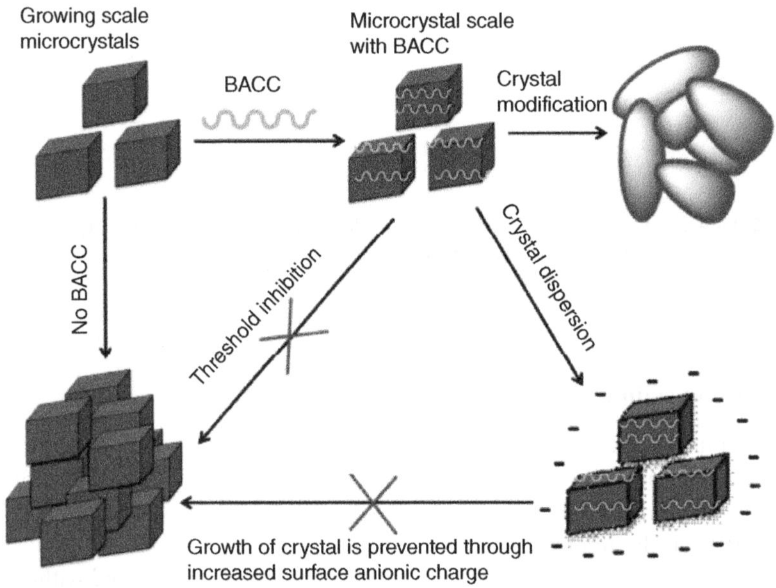

Figure 15.11 The scale inhibition mechanism of OCI in SCW. (Gan, T. et al., (2018) / With permission of American Chemical Society).

presence of 200 ppm BACC in SCW—a relatively considerable value. The scale inhibition mechanism of OCI in SCW is schematically represented in Figure 15.11.

As shown in this schematic, during the process of $CaCO_3$ formation on the electrode surface, the OCI can form different Ca(II)-BACC stable complexes leading to the formation of disordered crystalline precipitates with intensive surface stresses and inhibition of crystal growth by altering the shape of the microcrystals. Hence, it can be concluded that the OCI can obstruct the formation of calcium carbonate scale deposited by threshold inhibition and modification of the growth of microcrystals.

Hasanin et al. [52] employed some compounds based on cellulose niacin (i.e. ethyl cellulose-niacin composite (NEC), cellulose-niacin composite (NCMC), and microcrystalline cellulose-niacin composite (NMCC)) as green corrosion inhibitors for copper in 3.5% NaCl. The Nyquist and Bode-impedance plots of the studied samples are shown in Figure 15.12.

As can be seen in the figure, the semi-circle diameter in the Nyquist diagrams as well as the impedance at the lowest frequency in the Bode plots increased as follows: NEC>NCMC>NMCC>Blank. The EIS analysis results proved that the inhibition efficiency of NEC, NMCC, and NCMC were 94.7%, 33.2%, and 83.4%, respectively. In addition, the PP measurements suggested that all of the studied OCIs act as mixed-type inhibitors.

The inhibition performance of methyl-cellulose polysaccharide for magnesium dipped in HCl solution was evaluated by Hassan et al. [53]. Their results showed that the inhibition efficiency increased by increasing the OCI concentration as well as decreasing the electrolyte temperature. It was claimed that -OH functional groups in the alcoholic polysaccharide molecules can form bridges between the OCI and the immersed electrode surface, resulting in the formation of a protective layer on the magnesium surface. The experimental data were fitted well by Langmuir and Freundlich isotherm models.

15.1.6 Synergistic effect in cellulose-based inhibition systems

Although some cellulose-based inhibitors offer moderate to good inhibition performance, their inhibition activity generally decreases at high temperatures due to physical adsorption on metal surfaces. Synergism is one of the most popular ways to overcome this problem [54].In general, the synergistic effect is obtained when the inhibition efficiency of two or more inhibitors is greater than the sum of the efficiency generated by the same inhibitors individually [55]. Therefore, the synergism may occur through two mechanisms: (I) interaction between the chemical structures of the inhibitors, and (II) interaction between the inhibitor(s) and the species in the electrolyte. The synergistic effect can cause to improve the inhibition efficiency of the inhibition system and decrease the usage of valuable inhibitors. This effect is often expressed by Equation (15.1):

$$S = 1 - ((\theta_1 + \theta_2) - (\theta_1\theta_2)) / 1 - \theta'_{1+2} \quad (15.1)$$

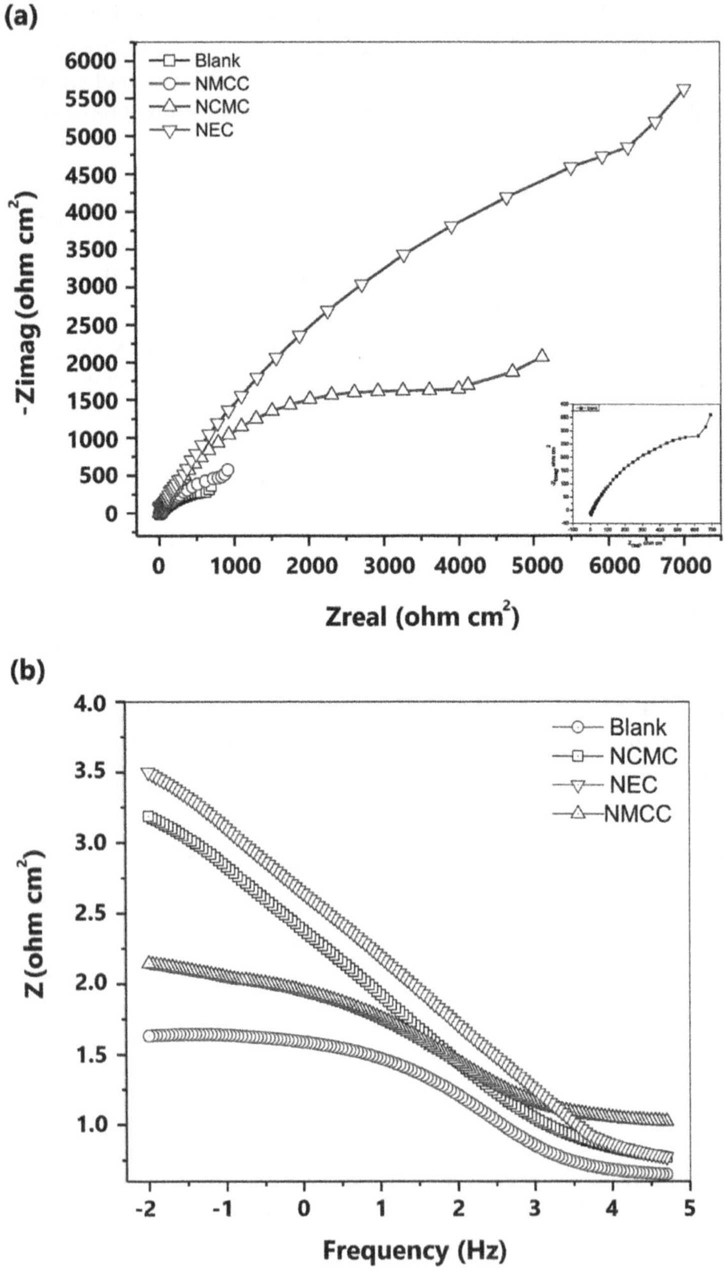

Figure 15.12 (a) Nyquist and (b) Bode-impedance curves of the blank, NEC, NCMC, and NMCC samples. (Hasanin, M. S. and Al, S. A. (2020) / With permission of Elsevier).

where θ_1 and θ_2 are the surface coverage of each inhibitor and θ_{1+2} is the surface coverage obtained when the inhibitors were used simultaneously. The S>1 and S<1 correspond to synergistic and antagonistic effects, respectively.

Different investigations have been conducted on the synergistic effect between cellulose derivative inhibitors and other species especially inorganic inhibitors and halide ions. One of the earliest studies published in the field of the synergistic effect between CMC and other corrosion inhibitors is the work of Rajender et al. [56]. Their investigation proved that the inhibition efficiency of 10 ppm Zn^{2+} and 300 ppm 1-hydroxyethanole-1, 1-diphosphonic acid (HEDP) for MS immersed in a neutral aqueous environment containing 60 ppm $Cl^-$was about 10%, while the addition of 50 ppm CMC to the electrolyte increased the inhibition efficiency by up to 80%. Their further characterization analyses from the electrode surface immersed in the inhibitor-containing solution revealed that the produced film on the electrode consists of Fe^{2+}-HEDP complex, Fe^{2+}-CMC complex, and $Zn(OH)_2$. In another work, Manimaran et al. [30] observed an inhibition efficiency of 250 ppm CMC and 50 ppm Zn^{2+} for carbon steel dipped in the groundwater solution reached 98%. SEM micrographs of the carbon steel surface before and after immersion in the corrosive solution without and with inhibitors are shown in Figure 15.13.

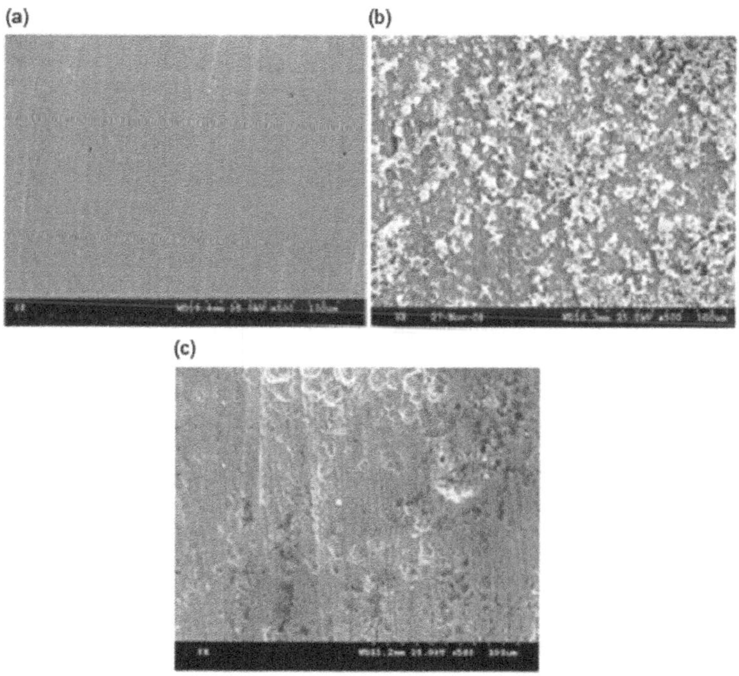

Figure 15.13 SEM micrographs of the carbon steel surface (a) before and after immersion in the corrosive solution (b) without and (c) with the CMC-Zn^{2+}. (Manimaran et al. (2013) / from European Chemical Bulletin).

The electrode surface after immersion in groundwater solution can be seen in the micrographs. The SEM micrograph of the carbon steel surface immersed in groundwater became completely rough due to the corrosive species attacking the exposed electrode surface. The relatively smoother surface in presence of 250 ppm CMC and 50 ppm Zn^{2+} suggests the formation of an insoluble compact layer on the metal surface, effectively inhibiting the corrosion reactions.

Umoren et al. [57] investigated the synergistic/antagonistic effects between CMC and some halide ions (Cl^-, Br^-, and I^-) and for MS in a 2 M H_2SO_4 solution. The variety of inhibition efficiency of these samples is shown in Figure 15.14.

According to Figure 15.14, it was found that after 6h of immersion, the presence of chloride and iodide halide ions had respectively antagonistic and synergistic effects on the corrosion inhibition of CMC. To find the effect of halide ions on the inhibition mechanisms of CMC some knowledge about the interactions between the corrosion inhibitors and the MS surface is required. The Fe dissolution mechanisms from the MS surface in the acidic electrolyte are as follows:

- Anodic interactions:

$$Fe + H_2O \leftrightarrow (FeOH)_{ads} + H^+ + e^- \qquad (15.2)$$

$$(FeOH)_{ads} \leftrightarrow (FeOH)^+ + e^- \qquad (15.3)$$

$$(FeOH)^+ + H^+ \leftrightarrow Fe^{2+} + H_2O \qquad (15.4)$$

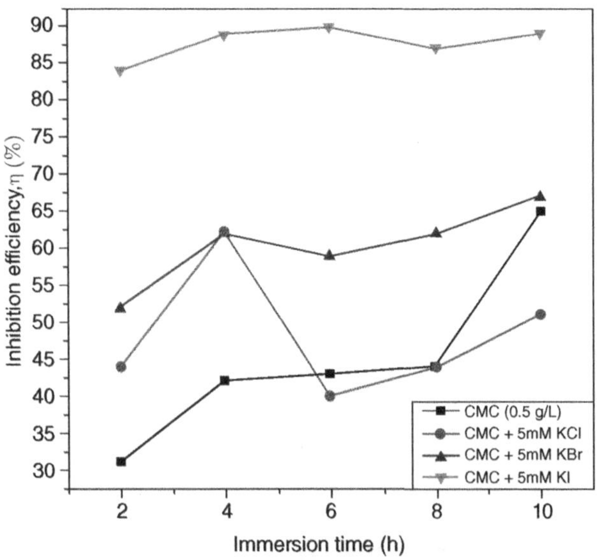

Figure 15.14 The inhibition efficiency of CMC without and with some halide ions (Cl^-,Br^-, and I^-) at different immersion times. (Umoren, S. A. et al. (2010) / With permission of Springer Nature).

- Cathodic interactions:

$$Fe + H^+ \leftrightarrow (FeH^+)_{ads} \quad (15.5)$$

$$(FeH^+)_{ads} + e^- \leftrightarrow (FeH)_{ads} \quad (15.6)$$

$$(FeH)_{ads} + H^+ + e^- \leftrightarrow Fe + H_2 \quad (15.7)$$

In the presence of a halide ion (X^-), the anodic interactions altered as follows:

$$Fe + H_2O + X^- \leftrightarrow (FeXOH)_{ads} + H^+ + e^- \quad (15.8)$$

$$(FeOH)^-_{ads} \leftrightarrow (FeXOH)^+ + e^- \quad (15.9)$$

$$(FeXOH)^+ + H^+ \leftrightarrow Fe^{2+} + X^- + H_2O \quad (15.10)$$

The combination of CMC molecules with the above-mentioned intermediates leads to the formation of CMC-Fe complexes that can increase or decrease the inhibition efficiency depending on the product solubility. Hence, the highly soluble CMC-based complex formed in presence of Cl^- and the low solubility of the formed complex in the presence of Br^- and I^- are the main reasons for the antagonistic and synergistic effects of these halide ions on the corrosion inhibition efficiency of CMC. In addition, the higher inhibition efficiency of I^- compared to Br^- can be due to the ease of chemisorption with MS free orbitals because of its high polarizability/hydrophobicity and low electronegativity of this ion.

Ethyl hydroxyethyl cellulose (EHEC) is another cellulose derivative OCI. Arukalam et al. [58] evaluated its synergistic effect with iodide ions on the corrosion inhibition performance for MS in 1 M H_2SO_4 solution. The EIS results showed that the inhibition efficiency of EHEC and EHEC+KI was about 72% and 91%, respectively. Indeed, I^- anions acted as intermediate bridges between the protonated EHEC molecules and the MS surface (see Equation 15.11).

$$I^-_{ads} + Inh^+_{sol} \rightarrow \left[I^- - Inh^+\right]_{ads} \quad (15.11)$$

Sovizi et al. [23] found that the addition of iodide halide ion to 2 M H_2SO_4 electrolyte containing CMC leads to a desirable synergistic effect against corrosion reactions of an aluminium-based alloy. It is well known that the isoelectric point of the Al_2O_3 layer covering Al-based alloys is pH≈. Therefore, below the isoelectric point, the aluminium oxide layer was positively charged and an electrostatic interaction with I^- was formed. On the other hand, the protonation of CMC in this low pH introduces a positive charge to this chemical structure. As a result, regarding the electrostatic interaction between CMC^+ and I^- on the surface of the electrode, the authors claimed that the adsorption of CMC on the surface of the Al-based alloy was stabilized in presence of KI, improving the CMC inhibition efficiency.

Synergistic corrosion inhibition of MS immersed in 0.5 M sulphuric acid containing hydroxypropyl cellulose and the iodine halide ion was another study in this field conducted by Nwanonenyi et al. [59]. They showed that the inhibition efficiency increased from 60.79% in presence of 800 ppm hydroxypropyl cellulose to 76.22% in presence of the same amount of hydroxypropyl cellulose and KI. In addition, their potentiodynamic polarization results demonstrated that when merely hydroxypropyl cellulose was used, the cathodic reactions predominantly controlled the corrosion rate of MS, whereas in presence of hydroxypropyl cellulose and KI, anodic reactions controlled the induced corrosion process.

Kalaivani et al. [60] showed that for aluminium dipped in an aqueous solution at pH 11, the inhibition efficiency of 250 ppm CMC without and with 25 ppm Zn^{2+} was about 62% and 95%, respectively. The increase in inhibition efficiency of the system with zinc cation can be related to the presence of the CMC-Zn^{2+} complex in the electrolyte. In fact, on the surface of the immersed electrode, the CMC-Zn^{2+} complex is converted to CMC- Al^{3+} complex via Equation (15.12):

$$\text{CMC-Zn}^{2+} + \text{Al}^{3+} \rightarrow \text{CMC-Al}^{3+} + \text{Zn}^{2+} \tag{15.12}$$

The CMC-Al^{3+}complex can block the anodic sites of the aluminium electrode and the produced zinc cation can transfer to the cathodic zones and precipitate as $Zn(OH)_2$, inhibiting the cathodic reactions of the aluminium corrosion process.

CMC-Ag complex is another corrosion inhibitor couple system designed by Solomon et al. [61] for corrosion inhibition of St37 steel in a 15% sulphuric acid solution. They demonstrated that the inhibition efficiency and the inhibitor stability on the immersed electrode surface significantly increased by incorporating of AgNPs into the CMC matrix. Their results revealed that the CMC-Ag complex was endothermically chemisorbed on the immersed steel obeying the Langmuir adsorption isotherm and a higher inhibition efficiency was obtained at higher temperatures. In another study, Abd El-Lateef et al. [62] used CMC-metal (Fe, Cu, and Ni) complex as a non-precious corrosion inhibitor for MS corrosion in 2 N HCl solutions at 50 °C. The electrochemical analyses revealed that the maximum inhibition efficiency for the CMC, CMC/Fe, CMC/Cu, and CMC/Ni complexes at 400 ppm were respectively about 77%, 95%, 96%, and 98%. The obtained value of Gibbs free energy ($\Delta G°_{ads}$) for the adsorption processes confirmed that the chemisorption is predominate, and this value of the parameter for CMC/Ni was less than that obtained for CMC/Fe and CMC/Cu complexes. The lower value of $\Delta G°_{ads}$ for the CMC/Ni sample may be considered the main reason for the higher protection power of this complex, since decreasing the $\Delta G°_{ads}$ value means a higher thermodynamic tendency to spontaneous reaction.

The synergistic inhibitory effect of hydroxyethyl cellulose mixed by some surfactants (triton X 100 (TX), cetyl pyridinium chloride (CPC), and sodium dodecyl sulphate (SDS) on the dissolution of A1020 carbon steel in 1 M HCl solution was

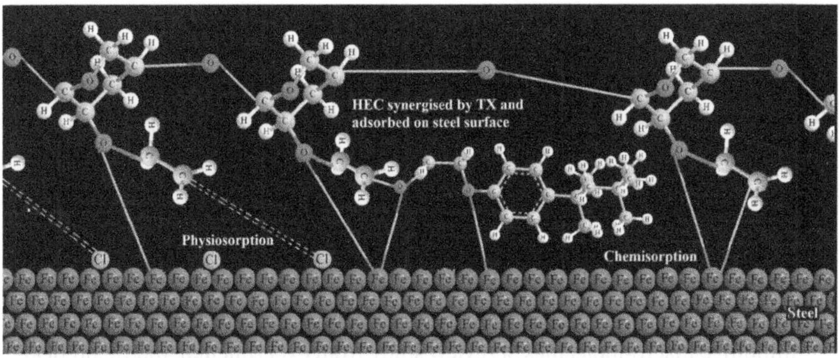

Figure 15.15 Schematic depiction of adsorption of hydroxyethyl cellulose mixed by TX on the c-steel surface (Mobin, M. and Rizvi, M. (2017) "Adsorption and corrosion inhibition behavior of hydroxyethyl cellulose and synergistic surfactants additives for carbon steel in 1 M HCl", *Carbohydrate Polymers*. Elsevier Ltd., 156, pp. 202–214).

investigated by Mobin et al. [63]. It was found from this study that the inhibition performance of hydroxyethyl cellulose mixed by TX (%IE = 93.5) is more than that obtained for the mixing with CPC (%IE = 91.5) and SDS (%IE = 89.3). A schematic of the adsorption mechanisms of hydroxyethyl cellulose mixed by TX on the A1020 carbon steel surface is shown in Figure 15.15.

According to Figure 15.15, the interaction between the cellulose-based inhibitor and the molecule structure of the surfactants causes the polymer backbone to expand and leads to a higher degree of surface coverage.

15.2 Conclusion

Biopolymers, especially cellulose-based biopolymers and their derivatives have always been considered natural and environmentally friendly compounds for use as corrosion inhibitors due to the presence of different heteroatoms, aromatic rings, and multiple bonds in their chemical structure. In this regard, these compounds have been added to corrosive solutions or anti-corrosion coatings after extraction from various parts of plants, either without further processing or after structural modification to prevent corrosion of immersed metals in electrolytes. As was reviewed in this chapter, CMC, cellulose acetate, methyl, ethyl, and propyl cellulose, and their derivatives—as well as some grafted cellulose compounds such as borated aminated cellulose citrate, cellulose niacin, and methyl-cellulose polysaccharide—have been used widely as green organic corrosion inhibitors for different metals and alloys immersed in various corrosive media. In addition, the studies on the synergistic effect between cellulose-based compounds and other corrosion inhibitors were reviewed in this chapter. Thus, it can be concluded that cellulose-based inhibitors, due to their eco-friendly nature, availability, low price, and significant inhibition efficiency—especially in acidic environments—can be considered a suitable candidate for applications as OCIs in large-scale industrial operations.

References

1 Shahini, M.H., Ramezanzadeh, B., and Mohammadloo, H.E. (2021). Recent advances in biopolymers/carbohydrate polymers as effective corrosion inhibitive macro-molecules: a review study from experimental and theoretical views. *Journal of Molecular Liquids*. Elsevier, 325: 115110. doi: 10.1016/J.MOLLIQ.2020.115110.

2 Umoren, S.A. and Eduok, U.M. (2016). Application of carbohydrate polymers as corrosion inhibitors for metal substrates in different media: a review. *Carbohydrate Polymers*. Elsevier, 140: 314–341. doi: 10.1016/J.CARBPOL.2015.12.038.

3 Vaidya, N.R., Aklujkar, P., and Rao, A.R. (2021). Modification of natural gums for application as corrosion inhibitor: a review. *Journal of Coatings Technology and Research*. Springer, 1–17. doi: 10.1007/s11998-021-00510-z.

4 Tamalmani, K. and Husin, H. (2020). Review corrosion inhibitors for oil and gas corrosion issues. *Applied Sciences*. Multidisciplinary Digital Publishing Institute, 10 (10): 3389. doi: 10.3390/app10103389.

5 Aftab, S., Ullah, S., and Anjum, F. (2021). Evaluation of chemo-microbial methods for cellulose hydrolysis in sugarcane bagasse and sawdust. *Sarhad Journal of Agriculture* 37 (3). doi: 10.17582/journal.sja/2021/37.3.1009.1015.

6 Sharma, A. et al. (2019). Commercial application of cellulose nano-composites – a review. *Biotechnology Reports*. Elsevier, 21: e00316. doi: 10.1016/J.BTRE.2019.E00316.

7 Wang, Y. et al. (2018). Functional nanomaterials through the esterification of cellulose: a review of chemistry and application. *Cellulose*. Springer, 25 (7): 3703–3731. doi: 10.1007/s10570-018-1830-3.

8 Onwukamike, K.N. et al. (2019). Critical review on sustainable homogeneous cellulose modification: why renewability is not enough. *ACS Sustainable Chemistry & Engineering*. American Chemical Society, 7 (2): 1826–1840. doi: 10.1021/acssuschemeng.8b04990.

9 Abushammala, H. and Mao, J. (2019). A review of the surface modification of cellulose and nanocellulose using aliphatic and aromatic mono- and di-isocyanates. *Molecules*. Multidisciplinary Digital Publishing Institute, 24 (15): 2782. doi: 10.3390/molecules24152782.

10 Yang, J. and Li, J. (2018). Self-assembled cellulose materials for biomedicine: a review. *Carbohydrate Polymers*. Elsevier, 181: 264–274. doi: 10.1016/J.CARBPOL.2017.10.067.

11 Gopi, S. et al. (2019). General scenarios of cellulose and its use in the biomedical field. *Materials Today Chemistry*. Elsevier, 13: 59–78. doi: 10.1016/J.MTCHEM.2019.04.012.

12 Tayeb, A. et al. (2018). Cellulose nanomaterials—binding properties and applications: a review. *Molecules*. Multidisciplinary Digital Publishing Institute, 23 (10): 2684. doi: 10.3390/molecules23102684.

13 Orasugh, J.T. et al. (2018). Synthesis of methylcellulose/cellulose nano-crystals nanocomposites: material properties and study of sustained release of ketorolac tromethamine. *Carbohydrate Polymers*. Elsevier, 188: 168–180. doi: 10.1016/J.CARBPOL.2018.01.108.

14 Sugai, K. et al. (2020). All carbohydrate-based nanocomposites are composed of sorbitol polyglycidyl ether, aminated trehalose, and cellulose nanofiber. *Carbohydrate Polymers*. Elsevier, 232: 115779. doi: 10.1016/J.CARBPOL.2019.115779.

15 Seidi, F. et al. (2022). Crystalline polysaccharides: a review. *Carbohydrate Polymers*. Elsevier, 275: 118624. doi: 10.1016/J.CARBPOL.2021.118624.

16 Yang, Y. et al. (2019). Preparation and applications of the cellulose nanocrystal. *International Journal of Polymer Science* 2019: 1–10. doi: 10.1155/2019/1767028.

17 Debnath, B., Haldar, D., and Purkait, M.K. (2021). A critical review of the techniques used for the synthesis and applications of crystalline cellulose derived from agricultural wastes and forest residues. *Carbohydrate Polymers*. Elsevier, 273: 118537. doi: 10.1016/J.CARBPOL.2021.118537.

18 Shahmoradi, A.R. et al. (2020). Studying the adsorption/inhibition impact of the cellulose and lignin compounds extracted from agricultural waste on the mild steel corrosion in HCl solution. *Journal of Molecular Liquids*. Elsevier, 304: 112751. doi: 10.1016/J.MOLLIQ.2020.112751.

19 Raghavendra, N. and Bhat, J.I. (2018). Chemical components of mature areca nut husk extract as potential corrosion inhibitor for mild steel and copper in both acid and alkali media. *Chemical Engineering Communications* 205 (2): 145–160. doi: 10.1080/00986445.2017.1370709.

20 Toghan, A., Gouda, M., and Shalabi, K. (2021). Preparation, characterization, and evaluation of macrocrystalline and nanocrystalline cellulose as potential corrosion inhibitors for SS316 alloy during acid pickling process: experimental and computational methods. *Polymers* 13 (14): 2275–2300.

21 Kanikireddy, V. et al. (2020). Carboxymethyl cellulose-based materials for infection control and wound healing: a review. *International Journal of Biological Macromolecules*. Elsevier, 164: 963–975. doi: 10.1016/J.IJBIOMAC.2020.07.160.

22 Javanbakht, S. and Shaabani, A. (2019). Carboxymethyl cellulose-based oral delivery systems. *International Journal of Biological Macromolecules*. Elsevier, 133: 21–29. doi: 10.1016/J.IJBIOMAC.2019.04.079.

23 Sovizi, M.R. and Abbasi, R. (2020). Effect of carboxymethyl cellulose on the corrosion behavior of aluminum in H2So4 solution and synergistic effect of potassium iodide. *Journal of Adhesion Science and Technology*. Taylor & Francis, 34 (15): 1664–1678. doi: 10.1080/01694243.2020.1717803.

24 Yang, X.H. and Zhu, W.L. (2007). Viscosity properties of sodium carboxymethylcellulose solutions. *Cellulose*. Springer, 14 (5): 409–417. doi: 10.1007/s10570-007-9137-9.

25 Benchabane, A. and Bekkour, K. (2008). Rheological properties of carboxymethyl cellulose (CMC) solutions. *Colloid and Polymer Science*. Springer, 286 (10): 1173–1180. doi: 10.1007/s00396-008-1882-2.

26 Fox, S.C. et al. (2011). Regioselective esterification and etherification of cellulose: a review. *Biomacromolecules*. American Chemical Society, 12 (6): 1956–1972. doi: 10.1021/bm200260d.

27 Taubner, T., Synytsya, A., and Čopíková, J. (2015). Preparation of amidated derivatives of carboxymethylcellulose. *International Journal of Biological Macromolecules*. Elsevier, 72: 11–18. doi: 10.1016/J.IJBIOMAC.2014.07.049.

28 Zennifer, A. et al. (2021). Key advances of carboxymethyl cellulose in tissue engineering & 3D bioprinting applications. *Carbohydrate Polymers*. Elsevier, 256: 117561. doi: 10.1016/J.CARBPOL.2020.117561.

29 Arancibia, C. et al. (2016). Application of CMC as thickener on nanoemulsions based on olive oil: physical properties and stability. *International Journal of Polymer Science* 2016: 1–10. doi: 10.1155/2016/6280581.

30 Manimaran, N. et al. (2013). Corrosion inhibition by carboxymethyl cellulose. *European Chemical Bulletin* 2 (7): 494–498.

31 Solomon, M.M. et al. (2010). Inhibitive and adsorption behavior of carboxymethyl cellulose on mild steel corrosion in sulphuric acid solution. *Corrosion Science*. Elsevier Ltd., 52 (4): 1317–1325. doi: 10.1016/j.corsci.2009.11.041.

32 Bayol, E., Gürten, A.A., and Dursun, M. (2008). Adsorption behavior and inhibition corrosion effect of sodium carboxymethyl cellulose on mild steel in acidic medium. *Acta Physico-Chimica Sinica* 24 (12): 2236–2243. doi: 10.1016/S1872-1508(08)60085-6.

33 Behpour, M. et al. (2011). Green approach to corrosion inhibition of mild steel in two acidic solutions by the extract of Punica granatum peel and main constituents. *Materials Chemistry and Physics* 131: 621–633. doi: 10.1016/j.matchemphys.2011.10.027.

34 Egbuhuzor, O.M. et al. (2020). Adsorption behavior and corrosion rate model of sodium carboxymethyl cellulose (Na-CMC) polymer on aluminum in HCl solution. *Nigerian Journal of Technology Search Articles For* 39 (2): 369–378.

35 Abdallah, M. and Fawzy, A. (2013). Inhibition of aluminum corrosion in hydrochloric acid by cellulose and chitosan. *Journal of American Science* 9 (4): 580–586.

36 Umoren, S.A. et al. (2018). International Journal of Biological Macromolecules evaluation of chitosan and carboxymethyl cellulose as eco-friendly corrosion inhibitors for steel. *International Journal of Biological Macromolecules*. Elsevier B.V., 117: 1017–1028. doi: 10.1016/j.ijbiomac.2018.06.014.

37 Bifari, E.N. et al. (2016). Cellulose acetate based nanocomposites for biomedical applications: a review. *Current Pharmaceutical Design*. Bentham Science Publishers, 22 (20). Available at: https://www.ingentaconnect.com/content/ben/cpd/2016/00000022/00000020/art00007 (accessed 19 December 2021).

38 Pandey, J.K. et al. (2012). An overview of cellulose-based conducting composites. *Composites Part B: Engineering*. Elsevier, 43 (7): 2822–2826. doi: 10.1016/J.COMPOSITESB.2012.04.045.

39 Tamborim, S.M. et al. (2011). Preparation and electrochemical characterization of amoxicillin-doped cellulose acetate films for AA2024-T3 aluminum alloy coatings. *Corrosion Science*. Elsevier Ltd., 53 (4): 1571–1580. doi: 10.1016/j.corsci.2011.01.034.

40 Konwarh, R., Karak, N., and Misra, M. (2013). Electrospun cellulose acetate nanofibers: the present status and gamut of biotechnological applications. *Biotechnology Advances*. Elsevier, 31 (4): 421–437. doi: 10.1016/J.BIOTECHADV.2013.01.002.

41 Kumar Gupta, P. et al. (2019). An update on overview of cellulose, its structure, and applications. In: *Cellulose*. IntechOpen. doi: 10.5772/intechopen.84727.

42 Andarany, K.S. et al. (2017). Cellulose acetate layer effect toward aluminum corrosion rate in hydrochloric acid media. *IOP Conference Series: Materials Science and Engineering*. IOP Publishing, 237 (1): 012042. doi: 10.1088/1757-899X/237/1/012042.

43 Arça, H.C. et al. (2018). Pharmaceutical applications of cellulose ethers and cellulose ether esters. *Biomacromolecules*. American Chemical Society, 19 (7): 2351–2376. doi: 10.1021/acs.biomac.8b00517.

44 Tosh, B. and Ranjan Routray, C. (1720). Chemical science review and letters grafting of cellulose-based materials: a review. *Chemical Science Review and Letters* 2014 (10): 74–92. Available at: https://chesci.com/wp-content/uploads/2016/10/V3i10_1_Review_Tosh-et-al_CS17204404_74-92.pdf (accessed 19 December 2021).

45 Eid, S., Abdallah, M., and Kamar, E.M. (2015). Corrosion inhibition of aluminum and aluminum silicon alloys in sodium hydroxide solutions by methylcellulose. *Journal of Materials and Environmental Science* 6 (3): 892–901.

46 Arukalam, I.O. et al. (2015). Cellulosic polymers for corrosion protection of aluminum. *International Journal of Engineering and Technical Research* 3 (1): 163–169.

47 Nwanonenyi, S.C., Obasi, H.C., and Eze, I.O. (2019). Hydroxypropyl cellulose as an efficient corrosion inhibitor for aluminum in acidic environments: experimental and theoretical approach. *Chemistry Africa*. Springer, 2 (3): 471–482. doi: 10.1007/s42250-019-00062-1.

48 Sobhi, M. and Eid, S. (2018). Chemical, electrochemical, and morphology studies on methyl hydroxyethyl cellulose as green inhibitor for corrosion of copper in hydrochloric acid solutions 1. *Physicochemical Problems of Materials Protection* 54 (5): 893–898. doi: 10.1134/S2070205118050210.

49 Sangeetha, Y., Meenakshi, S., and Sundaram, C.S. (2016). Corrosion inhibition of aminated hydroxyl ethyl cellulose on mild steel in acidic condition. *Carbohydrate Polymers*. Elsevier Ltd., 150: 13–20. doi: 10.1016/j.carbpol.2016.05.002.

50 Farhadian, A. et al. (2021). Modified hydroxyethyl cellulose as a highly efficient eco-friendly inhibitor for suppression of mild steel corrosion in a 15% HCl solution at elevated temperatures. *Journal of Molecular Liquids*. Elsevier, 338: 116607. doi: 10.1016/J.MOLLIQ.2021.116607.

51 Gan, T. et al. (2018). Synthesis, characterization, and application of a multifunctional cellulose derivative as an environmentally friendly corrosion and scale inhibitor in simulated cooling water systems. *Industrial & Engineering Chemistry Research* 57 (32): 10786–10797. doi: 10.1021/acs.iecr.8b02128.

52 Hasanin, M.S. and Al, S.A. (2020). Environmentally benign corrosion inhibitors based on cellulose niacin nano-composite for corrosion of copper in sodium chloride solutions. *International Journal of Biological Macromolecules*. Elsevier B.V., 161: 345–354. doi: 10.1016/j.ijbiomac.2020.06.040.

53 Hassan, R.M. and Ibrahim, S.M. (2021). Performance and efficiency of methyl-cellulose polysaccharide as a green promising inhibitor for inhibition of corrosion of magnesium in acidic solutions. *Journal of Molecular Structure*. Elsevier B.V., 1246: 131180. doi: 10.1016/j.molstruc.2021.131180.

54 Javidparvar, A.A., Naderi, R., and Ramezanzadeh, B. (2019). Manipulating graphene oxide nanocontainer with benzimidazole and cerium ions: application in epoxy-based nanocomposite for active corrosion protection. *Corrosion Science*. Pergamon, 108379. doi: 10.1016/J.CORSCI.2019.108379.

55 Umoren, S.A. and Solomon, M.M. (2016). Synergistic corrosion inhibition effect of metal cations and mixtures of organic compounds: a review. *Journal of Environmental Chemical Engineering*. Elsevier B.V., 5 (1): 246–273. doi: 10.1016/j.jece.2016.12.001.

56 Rajendran, S. et al. (2002). Corrosion inhibition by carboxymethyl cellulose-1-hydroxyethane-1, 1-phosphonic acid-Zn2+ system. *Bulletin of Electrochemistry* 18 (1): 25–28.

57 Umoren, S.A. et al. (2010). Synergistic and antagonistic effects between halide ions and carboxymethyl cellulose for the corrosion inhibition of mild steel in sulphuric acid solution. *Cellulose* 17: 635–648. doi: 10.1007/s10570-010-9409-7.

58 Arukalam, I.O. et al. (2014). Acid corrosion inhibition and adsorption behavior of ethyl hydroxyethyl cellulose on mild steel corrosion. *Journal of Materials* 2014: 1–11.

59 Nwanonenyi, S.C. et al. (2016). Inhibitive performance of hydroxypropyl cellulose and potassium iodide on the corrosion of mild steel in a sulphuric acid environment. *American Chemical Science Journal* 16 (2): 1–12. doi: 10.9734/ACSJ/2016/28250.

60 Transactions, C.S. et al. (2013). Inhibitive nature of carboxymethylcellulose with Zn2 + Ion. *Chemical Science Transactions* 2 (4): 1352–1357. doi: 10.7598/cst2013.618.

61 Solomon, M.M., Gerengi, H., and Umoren, S.A. (2017). Carboxymethyl cellulose/silver nanoparticles composite: synthesis, characterization, and application as a benign corrosion inhibitor for St37 steel in 15% H2SO4 medium. *ACS Applied Materials and Interfaces* 9 (7): 6376–6389. doi: 10.1021/acsami.6b14153.

62 Abd El-Lateef, H.M., Gouda, M., and Albokheet, W.A. (2020). Carboxymethyl cellulose/metal (Fe, Cu, and Ni) nanocomposites as non-precious inhibitors of C-steel corrosion in HCl solutions: synthesis, characterization, electrochemical and surface morphology studies. *Cellulose*. Springer Netherlands, 27: 8039–8057. doi: 10.1007/s10570-020-03292-6.

63 Mobin, M. and Rizvi, M. (2017). Adsorption and corrosion inhibition behavior of hydroxyethyl cellulose and synergistic surfactant additives for carbon steel in 1 M HCl. *Carbohydrate Polymers*. Elsevier Ltd., 156: 202–214. doi: 10.1016/j.carbpol.2016.08.066.

[illegible]

16

Sodium Alginate

Grafted Alginates as Sustainable Corrosion Inhibitors

Lakshmanan Muthulakshmi[1,*], *Shalini Mohan*[1], *Nellaiah Hariharan*[2], *and Jeenat Aslam*[3]

[1] *Department of Biotechnology, Kalasalingam Academy of Research and Education, Tamil Nadu, India*
[2] *2Bangalore Biotech Labs Private Limited (BiOZEEN), Bangalore – 560077, India*
[3] *Department of Chemistry, College of Science, Taibah University, Yanbu-30799, Al-Madina, Saudi Arabia*
[*] *Corresponding author*

16.1 Introduction

Polymers are materials composed of repeating units of monomers, either homogenously or heterogeneously polymerized based on the nature of the units being linked together. They have a wide range of applications in several fields of research and development, besides commercial uses. Biopolymers play a prominent role in the development of sustainable biodegradable and non-biodegradable production, and without disruption of the available resources. Biodegradable polymers are of the utmost importance because of their harmlessness to the environment [1]. With the advent of biodegradable polymers, bio-based polymers could be obtained from microbial sources like bacteria and fungi, plant sources, or algal sources. They can replace the currently available synthetic polymers that have hazardous effects on the environment [2]. They could be used for many applications like drug delivery, smart material production, anti-fouling, and metal chelation. Biopolymers also tend to possess desirable properties like increased water-holding capacity, viscoelasticity, and many other attributes, making them a better choice for commercial uses. One of the many biopolymers available is sodium alginate, which is derived from alginic acid that is extensively produced by both microbial and algal sources [3]. Sodium alginate finds use in processes like enzyme and cell encapsulation, and as an adhesive in many products. Alginates are a class of polysaccharides, with varying numbers of different monosaccharide units joined together, that are

Grafted Biopolymers as Corrosion Inhibitors: Safety, Sustainability, and Efficiency, First Edition.
Edited by Jeenat Aslam, Chandrabhan Verma, and Ruby Aslam.

widely used in various industries [4]. They are usually extracted in a complex with sodium as the latter enhances their molecular stability. Characterization techniques like XRD and FTIR are used to determine the presence of sodium alginate [5]. Sodium alginate is formulated and used in many forms like hydrogels, films, and any other forms suiting its applications. It is used as a corrosion inhibitor of various metals and in many pharmaceutical applications. Grafting is one of those techniques wherein hydrogels are co-polymerized and blended to increase the physical and mechanical properties of sodium alginate. Sodium alginate is found to play an important role in drug delivery as it helps in the encapsulation and controlled release of drug molecules. Further, it is also used to fabricate dressings for wounds as it possesses antibiotic properties [6]. In other words, grafting ensures *in situ* encapsulation of more than one polymer to enhance its desired properties. Control of bio-interfacial interactions is critical in many biomedical applications like diagnostics and biosensing. Biopolymers are known to have a potential role in minimizing or controlling the development of biofilms that may occur on implants and cause chronic infections [7]. A similar use of biopolymers is seen in naval vessels and many other clinical applications. The potential of biopolymers to act as corrosion inhibitors is thus a prospective field of research. Biopolymers could be produced, isolated, purified, and formulated as suitable products for application as corrosion inhibitors [8]. The selection of the production source, physical and mechanical properties of the biopolymer, cost of production, availability, non-toxicity, modification strategies if required, and ease of production scale-up determine the commercial potential of the biopolymer in question [9].

This chapter focuses on the types of biopolymers and their classification, sodium alginate and its characterization, formulations of sodium alginate, and the uses of sodium alginate as a corrosion inhibitor, and in biomedical applications.

16.2 Types of Biopolymers

The types of biopolymers that are conventionally extracted and used include the following:

16.2.1 Microbial Biopolymers

Several microbes are known to produce a wide range of biopolymers like glycogen, alginate, cellulose, and pullulan to enhance their infectivity as pathogens and as a means for their survival and persistence. Microbes synthesize these biopolymers as a response to biotic and abiotic stresses in the environment despite the requirement of energy for their synthesis. The biopolymers produced could act as virulence factors for the organisms as well as serve as ingredients for the food industry if they are produced by non-pathogenic bacteria. Therefore, microbial production of biopolymers could be considered an interdisciplinary activity involving microbial synthetic mechanisms, biopolymer characterization, and large-scale production, extraction and purification of the biopolymers to enhance their usability (Figure 16.1) [1].

Biopolymers have a wide array of applications: as vaccines, in the manufacture of smart implant materials, in the production of hydrogels, and as coating materials for preventing biofilm formation. The classification scheme for bio-based polymers is presented in Figure 16.2.

16.2.1.1 Bacterial Biopolymers

Bacterial biopolymers are produced by bacterial biofactories and have many useful applications. With the boom in genome sequencing and the development of novel molecular techniques, the metabolic machinery of bacteria came to be better understood. These advancements also provided insights into engineering the genetic material for large-scale production of several bio-based products with fair biocompatibility [2]. Bacterial biopolymers generally fall into one of the following classes:

16.2.1.1.1 Polysaccharides

These could be homopolymers or heteropolymers based on the types of repeating sugar units. Examples of polysaccharides include cellulose, alginate, and

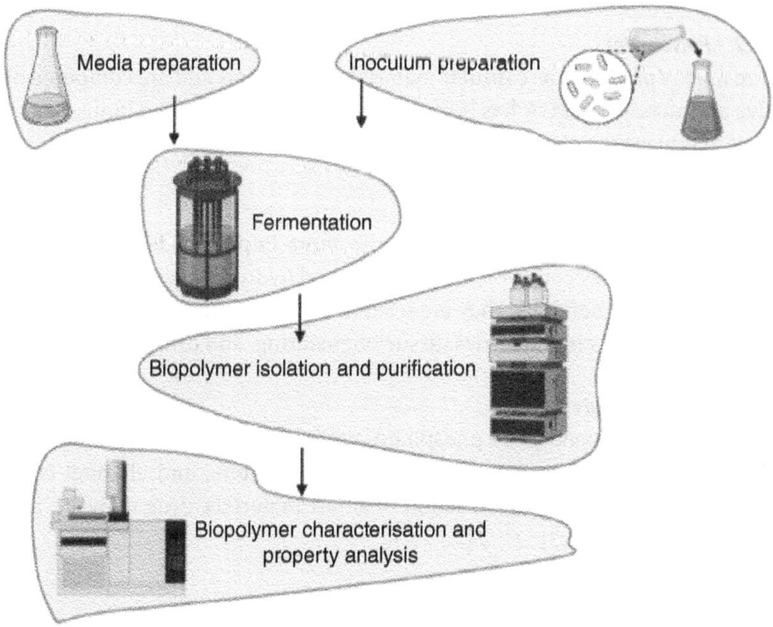

Figure 16.1 Production and development of biopolymers. *Source:* Muthulakshmi et al.

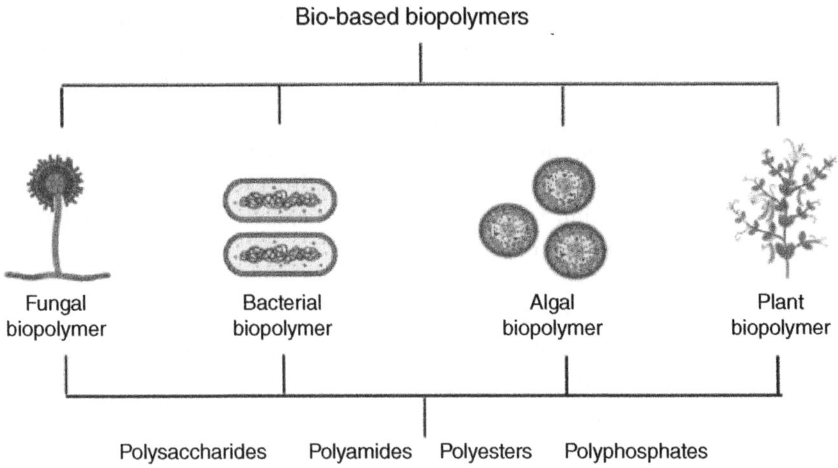

Figure 16.2 Classification of bio-based polymers. *Source*: Muthulakshmi et al.

hyaluronate. These materials possess high water-holding capacity, viscoelasticity, ionic strength, extensibility, gelation, stability, cross-linking, modifiability, biocompatibility, and biodegradability, making them suitable for the manufacture of hydrogels, films, microspheres, and 3D-printed organs.

16.2.1.1.2 Polyamides

These are mostly produced as capsular materials for biofilms and are composed of repetitive amino acids. γ–PGA has been extensively produced in the industry as a flocculant to replace the chemicals used in wastewater treatment.

16.2.1.1.3 Polyesters

Polyhydroxyalkanoates (PHA) are one of the most important bioplastics produced by microbes. They comprise linearly arranged hydrophobic structures used as storage units. Commercially, they are used in enzyme immobilization, several downstream operations in industries, tissue engineering, and drug delivery.

16.2.1.1.4 Polyphosphates

These are polymers of condensed phosphates with high-energy bonds, employed in many biosynthetic pathways as buffers, metal chelators, and channel compounds. They are used in the production of smart materials, drug delivery, and hydrogel production [10] [3].

16.2.1.2 Fungal Biopolymers

Fungal biopolymers have recently gained attention due to their increased yield. Chitin and chitosan are produced by many classes of fungi and have a plethora of applications in food, biomedicine, and agriculture due to their desirable properties like non-toxicity, biodegradability, bio-activity, bio-compatibility, bio-reabsorption and absorption [11, 12].

16.2.1.3 Algal Biopolymers

Algae require less water for their growth and also sequestrate CO_2 from the atmosphere by carbon fixation, thus playing a vital role in the maintenance of the carbon balance. Like microbes, algae are also known to produce improved PHA with desirable properties. Several genetic engineering techniques have been adopted for the production of biopolymers from algae by effectively increasing the production of PHA from the fixed carbon by activation of the particular pathway responsible for the production of the biopolymer. Moreover, algal bioplastics are believed to make a positive impact on the circular economy due to their extensive recyclable properties. The greenhouse gas credit implored by carbon sequestration signifies the environmental friendliness of algal biopolymers. Besides these, algal biopolymers could be used for 3D printing, in several biomedical applications, and as antioxidants and absorbents [13, 14].

16.2.2 Plant Biopolymers

These are produced by plants for maintaining their structural integrity, for storage of carbon, as a defensive component, and for resisting desiccation. Hence, the biopolymers produced are used for diverse purposes by humans in their daily lives. A vital example is rubber extracted from the rubber tree that is used for the manufacture of tyres, tubing, and many other consumer products. Production could be improved by adopting transgenic crop cultivation or growing other alternative crops. PHB produced by plants finds use as a chemical, in plant production, and as a feed supplement [15, 16, 17].

16.3 Biopolymers as Antifouling Agents

Some biopolymers are known to have an antifouling property i.e., the ability to prevent the accumulation of biological organisms over a surface. Examples of biopolymers with the anti-fouling property are shown in Table 16.1.

16.4 Sodium Alginate

Sodium alginate is a water-soluble linear polysaccharide obtained mostly from seaweed. It is composed of mannuronic and guluronic acids complexed together, and is, more precisely, one of the useful derivatives of alginic acids. It possesses

Table 16.1 Biopolymers with antifouling property.

S. No.	Polymer	Application in anti-fouling	References
1.	Chitosan	Marine vessel coating	[18]
2.	Heparin	Decreased catheter-associated colonization	[19]
3.	Cellulose acetate	Heavy metal removal	[20]
4.	Dextran	Marine coating	[21]
5.	Hyaluronic acid	Medical implant coating	[22]
6.	Pectin	Food packaging	[23]
7.	Collagen	Medical implant coating	[24]
8.	Polydopamine	Dental implants	[25]
9.	Polylactic acid	Marine vessel coating	[26]
10.	Gelatin	Antifouling paint	[27]

many desirable properties like non-toxicity, the ability to thicken, the ability to form films, effective bio-adhesion, and biocompatibility, among others, making it suitable for a wide range of industrial applications.

16.4.1 Producers of Sodium Alginate

The main producers of alginate are species of brown seaweed, namely, *Laminaria, Macrocystis*, and *Ascophyllum* [28]. A major fraction of the harvested *Laminaria* is used as food while the remaining is used for alginate production. The other species producing alginate include *Sargassum baccularia, S. siliquosum, S. binderi, S. muticum, Durvillaea antarctica, Laminaria digitata* and *Lobophora variegata* [4, 6].

16.4.2 Extraction and Characterization of Sodium Alginate

Crude extraction of sodium alginate is done by immersing the seaweed in an acid solution, followed by neutralization of the solution, and a reaction with sodium hydroxide. Since sodium alginate occurs mostly in insoluble forms as it is complexed with calcium or magnesium, a pH of 10 after crude extraction is necessary to obtain sodium alginate in pure form [29, 30].

The following properties of sodium alginate are analyzed for checking the suitability of the extracted sodium alginate for various applications [31].

16.4.2.1 Moisture Content

Evaluating the moisture content indicates the shelf-life and stability of the extracted sodium alginate. Therefore, a reduced level of moisture signifies greater rigidity of the sodium alginate.

16.4.2.2 Viscosity

The viscosity of the crude extract determines the stability and applications of sodium alginate. If the extract is to be used for gelling applications, then a high viscosity would be more appropriate.

16.4.2.3 Whiteness Degree

The brightness or the degree of whiteness directly correlates with the browning effect that could occur with the use of sodium alginate; therefore, this parameter further ensures the applicability of sodium alginate.

The characterization techniques for sodium alginate are as follows [6] and shown in Figure 16.3.

16.4.2.4 UV Spectroscopy

UV-VIS-NIR spectroscopy at room temperature in the wavelength range of 200–800 nm indicates the presence of sodium alginate in the sample.

16.4.2.5 FTIR Spectroscopy

The presence of sodium alginate in solid samples could be analyzed using FTIR in the range of 400–4000/cm, followed by baseline correction and normalization.

16.4.2.6 X-Ray Diffraction

The crystalline and amorphous nature of sodium alginate could be analyzed using XRD analysis.

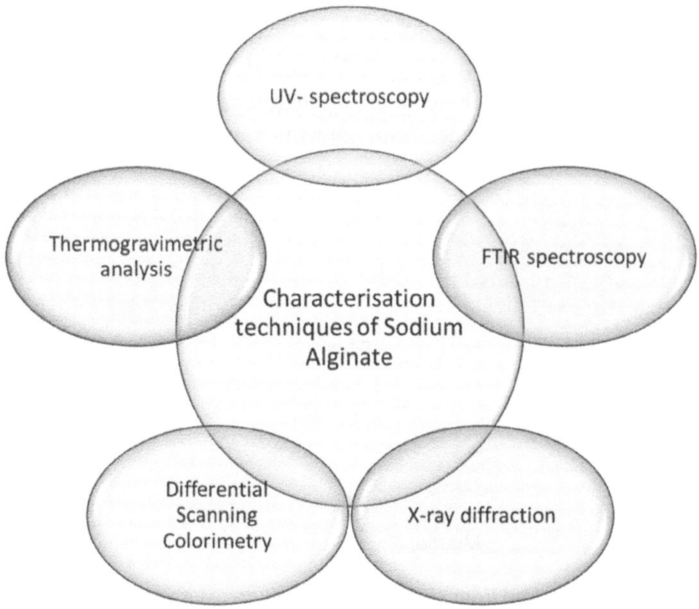

Figure 16.3 Characterization techniques for sodium alginate. *Source:* Muthulakshmi et al.

16.4.2.7 Differential Scanning Calorimetry

The properties of glass transition and melting behavior could be analyzed using DSC.

16.4.2.8 Thermogravimetric Analysis

This indicates the thermal degradation pattern of the sodium alginate to be analyzed.

Purification of the extracted sodium alginate is performed using ethanol precipitation, filtering, and drying [5, 32, 33].

16.4.3 Material Formulations of Sodium Alginate

The following are the biomaterials formulated using sodium alginate [34], as shown in Figure 16.4.

16.4.3.1 Hydrogels

These are three-dimensional polymers composed of complexly interlinked hydrophilic molecules. They are capable of absorbing huge volumes of body fluids without disrupting their intact structure, making them similar to living tissues. They are prepared by many methods like the interaction of ions, cross-linking by covalent bonds, gelation, polymerization using free radicals, and cross-linking within cells. They have various applications in medicine like drug delivery, wound dressing, and bioprinting [35].

16.4.3.2 Microspheres

These are small, enclosed spheres with materials trapped inside to promote their effective release or storage for a prolonged period. Sodium alginate microspheres are prepared by emulsification coupled with solvent evaporation, based on the

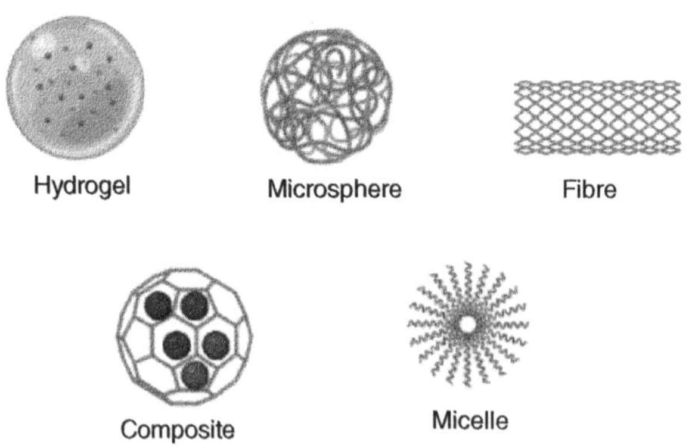

Figure 16.4 Biomaterials prepared from sodium alginate. *Source*: Muthulakshmi et al.

principle of complexation by polyelectrolytes. The viscosity of the sodium alginate should be adjusted to prevent the clumping of spheres [36].

16.4.3.3 Fibres

The concept of moist healing led to the development of products like gels and fibrous materials that retain water. Alginate fibers are mostly employed in the synthesis of highly absorbent materials that could be used as wound dressings. They are generally prepared by wet spinning using a spinneret and coagulation bath. Nanofibers are also produced because of their high surface area-to-volume ratio, which might promote the migration of cellular factors during wound healing [37].

16.4.3.4 Composites

Sodium alginate composites are formed by blending sodium alginate with other polymers, natural like chitosan or synthetic materials like polylactide, to obtain materials with enhanced physico-chemical properties. Plant-based cellulose nanofibers and hydroxyapatite are commonly used in various medical applications. However, sodium alginate composites also exhibit a fair degree of anti-microbial activity, thereby ensuring their applicability in topical treatment strategies. Further, they have a large surface area for loading ample amounts of drug preparations [38].

16.4.3.5 Films

Sodium alginate films are produced to obtain higher thermal stability and tensile strength. They could be used to develop materials that are biodegradable and edible, along with possessing certain biomedical properties. They are prepared by soaking sodium alginate powder in a solvent followed by drying of the solvent. These films could then be grafted to further increase their stability [39].

16.4.3.6 Other Formulations

The other formulations of sodium alginate include polyacrylamide, lactate oxidase, essential oil-based micelles, PEG (Polyethyleneglycol) or SPION (Superparamagnetic iron oxide nanoparticles)-based microcarriers, and glycolide- or lactide-based tubes, each with a unique range of applications [6].

16.4.4 Sodium Alginate Grafting

Natural polymers are susceptible to damage by many reactions; therefore, it is necessary to modify them to enhance their stability. Methods like blending, grafting, and curing are used for modifying the properties of polymers. Blending refers to the mixing of two or more properties for incurring the desired property curing involves filling the gaps over the surface of the polymer; and grafting refers to covalent bonding of the blended polymers by chemical reactions [40]. The methods of polymer modification are shown in Figure 16.5.

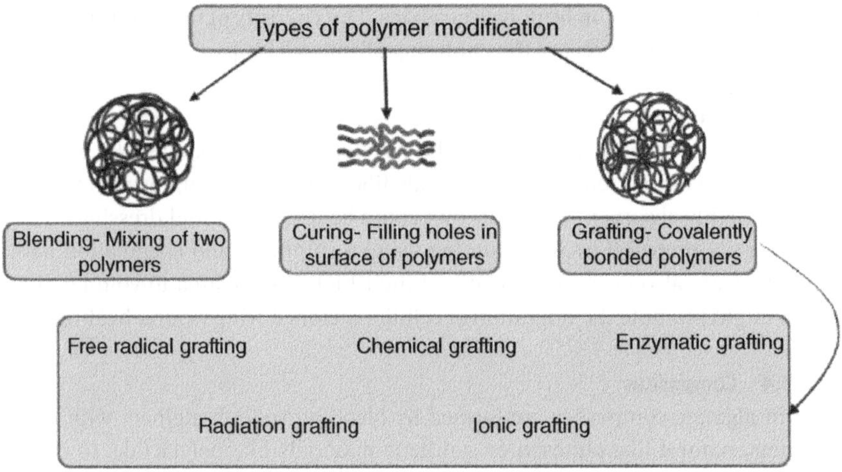

Figure 16.5 Types of polymer modification. *Source*: Muthulakshmi et al.

16.4.4.1 Methods of Grafting

Graft copolymer synthesis is generally of three types based on the methods involved in the synthesis, namely, "grafting to," which involves adherence of pre-polymer chains to the end groups of the polymer backbone, "grafting from," which involves functionalizing of the initial polymer chain using an initiator, and "grafting through," which involves the synthesis of the polymer after polymerization [41]. The different techniques used for grafting are as follows:

16.4.4.1.1 Free Radical Grafting

This method involves initiator molecules that can generate free radicals or transferring high-energy radicals to polymers, which eventually enhances the polymer modification. Free radicals could be generated by direct or indirect methods depending on their initiation process.

16.4.4.1.2 Chemical Grafting

Grafting by chemical methods involves using certain chemical substances that are capable of polymer modification.

16.4.4.1.3 Ionic Grafting

This is similar to free-radical grafting, but instead of free radical generation, ionic molecules (both cationic and anionic centers) are used for polymer modification.

16.4.4.1.4 Radiation Grafting

This type of grafting utilizes radiation for generating free radicals that could initiate polymer modification.

16.4.4.1.5 Enzymatic Grafting

Enzymes like tyrosinase modify the backbones of polymers, which enhances covalent bonding between polymer types [41, 42].

16.4.5 Sodium Alginate in Anti-Fouling

Biofouling is the phenomenon in which harmful microorganisms adhere to the surfaces of materials, effectively preventing the transport of nutrients in case of a surface with viable cells underneath, or corroding the surface of inorganic materials [43,44]. Sodium alginate is also known to possess antimicrobial activity against a wide range of microbial species [45]. It also has an inhibitory effect on both mucoid and non-mucoid strains of *Pseudomonas aeruginosa* [46], *Staphylococcus aureus, Lactococcus lactis* [47], and *E. coli* [48].

16.4.5.1 Anti-fouling of Medical Implants

The emerging need to develop anti-fouling agents to prevent infections from medical implants has prompted the discovery of many polymeric agents like polydopamine [49] which, on coating with chitosan and sodium alginate, acquires antibacterial properties against pathogens of dental implants [50]. Sodium alginate could also be used for functionalizing strontium coatings to minimize the orthopedic failure of implants [51].

16.4.5.2 Anti-fouling in Marine Vessels

Marine biofouling is often associated with severe economic losses. The conventional coating with tributyltin that is used to prevent fouling is toxic to marine life to a significant extent, thereby rendering coating with sodium alginate a sustainable replacement for it [8]. Combinations of sodium alginate with other polymers, fabricated layer by layer, were also found to be synergetic inhibitors of biofouling [52].

16.4.5.3 Sodium Alginate as a Corrosion Inhibitor of Metals

Corrosion occurring in many metals could be controlled by sodium alginate. API X60 carbon steel is protected from corrosion in saline conditions due to the physical absorption of the carboxylate oxygen present in it by sodium alginate. Therefore, the repercussions of wetting on the steel surface could be mitigated using sodium alginate [53, 54]. Even in acidic environments, sodium alginate proved to be a better inhibitor of corrosion [55]. Aluminium alloys complexed with chromium that are used in aircraft are found to be less corrosion-sensitive on treatment with sodium alginate isolated from brown seaweed. This might help reduce the maintenance costs of hardware used in the aerospace and defense industries. Sodium alginate reduced corrosion in the AZ91D magnesium alloy used in automobile and electronic applications [56].

16.4.6 Biomedical Applications of Sodium Alginate

The pharmacological applications of sodium alginate are as below [9].

16.4.6.1 Drug Delivery

Sodium alginate in the form of hydrogels is a potent carrier for drugs. In the case of oral delivery, sodium alginate is used in floating mucoadhesive systems. Sodium alginate grafted with polyglycidyl methacrylate hydrogels has been found to have increased bioavailability in intestinal fluids. Alginates as nanoparticles are used for drug delivery in injectables. Low-molecular-weight sodium alginate is effective for preparing ocular inserts with brimonidine tartrate. In addition, it is used for the delivery of many molecules and proteins [57].

16.4.6.2 Wound Dressing

Sodium alginate has proven to be a better alternative to other materials in moist wound healing processes. It can be loaded with soluble factors that could promote healing by reducing scars. It is also known to reduce bacterial and other microbial infections in connection with preventing chronic infections in the affected regions. Many fabrication types like films, hydrogels, foams, and wafers are used in wound dressing [58].

16.4.7 Sodium Alginate in Cell Culture

Sodium alginate could be used to model cell culture environments and is also capable of establishing stable 3D cultures. Matrix interactions between neighboring cells could be enhanced by using alginates coupled with peptides. Further, products based on alginate used as 3D cell culture systems could pave the way for replacing or regenerating tissues like blood vessels, nerve cells, or even bones [59].

Besides biomedical applications, sodium alginate has a number of other uses, some of which are presented in Table 16.2.

Table 16.2 Uses of sodium alginate.

S. No.	Formulation of Sodium alginate	Application	References
1.	Thin composite film	Desalination of brackish water	[60]
2.	Hydrogel	Tissue scaffold and implant devices	[43]
3.	Nanocomposite hydrogel	Water purification	[61]
4.	Colloid	Seawater desalination	[62]
5.	Composite hydrogel	Water purification	[63]
6.	Aerogel	Oil/ Water separation	[64]
7.	Emulsion	Antifouling paint	[65]
8.	Scaffold	Tissue engineering and drug delivery	[66]

16.5 Conclusion

This chapter discussed the basics of biopolymers and their classification based on their sources, while dwelling at length on sodium alginate, its applications, and its natural producers. Several characterization techniques for sodium alginate were described, along with their applications in pharmacology and anti-fouling. Thus, the eco-friendly and sustainable biopolymer sodium alginate could be exploited for a wide range of applications.

Acknowledgement

The authors gratefully acknowledge the financial support of DBT, Govt. of India, for this research.

References

1 Kreyenschulte, D., Krull, R., and Margaritis, A. (2014). Recent advances in microbial biopolymer production and purification. *Critical Reviews in Biotechnology* 34 (1): 1–15.

2 Moradali, M.F. and Rehm, B.H. (2020). Bacterial biopolymers: from pathogenesis to advanced materials. *Nature Reviews Microbiology* 18 (4): 195–210.

3 Ghosh, S., Lahiri, D., Nag, M. et al. (2021). Bacterial biopolymer: its role in pathogenesis to effective biomaterials. *Polymers* 13 (8): 1242.

4 Szekalska, M., Puciłowska, A., Szymańska, E. Ciosek, P., and Winnicka, K. (2016). Alginate: current use and future perspectives in pharmaceutical and biomedical applications. *International Journal of Polymer Science* 2016.

5 Aprilliza, M. (2017 April). Characterization and properties of sodium alginate from brown algae used as an eco-friendly superabsorbent. *In IOP Conference Series: Materials Science and Engineering* 188 (1): 012019. IOP Publishing.

6 Ahmad, A., Mubarak, N.M., Jannat, F.T., et al. (2021). A critical review on the synthesis of natural sodium alginate-based composite materials: an innovative biological polymer for biomedical delivery applications. *Processes* 9 (1): 137.

7 Tripathi, R. and Mishra, B. (2012). Development and evaluation of sodium alginate–polyacrylamide graft–co-polymer-based stomach targeted hydrogels of famotidine. *Aaps Pharmscitech* 13 (4): 1091–1102.

8 Abi Nassif, L., Rioual, S., Farah, W. et al. (2020). Electrophoretic deposition of zinc alginate coatings on stainless steel for marine antifouling applications. *Journal of Environmental Chemical Engineering* 8 (5): 104246.

9 Lee, K.Y. and Mooney, D.J. (2012). Alginate: properties and biomedical applications. *Progress in Polymer Science* 37 (1): 106–126.

10 Osman, Y., Abd Elrazak, A., and Khater, W. (2016). Microbial biopolymer production by Microbacterium WA81 in batch fermentation. *Egyptian Journal of Basic and Applied Sciences* 3 (3): 250–262.

11 Abo Elsoud, M.M. and El Kady, E.M. (2019). Current trends in fungal biosynthesis of chitin and chitosan. *Bulletin of the National Research Centre* 43 (1): 1–12.

12 Luft, L., Confortin, T.C., Todero, I., Zabot, G.L., and Mazutti, M.A. (2020). An overview of fungal biopolymers: bio-emulsifiers and biosurfactants compound production. *Critical Reviews in Biotechnology* 40 (8): 1059–1080.

13 Devadas, V.V., Khoo, K.S., Chia, W.Y. et al. (2021). Algae biopolymer towards a sustainable circular economy. In: *Bioresource Technology* 325: 124702.

14 Kartik, A., Akhil, D., Lakshmi, D. et al. (2021). A critical review on the production of biopolymers from algae biomass and their applications. *Bioresource Technology* 124868.

15 Madadi, R., Maljaee, H., Serafim, L.S., and Ventura, S.P. (2021). Microalgae as Contributors to Produce Biopolymers. *Marine Drugs* 19 (8): 466.

16 Snell, K.D., Singh, V., and Brumbley, S.M. (2015). Production of novel biopolymers in plants: recent technological advances and prospects. *Current Opinion in Biotechnology* 32: 68–75.

17 Van Beilen, J.B. and Poirier, Y. (2008). Production of renewable polymers from crop plants. *The Plant Journal* 54 (4): 684–701.

18 Dutta, J. (2019). Non-medical applications of chitosan nanocomposite coatings.

19 Wallace, A., Albadawi, H., Patel, N. et al. (2017). Anti-fouling strategies for central venous catheters. *Cardiovascular Diagnosis and Therapy* 7 (Suppl 3): S246–S257.

20 Ounifi, I., Guesmi, Y., Ursino, C. et al. (2021). Antifouling membranes based on cellulose acetate (CA) Blended with poly (acrylic acid) for heavy metal remediation. *Applied Sciences* 11 (10): 4354.

21 Xu, G., Liu, P., Pranantyo, D., Neoh, K.G., and Kang, E.T. (2018). Dextran-and chitosan-based antifouling, antimicrobial adhesion, and self-polishing multilayer coatings from the pH-responsive linkages-enabled layer-by-layer assembly. *ACS Sustainable Chemistry & Engineering* 6 (3): 3916–3926.

22 Gaetano, G., Giuseppe, P., Salvatore, P.F., Sara, S., and Luca, R.C. (2018). Hyaluronic-based antibacterial hydrogel coating for implantable biomaterials in orthopedics and trauma: from basic research to clinical applications. In: *Hydrogels*, 1e. 179–200. London: IntechOpen.

23 Calce, E., Mignogna, E., Bugatti, V., Vittoria, V., and De Luca, S. (2014). Pectin functionalized with natural fatty acids as an antimicrobial agent. *International Journal of Biological Macromolecules* 68: 28–32.

24 Utrata-Wesołek, A.L.I.C.J.A. (2013). Antifouling surfaces in medical applications. *Polimery* 58 (9).

25 Goh, S.C., Luan, Y., Wang, X. et al. (2018). Polydopamine–polyethylene glycol–albumin antifouling coatings on multiple substrates. *Journal of Materials Chemistry B* 6 (6): 940–949.

26 Kamtsikakis, A., Kavetsou, E., Chronaki, K. et al. (2017). Encapsulation of antifouling organic biocides in poly (lactic acid) nanoparticles. *Bioengineering* 4 (4): 81.

27 Szabó, T., Mihály, J., Sajó, I., Telegdi, J., and Nyikos, L. (2014). One-pot synthesis of gelatin-based, slow-release polymer microparticles containing silver nanoparticles and their application in antifouling paint. *Progress in Organic Coatings* 77 (7): 1226–1232.

28 McHugh, D.J. (1987). Production, properties, and uses of alginates: production and utilization of products from commercial seaweeds. *FAO. Fisheries Technical Paper* 288: 58–115.

29 Fertah, M., Belfkira, A., Dahmane, E. et al. (2017). Extraction and characterization of sodium alginate from Moroccan *Laminaria digitata* brown seaweed. *Arabian Journal of Chemistry* 10: S3707–S3714.

30 Sachan, N.K., Pushkar, S., Jha, A., and Bhattcharya, A. (2009). Sodium alginate: the wonder polymer for controlled drug delivery. *Journal of Pharmacy Research* 2 (8): 1191–1199.

31 Kusumawati, R., Basmal, J., and Utomo, B.B. (2018). Physicochemical characteristics of sodium alginate extracted from *Turbinaria* sp. and *Sargassum* sp. *Squalen Bulletin of Marine and Fisheries Postharvest and Biotechnology* 13 (2): 79–84.

32 Jana, S., Trivedi, M.K., Tallapragada, R.M. et al. (2015). Characterization of physicochemical and thermal properties of chitosan and sodium alginate after biofield treatment. *Pharmaceutica Analytica Acta* 6: 10.

33 Johnson, F.A., Craig, D.Q., and Mercer, A.D. (1997). Characterization of the block structure and molecular weight of sodium alginates. *Journal of Pharmacy and Pharmacology* 49 (7): 639–643.
34 Zhang, H., Cheng, J., and Ao, Q. (2021). Preparation of Alginate-based biomaterials and their applications in biomedicine. *Marine Drugs* 19 (5): 264.
35 Abasalizadeh, F., Moghaddam, S.V., Alizadeh, E. et al. (2020). Alginate-based hydrogels as drug delivery vehicles in cancer treatment and their applications in wound dressing and 3D bioprinting. *Journal of Biological Engineering* 14 (1): 1–22.
36 Nagpal, M., Maheshwari, D.K., Rakha, P. et al. (2012). Formulation development and evaluation of alginate microspheres of ibuprofen. *Journal of Young Pharmacists* 4 (1): 13–16.
37 Qin, Y. (2008). Alginate fibers: an overview of the production processes and applications in wound management. *Polymer International* 57 (2): 171–180.
38 Raus, R.A., Nawawi, W.M.F.W., and Nasaruddin, R.R. (2021). Alginate and alginate composites for biomedical applications. *Asian Journal of Pharmaceutical Sciences* 16 (3): 280–306.
39 Zaman, H.U. and Beg, M.D.H. (2013). Preparation and properties of sodium alginate films. *Journal of Polymer Engineering* 33 (9): 829–836.
40 Bhattacharya, A. and Misra, B.N. (2004). Grafting: a versatile means to modify polymers: techniques, factors, and applications. *Progress in Polymer Science* 29 (8): 767–814.
41 Maity, N. and Dawn, A. (2020). Conducting polymer grafting: recent and key developments. *Polymers* 12 (3): 709.
42 Koshy, T.M., Gowda, D.V., Tom, S. et al. (2016). Polymer grafting–An overview. *American Journal of Pharmtech Research* 6 (2): 1–13.
43 Zhang, J., Qian, S., Chen, L. et al. (2021). Highly antifouling double network hydrogel based on poly (sulfobetaine methacrylate) and sodium alginate with great toughness. *Journal of Materials Science & Technology* 85: 235–244.
44 Zhang, J., Peng, P., Chen, L., Zhao, L., and Feng, J. (2021). Antifouling poly (N-(2-hydroxyethyl) acrylamide)/sodium alginate double network hydrogels with eminent mechanical properties. *Polymer Testing* 95: 107087.
45 Kumar, L., Brice, J., Toberer, L. et al. (2019). Antimicrobial biopolymer formation from sodium alginate and algae extract using aminoglycosides. *Plos one* 14 (3): e0214411.
46 Baltimore, R.S., Cross, A.S., and Dobek, A.S. (1987). The inhibitory effect of sodium alginate on antibiotic activity against mucoid and non-mucoid strains of *Pseudomonas aeruginosa*. *Journal of Antimicrobial Chemotherapy* 20 (6): 815–823.
47 Ye, J., Ma, D., Qin, W., and Liu, Y. (2018). Physical and antibacterial properties of sodium alginate—sodium carboxymethylcellulose films containing *Lactococcus lactis*. *Molecules* 23 (10): 2645.

48 Fundador, E.O.V., Villanueva, J.M.A., Fundador, N.G.V., and de Cadiz, A.E. (2018). Antimicrobial property of sodium Alginate/TiO2 Nanocomposite film. *Philippine Journal of Science* 147 (3): 545–550.

49 Zeng, Q., Zhu, Y., Yu, B. et al. (2018). Antimicrobial and antifouling polymeric agents for surface functionalization of medical implants. *Biomacromolecules* 19 (7): 2805–2811.

50 Duan, Y., Wu, Y., Yan, R. et al. (2021). Chitosan-sodium alginate-based coatings for self-strengthening anticorrosion and antibacterial protection of titanium substrate in artificial saliva. *International Journal of Biological Macromolecules.*

51 Yuan, N., Jia, L., Geng, Z. et al. (2017). The incorporation of strontium in a sodium alginate coating on titanium surfaces for improved biological properties. *BioMed Research International* 2017.

52 Zhao, L., Chen, R., Lou, L. et al. (2020). Layer-by-Layer-Assembled antifouling films with surface microtopography inspired by *Laminaria japonica. Applied Surface Science* 511: 145564.

53 Obot, I.B., Onyeachu, I.B., and Kumar, A.M. (2017). Sodium alginate: a promising biopolymer for corrosion protection of API X60 high-strength carbon steel in a saline medium. *Carbohydrate Polymers* 178: 200–208.

54 Zhang, W., Nie, B., Li, H.J. et al. (2021). Inhibition of mild steel corrosion in 1 M HCl by chondroitin sulfate and its synergistic effect with sodium alginate. *Carbohydrate Polymers* 260: 117842.

55 Tawfik, S.M. (2015). Alginate surfactant derivatives as an ecofriendly corrosion inhibitor for carbon steel in acidic environments. *RSC Advances* 5 (126): 104535–104550.

56 Li, Y., Ba, Z.X., Li, Y.L., Ge, Y., and Zhu, X.C. (2017). Influence of sodium alginate inhibitor addition on the corrosion protection performance of AZ91D magnesium alloy in NaCl solution. *Anti-Corrosion Methods and Materials.*

57 Chaturvedi, K., Ganguly, K., More, U.A. et al. (2019). Sodium alginate in drug delivery and biomedical areas. In: *Natural Polysaccharides in Drug Delivery and Biomedical Applications* (ed. M.S. Hasnain and A.K. Nayak), 59–100. Academic Press.

58 Aderibigbe, B.A. and Buyana, B. (2018). Alginate in wound dressings. *Pharmaceutics* 10 (2): 42.

59 Andersen, T., Auk-Emblem, P., and Dornish, M. (2015). 3D cell culture in alginate hydrogels. *Microarrays* 4 (2): 133–161.

60 Kancherla, R., Vadeghar, R.K., Ginuga, P.R., and Sridhar, S. (2020). Antifouling membrane based on sodium alginate coated polyamide thin film composite for desalination of brackish water. *Polymer Engineering & Science* 60 (11): 2827–2840.

61 Amiri, S., Asghari, A., Vatanpour, V., and Rajabi, M. (2020). Fabrication and characterization of a novel polyvinyl alcohol-graphene oxide-sodium alginate nanocomposite hydrogel blended PES nanofiltration membrane for improved water purification. *Separation and Purification Technology* 250: 117216.

62 Melián-Martel, N., Alonso, J.J.S., and Ruiz-García, A. (2018). Combined silica and sodium alginate fouling of spiral-wound reverse osmosis membranes for seawater desalination. *Desalination* 439: 25–30.

63 Huang, Q., Liu, S., Li, K. et al. (2017). Sodium alginate/carboxyl-functionalized graphene composite hydrogel via neodymium ions coordination. *Journal of Materials Science & Technology* 33 (8): 821–826.

64 Li, Y., Zhang, H., Fan, M. et al. (2017). A robust salt-tolerant superoleophobic alginate/graphene oxide aerogel for efficient oil/water separation in marine environments. *Scientific Reports* 7 (1): 1–7.

65 Liu, Y. and Nisisako, T. (2020). Microfluidic encapsulation of hydrophobic antifouling biocides in calcium alginate hydrogels for controllable release. *ACS omega* 5 (40): 25695–25703.

66 Guarino, V., Caputo, T., Altobelli, R., and Ambrosio, L. (2015). Degradation properties and metabolic activity of alginate and chitosan polyelectrolytes for drug delivery and tissue engineering applications. *AIMS Materials Science* 2 (4): 497–502.

17

Grafted Dextrin as a Corrosion Inhibitor

M. Mobin[1,], K. Cial[1], J. Aslam[2], M. Parveen[1], and R. Aslam[1,*]*

[1] *Corrosion Research Laboratory, Department of Applied Chemistry, Aligarh Muslim University, Aligarh, Uttar Pradesh, India*
[2] *Department of Chemistry, College of Science, Taibah University, Yanbu, Al-Madina 30799, Saudi Arabia*
* *Corresponding authors*

17.1 Corrosion and Its Adverse Impact

Corrosion is the material's destructive reaction to its surrounding environment. During the past years, it has caused irreparable damage to human life [1]. Corrosion costs manifest in the form of premature deterioration or failure necessitating maintenance, repairs, and replacement of damaged parts. Corrosion has a vast environmental and economic impact on all the surfaces of national infrastructure like highways, bridges, buildings, chemical processing units, wastewater treatment, and virtually on all metallic objects in our day-to-day life use [2]. Other than material loss, corrosion interferes not only with the environment but also affects human safety and industrial operations severely. Awareness of corrosion and adaptation of timely and appropriate control measures hold the key to the abatement of corrosion failures.

The use of corrosion inhibitors [3, 4], cathodic protection [5, 6], different types of coatings on the surface [7], alloying elements, and additives [8] are several methods for controlling corrosion.

17.2 Corrosion Inhibitors and Factors Affecting Their Efficiency

An inhibitor is a substance that when added to an environment in a small concentration minimizes the loss of metal. Small amounts of corrosion inhibitors as external materials are inserted into aggressive media during descaling, acid pickling, and oil well refining to protect metal surfaces from corrosive damage [9]. Inhibitors

Grafted Biopolymers as Corrosion Inhibitors: Safety, Sustainability, and Efficiency, First Edition.
Edited by Jeenat Aslam, Chandrabhan Verma, and Ruby Aslam.

may be organic or inorganic compounds and they are usually dissolved in aqueous environments. Organic inhibitors, usually designated as film-forming, protect the metal by forming a hydrophobic film on the metal surface. Their effectiveness depends on the metal surface's chemical composition, molecular structures, and affinities. Because film formation is an adsorption process, the adsorption of inhibitors is related to the presence of heteroatoms such as nitrogen, oxygen, phosphorous, and sulfur, as well as a triple bond or an aromatic ring in their molecular structure through which they can adsorb on the metal surface [10]. Inhibitors are normally distributed from a solution or dispersion. Inhibitors slow corrosion processes by

- Increasing the anodic or cathodic polarization behavior (Tafel slopes)
- Reducing the movement or diffusion of ions to the metallic surface
- Increasing the electrical resistance of the metallic surface

Before choosing an inhibitor, it is important to consider the various factors that affect its properties, such as the presence of metals and the operating conditions of the fluid. In addition to these, it is also important to check the concentration of the inhibitor at regular intervals. For optimal results, it is recommended to add an inhibitor to the fluid or replace the whole fluid. Although continuous monitoring is generally recommended in certain applications, it is important to note that the results of this type of monitoring can be influenced by the behavior of certain components in the system.

17.3 Biopolymers as Corrosion Inhibitors: Advantages and Disadvantages

Polymers generated from petroleum have been used to make a wide range of goods used in daily life, agriculture, industry, and a variety of other purposes, but their widespread use has major environmental repercussions. To provide ecologically friendly items, biodegradable materials synthesis is attracting a lot of attention in the research and modern field. Biopolymer-based products are a potential option since they help the soil to lessen the detrimental consequences of plastic waste. Biodegradable polymeric materials derived from microbial, plant, synthetic, and marine sources have gotten a lot of attention in recent years for their use as diluents, binders, disintegrants, and gelling agents in tablets and gels, as well as protective colloids in suspension, corrosion inhibitors, and thickeners in oral liquids.

Instead of petroleum, which is the traditional source of polymers, biopolymers are polymers generated or derived from living things such as plants and bacteria. The primary sources of biopolymers are renewable. They are non-toxic and long-lasting. Furthermore, they have higher adsorption on metal surfaces than monomers and oligomers. Biopolymers have two advantages as corrosion inhibitors: a single polymeric chain expels more water molecules from the metal surface, and the presence of multiple adsorption sites makes desorption more difficult.

Type of Biopolymers – Figure 17.1 shows the various types of biopolymers based on their origin.

Biopolymers have attracted interest around the world for corrosion protection of industrial metal substrates, particularly as corrosion inhibitors, because they are environmentally friendly and have many adsorption centers [9]. However, most biopolymers tested are shown to be modest corrosion inhibitors, with poor protective effectiveness at higher concentrations.

One of the most challenging features of using biopolymers as corrosion inhibitors is their limited solubility in polar electrolytes due to their chemical structure, especially at higher concentrations. In most cases, organic extracts have been reported to precipitate in polar electrolytes. To address these difficulties, new techniques to reinforce their structures using biocompatibility have emerged, recognizing the urgent need to develop high-performance materials sourced from natural sources.

Copolymerization, metal ion addition, surfactant synergistic action, and the introduction of inorganic compounds in nano-size into the biopolymer matrix to improve corrosion inhibition capabilities of the biopolymers are some of the treatments that have been tried to minimize this disadvantage. Synergism is a phrase used to describe the practice of mixing external additions with material or organic inhibitors to increase their efficiency. Grafted biopolymer has shown promise in terms of enhancing inhibitory efficiency, which could protect metal surfaces from corrosive ions.

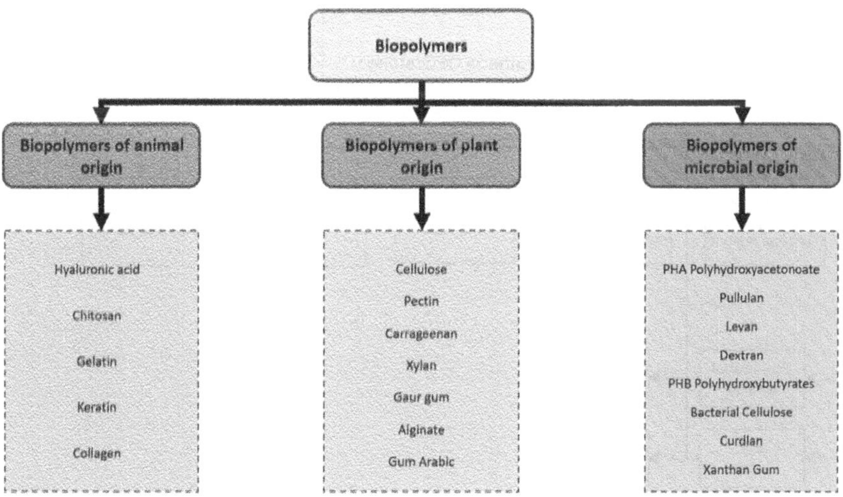

Figure 17.1 Types of biopolymers based on different origin.

17.4 Methods of Grafting the Biopolymers

Graft polymerization is a method of covalently linking monomers to the parent polymer chain [10]. Grafting is done to increase the biocompatibility, wettability, mechanical properties, and other properties of a surface polymer. The following sections examine several grafting techniques (Figure 17.2).

1-By chemical means-This process can take place through three different routes: free radical, ionic, and grafting via living polymerization. The initiators in Free Radicals generate free radicals, which are subsequently transferred to the substrate and react with the monomer to form graft copolymers.

(i) In the ionic method-Sodium naphthalenide and organometallic compounds are useful initiators. Copolymerization occurs when alkyl aluminum (R3Al) and the backbone polymer in the halide form (ACl) combine to generate carbonium ions along the polymer chain.

(ii) In living polymerization method-This approach couples the electrophilic groups of the backbone polymer and the propagation site of an anionic living polymer. This method would not be possible without the development of a backbone polymer with reactive groups.

2- By radiation method-In this procedure, the macromolecule is irradiated to produce free radicals and ions, which react with monomers to produce graft copolymer. Although there is no need for an initiator in this method, the medium is crucial.

3-Photochemical method-Photochemical radiations are used in this grafting method. When a macromolecule's chromophore absorbs light during the grafting process, it transforms into an excited state, from which it might further dissociate into

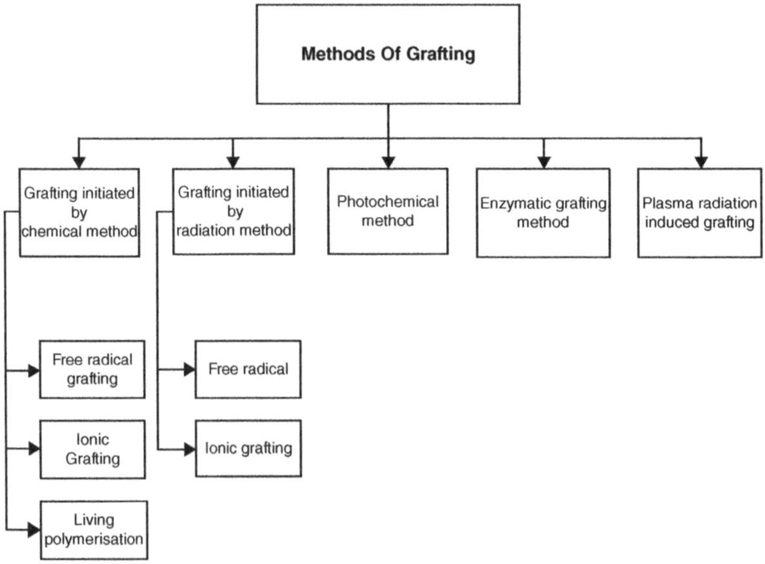

Figure 17.2 Methods of grafting.

reactive free radicals. If light absorption prevents bond breaking from resulting in the formation of free radical sites, the process can be hastened by adding photosensitizers.

4-Enzymatic grafting method-The grafting reaction is started by enzymes. Catalysts are used in uniting processes to provide a green technique by standardizing the use of responsive reagents for economic well-being and viability.

5- Plasma radiation induced grafting-Plasma polymerization is the process of activating monomers with an energetic plasma species to transform them into high-molecular-weight polymers. There are three steps to this method: When electrons and ions collide with gas monomers, atoms and free radicals develop in the initiation step. The polymeric chain is generated during the propagation stage, and the polymeric chain is closed during the termination stage. Plasma polymerization creates polymer films that are extensively branched and crosslinked.

Table 17.1 shows a few examples of grafted biopolymers employed as corrosion inhibitors [11–16].

Table 17.1 Shows a few grafted biopolymers employed in corrosion inhibition of metals. Adapted from [11–16]

Name of Polymer	Structure	Source of Origin	Grafted with Polymer	Reference
Guar Gum	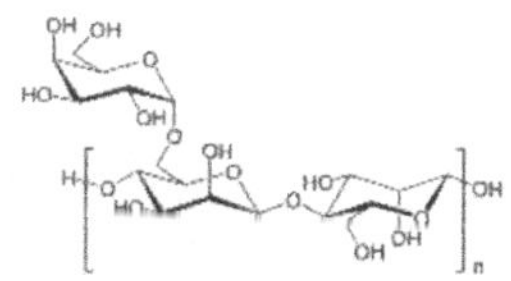	Cyamopsis tetragonolobus	Polyacrylamide	[11]
Gum Arabic	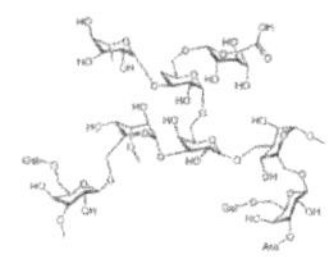	Stems and branches of Acacia tree	Polyacrylamide	[12]
Cellulose	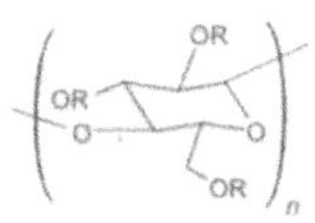	Plant tissue (tree, cotton, etc), bacteria (Acetobecter xylinum)	Polyurethane	[13]
Alginate	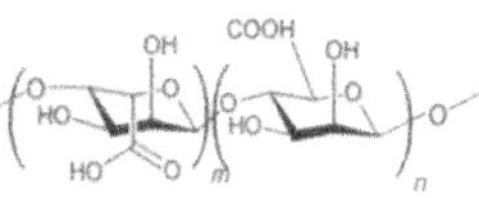	Brown algae of the genera Nacrocystis, Laminaria,	8-hydroxyquinoline	[14]
Pectin	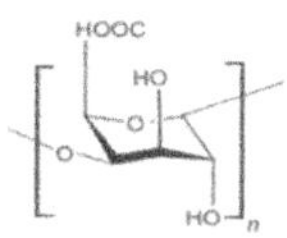	Plant cell walls, citrus peels, apple pomace	Polyacrylamide and polyacrylic acid	[15]
Lignin	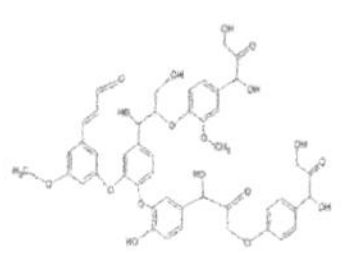	Red Algae	1,8-dihydroxyanthraquinone	[16]

17.5 Grafted Dextrin as Corrosion Inhibitor: A Literature Survey

Dextrin refers to a generic term for a variety of products that are created by heating starch with a small amount of moisture and acid. Maltodextrin, amylodextrin, α,β-dextrin, cyclic, and highly branched cyclic dextrin molecules are examples of dextrin compounds found in nature. Figure 17.3 shows the molecular structure of dextrin, which are carbohydrate with glucose (D) units linked by glycosidic linkages [1→4) or (1→6).

They are found in the human digestive tract as a hydrolysis product of starch due to amylase action. Pyroconversion is the heat treatment of a dry starch with or without the addition of acid. The three most common pyrodextrins are white dextrins, yellow dextrins, and British gums.

Dextrin and its derivatives are utilized in a variety of applications, including food, textiles, adhesives, pharmaceutical thickening, and binding agents, stabilizing agents, and more. Apart from these applications, Dextrin has long been employed as a corrosion inhibitor and coating to prevent metal from dissolving. In various corrosive situations, short-chain dextrin and its highly branched cyclic derivatives have been utilised as corrosion inhibitors. Grafted or modified dextrin and its derivatives are explored as corrosion inhibitors here.

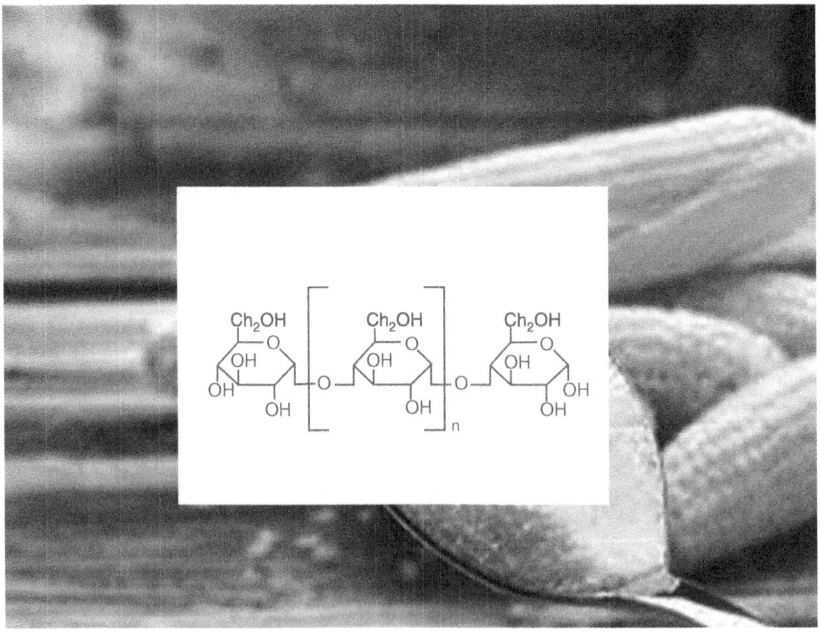

Figure 17.3 Molecular structure of dextrin.

Amrita Biswas et al. [17] used chemical, electrochemical, and surface analytical techniques to investigate the effectiveness of a graft copolymer of dextrin biopolymer, Dxt-g-pVAc (dextrin graft polyvinyl acetate), for mild steel corrosion in 15% HCl. The ATRP (atom transfer radical polymerization) approach was used to modify/graft the biopolymer. Corrosion inhibition by grafted dextrin was found to be 97.43%, while dextrin provided 82.73% at a concentration of 0.15 g/L. Both dextrin and its grafted biopolymer were mixed types in nature, according to polarisation data, and the adsorption process followed the Langmuir isotherm. Surface analytical tests confirmed the adsorption of the studied biopolymer which reduced mild steel surface damage.

Liu et al. [18] used PDP, EIS, weight loss, SEM, and EDS experiments to explore the β cyclodextrin (CD) modified chitosan as a corrosion inhibitor for Q235 carbon steel in 0.5 M HCl. Figure 17.3 shows the fluctuation in corrosion rate and corresponding inhibition efficiency at 298 K for varying concentrations of β-CDchitosan. With an increase in concentration, the corrosion rate reduces, and the efficiency changes in another manner (Figure 17.4). The improved version of cyclodextrin has a corrosion inhibition efficacy of 96.02% at 230 ppm, according to DC polarisation measurements.

The Langmuir adsorption isotherm was attributed to the adsorption of inhibitors and prevention corrosion. Thermodynamic characteristics have also been used to confirm the physisorption and chemisorption mechanisms. Because of the corrosive attack of the acid solution, the steel without inhibitor is heavily

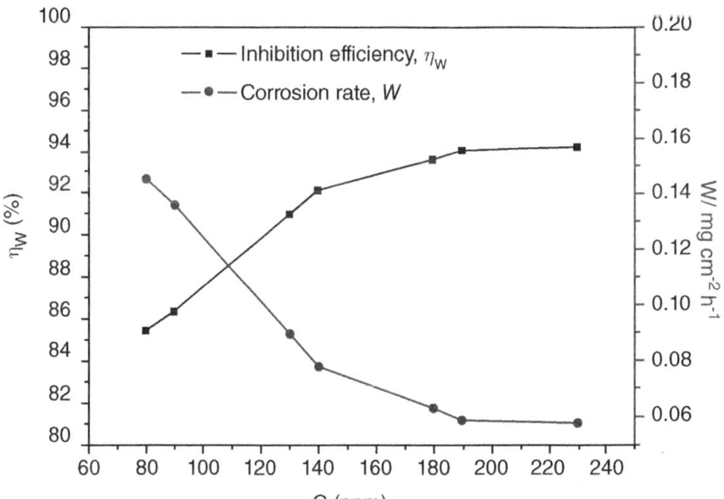

Figure 17.4 Variations of the corrosion rate calculated from weight loss measurements and corresponding inhibition efficiency at different concentrations of β-CD – chitosan in 0.5 M HCl. [18] / American Chemical Society.

corroded with deep localized pits, as expected. Even at higher magnifications, there is substantially less damage expected in the polishing lines in the picture when inhibitor molecules are present, confirming that the inhibitor molecules coated the active areas of metal steel to build a protective layer, but some portions remain uncovered. Furthermore, the proportion of Fe elements in carbon steel dipped in 0.5 M HCl in the absence and presence of polymer is 86.09% and 91.71%, respectively (Figure 17.5). The Fe peaks are greatly reduced in contrast to the unrestrained mild steel sample. The metal dissolution and hydrogen evolution reactions are hindered by the inhibitor coating on the mild steel surface, resulting in the suppression of Fe lines.

The inhibitory performance of dextrin grafted with caprolactam (Dxt-g-CPL) on J55 steel in 1 M HCl was recently discussed by Liu et al. [19]. The chemical synthesis approach was used to graft caprolactam with dextrin, which was validated by FTIR data. Figure 17.6 depicts the corrosion rate and inhibition efficiency of

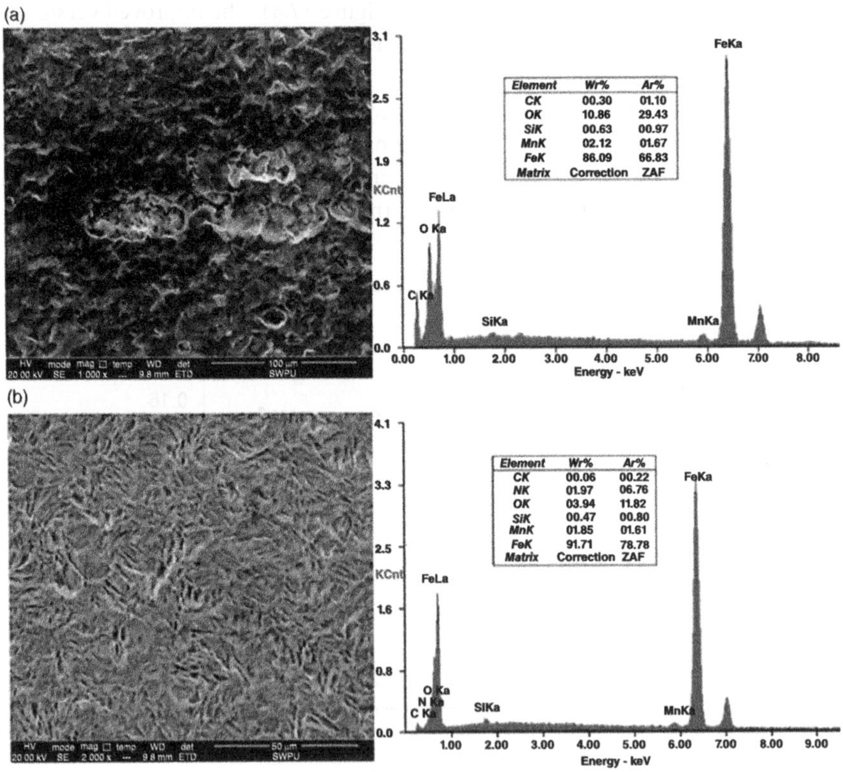

Figure 17.5 SEM images and EDS spectra of Q235 carbon steel exposed to 0.5 M HCl solution in the absence (a) and presence (b) of β-CD – chitosan. [18] / from American Chemical Society.

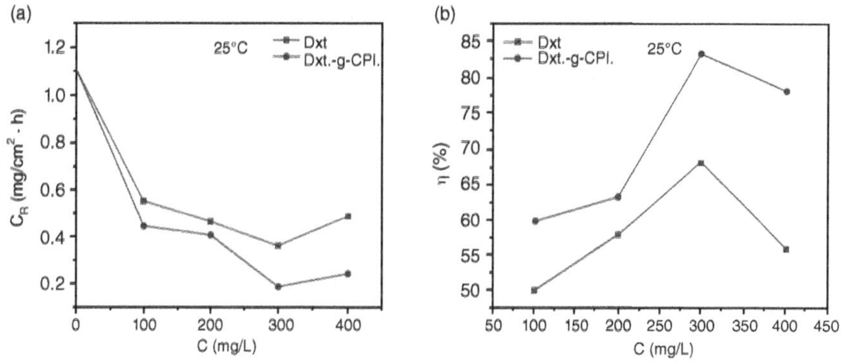

Figure 17.6 CR and η of J55 steel were added with different concentrations of corrosion inhibitor in 1 M HCl solution. (A) Dxt; (B) Dxt-g-CPL. [19] / MDPI / Licensed under CC BY 4.0.

J55 steel in 1 M HCl solution with various concentrations of Dxt and Dxt-g-CPL. The figure shows that when the concentration of the additional corrosion inhibitor increases, the corrosion rate decreases, and the corrosion inhibition efficiency increases at room temperature. The corrosion inhibition efficiency of Dxt and Dxt-g-CPL achieves the maximum level now, 67.30% and 83.27%, respectively when the additional corrosion inhibitor concentration reaches 300 mg/L. The results showed that the modified form of dextrin containing caprolactam inhibits corrosion more effectively than dextrin alone. The Langmuir isotherm is confirmed by the Tafel polarisation curve of grafted dextrin. When compared to the metal without the corrosion inhibitor, the morphology of the metal surface with Dxt and Dxt-g-CPL was dramatically enhanced, suggesting that the graft copolymer has a greater corrosion prevention performance.

Using potentiodynamic polarization, electrochemical impedance spectroscopy, and energy dispersive spectroscopy, Zou et al. [20] investigated the inhibitory impact of β-cyclodextrin modified acrylamide polymer on X70 steel in 0.5 M H_2SO_4. The polarization results show that the grafted polymer operates as a mixed-type inhibitor, with the highest inhibition efficacy (79.3%) at 150 mg/L. As the polymer concentration increases, the adsorption extent broadens, and the polymer forms a protective layer on the metal surface, limiting mass and charge transfer processes, according to the EIS data. Chemisorption was primarily responsible for corrosion prevention.

Cao et al. [21] produced ß cyclodextrins modified with xanthan gum (ß-CD-XG) to reduce corrosion. The grafted biopolymer was utilized to inhibit L80 steel corrosion in 1 M HCl. According to the findings, ß-CD-XG is a mixed-type inhibitor that reduces anode metal dissolution primarily by producing a tight adsorption layer. The formation of the film was attributed to the chemisorption of –OH,

–COO–, –CH2–O–, and –CH2–O–CH2– groups on the surface of L80 steel, which followed the Langmuir adsorption model. At 200 mg/L β-CD-XG and 293 K, the greatest corrosion inhibition efficiency was 94.74%. Based on weight loss measures, it can be concluded that the inhibitory performance of β-CD-XG in 1 M HCl is superior to that of β-CD and XG alone.

The synthesis of β-cyclodextrins grafted with polyacrylamide (β-CD-PAM) as a green corrosion inhibitor for X80 steel in 1 M H_2SO_4 was reported by Wang et al. [22]. EIS, EDX, SEM, weight loss, potentiodynamic polarisation, and FTIR are used to assess inhibitory performance. The polarization results revealed that ß-CD-PAM is a mixed type of inhibitor that is predominantly cathodic and that their adsorption followed a Langmuir isotherm with an IE of 92.48% at 200 mg/L. Adsorption increases with increasing the concentration of grafted cyclodextrin as more amide and hydroxyl groups increase, increasing surface coverage of inhibitor molecules and thus increasing inhibition. SEM measurements also revealed the creation of a dense layer on the inhibited steel surface, demonstrating that the modified polymer has been adsorbed on the steel surface. According to EDX analysis, as the grafted polymer was adsorbed, the content of Fe atoms on the metal surface increased dramatically (86.91%) while the content of O atoms on the metal surface decreased dramatically (only 1.31%), indicating that virtually no corrosion occurred on the metal surface, which could have been induced by the adsorption of β-CD-PAM.

Tang et al. [23] reported the synthesis of a novel scale and corrosion inhibitor namely β-MEA. This was synthesized from β-cyclodextrin-3-chloro-2-methylpropene, epoxysuccinic acid (ESA), and 2-acrylamido-2-methyl propane sulfonic acid (AMPS). The structure was confirmed by FTIR, mass spectrometry, TGA, and gel permeation chromatography. β-MEA has a greater scale inhibition efficiency (100%) for BaSO4 than poly(aspartic acid) (94.9%) at a concentration of 20 mg/L. A higher corrosion inhibition effectiveness i.e., 91.2% for N80 carbon steel in saline water at a concentration of 1 g/L was also observed. Both physisorption and chemisorption mechanisms have been demonstrated using thermodynamic characteristics.

In saline water, Fu et al. [24] studied the inhibitory efficiency of polyaspartic acid grafted β-cyclodextrin (PASP/β-CD) on N80 carbon steel. Figure 17.7 depicts the PASP/β-CD chemical production method. Weight loss measurements, electrochemical procedures, atomic absorption spectrophotometry (AAS), surface examination (SEM and EDS), and analyses were used by the authors to analyze corrosion and inhibition.

According to the data, grafted CD has a corrosion inhibition efficacy of 68.4% at 800 mg/L. The scale inhibition efficacy of the synthesized inhibitor against the $CaCO_3$ and $CaSO_4$ scale was also reported by the authors (Figure 17.8).

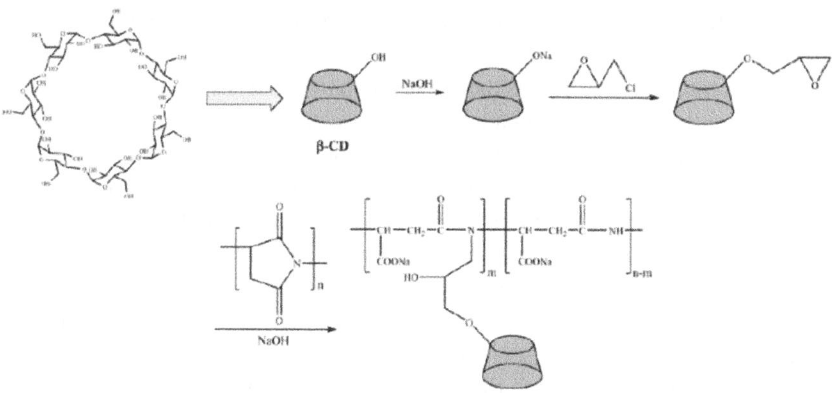

Figure 17.7 The synthesis of PASP/ ß-CD. [24] / With permission of Elsevier.

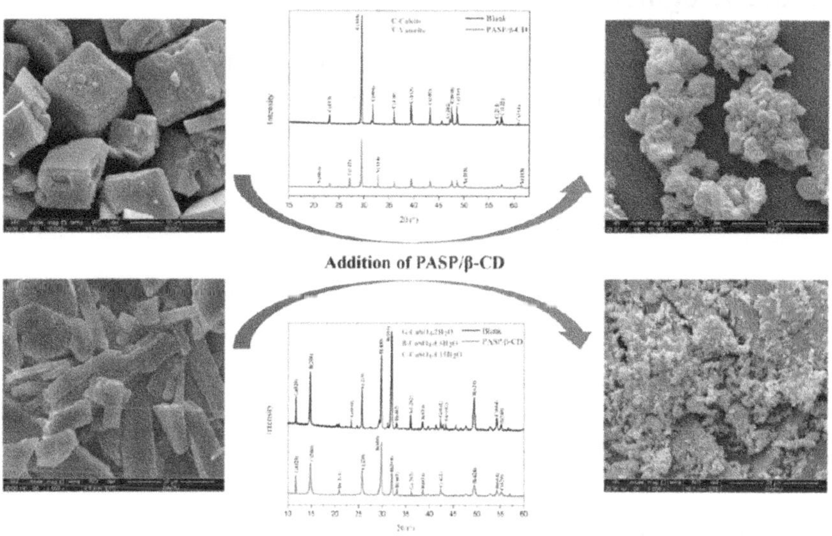

Figure 17.8 Analysis of scale inhibition mechanism of PASP/β-CD. [24] / from ELSEVIER.

17.6 Conclusion and Future Scope

Analysis of past research has revealed that minor changes to the structures of carbohydrate polymers can result in extremely high inhibitory efficacy in an aggressive environment with high mineral acid molarity. The only slight disadvantage of these carbohydrate polymers as corrosion inhibitors is that some of them are partially soluble and have lower stability when stored for days, i.e., they may not remain functional in extremely aggressive solutions of 1 or 2 M mineral acids after a defined duration.

Another interesting feature is the use of computer analysis to thoroughly investigate the chemical routes and adsorption processes of these carbohydrate polymers when used as inhibitors. A single molecule may be assigned to more than one type of functionality. To determine all possible bonding locations and the energy involved in the adsorption process, theoretical computations are required. Not only are bonding locations and bonding energies significant in corrosion research, but the spatial orientation of carbohydrate polymer molecules on metal surfaces is also crucial. It's worth noting that these carbohydrate polymers feature long molecular chains and high molecular weights, which researchers have often cited as a contributing factor to their efficacy as corrosion inhibitors.

Acknowledgments

Ruby Aslam acknowledges the Council of Scientific & Industrial Research, New Delhi, India for providing financial aid under the Research Associate fellowship (09/112(0616)2K19 EMR-I).

Conflict of Interest Statement

The authors declare that there is no conflict of interest in any way.

Competing Interest Statement

The authors declare that there is no competing interest in any way.

Author's Contributions

All authors collectedly contributed to the design and write-up of the book chapter.

References

1 Revie, R.W. and Uhlig, H.H. (2008). *Corrosion and Corrosion Control*. Hoboken, NJ: John Wiley and Sons. McCafferty, E. (2010). *Introduction to Corrosion Science*. New York: Springer.

2 Bhattacharya, A., Rawlins, J.W., and Ray, P. (2009). *Polymer Grafting and Crosslinking*. Hoboken, New Jersey: John Wiley & Sons, Inc.

3 Abdel Hameed, R.S. and Abdallah, M. (2018). Corrosion Inhibition of Carbon Steel in 1M Hydrochloric Acid Using Some Pyrazolo[3,4-d]Pyrimidinone Derivatives. *Protection of Metals and Physical Chemistry of Surfaces* 54: 113–121.
4 Abdel Hameed, R.S. and Abdallah, M. (2018). *Protection of Metals and Physical Chemistry of Surfaces* 54: 113.
5 Kuznetsova, E. and Remezkova, L. (2001). *Protection of Metals and Physical Chemistry of Surfaces* 37: 262.
6 Freiman, L. and Kasatkina, M. (2000). *Protection of Metals and Physical Chemistry of Surfaces* 36: 85.
7 Sanati, A., Raeissi, K., and Edris, H. (2017). *Protection of Metals and Physical Chemistry of Surfaces* 53: 902.
8 Mohammadnejad, M., Habibolahzadeh, A., and Yousefpour, M. (2016). *Protection of Metals and Physical Chemistry of Surfaces* 52: 100.
9 Zehra, S., Aslam, R., and Mobin, M. (2021). Corrosion Inhibitors: an Introduction. In: *Environmentally Sustainable Corrosion Inhibitors. Fundamentals and Industrial Applications* (ed. C.M. Hussain, C. Verma, and J. Aslam), 47–66. Elsevier. ISBN Number- 978-0-323-85405-4.
10 Aslam, R., Mobin, M., Zehra, S. et al. (2017). N,N'-Dialkylcystinegemini and monomeric N-alkyl cysteine surfactants as corrosion inhibitors on mild steel corrosion in 1M HCl solution – a comparative study. *ACS omega* 2: 5691.
11 Roy, P., Karfa, P., Adhikari, U., and Sukul, D. (2014). Corrosion inhibition of mild steel in acidic medium by polyacrylamide grafted Guar gum with various grafting percentage: effect of intramolecular synergism. *Corrosion Science* 88: 246–253.
12 Biswas, A., Mourya, P., Mondal, D. et al. (2018). Grafting effect of gum acacia on mild steel corrosion in acidic medium: gravimetric and electrochemical study. *Journal of Molecular Liquids* 251: 470–479.
13 Farhadian, A., Kashani, S.A., Rahimi, A. et al. (2021). Modified hydroxyethyl cellulose as a highly efficient eco-friendly inhibitor for suppression of mild steel corrosion in a 15% HCl solution at elevated temperatures. *Journal of Molecular Liquids* 338: 116607.
14 Fardioui, M., Rbaa, M., Benhiba, F. et al. (2021). Bio-active corrosion inhibitor based on 8-hydroxyquinoline-grafted-Alginate: experimental and computational approaches. *Journal of Molecular Liquids* 323: 114615.
15 Geethanjali, R., Fathima Sabirneeza, A.A., and Subhashini, S. (2014). Water-Soluble and Biodegradable Pectin-Grafted Polyacrylamide and Pectin-Grafted Polyacrylic Acid: electrochemical Investigation of Corrosion-Inhibition Behaviour on Mild Steel in 3.5% NaCl Media. *Indian Journal of Materials Science* 2014.
16 Hussin, M.H., Rahim, A.A., Ibrahim, M.N.M., and Brosse, N. (2015). Improved corrosion inhibition of mild steel by chemically modified lignin polymers from Elaeis guineensis agricultural waste. *Materials Chemistry and Physics* 163: 201–212.

17 Biswas, A., Das, D., Lgaz, H. et al. (2019). Biopolymer dextrin and poly (vinyl acetate) based graft copolymer as an efficient corrosion inhibitor for mild steel in hydrochloric acid: electrochemical, surface morphological and theoretical studies. *Journal of Molecular Liquids* 275: 867–878.

18 Liu, Y., Zou, C., Yan, X. et al. (2015). β-Cyclodextrin Modified Natural Chitosan as a Green Inhibitor for Carbon Steel in Acid Solutions. *Industrial & Engineering Chemistry Research* 54: 5664–5672.

19 Liu, M., Xia, D., Singh, A., and Lin, Y. (2021). Analysis of the Anti-Corrosion Performance of Dextrin and Its Graft Copolymer on J55 Steel in Acid Solution. *Processes* 9: 164.

20 Zou, C., Yan, X., Qin, Y. et al. (2014). Inhibiting evaluation of β-Cyclodextrin-modified acrylamide polymer on Alloy steel in sulfuric solution. *Corrosion Science*. http://dx.doi.org/10.1016/j.corsci.2014.04.046.

21 Cao, Y., Zou, C., Wang, C. et al. (2021). β-cyclodextrin modified xanthan gum as an eco-friendly corrosion inhibitor for L80 steel in 1 M HCl. *Cellulose* 28: 11133–11152.

22 Wang, C., Zou, C., and Cao, Y. (2021). Electrochemical and isothermal adsorption studies on corrosion inhibition performance of β-cyclodextrin grafted polyacrylamide for X80 steel in oil and gas production. *Journal of Molecular Structure* 1228: 129737.

23 Tang, M., Li, J., Ye, Z. et al. (2017). Scale and Corrosion Inhibitor Modified by β-Cyclodextrin. *Australian Journal of Chemistry* 70: 933–942.

24 Fu, L., Lv, J., Zhou, L. et al. (2020). Study on corrosion and scale inhibition mechanism of polyaspartic acid grafted β-cyclodextrin. *Materials Letters* 264: 127276.

18

Grafted Biopolymer Composites and Nanocomposites as Sustainable Corrosion Inhibitors

Syed Ali Abdur Rahman[1], P. Priyadharsini[2], R. V. Deeksha[2], and J. Arun[2,]*

[1] *Department of Biotechnology, Sathyabama Institute of Science and Technology, Jeppiaar Nagar (OMR), Chennai – 600 119, Tamil Nadu, India*
[2] *Centre for waste management – International Research Centre, Sathyabama Institute of Science and Technology, Jeppiaar Nagar (OMR), Chennai – 600 119, Tamil Nadu, India*
* *Corresponding author*

18.1 Introduction

Corrosion is a serious issue, mainly petroleum-based industrial sectors are affected because of the substantial amounts of metallic assets used across the supply chain. The annual expense of corrosion and its impacts are predicted to be 3 to 5% of the global GDP for wealthy countries [1]. The disadvantages of ordinary coatings include a greater risk of fracturing and collapse due to inadequate coupling between both the coatings' components and metallic substrates [2]. Choice of materials, coverings and liners, cathodic protection, and corrosion inhibitors are some of the corrosion mitigation measures used by several industries The utilization of biopolymeric or composite coatings intended to reduce corrosion protection to acceptable standards is among the most promising approaches for overcoming the limitations. Polymer coatings must increase their adhesion, mechanical characteristics, and barrier effect to improve their corrosion protection efficacy [1, 3].

Recent research efforts have focused on identifying sustainable corrosion inhibitors to substitute harmful organic and inorganic chemicals. This need has been met by naturally occurring substances. Natural occurring compounds are eco-friendly and ecologically acceptable, in addition to being widely available, inexpensive, and a renewable source of resources [4]. Xu et al. developed a polyaspartic acid-melamine grafted copolymer and investigated its $CaCO_3$ and $Ca_3(PO_4)_2$ scale reduction and ferric oxide dispersion capabilities [5].

Biopolymer nanocomposites and composites are increasingly being studied as corrosion inhibitors for a variety of substrates in a variety of extreme conditions. Compositing is a modifying process that includes the reactions of several components to form a chemically unique material with qualities that differ from the attributes of the specific elements. The components of a nanocomposite have a diameter of less than a micron (usually <0.1 pm) [6].

Grafted Biopolymers as Corrosion Inhibitors: Safety, Sustainability, and Efficiency, First Edition.
Edited by Jeenat Aslam, Chandrabhan Verma, and Ruby Aslam.

Several biopolymers have been studied as corrosion inhibitors. Jmiai et al. investigated the effect of di-biopolymers (including chitosan and sodium alginate) on copper corrosion inhibition [7]. The dynamic polarization test revealed that both inhibitors moved the E_{corr} further toward the negative values and greatly reduced the i_{corr}. For sodium alginate and chitosan, inhibition efficiencies of 82% and 87% were found. Furthermore, theoretical research supported the experimental results. For the steel surface, Shahmoradi et al. used a mixture of cellulose and lignin-derived from the pistachio nutshell. They discovered that the inhibitor is mixed-type and behaves according to the Langmuir isotherm [8].

18.2 Corrosion – Headache to Industrial Pipelines

Corrosion is often known as rust and is an undesired natural process that results in the progressive deterioration of objects, particularly metals, by chemical or electrochemical interactions with the surrounding environment. Corrosion is seen as a severe hazard to metal equipment during its lifetime, as degradation may result in leaks, faults, and harm to either operators or the environment [9]. Depending on whether CO_2 and H_2S are contained in the hydrocarbon fluid, two forms of corrosion can occur in the oil and gas pipeline system: sour corrosion and sweet corrosion. Sour corrosion happens in materials containing H_2S beyond a partial pressure of 0.05 psi (0.34 kPa) and CO_2. Sweet corrosion happens in materials containing mainly CO_2 or a trace amount of H_2S (H_2S partial pressure of 0.05 psi). Corrosion has an impact on a material's strength, mechanical capabilities, physical features, and operational condition. At the industrial level, the interaction of components and their surrounding environment causes a variety of corrosion issues [10]. Figure 18.1 shows the mechanistic process of corrosion formation on solid surfaces.

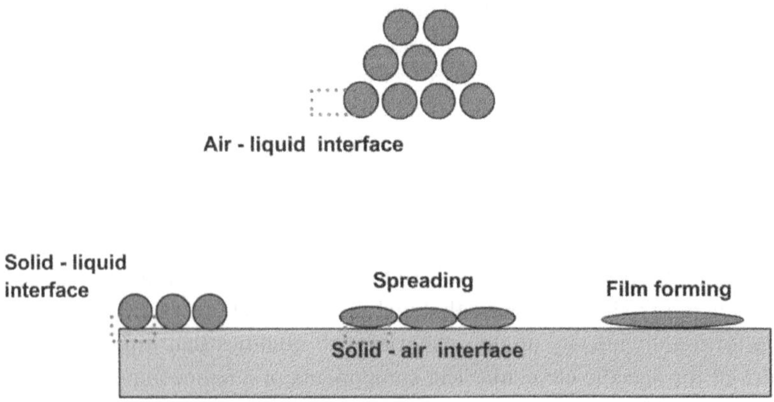

Figure 18.1 Mechanism of corrosion formation and its components.

In general, fluid transfer pipelines are classified as oil and gas (crude oil, natural gas, and products obtained from oil refinement), drinking water and sewage water (aqueduct and sewage), industrial materials (ammonia, hydrogen, carbon dioxide, and so on), are different types of fluid transfer pipelines. Water supply pipelines are pipes that deliver pressurized water from natural sources to treatment plants via open surface channels or enclosed basement tunnels and pipelines. Sewer lines, on the other hand, are return lines that collect and convey water, mostly at atmospheric pressure, to the natural environment or treatment plants. Water and sewage pipeline systems, in particular, are subject to sudden aging, degradation, and constant and cyclic strength decreases [11].

Pipelines are unavoidably subjected to many forms of corrosion, such as microbiologically caused corrosion, galvanic corrosion, and so on. The true rate of corrosion in sewage systems is difficult to quantify [12]. It necessitates a thorough understanding of the link between corrosion rate and environmental parameters such as H_2S concentration, humidity, and temperature, all of which are crucial for corrosion monitoring and mitigation. For example, moisture increases microbial corrosion, temperature changes impact sulfide oxidation, and sewage pH fluctuation affects biochemical process speeds [13]. Meanwhile, lifetime decreases, and maintenance or replacement requirements result in additional expenditures, and developing an effective approach for corrosion detection is essential [10].

18.3 Biopolymers – Production and Applications

Derived naturally from living organisms, the wide applications of biopolymers have been displayed time and again, since their first appearance in the market during the late 1980s [14]. The first generation of biopolymers that were introduced was composed of starch and polyhydroxyalkanoates (PHAs). Since then, numerous research experts have laid the foundation for the extraction of biopolymers from plants (corn, soybean), animals, agricultural wastes, and also microbes, just to name a few [15]. Biopolymers serve as a potent source of attraction due to their following desirable properties:

i) **Non-toxicity** – as these long chains of macromolecules are derived from living systems, the burning issues associated with their biodegradability and the consequential environmental hazards are eliminated [16]. They serve as novel alternatives to synthetic polymers such as plastics, which have emerged as potent sources of harmful greenhouse gas emissions in recent years.
ii) **Abundance in nature** – as research on this perspective continues to grow, various projects are aiming to synthesize biopolymers from numerous natural sources, thus providing wider scope for increasing their availability [15].

iii) **Renewability** – following up with the previous point, the abundance of biopolymers in the environment reflects on its renewability as well; a property that lacks in synthetic polymers. The growing need for synthetic polymeric materials in daily activities has resulted in scarcity and higher costs, owing to their non-renewability. Hence biopolymers step in as the cheaper and more renewable alternative to replace synthetic polymers [17].

18.3.1 Applications of Biopolymers

i) **Bone repair** – In a research paper published in 2016 [18], scientists used an approach to create a cellulose composite derived from the bacteria *Gluconacetobacterxylinus* (ATCC 700178), and fused it with degenerating bone tissue to repair the bone. Bacterial celluloses are hydrogels that are generally chosen for bone repair due to their excellent porous structure and fluid-exchange abilities, proving to be extremely biocompatible with animal bone [19].
ii) **Eco-friendly Plant Pots** – in recent years, the agriculture industry has started leaning towards the usage of biopolymers or materials derived from the same, to decrease the use of hazardous materials such as plastics at a large scale. One such application of biopolymers is the invention of biodegradable plant pots that are used to grow seedlings [20]. A major highlight of this method is that once the pot degrades, its contents would mix with the soil and provide better nutrition to the growing plant.
iii) **Drug-delivery systems** – due to their biodegradability and specific drug-delivery properties, biopolymers are widely used in delivering drugs to specific targets in a living system. Chitosan, a cationic polysaccharide derived from chitin, has been specifically highlighted for the same, due to its added beneficial medicinal properties, such as being bactericidal and fungicidal [21]. Some plant extracts such as Acacia are used as emulsifying agents and for micro-encapsulation in drug delivery systems [22]. Tragacanth gum, another plant extract isolated from *Astragalus gummifer*, a woody shrub, is used in drug manufacturing as a suspending agent as it helps increase the viscosity of the preparation.
iv) **Paper industry** – Guar gum, extracted from the plant *Cyamopsis tetragonolobus*, is used for the reduction of wood-resin depositions from paper, during manufacturing [23]. This increases the strength of the paper and has been employed by many paper manufacturers [24].
v) **Food packaging** – it goes without saying that the need for finding alternate solutions to the abundant usage of plastics for food wrapping and storage, has risen tremendously over the years. Biopolymers serve an excellent purpose by

providing a much-needed alternative to plastic wrappers. Researchers working on bio-polymeric containers focus on their mechanical stability and moisture-barrier properties [25]. Biopolymers based on starch, cellulose, and chitin have been utilized for creating food containers such as cups, plates, and cutlery [26].

vi) **Waste-treatment** – biopolymers such as Guar gum and Xanthan gum have been documented for their use in efficiently disposing of glass-wastes generated from discarded computers and television sets. These gums encapsulate the wastes, thus decreasing the leakage of toxic lead traces from them [27]. Lead extracts are known to have hazardous impacts on the environment. Additionally, if a certain biopolymer material has reached the end of its lifespan, it can be used as compost along with other kitchen waste, thus serving its purpose even beyond its usability [28].

vii) **Textile industry** – various "green-chemistry" approaches have been proposed to create *antimicrobial* textiles to enhance the quality of textile fibers. Biopolymers such as cyclodextrin, chitosan, and sericin have been reviewed to play roles as binding, levelling, and sizing agents in textile manufacturing [29, 30].

18.3.2 Production of Biopolymers

The method of production of biopolymers depends largely on the source of extraction; for example, microbial, plant-based, or animal-based. The broad methods of production based on the source are listed as follows:

i) **From seafood-wastes** – chitin is a major food waste that comes from sea animals such as crustaceans, crabs, shrimps, and prawns that are consumed as food. Chitin is an excellent biopolymer that can be extracted in two ways: enzyme-mediated and fermentation [31]. The former method makes use of several proteases that deproteinize the crustacean shell, thus eluting the chitin protein. Isolation of chitin from shrimp and prawn shells generally involves the usage of chemicals such as HCl and NaOH in correct proportions, during deproteinization and demineralization [32]. Similarly, collagen is extracted from disposed eggs, chicken meat, and fish scales using acid hydrolysis.

ii) **From bacteria** – bacterial systems are known to primarily produce four types of biopolymers: namely, polysaccharides, polyamides, polyesters, and polyphosphates. Bacteria can produce these biopolymers through the usage of various proteins and enzymes, and by specific gene regulation at the transcriptional level. Large-scale production of these biopolymers involves fermentation of these bacterial cultures at an industrial level, catering to all the physical, chemical, and nutritional requirements of the culture. The main precursor for biopolymer production from a bacterium is carbon [33]. Once carbon is metabolized into the system, depending on the metabolic pathway, various molecules help produce the particular biopolymer, such as alginate, glycogen, dextran, polyphosphate, etc. The well-known bacteria that are currently being used for the commercial production of biopolymers include *Ralstoniaeutropha, Pseudomonas putida, Alcaligenes latus, E. coli*, and *Bacillus subtilis* [34].

iii) **From fungi** – All biopolymers are synthesized by enzymatic processes in the cytoplasm. At an industrial level, the production of chitin and chitosan from fungi involves three steps [35]:

a) Alkaline treatment to remove protein and alkali-soluble polysaccharides;
b) Acid reflux to separate chitin and chitosan;
c) Precipitation of chitosan under alkaline conditions.

The use of the fungal genus *Mucor* for biopolymer production using glycerol as a carbon source has been recorded. The use of glycerol as a substrate was shown to be very promising, especially when using a fungal source [36].

iv) **From plants** – modern biotechnological techniques of biopolymer production from plants involve recombinant microbes in fermentation, using carbohydrates from plants. Transgenic plants are the current focus for producing better biopolymer materials. A major drawback with plant sources is the concern with their seasonal production. Hence crops that can grow throughout the year such as sugarcane are preferred over seasonal crops. Also, non-food crops such as tobacco and rubber are preferred over consumable crops in most parts of the world [37].

Several other living systems are currently being tested repeatedly, to improve biopolymer production. Better advanced biotechnological tools and approaches are being invented to further enhance this field. Only the main approaches to production have been highlighted above.

18.4 Nanocomposites – Production and Applications

A nanocomposite is a combination of two or more substances, out of which at least one is a nanomaterial. They comprise several phases, the significant one being the continuous phase called a "matrix," onto which other phases are loaded/mixed (reinforcement). The composition of the matrix material varies based on the application of the nanocomposite, the most popular being polymer matrices [38]. Other matrices include those made from ceramics and metals. The reinforcement phase may be made from fibers and other layered materials. Nanocomposites have gained recognition for their heightened qualities of tensile and mechanical strength when compared with traditional polymer composites. The nanocomposites exhibit a high surface-to-volume ratio for the loading of biomolecules [39]. They also have higher heat-distortion temperatures and noise dampening. They overpower traditional composites while assessing their properties such as increased toughness, higher optical clarity, and lower melt viscosity [38]. Table 18.1 shows the advantages and disadvantages of bio-based renewable sources.

Nanocomposites can be prepared from non-renewable sources such as petroleum products, and renewable sources such as chitin, lignin, hyaluronic acid, and natural rubber [40], among others. The various techniques involved in producing nanocomposites from renewable sources may be elaborated as follows:

i) **In-situ polymerization** – in general, polymerization is the process of chemically combining small building blocks called "monomers," to create a large chain of subunits, called a "polymer. Concerning nanocomposite production using renewable sources, the monomers that are polymerized, are bio-renewable. A certain additive is used during the reaction that aid in giving the novel properties of the nanocomposite. Examples of additives include cellulose, starch, and lignin.
ii) **Extrusion** – this method is industrially the most popular method used for nanocomposite production. Briefly, the technique involves molding the raw material into the desired shape by melting it through a melting channel,

Table 18.1 Summarized table of various bio-renewable sources along with their advantages and disadvantages. Adapted from [40].

Bio-renewable source	Advantages	Disadvantages
Chitin	Abundant in nature, renewable, non-toxic	Lacks mechanical stability
Lignin	High carbon content, low cost	Complex structure gives complications during production
Hyaluronic acid	Highly viscoelastic, renewable, biocompatible	May give rise to unwanted biological reactions
Starch	Combustible, non-abrasive, renewable	Poor mechanical stability
Collagen	Easily available, non-antigenic, renewable, non-toxic	Complex properties, expensive, quick release
Chitosan	Renewable, antibacterial properties, low toxicity	Low porosity
Alginate	Abundance, easy gelation process, renewable	Lacks mechanical stability

which has the same shape as desired for its final outlook. This method is popularly used for creating thermoplastic composites [40].

iii) **Melt-mixing** – this method is closely related to the extrusion method and thus is also popularly used for general composite production. This technique involves melting the raw material at a very high temperature and then adding certain additives into it (which are renewable), to change the properties of that raw material while it is still in its molten form. A certain study [41] revealed that this technique yields better results in fiber composition, compared to extrusion.

iv) **Injection-molding** – in this method, the raw material is initially chopped into very small pieces using a sharp pair of scissors, and these pieces are fed into an injection-molding unit at temperatures as high as 300 °C, at high pressure for a few minutes. Certain reinforcing elements are also added to the melt, such as felt and fabric [40]. Once the desired shape is obtained, the mold is allowed to cool and solidify and later removed [42].

v) **Other methods** – apart from the listed techniques, other methods incorporated for nanocomposite production include electro-spinning, nanofiber production, and centrifugal casting, which are especially used for production from renewable sources.

18.4.1 Applications of Nanocomposites in Biotechnology

i) **Drug-delivery systems** – nanostructures based on antimicrobial peptides or AMPs for use in drug delivery [43], are the hot topic of the pharmaceutical community, as an increasing number of drugs are rendered useless due to microbial resistance. Peptide molecules such as chitin exhibit certain antimicrobial activities and are actively used in selective drug delivery [21]. Such techniques help increase the bioavailability of the drug. Double-layered hydroxide nanocomposites for controlled drug release have also been presented as a cutting-edge methodology. Double-layered hydroxide polylactic acid has been used for the release of anti-inflammatory drugs into a system [40].

ii) **Tissue-regeneration** – in recent years, nanocomposite materials such as carbon nanotubes, TiO_2 nanoneedles, and Cu-Zn alloys [44] have gained a lot of recognition in tissue engineering applications. Such materials play important roles in creating substitutions of grafts which may otherwise face problems with host rejection. Additionally, chitin-poly (butylene succinate) (PBS) with chondroitin sulfate nanoparticles have been fabricated using the ionic gelation method for engineering skin tissue.

iii) **Biosensors** – biosensors based on polyaniline [45] and graphene [46] have been reviewed for their use in environmental monitoring and food quality control. Such materials encompass unique optical, mechanical, and electrical properties that make them suitable for roles in bio-sensing. Glucose biosensors based on a combination of chitosan, gold nanoparticles, and graphene have also been developed. Graphene-based nanocomposites have been shown to have extensive biomedical applications [47].

iv) **Food protectants** – some nanocomposites based on essential oils have been used to coat food items to protect them from microbes. Some research papers showed that these protective agents could protect food from spoilage for 12 days [48].

v) **Water-purification** – the porous structure of nanocomposites makes them an excellent agent for the purification of water. Plant-derived and protein-derived nanocomposites have the highest demand in this application. Nanocomposites have a high adsorption capacity that helps remove heavy-metal contaminants from water.

18.5 Role of Composites and Nanocomposites in Corrosion Inhibition

To protect metals with polymer nanocomposite technologies such as wet chemical reduction [49], photochemical reduction [50], laser mediation [51], hot water [52], and microwave support [53] are utilized. However, if you look closely, you can divide all these methods into two. Group of technologies, namely ex situ and in situ. Out-of-habitation technology can include all methods in which nanoparticles occur outside the host matrix. Pre-developed nano-sized fillers are added to this directly host polymer matrix [54]. The advantages of this method are size, shape, density, and particle size. Also suitable for industrial-scale synthesis. The basic problem with this method, the nanoparticles are evenly scattered in the matrix, and stability is maintained. For long-term aggregation [55]. The following technique simply means an in-situ technique Development of nanofillers in polymer matrices. This is very simple and in most cases stem. Ascending nanoparticles in the matrix suppress and then maintain component aggregation spatial expanse. For this reason, this technique is widely used by researchers. The method is that unreacted components can affect the properties of the developed composite [56].

Mohamed Gamal et al. [57]; reported that they applied solution dispersion and hot ring-opening polymerization techniques to synthesize a new group of bisbenzylidene-based polybenzoxazines (PBZ; CPBZ or CHBZ). A composite material of epoxidized soybean oil (ESBO; 10 or 20 wt%) and ESBO / bentonite (nanotone; 3 or 5 wt%) with a mild corrosion-resistant coating. Their study of the chemical composition of CPBZ and CHBZ was done by Fourier transform infrared (FTIR) spectroscopy and 1H and 13C nuclear magnetic resonance spectroscopy. It was from differential scanning calorimetry (DSC), it is clear that the thermal polymerization temperature of CHBZ before the experiment is very low, 1980 °C compared to monomer 3 phenyl 3,4. Dihydro 2H benzoxazine is 2360 °C. For corrosion protection application and test coating salt spray and electrochemical methods are used to prevent corrosion. 20% by weight epoxy Ingredients mixed in polybenzoxazine epoxidized soybean oil (PBZ / ESBO). Corrosion is more severe than pure PBZ. 20 wt% ESBO and 3 wt% Nano-Clay were added for reduction. Corrosion efficiency of 1 or 2 digits compared to pure alloy (CHBZ) an order of magnitude compared to uncoated mild steel (MS). You get 98.16% protection efficiency This reveals a significant improvement in the corrosion suppression properties of composite foils. Applications of nanocomposites on corrosion inhibition were elaborated in Table 18.2.

If we talk about salt spray testing, that's where the simple comparison from it is you need the expected actual corrosion resistance level. Neat poly (CPBZ) and bare poly (CHBZ) showed charring, and the corroded area was much larger than other ESBO resin-coated films. This means that the coating exhibits severe chemical fluctuations during exposure, leading to failure.

Table 18.2 Applications of Composites and Nanocomposites in corrosion inhibition. Adapted from [58].

Compounds	Type	Application
PET/clay	Nanocomposites	Nourishment bundling applications. Particular illustrations incorporate bundling for handled meats, cheese, confectionery, cereals, and boil-in-the-bag nourishments, natural product juice, dairy items, and lager and carbonated drinks bottles.
Fe/TiC MMCs	Composites	A determination of Ultra-hand and wear-resistant components.
Polycarbonate/ SiO_2	Nanocomposites	Scraped area safe coating.
Epoxy	Composites	Outlined to demonstrate solid padding and vitality putting away impact
Shape memory polymers/SiC	Nanocomposites	Therapeutic gadgets for grasping or discharging therapeutics inside blood vessels.
Reinforced nylon	Composites	Engine compartment optimizing system
PMMA/SiO2	Nanocomposites	Application of Dental, and optical gadgets.
Polyamide 6	Composite	Great chemical resistance, weakness perseverance

but, POLY (CPBZ)/ESBO10 or 20 wt% and Poly (CHBZ)/ESBO (10) or 20% by weight. It has no corrosion dispersion only when using a crosshatch and shows better resistance. Shows excellent corrosion suppression of film. The content of resin ESBO is increasing among the coatings that provide corrosion resistance, the reason for this is PBZ/ESBO epoxy composite material that is highly crosslinked during curing. Film barrier properties were strengthened by reducing the permeability and increasing the length of the through the path of O_2 and the water.

18.6 Green Chemistry in Corrosion Inhibition

Green corrosion inhibitors are biodegradable and contain a negligible amount of heavy metals or other toxic compounds. Some research group reports the successful use of natural substances generated to suppress metal corrosion in acidic and alkaline environments. Figure 18.2 shows the role of ceramic coatings on corrosion inhibition. Royal poinciana extract suppressed the corrosion of aluminum in a hydrochloric acid solution [59]. Rosemary leaves were examined as a corrosive inhibitor of alloy Al and 2.5 Mg in a 3% NaCl solution at 25 °C [60], and El Etre studied and studied natural honey as a copper corrosion inhibitor [61]. Aluminum parrot extract [62]. The suppression effect of Khillah (*Ammi Visnaga*) seed extract corrosion of SX 316 steel in HCl solution was determined by weight loss measurements and potentiticstechnology. The mechanism of action is attributable to the formation of the insoluble complex by an interaction between the iron cation and khellin [63] and levels. Indicated inhibition of corrosion with ethanol extract Overview of African Bush Pepper (*Piper-Guinsens*) on Steel; Karika Papaya's

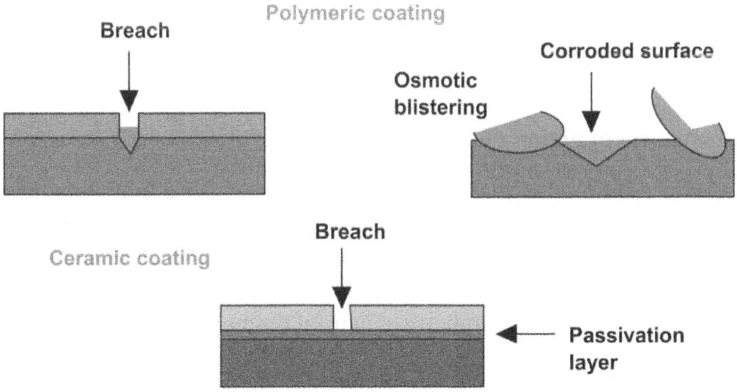

Figure 18.2 Role of ceramic coating on inhibition of corrosion.

escape [64]; Neem leaves extract. H_2SO_4 on light steel (*Azadirachta* ISMICA) [65]. Omagingy studied Papira plant extracts, *PopaniaPulcherrima, Cassia Occidentalis,* and *Datura StramoniumAzadirachta Indica, Caro Tropis Professional B, Carotropis Professional B, Papaya B* and for the corrosion suppression of *PorpoioTurquisal* Potential and all extracts except *AuhorpioTurkiale* and *AzadirachtaIndica* reduced corrosion of 1 N HCl 88% to 96% efficient steel Slightly lower efficiency with 2 N HCl. You start effectiveness to the product of hydrolysis of the protein content of these plants [66]. Yee et al. [67] have determined the inhibitory effect on organic matter compounds, namely honey and *Rosmarinus officinalis L.* Four different metals: aluminum, copper, iron, and zinc, each polarized in two different solutions (sodium chloride and sodium sulfate). The experimental approach uses the electro-potential polarization method. Best inhibitory effect in the presence of zinc polarizes honey with both sodium chloride and sodium sulfate solution.

Rosemary extract showed some cathode activity suppression when metal is polarized with sodium chloride solution. But this organic compound is less anode suppression compared to honey. The most important Rosemary's chemical composition includes borneol and bornyl acetate, camphor, cineole, camphene, and alpha-pinene. Chalchat et al. [68] reported that the oil was from rosemary 1,8-rich in cineole, camphor, and bornyl acetate, Okafor et al. examined onion extract (*Allium sativum*), Carica papaya extract, Garcinia kola, Phyllanthus Amarus [69]. ElEtre, Abdallah M used Natural Honey as a corrosion inhibitor for metals and alloys [70]. *Telfaria occidentalis, Occinumviridis, Azadirachta* were rated Trifaciata extract of Indica and Sansevieria [71]. Recent research about the use of some medicines as corrosion inhibitors reported [72]. Most of these drugs are heterocyclic compounds and are so recognized. It's eco-friendly, so it has great potential. The medicine is not so eco-friendly and contains heavy metals or other toxic compounds.

In view of this adsorption and suppression effect of ACPDQC (fifth) Amino 1 cyclopropyl 7 [(3R, 5S) 3,5 dimethylpiperazine 1YL] 6,8 (difluoro-4-oxowinolin 3 carboxylic acid), mild Steel corrosion has been studied and found to be effective.

18.6.1 Mechanism of Action of Green Inhibitors

Many theories to substantiate the mode of action of these green inhibitors have been put forth by several workers. Organic substances, which form onium ions in acidic solutions, are adsorbed on the cathodic sites of the metal surface and interfere with the cathodic reaction. Various mechanisms of action have been postulated for the corrosion inhibition property of natural products.

i) **Argemone mexicana.** It is a contaminant of mustard seeds that contain an alkaloid berberine which has a long chain of aromatic rings, an N atom in the ring, and, at several places H atoms joined to C are supplanted by bunches,

-CH, $-OCH_3$, and -O. The free electrons on the O and N form bonds with the electrons on the metal surface. Berberine in water ionizes to discharge a proton, hence the presently negatively charged O iota makes a difference to free an electron on the N atom and shapes a more grounded bond with the metallic electrons. These properties bestow great restraint properties to *Argemone mexicana*.

ii) **Garlic.** It contains allyl propyl disulfide. Likely, this S-containing unsaturated compound influences the potential cathodic preparation of steel.
iii) **Carrot.** It contains pyrrolidine in watery media, pyrrolidine ionizes, and the N molecule procures a negative charge, and the free electrons on N have still higher charge, resulting in the more grounded bond arrangement at N Carrot does not ionize in acidic media and hence does not ensure in acids.
iv) **Castor Seed.** They contain the alkaloid ricinine. The N atom is within the ring connection of the OCH_3.
v) **Dark Pepper.** Quraishi et al. [73] examined erosion restraint of mellow steel in a hydrochloric arrangement by dark pepper extract (Flute player nigrum family: Piperaceae) by mass misfortune estimations, potentiodynamic polarization, and electrochemical impedance spectroscopy (EIS). Dark pepper extricates gave maximum restraint productivity (98%) at 120 ppm. at 35 °C for mellow steel in a hydrochloric corrosive medium. The electrochemical evaluation uncovered it to be a mixed-type inhibitor and that charge exchange controls the erosion preparation. The corrosion inhibition property was credited to an alkaloid "Piperine."

18.7 Environmental Safety Aspects on Industrial Outlooks

The nanotechnology insurgency is making a ground-breaking impact on assorted science, designing, and commercial segments, counting the development industry. The physical and chemical properties interesting to the nanoscale can lead to exceptional viability improvement in (photo)catalysis, (warm and electrical) conductivity, mechanical quality, and optical affectability, empowering applications such as catalysts, electronic and vitality storage devices, progressed mechanical materials, and sensors. Tailing after rising nanotechnology applications in biomedical and electronic industries, the development industry has of late begun looking for out a way to progress customary development materials using an assortment of fabricated nanomaterials. Different MNMs can progress crucial characteristics of development materials such as quality, toughness, and lightness, and invest valuable property to screen development security and basic health. Despite the current, moderately tall fetched of nano-enabled items, their utilization in construction materials is likely to increase because of (1) exceedingly important properties imparted at generally added substance ratios, (2) fast improvement of modern applications harnessing of a nanoscale property, and (3) diminishing taken a toll of base nanomaterials as they are delivered in bigger quantities. The benefits of consolidating MNMs in construction materials can be balanced by concerns almost their potential to behave as destructive natural contaminants after their accidental or coincidental release. This underscores the requirement for proactive risk assessment and administrative rules to guarantee the secure utilization and transfer of products containing MNMs (e.g., heat-insulating, self-cleaning, and antifogging) and work as key sensing components to screen development security and basic health.

18.8 Conclusion and Future Directions

Corrosion is still a major issue for the industrial sector, ranging from oil and gas to wastewater. The future appears to be primarily focused on the development of various corrosion prevention strategies [9]. Current anti-corrosion surfaces confront various challenges, such as significant expense, weak super hydrophobicity qualities, anti-fouling, and a lack of endurance. In the effort against corrosion, graphene has shown promising results [74]. Fiber-reinforced polymer (FRP) composites are another intriguing material that has recently been employed as an anti-corrosion layer in sewage treatment facilities. They have good anti-corrosion capabilities, although their long-term behavior in this usage is still being studied [75].

Because of its correlation between efficiency, ecology, economics, and environmental friendliness, organic compounds are now recognized as one of the most effective and efficient techniques for corrosion prevention. Organic corrosion inhibitors are utilized in a variety of applications in various industries [76]. However, there are a number of drawbacks to employing these substances. The restricted solubility of organic corrosion inhibitors, particularly in polar electrolytes, is one of the most difficult aspects of employing them. Organic corrosion inhibitors, particularly those containing functional groups and non-polar hydrocarbon chains, have restricted solubility due to their hydrophobic nature, which reduces their protective efficacy [77].

Corrosion modeling, which includes semi-empirical, empirical, and principal techniques, has been shown to be a viable tool for combining atmospheric, susceptible, aspect, rate of corrosion, structural impacts modeling, and presenting integrated model findings in corrosion prediction [78]. Extensive research has been conducted to anticipate and prevent corrosion in sewage using a statistical model [9].

Acknowledgment

The author wishes to thank Sathyabama Institute of Science and Technology for their support.

Conflict of Interest

None

References

1 Umoren, S.A. and Solomon, M.M. (2021). Biopolymer composites and nanocomposites for corrosion protection of industrial metal substrates. In: *Corrosion Protection of Metals and Alloys Using Graphene and Biopolymer Based Nanocomposites*, (ed. H.M.A. Amin and A. Galal) 16–31. CRC Press.
2 Smith, L. (1999). Control of corrosion in oil and gas production tubing. *British Corrosion Journal* 34: 247–253.
3 Umoren, S.A. and Solomon, M.M. (2017). Application of polymer composites and nanocomposites as corrosion inhibitors. *Corrosion Inhibitors, Principles, Mechanisms, and Applications* 27–58.
4 Geethanjali, R., Sabirneeza, A., and Subhashini, S. (2014). Water-soluble and biodegradable pectin-grafted polyacrylamide and pectin-grafted polyacrylic acid: electrochemical investigation of corrosion-inhibition behavior on mild steel in 3.5% NaCl media. *Indian Journal of Materials Science* 2014.
5 Xu, Y., Zhao, L., Wang, L. et al. (2012). Synthesis of polyaspartic acid–melamine grafted copolymer and evaluation of its scale inhibition performance and dispersion capacity for ferric oxide. *Desalination* 286: 285–289.
6 Umoren, S.A. and Eduok, U.M. (2016). Application of carbohydrate polymers as corrosion inhibitors for metal substrates in different media: a review. *CarbohydrPolym* 140: 314–341.
7 Jmiai, A., El Brahimi, B., Tara, A. et al. (2020). The effect of the two biopolymers "sodium alginate and chitosan" on the inhibition of copper corrosion in 1 M hydrochloric acid. *Materials Today: Proceedings* 22: 12–15.
8 Shahini, M.H., Ramezanzadeh, B., and Mohammadloo, H.E. (2021). Recent advances in biopolymers/carbohydrate polymers as effective corrosion inhibitive macro-molecules: a review study from experimental and theoretical views. *Journal of Molecular Liquids* 325: 115110.
9 Foorginezhad, S., Mohseni-Dargah, M., Firoozirad, K. et al. (2021). Recent advances in sensing and assessment of corrosion in sewage pipelines. *Process Safety and Environmental Protection* 147: 192–213.

10 Ahmad, Z. (2006). *Principles of Corrosion Engineering and Corrosion Control*. Elsevier.

11 Glisic, B. (2014). Sensing solutions for assessing and monitoring pipeline systems. In: *Sensor Technologies for Civil Infrastructures*, (ed. M. L.Wang and J. P. Lynchand H. Sohn) 422–460. Elsevier.

12 Cosham, A. and Hopkins, P. (2004). The assessment of corrosion in pipelines–Guidance in the pipeline defect assessment manual (PDAM). In: *Pipeline Pigging and Integrity Management Conference*, 17–18. The Netherlands: Amsterdam.

13 Sharma, K., Ganigue, R., and Yuan, Z. (2013). pH dynamics in sewers and its modeling. *Water Research* 47: 6086–6096.

14 Jayan, J.S., Deeraj, B.D.S., Saritha, A., and Joseph, K. (2021). Biopolymer-derived carbonaceous composites and their potential applications. In: *Hybrid Natural Fiber Composites*, (ed. A. Khan, S.M. Rangappa, S. Siengchin, M. Jawaid, A.M. Asiri) et al, 253–280. Elsevier.

15 Dassanayake, R.S., Acharya, S., and Abidi, N. (2018). Biopolymer-based materials from polysaccharides: properties, processing, characterization and sorption applications. *Advanced Sorption Process Applications*, (ed. Serpil Edebali) 1–24.

16 Kaplan, D.L. (1998). Introduction to biopolymers from renewable resources. In: *Biopolymers from Renewable Resources*, (ed. David L. Kaplan) 1–29. Springer.

17 Gowthaman, N.S.K., Lim, H.N., Sreeraj, T.R. et al. (2021). Advantages of biopolymers over synthetic polymers: social, economic, and environmental aspects. In: *Biopolymers and Their Industrial Applications*, (ed. S. Thomas S. Gopi and A. Amalraj) 351–372. Elsevier.

18 Hu, Y., Zhu, Y., Zhou, X. et al. (2016). Bioabsorbable cellulose composites prepared by an improved mineral-binding process for bone defect repair. *Journal of Materials Chemistry B* 4: 1235–1246.

19 Hu, Y., Catchmark, J.M., Zhu, Y. et al. (2014). Engineering of porous bacterial cellulose toward human fibroblast ingrowth for tissue engineering. *Journal of Materials Research* 29: 2682–2693.

20 Râpă, M., Popa, M.E., Cinelli, P. et al. (2011). Biodegradable alternative to plastics for agriculture application. *Romanian Biotechnological Letters* 16: 59–64.

21 Saikia, C., Gogoi, P., and Maji, T.K. (2015). Chitosan: a promising biopolymer in drug delivery applications. *Journal of Molecular and Genetic Medicine S.* 4: 899–910.

22 Adikwu, M.U., and Esimone, C.O., Biopolymers in drug delivery: recent advances and challenges, (2010).

23 McLean, D., Agarwal, V., Stack, K.R. et al. (2011). Synthesis of guar gum-graft-poly (acrylamide-co-diallyl dimethylammonium chloride) and its application in the pulp and paper industry. *BioResources* 6: 4168–4180.

24 Chudzikowski, R.J. (1971). Guar gum and its applications. *Journal of the Society of Cosmetic Chemists* 22: 43.

25 Tang, X.Z., Kumar, P., Alavi, S., and Sandeep, K.P. (2012). Recent advances in biopolymers and biopolymer-based nanocomposites for food packaging materials. *Critical Reviews in Food Science and Nutrition* 52: 426–442.

26 Grujić, R., Vujadinović, D., and Savanović, D. (2017). Biopolymers as food packaging materials. *Advances in Applications of Industrial Biomaterials*, (ed. E. Pellicer, D. Nikolic, J. Sort, M. Baró, F. Zivic, N. Grujovic, R. Grujic, S. Pelemis) 139–160.

27 Kim, D., Quinlan, M., and Yen, T.F. (2009). Encapsulation of lead from hazardous CRT glass wastes using biopolymer cross-linked concrete systems. *Waste Management* 29: 321–328.

28 Wojnowska-Baryła, I., Kulikowska, D., and Bernat, K. (2020). Effect of bio-based products on waste management. *Sustainability* 12: 2088.

29 Shahid, M. and Mohammad, F. (2013). Green chemistry approaches to develop antimicrobial textiles based on sustainable biopolymers–a review. *Industrial & Engineering Chemistry Research* 52: 5245–5260.

30 Jahandideh, A., Ashkani, M., and Moini, N. (2021). Biopolymers in textile industries. In: *Biopolymers and Their Industrial Applications*, (ed. S. Thomas, S. Gopi, A. Amalraj) 193–218. Elsevier.

31 Ranganathan, S., Dutta, S., Moses, J.A., and Anandharamakrishnan, C. (2020). Utilization of food waste streams for the production of biopolymers. *Heliyon* 6: e04891. https://doi.org/10.1016/j.heliyon.2020.e04891.

32 Pandharipande, S. and Bhagat, P.H. (2016). Synthesis of chitin from crab shells and its utilization in preparation of the nanostructured film. *Synthesis (Stuttg)* 5: 1378–1383.

33 Rehm, B.H.A. (2010). Bacterial polymers: biosynthesis, modifications, and applications. *Nature Reviews. Microbiology* 8: 578–592.

34 Jose, A.A., Hazeena, S.H., Lakshmi, N.M. et al. (2022). Bacterial biopolymers: from production to applications in biomedicine. *Sustainable Chemistry and Pharmacy* 25: 100582.

35 Akila, R.M. (2014). Fermentative production of fungal chitosan, a versatile biopolymer (perspectives and its applications). *Advances in Applied Science Research* 5: 157–170.

36 Araújo, T.R., Petkowicz, C.L.D.O., Cardoso, V.L. et al. (2016). Biopolymer production using fungus Mucor racemosus Fresenius and glycerol as substrate. *Polímeros* 26: 144–151.

37 van Beilen, J.B. and Poirier, Y. (2008). Production of renewable polymers from crop plants. *The Plant Journal* 54: 684–701.

38 Armstrong, G. (2015). An introduction to polymer nanocomposites. *European Journal of Physics* 36: 63001.

39 Malhotra, B.D. and Ali, M.A. (2018). Nanomaterials in biosensors: fundamentals and applications. *Nanomaterials for Biosensors* 1.

40 Ates, B., Koytepe, S., Ulu, A. et al. (2020). Chemistry, structures, and advanced applications of nanocomposites from renewable resources. *Chemical Reviews* 120: 9304–9362.

41 Jaafar, J., Siregar, J.P., Tezara, C. et al. (2019). A review of important considerations in the compression molding process of short natural fiber composites. *The International Journal of Advanced Manufacturing Technology* 105: 3437–3450.

42 Asmatulu, R., Khan, W.S., Reddy, R.J., and Ceylan, M. (2015). Synthesis and analysis of injection-molded nanocomposites of recycled high-density polyethylene incorporated with graphene nanoflakes. *Polymer Composites* 36: 1565–1573.

43 Do Céu Teixeira, M., Santini, A., and Souto, E.B. (2017). Delivery of antimicrobials by chitosan-composed therapeutic nanostructures. In: *Nanostructures for Antimicrobial Therapy*, (ed. A. Ficai, A. M. Grumezescu) 203–222. Elsevier.

44 Narayanan, N., Kuang, L., Del Ponte, M. et al. (2016). Design and fabrication of nanocomposites for musculoskeletal tissue regeneration. In: *Nanocomposites for Musculoskeletal Tissue Regeneration*, (ed. H. Liu) 3–29. Elsevier.

45 Shoaie, N., Daneshpour, M., Azimzadeh, M. et al. (2019). Electrochemical sensors and biosensors based on the use of polyaniline and its nanocomposites: a review on recent advances. *Microchimica Acta* 186: 1–29.

46 Lei, W., Si, W., Xu, Y. et al. (2014). Conducting polymer composites with graphene for use in chemical sensors and biosensors. *Microchimica Acta* 181: 707–722.

47 Shan, C., Yang, H., Han, D. et al. (2010). Graphene/AuNPs/chitosan nanocomposites film for glucose biosensing. *Biosensors & Bioelectronics* 25: 1070–1074.

48 Valencia-Chamorro, S.A., Palou, L., Del Río, M.A., and Perez-Gago, M.B. (2011). Antimicrobial edible films and coatings for fresh and minimally processed fruits and vegetables: a review. *Critical Reviews in Food Science and Nutrition* 51: 872–900.

49 Shameli, K., Bin Ahmad, M., Yunus, W.Z.W., et al. (2010). Synthesis and characterization of silver/talc nanocomposites using the wet chemical reduction method. *International Journal of Nanomedicine* 5: 743.

50 Marin, M.L., McGilvray, K.L., and Scaiano, J.C. (2008). Photochemical strategies for the synthesis of gold nanoparticles from Au (III) and Au (I) using photoinduced free radical generation. *Journal of the American Chemical Society* 130: 16572–16584.

51 Castillejo, M., Ossi, P.M., and Zhigilei, L. (2014). *Lasers in Materials Science*. Springer.

52 V Novais, S.M., Monteiro, T.J., Teixeira, V.C. et al. (2018). Hydrothermal synthesis of CdWO4 for scintillator-polymer composite films development. *Journal of Luminescence* 199: 225–231.

53 Chen, J., Wang, J., Zhang, X., and Jin, Y. (2008). Microwave-assisted green synthesis of silver nanoparticles by carboxymethyl cellulose sodium and silver nitrate. *Materials Chemistry and Physics* 108: 421–424.

54 Salehoon, E., Ahmadi, S.J., Razavi, S.M., and Parvin, N. (2017). Thermal and corrosion resistance properties of unsaturated polyester/clay nanocomposites and the effect of electron beam irradiation. *Polymer Bulletin* 74: 1629–1647.

55 Quadri, T.W., Olasunkanmi, L.O., Fayemi, O.E. et al. (2017). Zinc oxide nanocomposites of selected polymers: synthesis, characterization, and corrosion inhibition studies on mild steel in HCl solution. *ACS Omega* 2: 8421–8437.

56 Solomon, M.M., Gerengi, H., Kaya, T., and Umoren, S.A. (2017). Enhanced corrosion inhibition effect of chitosan for St37 in 15% H2SO4 environment by silver nanoparticles. *International Journal of Biological Macromolecules* 104: 638–649.

57 Mohamed, M.G., Kuo, S.W., Mahdy, A. et al. (2020). Bisbenzylidene cyclopentanone and cyclohexanone-functionalized polybenzoxazine nanocomposites: synthesis, characterization, and use for corrosion protection on mild steel. *Materials Today Communications* 25: 101418.

58 Camargo, P.H.C., Satyanarayana, K.G., and Wypych, F. (2009). Nanocomposites: synthesis, structure, properties and new application opportunities. *Materials Research* 12: 1–39.

59 Abiola, O.K., Oforka, N.C., Ebenso, E.E., and Nwinuka, N.M. (2007). Eco-friendly corrosion inhibitors: the inhibitive action of Delonix Regia extract for the corrosion of aluminium in acidic media. *Anti-Corrosion Methods and Materials*, (ed. Yu Yan).

60 Kliškić, M., Radošević, J., Gudić, S., and Katalinić, V. (2000). Aqueous extract of Rosmarinus officinalis L. as an inhibitor of Al-Mg alloy corrosion in chloride solution. *Journal of Applied Electrochemistry* 30: 823–830.

61 El-Etre, A.Y. (1998). Natural honey in a corrosion inhibitor for metals and alloys. I. Copper in neutral aqueous solution. *Corrosion Science* 40: 1845–1850.

62 El-Etre, A.Y. (2003). Inhibition of aluminum corrosion using Opuntia extract. *Corrosion Science* 45: 2485–2495.

63 El-Etre, A.Y. (2006). Khillah extract as an inhibitor for acid corrosion of SX 316 steel. *Applied Surface Science* 252: 8521–8525.

64 Ebenso, E.E. and Ekpe, U.J. (1996). Kinetic study of corrosion and corrosion inhibition of mild steel in H2SO4 using Carica papaya leaves extract. *West African Journal of Biological and Applied Chemistry* 41: 21–27.

65 Ekpe, U.J., Ebenso, E.E., and Ibok, U.J. (1994). Inhibitory action of Azadirachta indica leaves extract on the corrosion of mild steel in H2SO4. *JW Afri. Sci. Assoc.* 37: 13–30.

66 Zucchi, F. and Omar, I.H. (1985). Plant extracts as corrosion inhibitors of mild steel in HCl solutions. *Surface Technology* 24: 391–399.

67 Yee, Y.J., Green inhibitors for corrosion control: a Study on the inhibitive effects of extracts of honey and rosmarinus officinalis L. (Rosemary), MS thesis, University of Manchester, Institute of Science and Technology, 2004.

68 Chalchat, J.-C., Garry, R.-P., Michet, A. et al. (1993). Essential oils of rosemary (Rosmarinus officinalis L.). The chemical composition of oils of various origins (Morocco, Spain, France). *Journal of Essential Oil Research* 5: 613–618.

69 Umoren, S.A. and Ebenso, E.E. (2008). Studies of the anti-corrosive effect of Raphia hookeri exudate gum-halide mixtures for aluminum corrosion in acidic medium. *Pigment & Resin Technology*, (ed. Long Lin).

70 Okafor, P.C., Osabor, V.I., and Ebenso, E.E. (2007). Eco-friendly corrosion inhibitors: inhibitive action of ethanol extracts of Garcinia kola for the corrosion of mild steel in H2SO4 solutions. *Pigment & Resin Technology*, (ed. Long Lin).

71 Oguzie, E.E. (2006). Studies on the inhibitive effect of Occimumviridis extract on the acid corrosion of mild steel. *Materials Chemistry and Physics* 99: 441–446.

72 Bendahou, M., Benabdellah, M., and Hammouti, B. (2006). A study of rosemary oil as a green corrosion inhibitor for steel in 2 M H3PO4. *Pigment & Resin Technology*, (ed. Long Lin).

73 Quraishi, M.A., Kumar Yadav, D., and Ahamad, I. (2009). Green approach to corrosion inhibition by black pepper extract in hydrochloric acid solution. *The Open Corrosion Journal* 2.

74 Chauhan, D.S., Quraishi, M.A., Ansari, K.R., and Saleh, T.A. (2020). Graphene and graphene oxide as a new class of materials for corrosion control and protection: present status and future scenario. *Progress in Organic Coatings* 147: 105741.

75 Castelo, A., Correia, J.R., Cabral-Fonseca, S., and de Brito, J. (2020). In-service performance of fiber-reinforced polymer constructions used in water and Sewage treatment plants. *Journal of Performance of Constructed Facilities* 34: 04020059.

76 Alhaffar, M.T., Umoren, S.A., Obot, I.B., and Ali, S.A. (2018). Isoxazolidine derivatives as corrosion inhibitors for low carbon steel in HCl solution: experimental, theoretical and effect of KI studies. *Rsc Advances* 8: 1764–1777.

77 Verma, C., Ebenso, E.E., Quraishi, M.A., and Hussain, C.M. (2021). Recent developments in sustainable corrosion inhibitors: design, performance and industrial scale applications. *Materials Advances* 2: 3806–3850.

78 Vachtsevanos, G., Natarajan, K.A., Rajamani, R., and Sandborn, P. (2020). *Corrosion Processes*. Springer.

19

Industrially Useful Corrosion Inhibitors

Grafted Biopolymers as Ideal Substitutes

Farhat A. Ansari[1,] and Hariom K. Sharma[2]*

[1] *Faculty of Pharmaceutical Chemistry, Hygia Institute of Pharmaceutical Education and Research, Faizullahganj, Lucknow, 226020, Uttar Pradesh, India*
[2] *Engineering Department, University of Technology and Applied Sciences (UTAS), Salalah, Dhofar, Sultanate of Oman*
* *Corresponding author*

19.1 Introduction

From the very beginning of the Industrial Era in the 1800s, corrosion has been a serious issue because most metals in use since then are exposed to moisture in the environment, resulting in unavoidable atmospheric corrosion. In 1938, Wanger et al. [1] concluded that the cause of corrosion took place largely at the metal-electrode interface. Corrosion generally occurs when metal is subjected to an atmosphere where the outermost portion of the metal thins away due to chemical or electrochemical action or exposure and the metals or alloys become rusted as a result [2]. Multiple methods are used to safeguard metals and their alloys [3], such as utilizing substances to generate a protective coating of films, and transferring impressed current to shield the cathodes—just to name a few. Corrosion can be reduced in some cases due to the auto electro-chemical process.

Corrosion inhibitors are substances known to prevent corrosion from occurring and have been in use for decades [4]. In general, there are four kinds of environments in which corrosion inhibitors are used:

- Water with a pH scale of 5 to 9
- Acidic aqueous solutions
- Oil refinery atmosphere (direct and indirect production)
- Numerous other environments

Organic substances are considered to be one of the most suitable and value-effective techniques of corrosion control compared to the alternative options available. The widespread use of organic substances as corrosion inhibitors can also be linked to their reasonable efficiency and convenience of use. Through their non-bonding and pi-electrons, these compounds adhere and form a corrosion-protective shield. Prior to 1960 the use of inorganic substances—primarily nitrites, silicates, zinc, and

Grafted Biopolymers as Corrosion Inhibitors: Safety, Sustainability, and Efficiency, First Edition.
Edited by Jeenat Aslam, Chandrabhan Verma, and Ruby Aslam.

molybdates—have been some of the earliest means of corrosion prevention [5]. As environmental concerns grew in prominence from 1980–1995 harmful chemicals were largely replaced with alternative substitutes such as biodegradable biopolymers, biosurfactants, vitamin supplements, polyphenols, and altogether more natural substances. The most recent research into corrosion inhibition has focused on the usage of rare metals [6], heterocyclic compounds, the synergistic effect of organic substances employing REM (rare earth metal), and the absorption of inhibitors.

Between the growing environmental consciousness and stringent government regulations, the use of naturally derived substances and organic resources came into existence for a variety of uses. Chemical synthesis of organic substances is not only costly to process initially, but it also releases various harmful substances into the air which are now known to have long-term negative impacts on living things which can be even more costly in the long run. With the catastrophic situation of the environment increasingly feeling the effects of global warming as a direct consequence of climate change, the development of systems that employ sustainable materials has soared in recent years which includes responsibly managing harmful fossil resources and preserving limited natural resources. The pricing of fossil fuel resources and the consumer products that rely on them for production or transportation of goods has increased sharply and has contributed to the growing public interest in adopting greener solutions that are more eco-friendly.

In the recent decade, corrosion scientists and engineers have been working together to develop and design several eco-friendly corrosion inhibition systems [7]. Green corrosion inhibitors are also known as sustainable corrosion inhibitors. Green corrosion inhibitors are designed by modifying the chemical process which decreases the formation of toxic material waste. Green chemistry, in general, is a set of concepts that encompasses the production, development, application, and discharge of synthetic chemicals [8]. The guiding principles of green chemistry include:

- Reducing environmental contamination.
- Decreasing the negative impact of procedures and chemicals on human health and the environment.
- Harmful chemicals are sometimes minimized by applying pre-existing techniques and practices in new ways.
- It also entails modifying and redesigning current products and practices to reduce inherent risks.

Green corrosion inhibitors were developed based on the following criteria:

- Compounds that have a lower impact on the environment and health.
- Preparing solvent, chemical, and reagent feedstocks that have reduced environmental impacts.
- Preparation activities are designed to consume as little energy as possible.
- Emphasizing the use of items naturally derived and from renewable resources.

- Utilizing new technology that produces less chemical waste.
- Chemicals are utilized through trash recycling.

There are a variety of natural compounds being used as corrosion inhibitors including pharmaceutical drugs that are derived from plants or ionic liquids as well as other synthesized inhibitors created from multicomponent reactions (MCRs) and mechanical alloying [9]. Biopolymers are non-biologically persistent and renewable. Oligosaccharides (carbohydrates, cellulose, glucosamine, etc.), tripeptides, nucleotides (RNA & DNA), rubber, tannin, and other biopolymers are some of the most prevalent biopolymers. Biopolymers, especially chitosan derivatives, have made significant progress. Chitosan and its compounds have been successfully used as corrosion inhibitors for metallic materials in a variety of extremely corrosive environments. The main attractive feature is that chitosan and its compounds contain multiple functional groups such as amine (–NH_2), hydroxy (–OH), and acylated (–$COCH_3$) groups that easily bond with metallic substrates and making them effective corrosion inhibitors. Chitosan is an n – acetyl – d chitin polymer having immunologic characteristics, excellent biocompatibility, tissue repair, and renewability [10, 11]. Figure 19.1 illustrates the molecular structure of various biopolymers [12] used as corrosion inhibitors.

Chitosan—similar to polysaccharides—is enriched with -OH and -NH_2 functional groups which could be used as an effective corrosion inhibitor. Chitin is a naturally occurring substance of chitosan that can be found in the carapace of

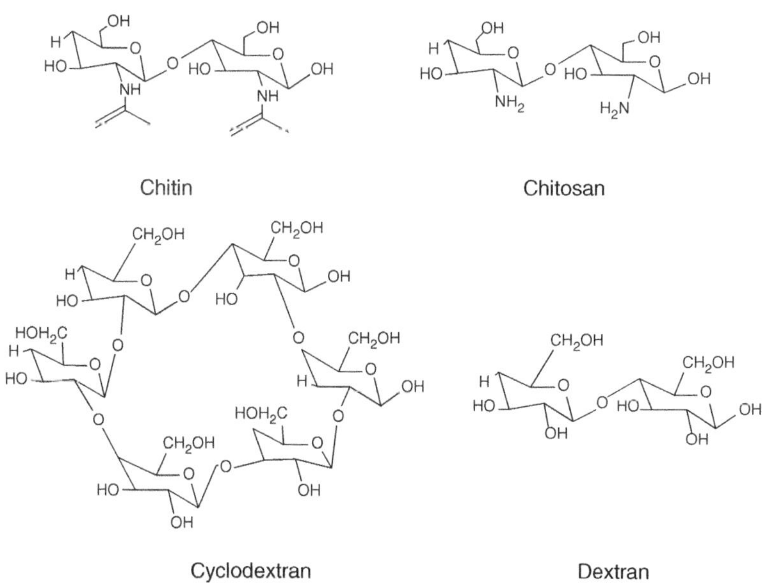

Figure 19.1 Molecular structure of biopolymers used as corrosion inhibitors [12] / American Chemical Society.

crustaceans, fungus cells, and insect coverings [10, 11]. N-deacetylation of chitin with a variety of deacetylated agents will also yield chitosan. Chitosan and its derivatives are reliably used due to their beneficial biological properties, non-toxic nature, superior renewability, elevated tissue regeneration capability, and antigenic assets [13–17].

19.2 Corrosion Inhibitors

19.2.1 Sustainable or Green Corrosion Inhibitors

With the growing need for green chemistry worldwide, the production of non-toxic and environmentally sustainable corrosion inhibitors is extremely desirable. Environmentally friendly substances are being synthesized, designed, developed, and consumed in science and engineering research investigations to yield sustainable alternatives to toxic substances [18–20]. The increasingly stringent environmental regulations are driving the current market demand for conservation knowledge and research. In this regard, the usage of conventional, synthetic organic compounds has been restricted and they have been replaced with green organic compounds which are usually extracted using conventional or novel processing technologies.

Green corrosion inhibitors are biodegradable, non-toxic, and sustainable, making them a better corrosion inhibitor alternative than their synthetic competitors. They are used as inhibitors for numerous metallic materials in several corrosive environments [21, 22]. Various types of green corrosion inhibitors are being employed as shown in Figure 19.2, including medicines or drugs, plant extracts, ionic liquids, and synthetic inhibitors created from multi-component reactions (MCRs) and mechanochemical mixing. MCRs—in combination with microwave and ultrasound irradiation—are currently considered to be among the most environmentally friendly techniques available today.

Despite the numerous environmental benefits, there have been several constraints of green corrosion inhibitors that promote tackling obstacles while employing such inhibitors. When it comes to green corrosion inhibitors, it's important to consider if the inhibitor remains non-toxic after 28 days. The bioaccumulation of the inhibitor must therefore be thoroughly investigated before being employed to prevent corrosion. The partition coefficient method is used to calculate bioaccumulation among 1-octanol and water. The bioaccumulation will be larger if the partition coefficient is higher. Another significant disadvantage is their short lifespan and they are often extremely biodegradable which makes it challenging to maintain them for an extended period. It is also important to remember that when extracting plant components for the chosen solvent to be as safe as possible. Furthermore, several processes necessitate extended processing times and elevated heat, both of which are unfavourable to green corrosion

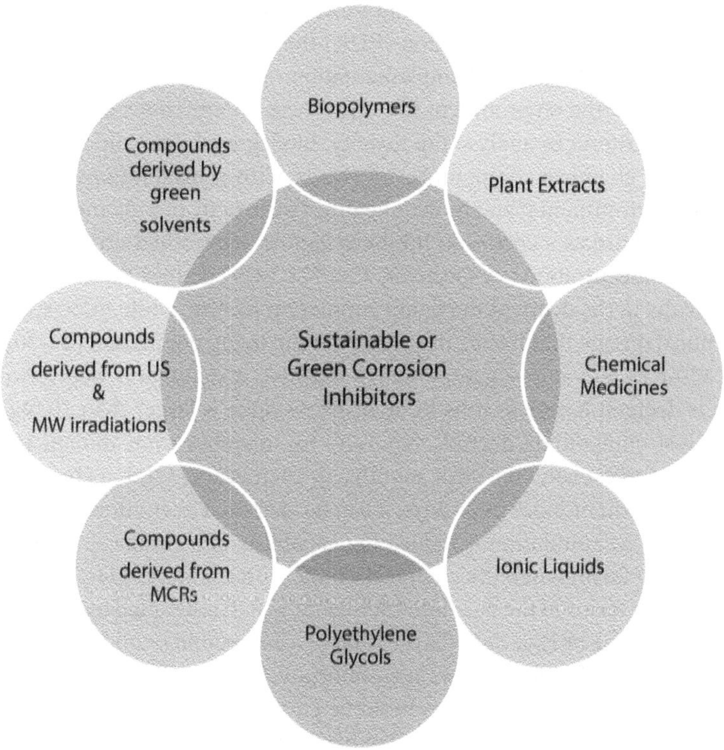

Figure 19.2 Types of green and sustainable corrosion inhibitors.

inhibition. Now, modern methods of supercritical fluid extraction are used as an alternative approach to combat many of these issues. However, the impact of physicochemical factors on the mechanism are now being researched. Despite these drawbacks, green corrosion inhibitors remain the most promising alternatives since they are commonly produced from natural ingredients which increases their readily available nature and economic viability [23].

19.2.2 Guidelines and Assessment of the Sustainability of Corrosion Inhibitors

Green corrosion inhibitors could be used instead of traditional inhibitors driven by environmental concerns, but the impacts of green inhibitors on the environment must still be considered. Biodegradability, Bioaccumulation ability, and the effect of toxicity on the ecosystem can all be used to establish how ecologically friendly a chemical compound is. The international commissions for assessing the above parameters are the Oslo and Paris Commission (OSPAR) and the

Registration, Evaluation, Authorization, and Restriction of Chemicals (REACH). REACH is a European Union Regulation (EUR) that focuses on chemical substance manufacturing, design, and implementation, as well as their influence on human health and the environment [24, 25]. These commissions established guiding rules and criteria to analyse the toxicity, biodegradability, and bioaccumulation ability of chemical compounds—of course with considerations for the impacts to both environmental and human health. The toxicity of a chemical substance intended for use as a corrosion inhibitor can be determined utilizing the LC50 and EC50 concentrations, respectively [26, 27]. LC50 defines the chemical dosage that results in the death of the living population by 50%, while EC50 specifies the chemical dosage that harms the growth of the living population [28–30]. A lower LC50 or EC50 value suggests higher sensitive toxicity, and vice versa. A substance is deemed harmless if its LC50 or EC50 value is larger than 10 mg/kg [28–30]. Most compounds degrade naturally by microbes even though the time-consuming process can take days, months, or even years [31]. If a chemical decomposes 60 percent or more in 28 days, it can be classified as environmentally benign [32, 33]. Bioaccumulation is indeed the capability of a chemical to accumulate in a live organism when the chemical's only source is water [34–36]. It is usually expressed in terms of the partition coefficient which is denoted by log K_{OW} or D_{OW}. In general, K_{OW} or D_{OW} denotes the ratio of a compound's concentrations in a mixture of two immiscible solvents at equilibrium. The K_{OW} or D_{OW} value for an environmentally friendly chemical should be equal to or less than 3.52. The partition coefficient of a chemical distributed between two immiscible phases of octanol and water is used to calculate bioaccumulation.

19.2.3 Industrial Uses of Corrosion Inhibitors

Green corrosion inhibition compounds are suggested for future corrosion exploration because they are cheap, widely available, bio-renewable, permeable, non-toxic, recyclable substances, incredibly effective, and ecologically friendly. The key technique utilized in preventing electrochemical corrosion in metals and their alloys is to adequately separate the metal from corrosive chemicals—corrosion inhibitors can be used to accomplish this. One of the most well-known means of corrosion control is the use of inhibitors. Figure 19.3 shows various corrosion inhibitors that have been efficiently employed in a variety of industrial settings for the protection of metals and alloys. Acidic solutions with high concentrations are used in a variety of industrial processes [37–39]. Acid pickling—which involves the elimination of surface contaminants including rusts from metallic sheets, wires, pipes, or any other metal instruments across almost6 every major industry such as cooling systems, heat transfer, heat exchange, and boilers—is one of the well-known and most significant acidic procedures. Different acidic solutions are

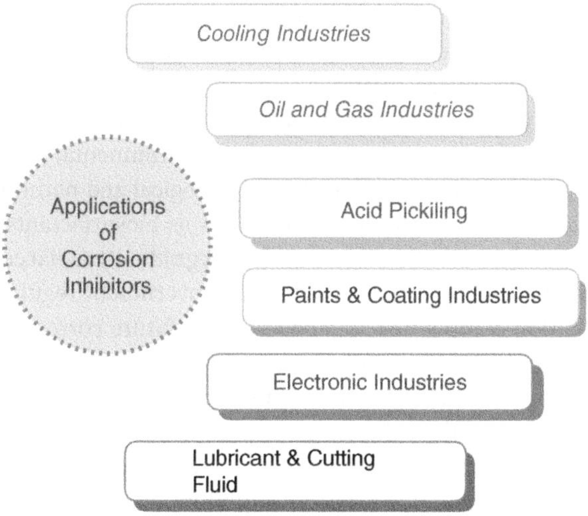

Figure 19.3 Application and industrial uses of various corrosion inhibitors.

often employed in descaling and acid pickling procedures in order to remove scales and metal-oxide rust from the metal surface [40]. Due to their extremely aggressive nature, acidic solutions promote corrosion-related deterioration of metallic structures throughout these processes. Furthermore, H_2-gas is generated during these operations, which could also reach inside the metallic structures leading to hydrogen embrittlement. As a result, corrosion inhibitors—which are chemical species that may be added to certain industrial processes—are required. It is worth noting that the majority of the inhibitors used in the acid pickling process are organic and are particularly heterocyclic as well [41]. These organic inhibitors not only prevent the metals from corrosion, but they can also increase the metallic equipment's life and reduce the quantity of acid consumed. Various organic compounds have been efficiently utilized as inhibitors in acid solutions for the acid pickling process of metallic alloys [42–49].

Another industrial process in the petroleum sector is oil-well acidification which necessitates the use of a highly concentrated acidic solution [50, 51]. To improve the flow of oil, an acidic solution of HCl with a high concentration (usually 15 to 28 percent) is injected into the well through metallic tubing. The tubing of such extremely acidic solutions causes substantial corrosion to the surface of the metal during the acidification process. As a result, inhibitors are introduced to the acidizing solution as an external component. According to a literature review, the majority of formerly utilized acidization inhibitors are heteroatoms (particularly nitrogen, and oxygen) containing heterocyclic compounds.

Organic inhibitors are one of the most efficient and cost-effective techniques for protection against corrosion among the many options available. The widespread use of organic compounds as corrosion inhibitors can also be linked to their high efficiency and simplicity of use. However, synthesis method of organic compounds is not only costly, but it also releases various harmful substances into the environment which can have a variety of negative impacts on living organisms.

Recent increases in ecological consciousness and growing environmental concerns have prompted the use of substances produced from biological and natural resources whenever possible. Thus, natural alternatives such as biosurfactants, natural biopolymers, tannins, vitamins, and natural substances gradually replaced the toxic chemicals as a result of public concerns and strict government regulations. Green corrosion inhibition compounds are suggested for future corrosion exploration because they are cheap, widely available, bio-renewable, permeable, non-toxic, recyclable substances, incredibly effective, and ecologically friendly. Due to their environmental compatibility, the usage of plant-based products is growing rapidly. Plant-based products—particularly plant extracts—are environmentally beneficial alternatives to metallic corrosion inhibitors for a variety of biological and industrial uses due to their natural origins [52–56].

Biopolymers are another class of green inhibitors that are considered environmentally friendly alternatives and employed in a variety of biological and industrial applications [57–59]. They are also biodegradable and non-bioaccumulative, contrary to synthetic polymers. Because of the numerous ecologically favourable qualities—such as minimal toxic effects, high polarity, low melting point, low vapor pressure, and high resistivity for chemical and thermal treatment—ionic liquids are an acceptable environmentally safe alternative to conventional corrosion inhibitors [60–62]. Ionic liquids are also referred to as fused salts, ionic melts, ionic fluids, ionic glasses, and ionic electrolytes. Ionic liquids have been produced and employed in a variety of biological and industrial applications [63–66]. Drugs or chemical medicines are other types of environmentally friendly corrosion inhibitors that can be used instead of standard harmful corrosion inhibitors [67–70]. Chemical medicines or drugs—which are produced from natural and biological resources—are perfect environmentally friendly options. Furthermore, because the majority of the medicines are biocompatible, bio-tolerable, and non-bioaccumulative, they provide environmentally friendly options [71, 72].

19.3 Biopolymers

Biopolymers are employed in a variety of industries including packaged foods, beauty products, and healthcare sectors [73]. They can be used to replace petrol-based polymers in a spectrum of uses. Some biomaterials were often used in applications where traditional plastic material would be ineffective, such as the manufacture of synthetic tissue. These applications may necessitate biocompatibility with polymeric materials that are sensitive to pH, chemical, and thermal changes. When compared to commercial polymers, biopolymers typically have inferior tensile characteristics, chemical stability, and ease of processing. They could be strengthened with fillers to increase their qualities and make materials better suited for highly specified applications. Graft-biopolymer composites are biopolymers that have been strengthened in this manner.

19.3.1 Classification of Biopolymers

Biopolymers are categorized in a variety of ways characteristics. Biomaterials can sometimes be classified into broader classes based on their processability as either biodegradable or non-biodegradable biopolymers. The nature of the monomeric unit also serves as a basis for biopolymers' classification.

Carbohydrates are a prime example of biopolymers that are made up of simple sugars connected by O-glycosidic bonds. Polysaccharides found as the largest natural polymeric are cellulose, as the second most prevalent biopolymer, chitin [74–76]. Physical and chemical properties such as mechanical characteristics, solubility, flowability, gelation potential, surface and surface traits are governed by monosaccharide structure, degree of polymerization, linking types, and trends which give polysaccharides a high degree of adaptability in the selection of materials for diverse applications.

Biopolymers—which include nucleotides, peptides, polysaccharides, triglycerides, and the form of non-polymeric molecules such as macrocycles—are the most prevalent macro-molecules [77]. Synthetic polymers include composites, fibres, and novel materials like nanomaterials, etc. Figure 19.4 depicts the categorization of biopolymers based on their sources.

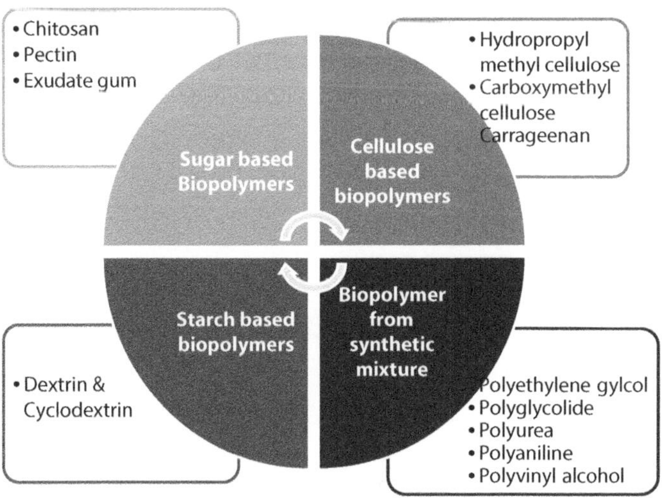

Figure 19.4 Classification of biopolymers based on their source of origin.

19.3.2 Applications of Biopolymers

Grafted synthesized polymers were not just for the purposes stated. Recent research indicates the growing possibilities for new applications of biopolymers as manufacturing materials from medicinal formulas with complexation; commodities (edible wrappers and emulsifying agents) [78]; sustainable agricultural production including methods of purifying water and soil repair; the cosmetics production houses [79], biosensing electronic devices, and even digital storage components are all examples of biopolymer application domains. Figure 19.5 shows the various industrial application of biopolymers.

Starch-based graft-polymers were utilized to wrap conservatories, mulching film, and fertilizers, according to Bohlmann [80]. Dave et al. [81] reported the application of polymer-based devices for the regulated delivery of synthetic nitrogen fertilizers.

A novel biomedical approach that Boesel et al. investigated the use of SCA (starch, cellulose acetate) mixtures for the manufacture of acrylic bone blocks of cement [82]. In rats, Gomes et al. investigated the parameters that influence the bone development of neural stem cells [83]. Bal mayor et al. studied the utilization of metabolic enzymes' degradable transport mechanism for the transport of differentiated chemicals for application in bone-tissue regeneration [84].

Carbon nanotubes have been used in aerospace components, according to Belluccia et al. [85]. Kang et al. looked at using carbon nanomaterials as a piezoresistive sensing material for healthcare monitoring that could also be employed with electrolytes to generate power from mechanical movements [86]. Wang et al. investigated using solid poly acrylic nitrile fibres and carbon nanotubes to create

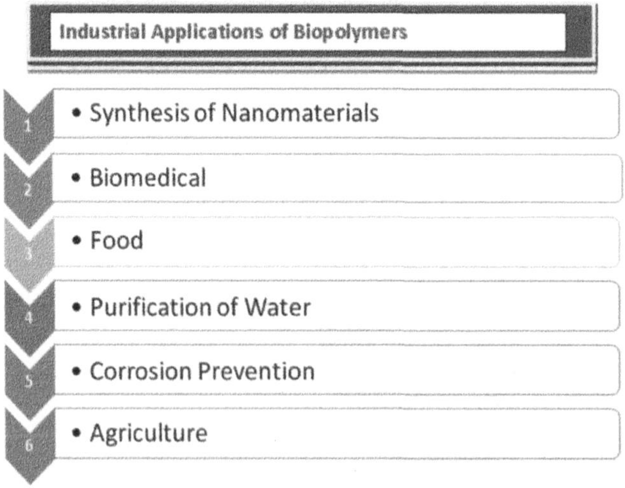

Figure 19.5 Industrial application of biopolymers.

novel rechargeable batteries [87]. Some conducting graft-polymers, such as PANi and CNTs were used as prosthetic muscle fibres as reported by another group of researchers [88]. Carbon nanoparticles were also investigated for sewage treatment, air quality control, bioengineering, renewable energy technology, superconductors, and eco-friendly nanocomposites by Ong et al. [89].

Grafted polymers are a type of biocompatible compound that have a multitude of uses in biomedical engineering. Scientists, for instance, used grafted poly-oligosaccharide to build a colon-targeted medication distribution system for amoebiasis treatments [90]. Silva et al. interrogated applications of starch-grafted copolymers as pharmaceutical and peptide delivery matrices as well as hydrodynamic matrices for colon-specific drug carrier systems [91]. Although work in this field has transformed the biological and biomedical sectors, much work is still needed to impact improvements, not just at an individual scale, but at a sustainable and economical level as well. While research is ongoing, biopolymer-based techniques are immensely beneficial for providing incredibly advantageous, potent, and secure compounds.

19.4 Biopolymers as Corrosion Inhibitors

Biopolymers are macro-biomolecules that are made up of oligomeric units that are covalently bound together to form bigger molecules. The word "bio" denotes that the components are biocompatible and created by life forms. The terminology "biopolymer" covers a broad spectrum of polymers that are generally produced from natural sources like microbes, plants, or trees. Biodegradable polymers are sophisticated chemical complexes with exact and definite 3D structures which is in contrast to the manufactured polymeric material that has a simple and more arbitrary shape. Their role is ultimately dependent on their specified composition and form. Haemoglobin, for instance, would have been unable to transport oxygen around the body if it was not rolled up into a quaternary structure.

Biopolymers, because of their underlying stability, accessibility, low cost, and nontoxic biomaterials are gaining a lot of interest as corrosion inhibitors [92, 93]. The incredible advantages of polysaccharides and their derivatives like sorption [94, 95], medicine delivery [96, 97], corrosion inhibitors [98], and surface coating [99] have been extensively used for decades. Polymeric compounds have subsequently received much interest in antirust studies [100–103]. The macro-molecules effectively blanket significant surface area regions of metallic materials in aqueous solutions which is necessary to protect it from degradation by corrosive chemicals and particles. The effectiveness of carbohydrate-biopolymers as inhibitors depends greatly upon respective molecular weights, heterocyclic rings, bond-forming sites, and the abundance of heteroatom centres.

19.4.1 Sugar-Based Biopolymers

19.4.1.1 Chitosan as Green Corrosion Inhibitors

Chitosan is a polysaccharide polymer formed from chitin which can be produced in a variety of natural materials including crabs, fungi, insects, and microalgae. Chitosan is produced by removing most of the acyl groups from chitin using a strong alkali [104]. Chitosan is chitin that has been deacetylated to a level greater than 70%. Chitosan resembles cellulose and chitin in appearance. Chitin is a straight strand of acetyl glucosamine groups, whereas chitosan is created through a procedure called deacetylation whereby sufficient acyl groups (CH_3–CO) are deleted to make the compound soluble in just about all acidic solutions. The acetyl component of a polymeric chain is what distinguishes chitin from chitosan. Chitosan [105], which has a free –NH_2 group, is perhaps the most valuable chitin derivative. The chemical structure of chitosan has polar groups such as hydroxymethyl, acyl, hydroxyl, amine, and ether which effectively form interaction with the metal substrate. Chitosan's corrosion protection action on metals and alloys has been thoroughly documented [106].

Chitosan-based anti-corrosive compounds are regarded as eco-friendly and beneficial due to their natural sources. In hydrochloric acid (1 M HCl), pure chitosan compounds are commonly used for ferrous alloys such as mild steel, carbon steel, and 316 austenitic steel [58, 107–109]. This is attributed as the reason why all these metal materials are commonly employed in a number of engineering applications where the degradation or erosion is much more severe. The effect of chitosan on Cu (copper) metal was also tested in acidic as well as synthetic seawater environments [110, 111]. The biopolymer was found to be quite an effective corrosion protection compound with 87 and 89% IE respectively.

Generally, chitosan that has been treated with different organic materials inhibits rusting better than chitosan alone. The preparation of chitosan-Schiff base compounds is carried out through a polymerization reaction or microwave radiation. It was found that adding Schiff bases substituents to the chitosan framework improves the inhibitory ability and polymeric layer adherence on metals. The inhibition efficiency of the chitosan-Schiff-based derivatives was greater than 80%, indicating that the chitosan-Schiff base may be a good option for use as an efficient anti-corrosion compound for metal surfaces [112].

The mixture of carboxymethyl-chitosan and benzaldehyde were refluxed at 60 °C for 5 hours. Similarly, urea and glutaric acid were refluxed at 100 °C for 3 hours. Then carboxy methyl chitosan was added to form Carboxymethyl-chitosan-benzaldehyde (CMChi-B) and Carboxymethyl chitosan-urea glutaric acid (CMChi-UGLU) copolymer derivatives [113]. With the fluidization approach, the percentage inhibition against corrosion on steel in 2% NaCl medium CMChi-B and CMChi-UGLU was 80.82% and 80.62%, respectively [Fig. 19.6]. The corrosion inhibition mechanism of Carboxymethyl-chitosan-benzaldehyde is shown in Fig 19.6.

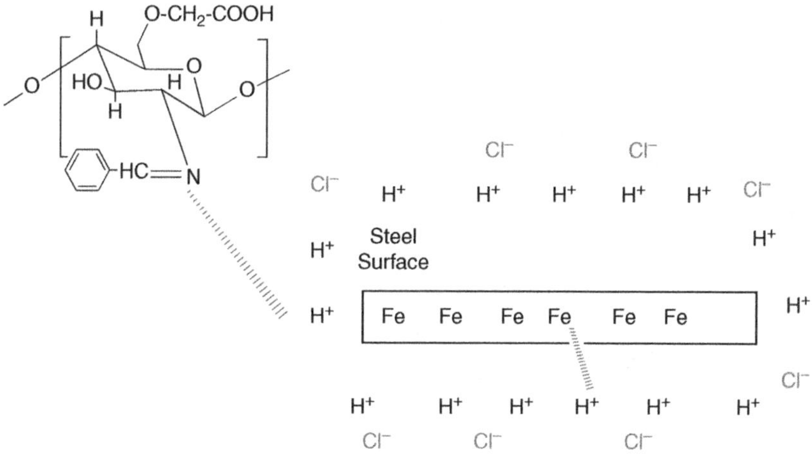

Figure 19.6 The mechanism of inhibition of carbon steel of Carboxymethyl-chitosan-benzaldehyde (CMChi-B) in the hydrochloric acid medium [113] With permission of Elsevier.

Every particle of the inhibitor changed into a fluid by dispersion in gas and liquid in the fluid flow process [114, 115]. Each particulate floats spins, and rotates on the surface, leading to enhanced particle-solution interaction and enhancing inhibition performance. Using microwave irradiation, cinnamaldehyde-chitosan (Cinn-Cht) was produced sequentially and studied using spectroscopic methods. Weight loss studies, EIS measurements, and potentiodynamic polarization investigations were used to assess corrosion. With increasing inhibitor concentration, inhibitive efficiency reached a peak of 85.16 percent at 400 mg/L. The protonated species of the biomolecule have greater adsorption energy than the normal inhibitor, according to Monte Carlo modeling [116].

Even though chitosan has excellent corrosion protection properties in sodium chloride and acid solutions its corrosion-resistant impact and absorption in electrolytes stands to be improved by increasing its functionality with a range of species. Novel chitosan compounds of polyamine-grafted chitosan composites have been designed to protect carbon steel from deterioration in acid conditions as shown in Table 19.1. Methyl acrylate-chitosan composite (CS-MAA) was synthesized followed by ethylene diamine (EN) and trimethylene tetraamine which was grafted into the matrix to form chitosan-methyl acrylamide-ethylenediamine (CS-MAA-EN) and chitosan-methyl acrylamide-trimethylene tetraamine (CS-MAA-TN) composite. In a 5% hydrochloric acid solution at 298 K, the composites of "CS-MAA-EN" and "CS-MAA-TN" show a significant corrosion prevention property against the Q235 carbon steel corrosion process [117].

PANI (polyaniline) is a conductive polymeric compound that has shown potential as an environmentally benign mixed-type inhibitor. Chitosan and polyaniline

Table 19.1 Chitosan and grafted chitosan as corrosion inhibitors.

Inhibitors	Name	Structure	Metal/ corrosive medium	Inhibition efficiency (%)/ Concentration	Reference
Chitosan as inhibitor					
1	Chitosan derived from Archachatina marginata snail shell	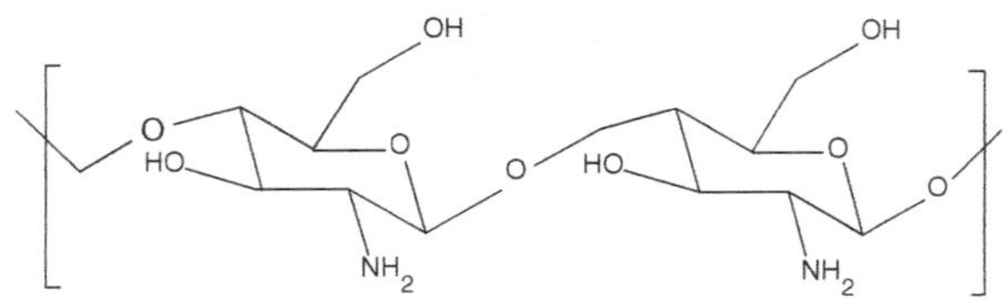	Mild steel/ 0.1 Molar HCl	93/1.8 mm	[107]
2	Chitosan nanoparticle		Carbon steel/ 1.0 Molar HCl	93/500 ppm	[58]
3	Chitosan		316 austenitic steel/ 0.1 MHCl	71/11 mm	[108]
4	Chitosan		Mild steel/ 0.1 MHCl	69/4 μM	[109]
5	Chitosan		Copper/1.0 Molar HCl	87/0.1 mg/l	[110]
6	Chitosan		Copper/ Simulated seawater + 20 ppm Sodium sulphide	89/800 ppm	[111]

Grafted-Chitosan based inhibitors					
1	**CSB-1** CH-Benzaldehyde **CSB-2** CH-4(dimethylamino) benzaldehyde & **CSB-3** 4-hydroxy-3-methoxy benzaldehyde	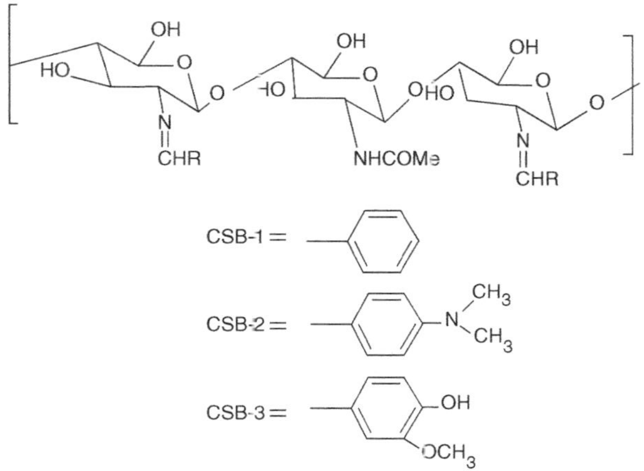	Mild steel/ 1 Molar HCl	91.43 /100 ppm 89.87/100 ppm 88.63/100 ppm	[112]

(Continued)

Table 19.1 (Continued)

Inhibitors	Name	Structure	Metal/ corrosive medium	Inhibition efficiency (%)/ Concentration	Reference
2	**CMCHT-B** Carboxymethyl-chitosan benzaldehyde	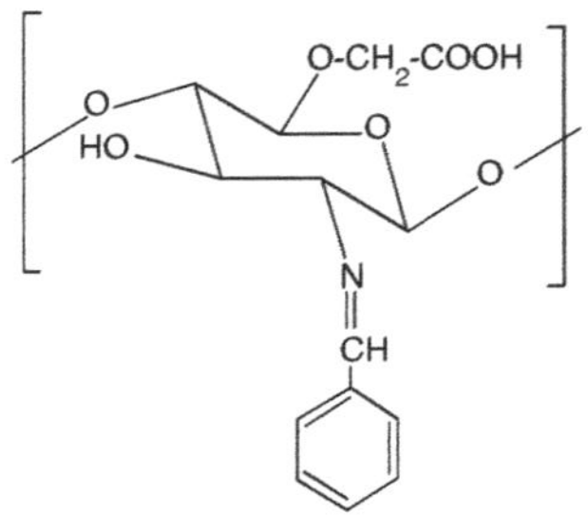	Steel/2% NaCl	80.82 CMCHT-B 80.62 CMCNT-UGLU	[113]
	CMCNT-UGLU Carboxymethyl-chitosan-urea-glutaric acid	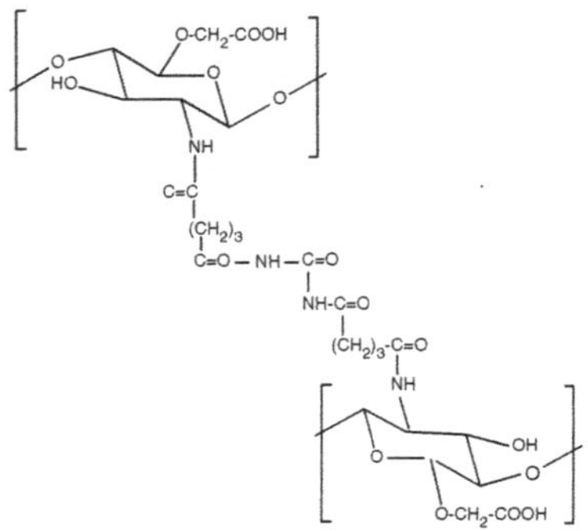			

3	**C-CHT** Cinnamaldehyde-chitosan	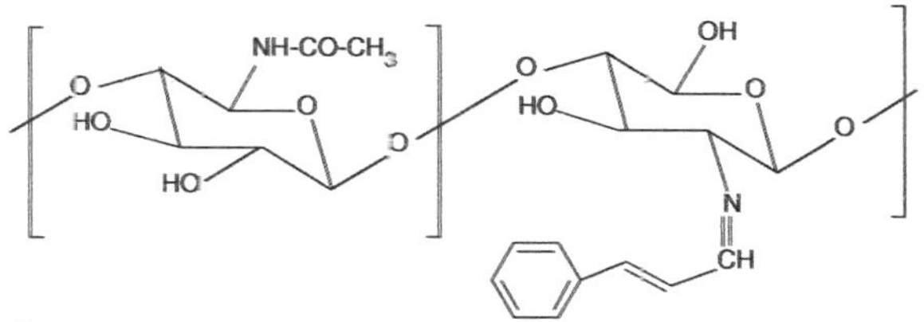	Carbon steel/15% HCl	87.72/600 ppm 92.67 + KI/600 ppm	[116]
4	**CHT-MEA-EN** Chitosan- methyl-acrylate-ethylene diamine	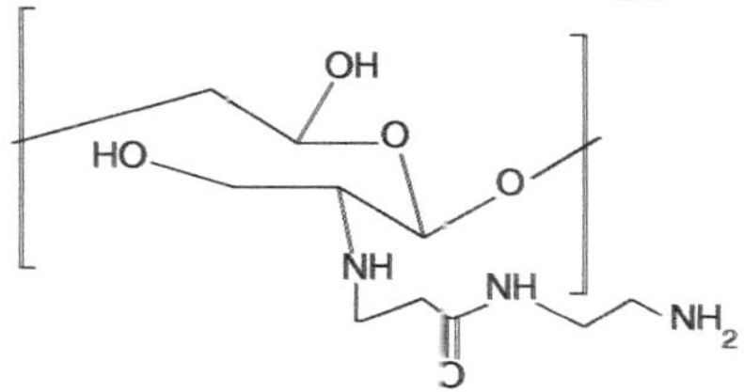	Carbon steel/5% HCl	88.06%/0.3% 69.46%/0.3%	[117]
	CHT-MEA-TN Chitosan- methyl-acrylate tri-ethylene tetraamine	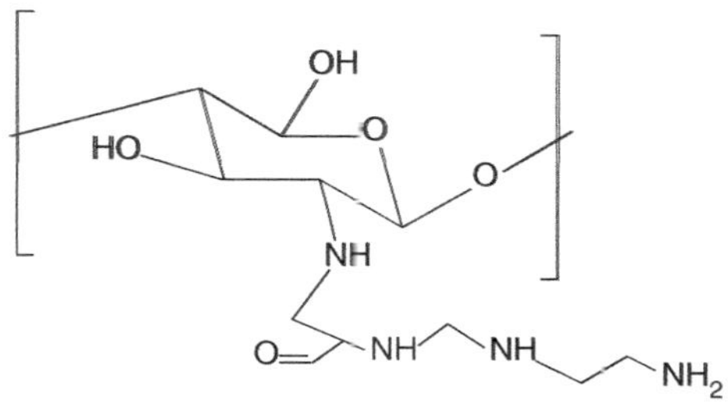			

(Continued)

Table 19.1 (Continued)

Inhibitors	Name	Structure	Metal/ corrosive medium	Inhibition efficiency (%)/ Concentration	Reference
5	**CHT-PA** Chitosan Polyaniline	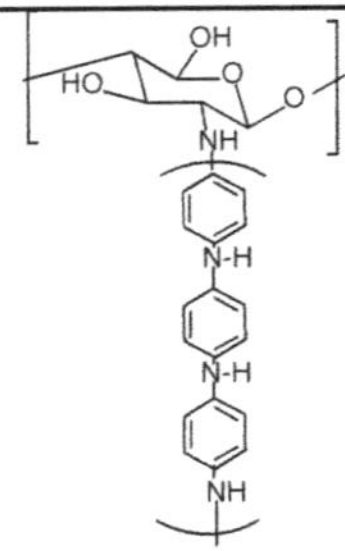	Q235 steel/0.5 Molar HCl	79.02%/200 ppm	[118]
6	**CHT-PAA** Chitosan- polyaspartic acid	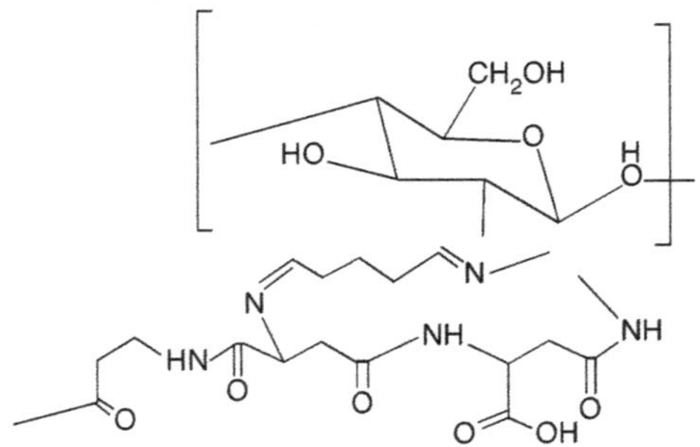	Carbon steel/3.5% NaCl	87.56/20 ppm	[119]

7	**CHT-Van** Chitosan Vanillin	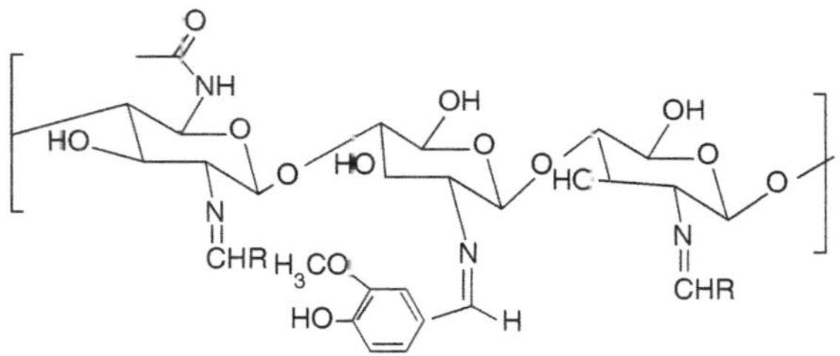	Carbon steel/15% HCl	92.72%/500 mg/L	[120]
8	**CHT-SA** Chitosan Salicylaldehyde	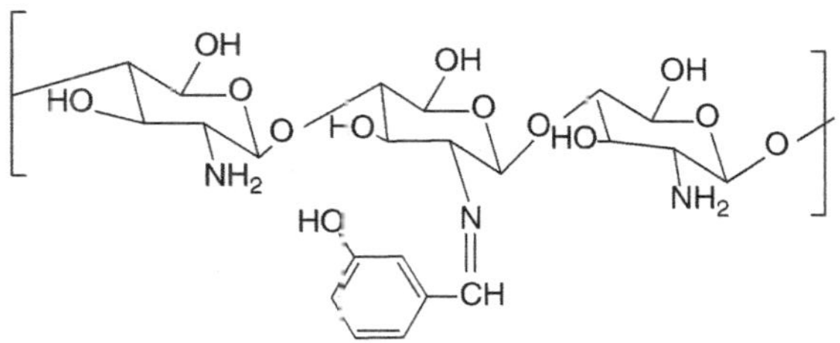	Mild steel/ 1 Molar HCl	70.08%/1500 ppm	[15]

(Continued)

Table 19.1 (Continued)

Inhibitors	Name	Structure	Metal/ corrosive medium	Inhibition efficiency (%)/ Concentration	Reference
9	**CHT-TSC** Chitosan Thiosemicarbazide	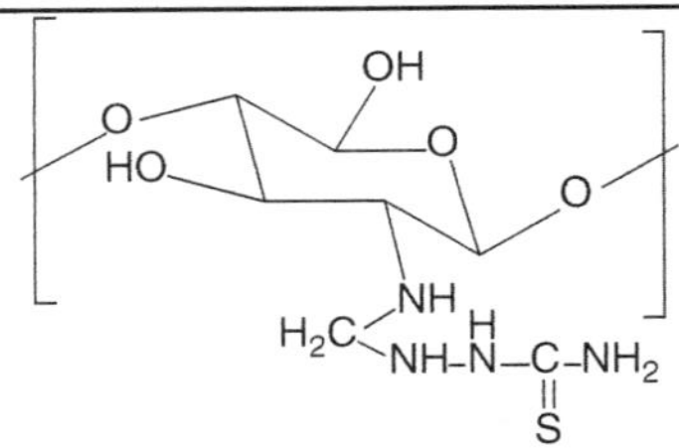 CHT-TSC	304 steel/ 2% acetic acid	92%/60 mg/L	[121]
	CHT-TCH Chitosan Thiocarbohydrazide	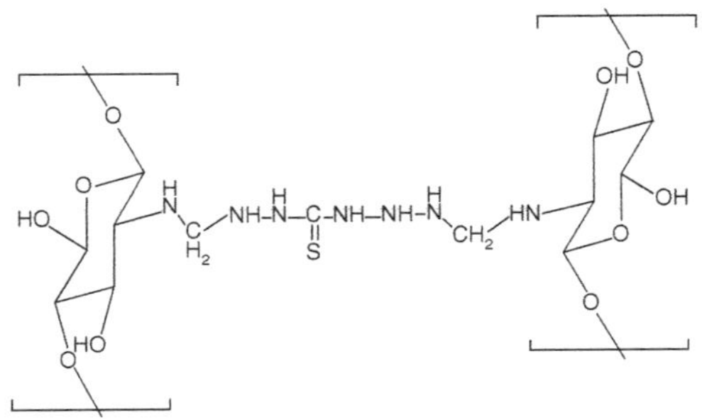 CHT-TCH			

10	**CHT-PVB** Chitosan Poly-vinyl butyral	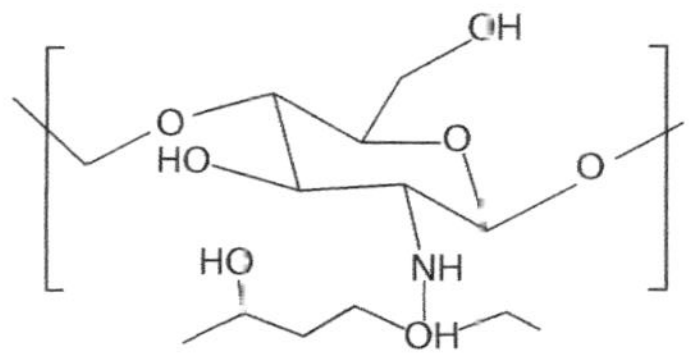	Carbon steel/ 0.3 M salt solution	-	[122]
11	**CHT-PEG** Chitosan Poly-ethylene glycol	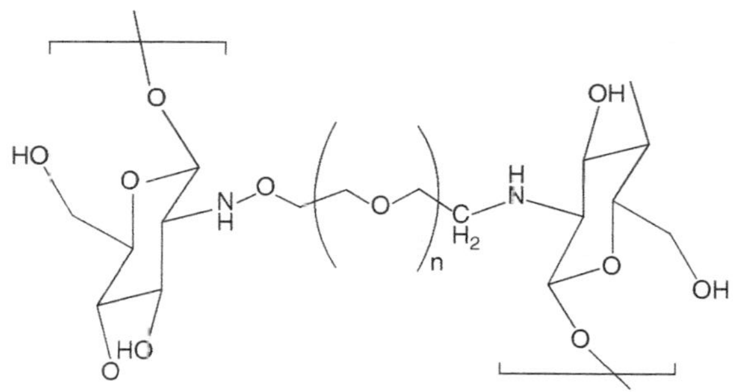	Mild steel/ 1 Molar sulphuric acid	93.9%/ 200 ppm	[123]

were found to be effective inhibitors of steel corrosion in acidic solutions by P. Kong et al [118]. Electrochemical impedance studies (EIS) and Potentiodynamic studies (PDP) showed that polyaniline grafted chitosan performed as a mixed-type inhibition for Q235 steel in 0.5 molar HCl. Several Chitosan-complex [15, 119–122] showed enhanced corrosive inhibition by the addition of other compounds that are species rich in heteroatoms. The heteroatoms act as active sites for interaction with the metal surface and the increased molecular size provides better surface coverage.

Further chitosan-composites include chitosan–Zn, Ag, Cu, Ni, Au, and F nanoparticles [123–131], chitosan TiO_2 nanocomposites [132–134], hydroxyapatite [135–139] – polyamines [115]. Some grafted-chitosan as corrosion-resistant compounds include PMVI (poly-2-methyl-1-vinyl imidazole) [140, 141], sulfonated chitosan [142, 143], chitosan-polymer blends [144], and others. In general, the grafted chitosan-based compounds outperformed pure chitosan in terms of protection. As a result, it is suggested that in future investigations, modified chitosan can be used. The adsorption-isotherm model predicts that these substances would bind on the metallic substrate and become efficient as inhibitors. Usually, chitosan-based compounds suppress both anodic-cathodic polarization reactions and behave as mixed-type inhibitors.

19.4.1.2 Pectin as Corrosion Inhibitors

Pectin is a unique combination of hetero-polysaccharides whose constituents vary depending on the plants as well as the parts of plants after ripening. Pectin is often derived from fresh fruits, apples, and other fruits to be utilized in the food manufacturing industries as a gel-forming or thickening agent as well as a stabilizer in various confections. Pectin's capacity to prevent corrosion of metals, like that of the other poly-saccharides research, is based on the structure. Pectin carbohydrates framework has acidic (COOH) and carboxymethyl ($COOCH_3$) substituents, making it a potential choice chemical for reducing the deterioration and scaling of metals in various environments.

Pectin is a bio-disposable, non-toxic, and environmentally friendly rust inhibitor [145] is presented in Table 19.2. The physio-electrochemical approaches recently showed the anti-corrosion capacity of pectin, which was generated from apples against X60 steel in hydrochloric acid media [146]. Khairou et al. [147] performed inhibitive experiments on cadmium electrodes in 0.5 Molar hydrochloric acids using EIS and Tafel polarization techniques. In a corrosive environment of 2% NaCl, 1% CH_3COOH, and 0.5% citric acid mixture, pectin concentrate from tomato peel was used to explore corrosion protection resulting in the value of % IE of 73 reported at very small concentrations (4 g/L) utilizing EIS (Electrochemical Impedance Spectroscopy) combined with the PDP (Potentiodynamic Polarization) technique [148].

Table 19.2 Pectin and grafted pectin as corrosion inhibitors.

Inhibitors	Name	Structure	Metal/ corrosive medium	Inhibition efficiency (%)/ Concentration	Reference
Pectin as Corrosion Inhibitor					
1	Pectin	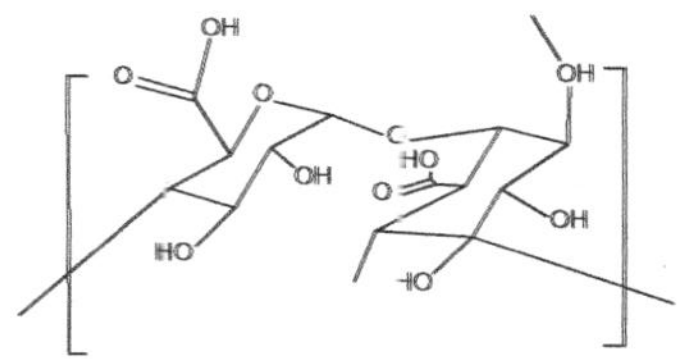	Aluminium/2.0 Molar HCl	91%/8.0 g/L	[145]
2	Pectin extracted from apple fruit		X 60 pipeline steel/0.5 Molar HCl		[146]
3	Pectin		Cadmium/0.5 Molar HCl	67.5%/2×10^{-1} g/ dm^{-3}	[147]
4	Pectin from tomato peel.		Tin/2% NaCl + 1% CH_3COOH + 0.5% citric acid	73%/4 g/L	[148]
Grafted-Pectin as Corrosion inhibitors					
1	Pectin-ascorbic acid binary system		Tin/Neutral NaCl solution		[149]
2	Pectinpropyl phosphonic acid + Zn^{2+} ion		Carbon steel/Neutral aqueous environment	84%/250 ppm + 50 ppm Zn^{2+}	[150]
3	Pectin-grafted Polyacrylamide Pectin-grafted Polyacrylic acid	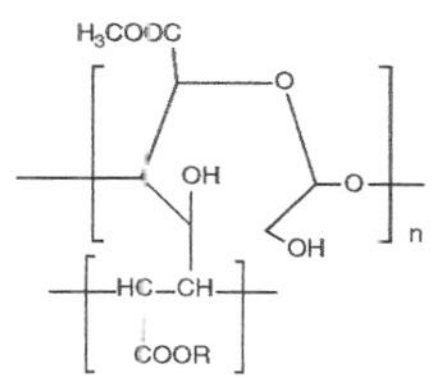	Mild steel/3.5 M NaCl	89.7%/700 ppm 82. 8%/400 ppm	[151]

Nada et al. [149] studied the pectin-ascorbic acid binary system which according to the researchers the protective film on the metal surface was made of Fe^{3+}/Fe-propyl phosphonic acid complex, $Zn(OH)_2$, and oxides. The creation of the metal-inhi at the tin/solution junction was related to the inhibiting impact of this inhibition mechanism. Using chemical-electrochemical approaches, Prabakaran et al. [150] discovered synergism in pectin- propyl phosphonic acid (PA) and Zn^{2+} ions in the neutral, alkaline medium for carbon steel. In this investigation, the addition of pectin improved the capability of the secondary compounds (PA and Zn^{2+} additives) to inhibit the corrosive system. The mechanism of inhibition was confirmed by FTIR, AFM, SEM, and X-ray photoelectron spectroscopy.

Pectin, like other carb-biomolecules, can be tailored for a variety of uses by tweaking its function chemistry. Some grafted biopolymers were created from pectin, acrylamide, and acrylic acid as substrates. Pectin-g polyacrylamide and Pectin-g polyacrylic acid grafted polymers were generated in the process. The overall performance of such corrosion protection polymers is enhanced by linking several active sites suitable for metal substrate attachment and physical displacement of corrosive substances like water all over the metal/solution boundary. Grafted polymer with polyacrylamide and polyacrylic acid protected mild steel corrosion in 3.5 wt% NaCl aqueous solution, and the resultant materials guarded steel as IE% (inhibition efficiency) was 85% [151]. Except for the modification of the final pH-sensitive gelatine compositions with acrylic and acrylamide acids, and the use of pectin as a predecessor, the graft polymerization technique differs significantly from that revealed by Mishra et al. [152].

Antiscalants made from pectin alteration compounds have also been observed to be employed in water purification [153]. This process was created solely to decrease the molecular mass of pectin before grafting to acrylamide resulting in either a copolymer graft having an antiscalant capable of inorganic salts (carbonates, sulphates, and phosphates). Pectates are D-Galacturonic (pectic) acid salts or esters.

Pectic acids (also known as D-Galacturonic acid) are found in plants, but primarily in ripening fruit [154]. Pectates, like all other pectic compounds, were used to limit corrosion due to their free alcohol-containing groups which bind with the metal surface. Zaafarany [155] used chemical approaches to limit the corrosion of pectates along with alginates and anionic polyelectrolytes for Aluminium in a 4M sodium hydroxide (NaOH) solution. The addition of pectate in the liquid electrolyte considerably lowered the corrosion process of aluminium, with an inhibitory efficacy of approximately 88% (at 1.6% of pectate). The value was measured by using the gravimetric weight loss methodology.

In the same corrosive environment, Zaafarany and his co-worker [156] repeated this study using alginate and pectate of sodium salt for the pure aluminium substrate. The results of the investigation were computed and revealed that there was no notable change. Corrosion behaviour by sodium pectate salt was attributed to physical molecule binding on the metal surface, as proven by kinetic and thermodynamic processes. The adsorption behaviour followed Langmuir and Freundlich isotherms.

19.4.1.3 Exudate Gum as Corrosion Inhibitors

Exudate gums are thick polysaccharide fluids produced by trees that seem wet or viscous when fresh and harden once dry. Because of their particular physical qualities, they are utilized as oil/gas commercial adhesives and bindings [157], and also in food products as gelling agents [158].

Gum Arabic (GA) was used as a mixed-type inhibitor for several metallic substrates [159–161] and published studies demonstrate that it prevents metals from corrosion in both acidic and neutral conditions to a significant extent. Guar gum (GG) proved to show anti-corrosive properties. Using Tafel polarization and gravimetric weight loss methods, Abdallah [162] was the one to demonstrate guar gum corrosion protection performance for L-52 grade-carbon steel in a sulfuric acid environment with NaCl which also caused pitting.

Grafted polyacrylamide with xanthan exudates gum showed a remarkable inhibitory effect on mild steel at 15% hydrochloric acid (HCl) concentration by Biswas et al. [163]. The IE% (inhibition efficiency) increased with increasing inhibitor concentration with all of the corrosion techniques such as EIS, PDP, and Surface studies which are included in this investigation. The maximum IE% was 92 which is considered significant for 15% HCl. It was revealed that the gum behaved as a mixed-type inhibitor with molecule binding at the metal interface commencing corrosion inhibition. The inclusion of polyacrylamide enhanced the inhibition phenomena. The inhibitory action was discovered using physical/chemical characteristics as well as xanthan gum binding on the surface of a metal in association with the polymeric matrix.

The synthesis and evaluation of the Polyaniline (PANI) grafted-Xanthan Gum (XG) framework as a conductive biopolymer was also described by G. Babaladimath et al. [164]. The graft was accomplished by using the microwave to polymerize aniline under an acid medium. The inclusion of polyaniline appears to increase the parental polymer's thermal-breakdown and electrical conductance. In an acidic environment (1molar HCl), both xantham-gum and xanthan-gum-polyaniline showed potential inhibiting efficacy against aluminium degradation.

Gum Arabic (GA) was incorporated into cerium oxide nanomaterial (CONP's) for corrosion protection in epoxy-covered metal parts. With the addition of Gum

Arabica, there had been a significant improvement in the coating barrier characteristics. The increased barrier qualities of the amended coating are attributable to the improved bridging of the cured epoxy, which decreases the number of gaps and conductive pathways. Gum Arabic generated from Cerium Oxide nanomaterial and deposited onto steel surface serves to minimize corrosion's active region to provide corrosion protection [165].

A greener anticorrosive silver composite [166], (GA-AgNPs) for steel protection has been developed for HCl (15%) and H_2SO_4 (15%) corrosive medium. Fourier transform infrared and UV-visible surface analysis (EDAX and SEM) are all used to characterize GA-AgNPs. All of the methods employed show GA-AgNPs to be an efficient corrosion protection substance. The inhibition mechanism follows the Langmuir adsorption isotherm. In sulphuric acid corrosive solution, GA-AgNPs behave as a mixed corrosion inhibitor, however, in a hydrochloric medium, it functions like an anodic-inhibitor.

The base materials guar gum and acrylamido-methyl-sulphonic acid (AMPS) were used to make grafted Guar gum acrylamide-methyl, propane-sulphonic acid (GG-AMPS). The grafted inhibitor GG-AMPS performed well in gravimetric weight loss tests at the optimum conc. of inhibitor showed the best result at 600 mg/L. The effectiveness of the system reduced as the temperature increased and the acid media percentage increased. The effectiveness of the inhibitor, on the other hand, rose as the immersion time was prolonged [167].

19.4.2 Dextrin and Cyclodextrin

There have been other research efforts related to the corrosion protection action of dextrin and cyclodextrin on metals/ alloys in various environments which have been published [168–172]. Dextrin is a less mol. weight natural polymer with carbohydrate (D) units linked by glycosidic bonds [–(14) or –(16)], as shown in Figure 19.7. They are produced in the human gastrointestinal system by hydrolytic enzymes catalysing the hydrolysis of starch in the mouth and they can be synthesized by thermal treatment in an acidic medium.

Dxt-g-pVAc (dextrin-based polyvinyl acetate), copolymer dextrin-based graft composites, were produced via atom transfer radical polymerization (ATRP) and used mixed-type inhibitors for mild steel in a 15% HCl media [168]. The protective characteristics of dextrin and its polymer were compared using electrochemical techniques. When compared to pure dextrin (84.56%), the

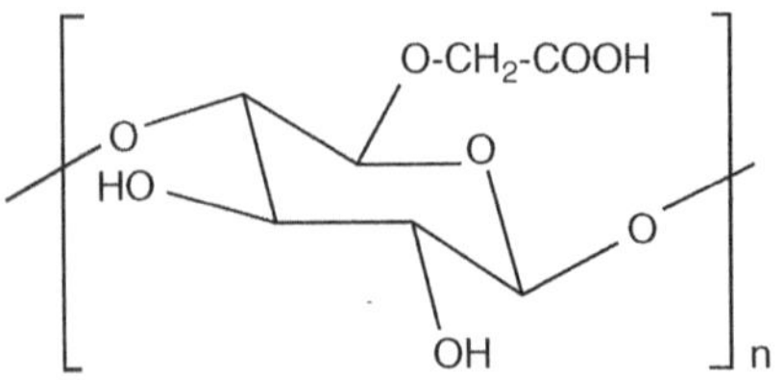

Figure 19.7 Structure of dextrin.

produced graft copolymer demonstrated outstanding corrosion protective efficacy (98.39%).

To improve the inhibitory property of dextrin against the dissolving of mild steel in acid media, the Floxacin nucleus (Cipro, Levo, and Norfloxacin) was utilized. The inclusion of ofloxacin in mild steel resulted in a significant increase in inhibitory properties compared to the native dextrin [169].

19.4.3 Cellulose as Corrosion Inhibitors

As corrosion inhibitors, cellulose and its variants might be considered eco-sustainable options. The natural origin of cellulose and its compounds, as well as its degradability and non-bioaccumulation qualities, are credited with their environmental friendliness.

19.4.3.1 Carboxymethyl-cellulose

The derivatives of cellulose most commonly used as corrosion inhibitors are carboxy-methyl cellulose [173] and hydroxy-methyl cellulose. Umoren et al. [174] extensively studied the influence of carboxy-methyl cellulose on mild steel corrosion in 2 M sulfuric acid. They also reported the role of synergents i.e potassium halides (i.e, KCl, KBr, and KI at 5mm concentration) on inhibition. Carboxymethyl cellulose along with KI showed the best performance and attained IE (inhibition efficiency) 85%.

Solomon et al. [175]. synthesized Carboxy-methyl cellulose-silver nanoparticle composites (CMC/AgNPs) and tested its inhibition proficiency for St37 type steel in 15% sulfuric acid. The corrosion studies were analysed using molecular, electro-chemical, and surface studies. The scanning electron microscopy (SEM) investigation revealed that with the addition of CMC/AgNPs, the metal's surface improves dramatically, confirming the adsorption capability of corrosion inhibitor composites.

The group of researchers [176] studied the effects of carboxy-methyl cellulose-1-hydroxyethane-1, 1-diphosphonic acid-Zn^{2+} (CMC-HEDP) on the inhibitory action for mild steel in neural Cl– ion (60 ppm Cl). In reviewing the combination of (HEDP: 300 ppm+Zn^{2+}: 10 ppm), IE was found to be 40%, but improved to 80% when 50 ppm CMC was added as the maximum concentration.

Certain nanoparticles, including Copper (Cu), Iron (Fe), and Nickel (Ni) were synthesized in a carboxymethyl cellulose (CMC) framework and subsequently employed as metal (steel) inhibitors in a hydrochloric acid solution. In comparison to the blank solution, the acid PDP result demonstrated improved behaviour in the addition of the inhibitor or without the nanomaterials. The corrosion reduction achieved in the presence of composites was in the sequence of CMC/Ni > CMC/Cu > CMC/Fe > CMC [177].

Three types of corrosion inhibitors for the copper in the NaCl solution were tested using the cellulose-polymeric niacin nanocomposite including ethyl, micro-crystalline, and carboxy-methyl cellulose-polymeric niacin composites [178]. The results revealed that all three composites reduced corrosion current (I_{corr}); nevertheless, the ethyl cellulose-niacin dextras inhibitor acted more consistently and had the maximum inhibition efficacy. The addition of inhibitors enhanced the impedance in the EIS test. Furthermore, in the case of ethyl cellulose-polymeric niacin, the manifestation of this action was more severe.

19.4.3.2 Hydroxy-methyl, Hydroxy-ethyl Cellulose

Another derivative of cellulose is hydroxyethyl-cellulose (HEC) which is utilized as a thickener, adhesive, and gel-forming agent as well as in the healthcare field for solubility of drugs in gastric fluid. Liao et al. [179] studied, microwave synthesized polymer hydroxyethyl cellulose, acrylamide/bentonite (HEC-AM) graft polymer, and polymer concentration permeation was used to make the composite. X-ray diffraction (XRD) investigation, Fourier-transform infrared (FTIR) spectroscopy, and Scanning electron microscopy (SEM) were used to examine the polymer/bentonite composite.

In a salt solution, Shih and Chieh [180] assess the overall corrosion protection effectiveness of bio-based polymer hydroxypropyl methylcellulose (HPMC) compounds. The anticorrosive abilities of hydroxypropyl methylcellulose (HPMC), hydroxypropyl methylcellulose phthalate (HPMCP), and hydroxypropyl methylcellulose acetate succinate (HPMCAS) film variants were studied. HPMC and HPMCP have shown anti-corrosive properties which has shown to improve with the increasing layer thickness in both substances. HPMCP also outperforms HPMCAS in terms of corrosion protection due to its hydrophobic surface and low water content. The research is significant in terms of developing sustainable corrosion inhibitors for steel surfaces using biopolymer-based derivatives.

Carrageenan, like the other biopolymers addressed in this series, is a greener and more sustainable carbohydrate polymeric molecule with corrosion-prevention potential. They are non-toxic and, owing to their distinct chemical compositions and molecular structure (Figure 19.8), can combine metallic ions at

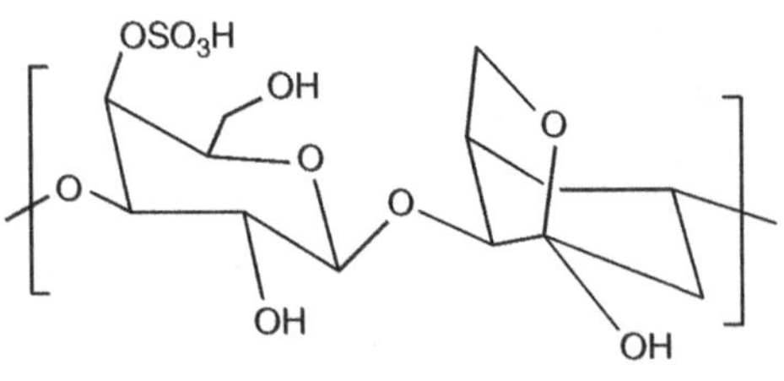

Figure 19.8 Structure of carrageenan.

interfaces to prevent corrosion. The –SO3H groups on these natural polymers have the pi-orbital property of giving electrons to the metal substrates' having vacant d orbital.

The red seaweed "Acanthophora spicifera-Rhodophyceae" produces the k-Carrageenan which has anticorrosive properties [181]. The k-Carrageenan hybrid composites with gelatine was synthesized using the sol-gel technique to provide the greatest corrosion prevention abilities. According to the electrochemical data, the covered SS 304 showed a promising corrosion inhibition with an efficiency of up to 97% after twenty-four hours and up to 65% after two hundred twenty-seven hours.

19.4.4 Starch as Corrosion Inhibitors

Starch, which is made up of straight-chain amylose and branched-chain amylopectin, is also used to protect metals and alloys against corrosion. Its applicability as a corrosion inhibitor is limited since it is poorly soluble. As a result, physiologically modified starch is commonly used for metallic corrosion protection.

A novel graft inhibitor of glycerine-starch copolymers was synthesized and tested for carbon-manganese protection [182]. The corrosion inhibition studies of Carbon-Manganese (C-Mn) steel in 1 molar hydrochloric acid (HCl) solution were investigated between 25 °C–50 °C in the absence and presence of glycerine-grafted starch as a bio-graft polymer that served as a corrosion inhibitor. According to the findings, the inhibition process can be more complicated than the physiosorption process on metal surfaces. It could be explained based on copolymer chemical adsorption on metal substrates.

The triple co-graft polymer of cassava starch-sodium allyl-sulfonate-acryl amide (CS-SAS-AAGC) was made by cross-linking two polymeric materials of sodium allyl sulfonate (SAS) and acryl amide (AA) concurrently onto cassava starch [183]. Weight loss, potentiodynamic polarization/electrochemical impedance methods, and surface characterization methods were used to investigate the inhibitory efficacy of triple graft starch on cold-rolled steel in a1 molar mineral acid (HCl) solution. At an inhibitor concentration (triple-graft) of 50 mg/L, the % IE achieved was 97.2%, and its corrosion inhibition capacity is better than casava starch, sodium allyl sulphonate, or acryl amide separately.

The description of exudates gum, dextrin, carboxy methylcellulose, hydroxyethyl-cellulose, starch, and their graft polymers are shown in Table 19.3.

Table 19.3 Grafted exudate gum, grafted dextrin, and grafted starch as corrosion inhibitors.

Inhibitors	Name	Structure	Metal/ corrosive medium	Inhibition efficiency (%)/Concentration	Reference
Exudate Gum as Corrosion Inhibitor					
1	Polyacrylamide – xantham exudates gum	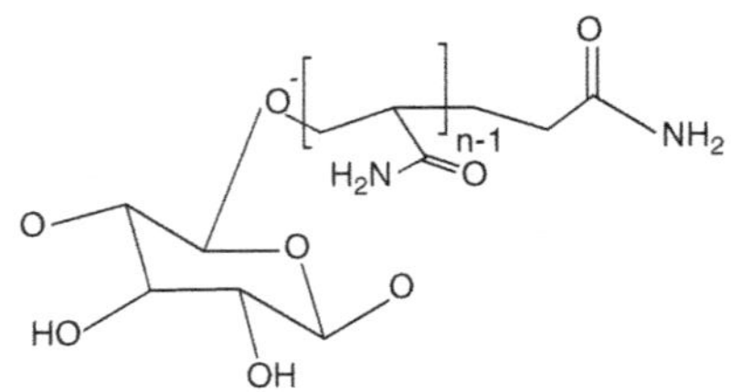	Mild steel/ 15% HCl	92%	[163]
2	Polyaniline grafted-Xanthan Gum	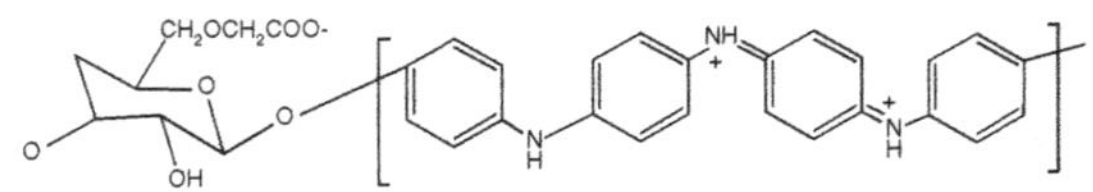	Aluminium/ 1 Molar HCl	95%/0.4g/L	[164]
3	Guar gum-acryl amido-methy lpropanesulphonic acid (GG-AMPS)	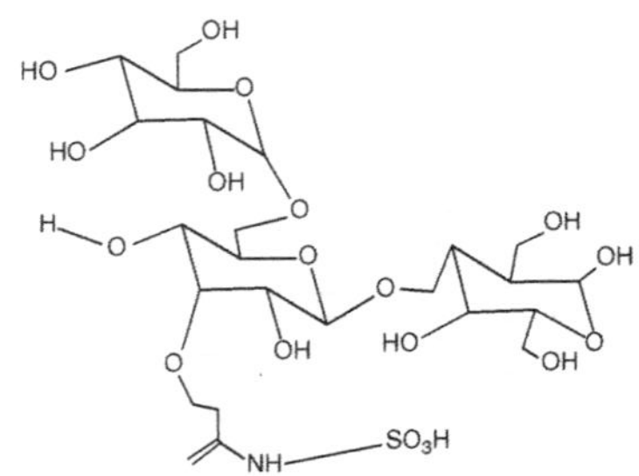	Copper/ 3.5% NaCl	90.3%/600 mg/L	[167]

Grafted Dextrin & Cyclodextrin					
1	Dextrin-PVA	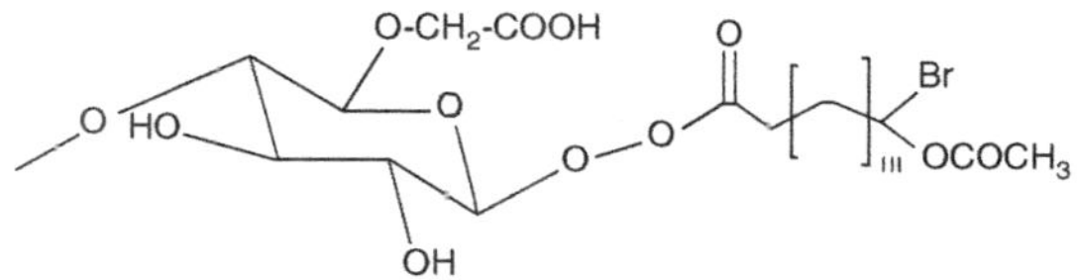	Mild steel/1 M HCl	98.39%/0.15 g/L	[168]
2	Dextrin-floxacin Graft	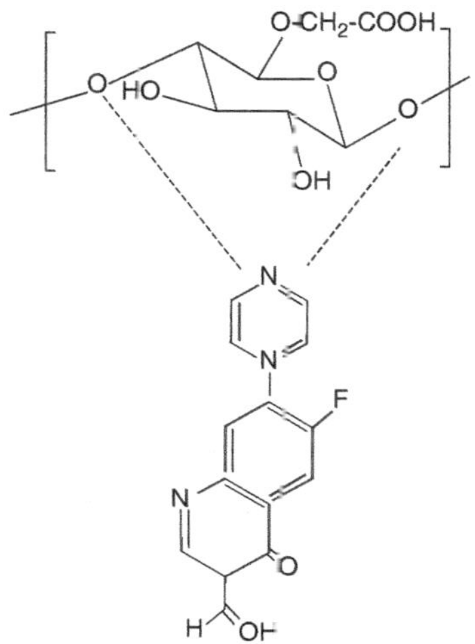	Mild steel/ 0.5 molar H_2SO_4	78.2%/50 ppm	[169]

(Continued)

Table 19.3 (Continued)

Inhibitors	Name	Structure	Metal/ corrosive medium	Inhibition efficiency (%)/Concentration	Reference
Starch					
1	Glycerine-Grafted Starch	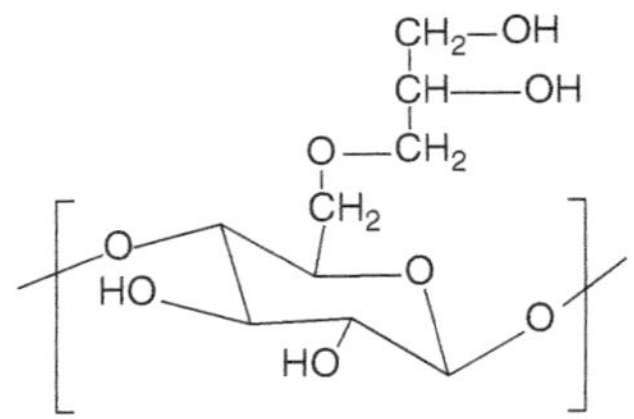	C-Mn Carbon-Manganese steel	93%/300 mg/L	[176]
2	Cassava starch-sodium allyl sulfonate-acryl amide graft copolymer ALDS-SAS-AA	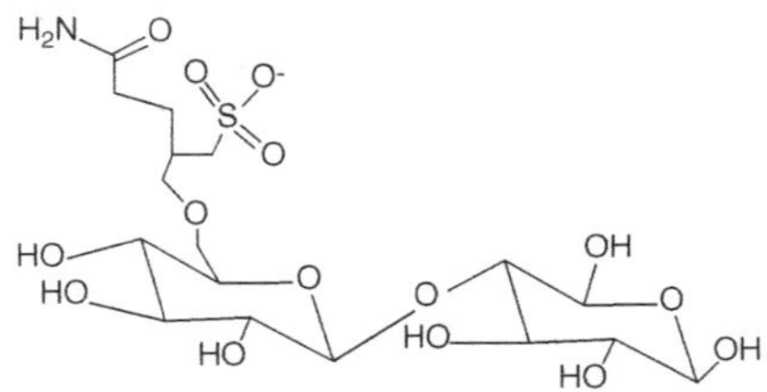	Cold-roll steel/ 1 molar HCl	97.2%/50 mg/L	[177]
	ALDS-AA- SAS	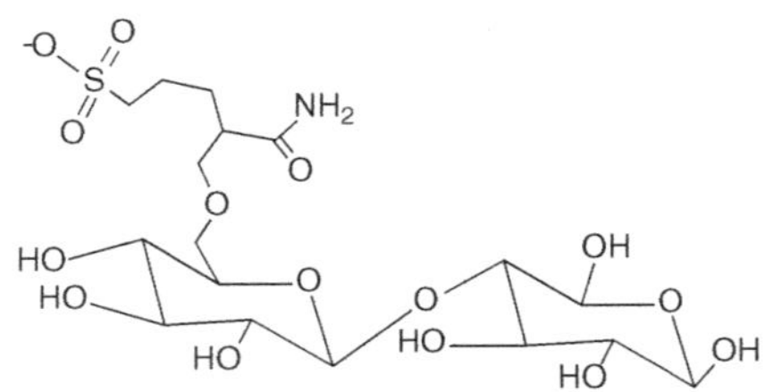			

19.5 Conclusion

The present chapter represents environmentally friendly alternative options to conventional hazardous corrosion inhibitors. According to a review of the literature, numerous categories of classical and upgraded natural or biopolymers have been reported widely as eco-friendly alternative options to conventional harmful corrosion inhibitors. The use of biopolymers in the production of formulations of inhibitors for diverse potential applications is growing rapidly and their chemistry is straightforward because the adsorption process is either chemisorption or physical adsorption at the surface of the metals. Some carbohydrate polymer combinations have been shown to have chemical ingredients capable of creating passivation layers, which prevent corrosive ions and molecules from passing through the metal/solution contact.

Biopolymers, because of their underlying stability, accessibility, low cost, and nontoxicity are considered environmentally friendly corrosion inhibitors for various metals in corrosive media. Dextrin and poly (vinyl acetate) based graft copolymers, and glycerine-grafted starch polymers have been used as efficient corrosion inhibitors for the protection of metals in aqueous solutions. These compounds by covering the huge metallic surface in aqueous solutions from degradation by corrosive chemicals and particles protect the metals.

Chitosan, a new versatile biopolymer, is a polysaccharide obtained from chitin. Chitosan's corrosion protection action on metals and alloys has been thoroughly documented by several researchers. Their performance as corrosion inhibitors is directly based on their chemical composition, macromolecular weights, and unique electronic as well as molecular structures. With an increasing need for sustainable and green technology, the use of environmentally friendly materials should then be investigated further. Biopolymers are a natural, non-toxic, sustainable, and environmentally beneficial replacement for certain other corrosion inhibitors with harmful effects. They are environmentally beneficial replacements for certain other corrosion inhibitors with harmful effects. Graft-polymer compounds are effective corrosion inhibitors for different metals in a variety of corrosive environments.

References

1 Wagner, C. and Traud, W. (1938). Über die Deutung von Korrosionsvorgängen durch Überlagerung von elektrochemischen Teilvorgängen und über die Potentialbildung an Mischelektroden. *Zeitschrift für Elektrochemie und angewandte physikalische Chemie* 44 (7): 391–402.

2 Saha, J.K. (2012). *Corrosion of Constructional Steels in Marine and Industrial Environment: Frontier Work in Atmospheric Corrosion*. Springer Science & Business Media.

3 Sastri, V.S. (2012). *Green Corrosion Inhibitors: Theory and Practice*. John Wiley & Sons.

4 Dunlop, A.K., Howard, R.L., and Raifsnider, P.J. (1969). *Material Protection* 8 (27).

5 De Damborenea, J., Conde, A., and Arenas, M. (2014). Corrosion inhibition with rare earth metal compounds in aqueous solutions. In: *Rare Earth based Corrosion Inhibitors* 84–116. Woodhead Publishing Series in Metals and Surface Engineering. Elsevier.

6 Verma, C., Obot, I.B., Bahadur, I. et al. (2018). Choline-based ionic liquids as sustainable corrosion inhibitors on mild steel surface in acidic medium: gravimetric, electrochemical, surface morphology, DFT and Monte Carlo simulation studies. *Applied Surface Science* 457: 134–149.

7 Marzorati, S., Verotta, L., and Trasatti, S.P. (2019). Green corrosion inhibitors from natural sources and biomass wastes. *Molecules* 24 (1): 48.

8 Rani, B.E. and Basu, B.B.J. (2012). Green inhibitors for corrosion protection of metals and alloys: an overview. *International Journal of Corrosion* 2012.

9 Verma, C., Kumar, A.M., Mazumder, M.A.J., and Quraishi, M.A. (2018). Chitosan-based green and sustainable corrosion inhibitors for carbon steel. *Chitin-Chitosan: Myriad Functionalities in Science and Technology* 143.

10 Cheng, S., Chen, S., Liu, T. et al. (2007). Carboxymenthylchitosan as an ecofriendly inhibitor for mild steel in 1 M HCl. *Materials Letters* 61 (14–15): 3276–3280.

11 Cheng, S., Chen, S., Liu, T. et al. (2007). Carboxymethylchitosan+ Cu2+ mixture as an inhibitor used for mild steel in 1 M HCl. *Electrochimica Acta* 52 (19): 5932–5938.

12 Varma, R.S. (2019). Biomass-derived renewable carbonaceous materials for sustainable chemical and environmental applications. *ACS Sustainable Chemistry & Engineering* 7 (7): 6458–6470.

13 Fekry, A.M. and Mohamed, R.R. (2010). Acetyl thiourea chitosan as an eco-friendly inhibitor for mild steel in a sulfuric acid medium. *Electrochimica Acta* 55 (6): 1933–1939.

14 Muzzarelli, R.A., Guerrieri, M., Goteri, G. et al. (2005). The biocompatibility of dibutyryl chitin in the context of wound dressings. *Biomaterials* 26 (29): 5844–5854.

15 Li, M., Xu, J., Li, R. et al. (2014). Simple preparation of aminothiourea-modified chitosan as a corrosion inhibitor and heavy metal ion adsorbent. *Journal of Colloid and Interface Science* 417: 131–136.

16 Jayakumar, R., Nwe, N., Tokura, S., and Tamura, H. (2007). Sulfated chitin and chitosan as novel biomaterials. *International Journal of Biological Macromolecules* 40 (3): 175–181.

17 Balakrishnan, B., Mohanty, M., Umashankar, P.R., and Jayakrishnan, A. (2005). Evaluation of an in situ forming hydrogel wound dressing based on oxidized alginate and gelatin. *Biomaterials* 26 (32): 6335–6342.

18 Wang, H. and Zheng, J. (2002). Progress of research on environmental-friendly corrosion inhibitors. *Corrosion Science and Protection Technology* 14 (5): 275–279.

19 Bethencourt, M., Botana, F.J., Calvino, J.J., et al. (1998). Lanthanide compounds as environmentally-friendly corrosion inhibitors of aluminum alloys: a review. *Corrosion Science* 40 (11): 1803–1819.

20 Powell, S.M., McMurray, H.N., and Worsley, D.A. (1999). Use of the scanning reference electrode technique for the evaluation of environmentally friendly, non-chromate corrosion inhibitors. *Corrosion* 55: 11.

21 Marzorati, S., Verotta, L., and Trasatti, S.P. (2018). Green corrosion inhibitors from natural sources and biomass wastes. *Molecules* 24 (1): 48.

22 Ninčević Grassino, A. (2017). Plant extracts as a natural corrosion inhibitors of metals and its alloys used in food preserving industry. In: *Science within Food: Up to date Advances on Research and Education Ideas, Food Science BookSeries*, 1 (ed. A. Mendez-Vilas), 185–193. Badajoz: Formatex Research Center.

23 Palou, R.M., Olivares-Xomelt, O., and Likhanova, N.V. (2014). Environmentally friendly corrosion inhibitors. *Developments in Corrosion Protection* 19 (1): 431–432.

24 Greenwood, J. (2007). Organized civil society and democratic legitimacy in the European Union. *British Journal of Political Science* 37 (2): 333–357.

25 Dehousse, R. (2008). Delegation of powers in the European Union: the need for a multi-principals model. *West European Politics* 31 (4): 789–805.

26 Castano, A., Cantarino, M.J., Castillo, P., and Tarazona, J.V. (1996). Correlations between the RTG-2 cytotoxicity test EC50 and in vivo LC50 rainbow trout bioassay. *Chemosphere* 32 (11): 2141–2157.

27 Hermens, J., Canton, H., Steyger, N., and Wegman, R. (1984). Joint effects of a mixture of 14 chemicals on mortality and inhibition of reproduction of Daphnia Magna. *Aquatic Toxicology* 5 (4): 315–322.

28 Zhang, M., Aguilera, D., Das, C. et al. (2007). Measuring cytotoxicity: a new perspective on LC50. *Anticancer Research* 27 (1A): 35–38.

29 Zakari, A. and Kubmarawa, D. (2016). Acute toxicity (LD50) studies using Swiss albino mice and brine shrimp lethality (LC50 and LC90) determination of the ethanol extract of stem bark of Echinacea Angustifolia DC. *Natural Products Chemistry & Research* 4 (2).

30 Parra, A.L., Yhebra, R.S., Sardiñas, I.G., and Buela, L.I. (2001). Comparative study of the assay of Artemia salina L. and the estimate of the medium lethal dose (LD50 value) in mice, to determine oral acute toxicity of plant extracts. *Phytomedicine* 8 (5): 395–400.

31 Chomel, M., Fernandez, C., Bousquet-Mélou, A. et al. (2014). Secondary metabolites of P inus halepensis alter decomposer organisms and litter decomposition during the afforestation of abandoned agricultural zones. *Journal of Ecology* 102 (2): 411–424.

32 Summers, M.D., Blunk, S.L., and Jenkins, B.M. (2003). How straw decomposes: implications for straw bale construction. *Ecological Building Network*, San Rafael, CA: USA.

33 Wedmann, R., Ivanovic-Burmazovic, I., and Filipovic, M.R. (2017). Nitrosopersulfide (SSNO–) decomposes in the presence of sulfide, cyanide, or glutathione to give HSNO/SNO–: consequences for the assumed role in cell signaling. *Interface Focus* 7 (2): 20160139.

34 van der Heijden, S.A. and Jonker, M.T. (2009). Evaluation of liposome– water partitioning for predicting bioaccumulation potential of hydrophobic organic chemicals. *Environmental Science & Technology* 43 (23): 8854–8859.

35 Conder, J.M., Hoke, R.A., Wolf, W.D., et al. (2008). Are PFCAs bioaccumulative? A critical review and comparison with regulatory criteria and persistent lipophilic compounds. *Environmental Science & Technology* 42 (4): 995–1003.

36 Williamson, R.L., Jr, Burton, D.T., Clarke, J.H., and Fleming, L.E. (1993). Gathering danger: the urgent need to regulate toxic substances that can bioaccumulate. *Ecology LQ* 20: 605.

37 Bahri, M., Mahdavi, A., Mirzaei, A. et al. (2018). Integrated oxidation process and biological treatment for highly concentrated petrochemical effluents: a review. *Chemical Engineering and Processing-Process Intensification* 125: 183–196.

38 Gueccia, R., Aguirre, A.R., Randazzo, S. et al. (2020). Diffusion dialysis for separation of hydrochloric acid, iron, and zinc ions from highly concentrated pickling solutions. *Membranes* 10 (6): 129.

39 Agrawal, A. and Sahu, K.K. (2009). An overview of the recovery of acid from spent acidic solutions from steel and electroplating industries. *Journal of Hazardous Materials* 171 (1–3): 61–75.

40 Finšgar, M. and Jackson, J. (2014). Application of corrosion inhibitors for steels in acidic media for the oil and gas industry: a review. *Corrosion Science* 86 17–41.

41 Chen, Y. and Yang, W. (2019). Formulation of corrosion inhibitors. *Water Chemistry*, 1–22.

42 Quraishi, M.A. and Ansari, F.A. (2003). Corrosion inhibition by fatty acid triazoles for mild steel in formic acid. *Journal of Applied Electrochemistry* 33 (3): 233–238.

43 Quraishi, M.A. and Ansari, F.A. (2006). Fatty acid oxadiazoles as corrosion inhibitors for mild steel in formic acid. *Journal of Applied Electrochemistry* 36 (3): 309–314.

44 Ansari, F.A., Sudheer, Chauhan, D.S., and Quraishi, M. A. (2022). Oleochemicals as corrosion inhibitors, Part III, Chapter 15. In: *Organic Corrosion Inhibitors: Synthesis, Characterization, Mechanism, and Applications*, 343–369. John Wiley & Sons, Inc.

45 Kumar, S., Sharma, D., Yadav, P., and Yadav, M. (2013). Experimental and quantum chemical studies on corrosion inhibition effect of synthesized organic

compounds on N80 steel in hydrochloric acid. *Industrial & Engineering Chemistry Research* 52 (39): 14019–14029.
46 Yadav, M., Behera, D., Kumar, S., and Sinha, R.R. (2013). Experimental and quantum chemical studies on the corrosion inhibition performance of benzimidazole derivatives for mild steel in HCl. *Industrial & Engineering Chemistry Research* 52 (19): 6318–6328.
47 Yadav, M., Sinha, R.R., Kumar, S., and Sarkar, T.K. (2015). Corrosion inhibition effect of spiropyrimidinethiones on mild steel in 15% HCl solution: insight from electrochemical and quantum studies. *RSC Advances* 5 (87): 70832–70848.
48 Umoren, S.A., Solomon, M.M., Obot, I.B., and Suleiman, R.K. (2018). Comparative studies on the corrosion inhibition efficacy of ethanolic extracts of date palm leaves and seeds on carbon steel corrosion in 15% HCl solution. *Journal of Adhesion Science and Technology* 32 (17): 1934–1951.
49 Sharma, H.K. and Quraishi, M.A. (2007). Inhibitive effect of some Gemini surfactants on mild steel corrosion in acid solutions.
50 Haruna, K., Saleh, T.A., and Quraishi, M.A. (2020). Expired metformin drug as green corrosion inhibitor for simulated oil/gas well acidizing environment. *Journal of Molecular Liquids* 315: 113716.
51 Singh, A. and Quraishi, M.A. (2015). Acidizing corrosion inhibitors: a review. *Journal of Materials and Environmental Science* 6 (1): 224–235.
52 Mo, S., Luo, H.Q., and Li, N.B. (2016). Plant extracts as "green" corrosion inhibitors for steel in sulphuric acid. *Chemical Papers* 70 (9): 1131–1143.
53 Miralrio, A. and Espinoza Vázquez, A. (2020). Plant extracts as green corrosion inhibitors for different metal surfaces and corrosive media: a review. *Processes* 8 (8): 942.
54 Fateh, A., Aliofkhazraei, M., and Rezvanian, A.R. (2020). Review of corrosive environments for copper and its corrosion inhibitors. *Arabian Journal of Chemistry* 13 (1): 481–544.
55 Hamilton-Amachree, A. and Iroha, N.B. (2020). Corrosion inhibition of API 5L X80 pipeline steel in an acidic environment using aqueous extract of Thevetia peruviana. *Chem International* 6 (3): 110.
56 Izionworu, V., Ukpaka, C., and Oguzie, E. (2020). Green and eco-benign corrosion inhibition agents: alternatives and options to chemical-based toxic corrosion inhibitors. *Chemistry International* 6 (4): 232–259.
57 Umoren, S.A., Banera, M.J., Alonso-Garcia, T. et al. (2013). Inhibition of mild steel corrosion in HCl solution using chitosan. *Cellulose* 20 (5): 2529–2545.
58 Rabizadeh, T. and Khameneh Asl, S. (2019). Chitosan as a green inhibitor for mild steel corrosion: thermodynamic and electrochemical evaluations. *Materials and Corrosion* 70 (4): 738–748.
59 Gupta, N.K., Joshi, P.G., Srivastava, V., and Quraishi, M.A. (2018). Chitosan: a macromolecule as a green corrosion inhibitor for mild steel in sulfamic acid

useful for the sugar industry. *International Journal of Biological Macromolecules* 106: 704–711.

60 Quraishi, M.A., Chauhan, D.S., and Ansari, F.A. (2021). Development of environmentally benign corrosion inhibitors for organic acid environments for the oil-gas industry. *Journal of Molecular Liquids* 329: 115514.

61 Deyab, M.A. (2020). Understanding the anti-corrosion mechanism and performance of ionic liquids in desalination, petroleum, pickling, de-scaling, and acid cleaning applications. *Journal of Molecular Liquids* 309: 113107.

62 Ardakani, E.K., Kowsari, E., and Ehsani, A. (2020). Imidazolium-derived polymeric ionic liquid as a green inhibitor for corrosion inhibition of mild steel in 1.0 M HCl: an experimental and computational study. *Colloids and Surfaces. A Physicochemical and Engineering Aspects* 586: 124195.

63 Plechkova, N.V. and Seddon, K.R. (2008). Applications of ionic liquids in the chemical industry. *Chemical Society Reviews* 37 (1): 123–150.

64 MacFarlane, D.R., Tachikawa, N., Forsyth, M. et al. (2014). Energy applications of ionic liquids. *Energy & Environmental Science* 7 (1): 232–250.

65 Lu, J., Yan, F., and Texter, J. (2009). Advanced applications of ionic liquids in polymer science. *Progress in Polymer Science* 34 (5): 431–448.

66 Wei, D. and Ivaska, A. (2008). Applications of ionic liquids in electrochemical sensors. *Analytica chimica acta* 607 (2): 126–135.

67 Gece, G. (2011). Drugs: a review of promising novel corrosion inhibitors. *Corrosion Science* 53 (12): 3873–3898.

68 Pathak, R.K. and Mishra, P. (2016). Drugs as corrosion inhibitors: a review. *International Journal of Science and Research* 5 (4): 671–677.

69 Liangtian, Y., Zhang, M., Shidong, C. et al. (2020). Investigation of corrosion inhibition effect of enprofylline drug on mild steel corrosion in sulphuric acid solution. *International Journal of Electrochemical Science* 15: 5102–5114.

70 Alfakeer, M., Abdallah, M., and Fawzy, A. (2020). Corrosion inhibition effect of expired ampicillin and flucloxacillin drugs for mild steel in aqueous acidic medium. *International Journal of Electrochemical Science* 15: 3283–3297.

71 Verma, C., Chauhan, D.S., and Quraishi, M.A. (2017). Drugs as environmentally benign corrosion inhibitors for ferrous and nonferrous materials in the acid environment: an overview. *Journal of Materials and Environmental Science* 8: 4040–4051.

72 Abdallah, M. (2004). Antibacterial drugs as corrosion inhibitors for corrosion of aluminum in hydrochloric solution. *Corrosion Science* 46 (8): 1981–1996.

73 Hassan, M.E.S., Bai, J., and Dou, D.Q. (2019). Biopolymers; definition, classification, and applications. *Egyptian Journal of Chemistry* 62 (9): 1725–1737.

74 Klemm, D., Heublein, B., Fink, H.P., and Bohn, A. (2005). Cellulose: fascinating biopolymer and sustainable raw material. *Angewandte Chemie International Edition* 44 (22): 3358–3393.

75 Rinaudo, M. (2006). Chitin and chitosan: properties and applications. *Progress in Polymer Science* 31 (7): 603–632.

76 Lenz, D.M., Tedesco, D.M., Camani, P.H., and Dos Santos Rosa, D. (2018). Multiple reprocessing cycles of corn starch-based biocomposites reinforced with Curauá fiber. *Journal of Polymers and the Environment* 26 (7): 3005–3016.

77 Jenkins, A.D., Kratochvíl, P., Stepto, R.F.T., and Suter, U.W. (1996). Glossary of basic terms in polymer science (IUPAC Recommendations 1996). *Pure and Applied Chemistry* 68 (12): 2287–2311.

78 Innovative Packaging of Medicines-IJAHR. Available online: https://www.ipinnovative.com/journals/IJAHR/article-full-text/10648 (accessed 27 June 2021).

79 Zadbuke, N., Shahi, S., Gulecha, B. et al. (2013). Recent trends and future of pharmaceutical packaging technology. *Journal of Pharmacy & Bioallied Sciences* 5 (2): 98.

80 Bohlmann, G. and Toki, G. (2004). *Chemical Economics Handbook*. Menlo Park: SRI International.

81 Dave, A.M., Mehta, M.H., Aminabhavi, T.M. et al. (1999). A review on controlled release of nitrogen fertilizers through polymeric membrane devices. *Polymer-Plastics Technology and Engineering* 38 (4): 675–711.

82 Boesel, L.F., Mano, J.F., and Reis, R.L. (2004). Optimization of the formulation and mechanical properties of starch-based partially degradable bone cements. *Journal of Materials Science. Materials in Medicine* 15 (1): 73–83.

83 Gomes, M.E., Sikavitsas, V.I., Behravesh, E. et al. (2003). Effect of flow perfusion on the osteogenic differentiation of bone marrow stromal cells cultured on starch-based three-dimensional scaffolds. *Journal of Biomedical Materials Research Part A: An Official Journal of the Society for Biomaterials, the Japanese Society for Biomaterials, and the Australian Society for Biomaterials and the Korean Society for Biomaterials* 67 (1): 87–95.

84 Balmayor, E.R., Tuzlakoglu, K., Marques, A.P. et al. (2008). A novel enzymatically mediated drug delivery carrier for bone tissue engineering applications: combining biodegradable starch-based microparticles and differentiation agents. *Journal of Materials Science. Materials in Medicine* 19 (4): 1617–1623.

85 Bellucci, S., Balasubramanian, C., Micciulla, F., and Rinaldi, G. (2007). CNT composites for aerospace applications. *Journal of Experimental Nanoscience* 2 (3): 193–206.

86 Kang, I., Heung, Y.Y., Kim, J.H. et al. (2006). Introduction to carbon nanotube and nanofiber smart materials. *Composites Part B: Engineering* 37 (6): 382–394.

87 Wang, C.Y., Mottaghitalab, V., Too, C.O. et al. (2007). Polyaniline and polyaniline–carbon nanotube composite fibers as battery materials in ionic liquid electrolytes. *Journal of Power Sources* 163 (2): 1105–1109.

88 Mottaghitalab, V., Xi, B., Spinks, G.M., and Wallace, G.G. (2006). Polyaniline fibres containing single-walled carbon nanotubes: enhanced performance artificial muscles. *Synthetic Metals* 156 (11–13): 796–803.

89 Ong, Y.T., Ahmad, A.L., Zein, S.H.S., and Tan, S.H. (2010). A review on carbon nanotubes in environmental protection and green engineering perspective. *Brazilian Journal of Chemical Engineering* 27: 227–242.

90 Mundargi, R.C., Patil, S.A., Agnihotri, S.A., and Aminabhavi, T.M. (2007). Development of polysaccharide-based colon targeted drug delivery systems for the treatment of amoebiasis. *Drug Development and Industrial Pharmacy* 33 (3): 255–264.

91 Silva, I., Gurruchaga, M., and Goñi, I. (2009). Physical blends of starch graft copolymers as matrices for colon targeting drug delivery systems. *Carbohydrate Polymers* 76 (4): 593–601.

92 Obot, I.B., Onyeachu, I.B., and Kumar, A.M. (2017). Sodium alginate: a promising biopolymer for corrosion protection of API X60 high strength carbon steel in saline medium. *Carbohydrate Polymers* 178: 200–208.

93 Christopher, G., Kulandainathan, M.A., and Harichandran, G. (2016). Biopolymers nanocomposite for material protection: enhancement of corrosion protection using waterborne polyurethane nanocomposite coatings. *Progress in Organic Coatings* 99: 91–102.

94 Pal, A., Das, D., Sarkar, A.K. et al. (2015). Synthesis of glycogen and poly (acrylic acid)-based graft copolymers via ATRP and its application for selective removal of Pb2+ ions from aqueous solution. *European Polymer Journal* 66: 33–46.

95 Sarkar, A.K., Saha, A., Panda, A.B., and Pal, S. (2015). pH Triggered superior selective adsorption and separation of both cationic and anionic dyes and photocatalytic activity on a fully exfoliated titanate layer–natural polymer-based nanocomposite. *Chemical Communications* 51 (89): 16057–16060.

96 Das, D., Patra, P., Ghosh, P. et al. (2016). Dextrin and poly (lactide)-based biocompatible and biodegradable nanogel for cancer targeted delivery of doxorubicin hydrochloride. *Polymer Chemistry* 7 (17): 2965–2975.

97 Das, D. and Pal, S. (2015). Modified biopolymer-dextrin based crosslinked hydrogels: application in controlled drug delivery. *RSC Advances* 5 (32): 25014–25050.

98 Zhang, H., Wang, D., Wang, F. et al. (2015). Corrosion inhibition of mild steel in hydrochloric acid solution by quaternary ammonium salt derivatives of corn stalk polysaccharide (QAPS). *Desalination* 372: 57–66.

99 Carneiro, J., Tedim, J., and Ferreira, M.G.S. (2015). Chitosan as a smart coating for corrosion protection of aluminum alloy 2024: a review. *Progress in Organic Coatings* 89: 348–356.

100 Rajeswari, V., Kesavan, D., Gopiraman, M., and Viswanathamurthi, P. (2013). Physicochemical studies of glucose, gellan gum, and hydroxypropyl cellulose—Inhibition of cast iron corrosion. *Carbohydrate Polymers* 95 (1): 288–294.

101 Fares, M.M., Maayta, A.K., and Al-Mustafa, J.A. (2012). Corrosion inhibition of iota-carrageenan natural polymer on aluminum in presence of zwitterion mediator in HCl media. *Corrosion Science* 65: 223–230.

102 Zou, C., Yan, X., Qin, Y. et al. (2014). Inhibiting evaluation of β-Cyclodextrin-modified acrylamide polymer on alloy steel in sulfuric solution. *Corrosion Science* 85: 445–454.

103 Kumar, S., Vashisht, H., Olasunkanmi, L.O. et al. (2016). Experimental and theoretical studies on inhibition of mild steel corrosion by some synthesized polyurethane tri-block co-polymers. *Scientific Reports* 6 (1): 1–18.

104 Li, J., Revol, J.F., and Marchessault, R.H. (1997). Effect of degree of deacetylation of chitin on the properties of chitin crystallites. *Journal of Applied Polymer Science* 65 (2): 373–380.

105 Li, Q., Dunn, E.T., Grandmaison, E.W., and Goosen, M.F.A. (1992). Applications and properties of chitosan. Journal of Bioactive Compatable Polymer 7: 370–397.

106 Verma, C., Ebenso, E.E., Quraishi, M.A., and Hussain, C.M. (2021). Recent developments in sustainable corrosion inhibitors: design, performance and industrial scale applications. *Materials Advances* 2 (12): 3806–3850.

107 Haque, J., Srivastava, V., Chauhan, D.S. et al. (2018). Microwave-induced synthesis of chitosan Schiff bases and their application as novel and green corrosion inhibitors: experimental and theoretical approach. *ACS Omega* 3 (5): 5654–5668.

108 Okoronkwo, A.E., Olusegun, S. J., and Oluwasina, O.O. (2015). The inhibitive action of chitosan extracted from Archachatina marginata shells on the corrosion of plain carbon steel in acid media. *Anti-Corrosion Methods and Materials* 62: 13–18.

109 Fayomi, O.S.I., Akande, I.G., Oluwole, O.O., and Daramola, D. (2018). Effect of water-soluble chitosan on the electrochemical corrosion behaviour of mild steel. *Chemical Data Collections* 17: 321–326.

110 Jmiai, A., El Ibrahimi, B., Tara, A. et al. (2017). Chitosan as an eco-friendly inhibitor for copper corrosion in acidic medium: protocol and characterization. *Cellulose* 24 (9): 3843–3867.

111 El Mouaden, K., El Ibrahimi, B., Oukhrib, R. et al. (2018). Chitosan polymer as a green corrosion inhibitor for copper in sulfide-containing synthetic seawater. *International Journal of Biological Macromolecules* 119: 1311–1323.

112 Suyanto, D., Darmokoesemo, H., Ruriyanti, R., and Anggara, L.S. (2015). Application chitosan derivatives as inhibitor corrosion on steel with fluidization method. *Journal of Chemical and Pharmaceutical Research* 2 (7): 260–267.

113 Kunii, D. and Levenspiel, O. (1991). *Fluidization Engineering*. Butterworth-Heinemann.

114 Gibilaro, L.G. (2001). *Fluidization Dynamics*, 1. Oxford, Auckland, Boston, Johannesburg, Melbourne, New Delhi: Butterworth Heinemann, Elsevier.

115 Chauhan, D.S., Mazumder, M.J., Quraishi, M.A., and Ansari, K.R. (2020). Chitosan-cinnamaldehyde Schiff base: a bioinspired macromolecule as corrosion inhibitor for oil and gas industry. *International Journal of Biological Macromolecules* 158: 127–138.

116 Li, H., Li, H., Liu, Y., and Huang, X. (2015). Synthesis of polyamine grafted chitosan copolymer and evaluation of its corrosion inhibition performance. *Journal of the Korean Chemical Society* 59 (2): 142–147.

117 Kong, P., Feng, H., Chen, N. et al. (2019). Polyaniline/chitosan as a corrosion inhibitor for mild steel in acidic medium. *RSC Advances* 9 (16): 9211–9217.

118 Chen, T., Zeng, D., and Zhou, S. (2018). Study of polyaspartic acid and chitosan complex corrosion inhibition and mechanisms. *Polish Journal of Environmental Studies* 27 (4).

119 Quraishi, M.A., Ansari, K.R., Chauhan, D.S. et al. (2020). Vanillin modified chitosan as a new bio-inspired corrosion inhibitor for carbon steel in oil-well acidizing relevant to petroleum industry. *Cellulose* 27 (11): 6425–6443.

120 Menaka, R. and Subhashini, S. (2016). Chitosan Schiff base as eco-friendly inhibitor for mild steel corrosion in 1 M HCl. *Journal of Adhesion Science and Technology* 30 (15): 1622–1640.

121 Luckachan, G.E. and Mittal, V. (2015). Anti-corrosion behavior of layer by layer coatings of cross-linked chitosan and poly (vinyl butyral) on carbon steel. *Cellulose* 22 (5): 3275–3290.

122 Chauhan, D.S., Srivastava, V., Joshi, P.G., and Quraishi, M.A. (2018). PEG cross-linked chitosan: a biomacromolecule as corrosion inhibitor for sugar industry. *International Journal of Industrial Chemistry* 9 (4): 363–377.

123 Sanmugam, A., Vikraman, D., Karuppasamy, K. et al. (2017). Evaluation of the corrosion resistance properties of electroplated chitosan-Zn1− xCuxO composite thin films. *Nanomaterials* 7 (12): 432.

124 John, S., Joseph, A., Jose, A.J., and Narayana, B. (2015). Enhancement of corrosion protection of mild steel by chitosan/ZnO nanoparticle composite membranes. *Progress in Organic Coatings* 84: 28–34.

125 Rasheed, P.A., Jabbar, K.A., Rasool, K. et al. (2019). Controlling the biocorrosion of sulfate-reducing bacteria (SRB) on carbon steel using ZnO/chitosan nanocomposite as an eco-friendly biocide. *Corrosion Science* 148: 397–406.

126 Liu, Y., Zou, C., Yan, X. et al. (2015). β-Cyclodextrin modified natural chitosan as a green inhibitor for carbon steel in acid solutions. *Industrial & Engineering Chemistry Research* 54 (21): 5664–5672.

127 Jena, G., Anandkumar, B., Vanithakumari, S.C. et al. (2020). Graphene oxide-chitosan-silver composite coating on Cu-Ni alloy with enhanced anticorrosive and antibacterial properties suitable for marine applications. *Progress in Organic Coatings* 139: 105444.

128 Solomon, M.M., Gerengi, H., Kaya, T., and Umoren, S.A. (2017). Performance evaluation of a chitosan/silver nanoparticles composite on St37 steel corrosion in a 15% HCl solution. *ACS Sustainable Chemistry & Engineering* 5 (1): 809–820.

129 Tabesh, E.S.H.R., Salimijazi, H.R., Kharaziha, M. et al. (2019). Development of an in-situ chitosan-copper nanoparticle coating by electrophoretic deposition. *Surface & Coatings Technology* 364: 239–247.

130 Karimi, S., Salahinejad, E., Sharifi, E. et al. (2018). Bio performance of chitosan/fluoride-doped diopside nanocomposite coatings deposited on medical stainless steel. *Carbohydrate Polymers* 202: 600–610.

131 Zhai, X., Ren, Y., Wang, N. et al. (2019). Microbial corrosion resistance and antibacterial property of electrodeposited Zn–Ni–Chitosan coatings. *Molecules* 24 (10): 1974.

132 John, S., Salam, A., Baby, A.M., and Joseph, A. (2019). Corrosion inhibition of mild steel using chitosan/TiO2 nanocomposite coatings. *Progress in Organic Coatings* 129: 254–259.

133 Cordero-Arias, L., Cabanas-Polo, S., Gao, H. et al. (2013). Electrophoretic deposition of nanostructured-TiO 2/chitosan composite coatings on stainless steel. *RSC Advances* 3 (28): 11247–11254.

134 Ledwig, P., Kot, M., Moskalewicz, T., and Dubiel, B. (2017). Electrophoretic deposition of nc-TiO2/chitosan composite coatings on X2CrNiMo17-12-2 stainless steel. *Archives of Metallurgy and Materials* 62 (1): 405–410.

135 Hahn, B.D., Park, D.S., Choi, J.J. et al. (2011). Aerosol deposition of hydroxyapatite–chitosan composite coatings on biodegradable magnesium alloy. *Surface & Coatings Technology* 205 (8–9): 3112–3118.

136 Sutha, S., Kavitha, K., Karunakaran, G., and Rajendran, V. (2013). In-vitro bioactivity, biocorrosion and antibacterial activity of silicon integrated hydroxyapatite/chitosan composite coating on 316 L stainless steel implants. *Materials Science and Engineering: C* 33 (7): 4046–4054.

137 Molaei, A., Yari, M., and Afshar, M.R. (2017). Investigation of halloysite nanotube content on electrophoretic deposition (EPD) of chitosan-bioglass-hydroxyapatite-halloysite nanotube nanocomposites films in surface engineering. *Applied Clay Science* 135: 75–81.

138 Sousa, L.L.D., Ricci, V.P., Prado, D.G. et al. (2018). Titanium coating with hydroxyapatite and chitosan doped with silver nitrate. *Materials Research* 20: 863–868.

139 Stevanovic, M., Đošić, M., Jankovic, A. et al. (2018). Gentamicin-loaded bioactive hydroxyapatite/chitosan composite coating electrodeposited on titanium. *ACS Biomaterials Science & Engineering* 4 (12): 3994–4007.

140 Eduok, U., Ohaeri, E., and Szpunar, J. (2019). Conversion of imidazole to N-(3-Aminopropyl) imidazole toward enhanced corrosion protection of steel in combination with carboxymethyl chitosan grafted poly (2-methyl-1-vinyl imidazole). *Industrial & Engineering Chemistry Research* 58 (17): 7179–7192.

141 Eduok, U. and Szpunar, J. (2019). Biocorrosion reduction of pipeline steel in Desulfovibrio ferrophilus culture in the presence of Carboxymethyl Chitosan

Grafted Poly (2-methyl-1-vinylimidazole)/Cerium Molybdate Nanocomposite. *ECS Meeting Abstracts*. IOP Publishing, 853.

142 Farhadian, A., Varfolomeev, M.A., Shaabani, A. et al. (2020). Sulfonated chitosan as green and high cloud point kinetic methane hydrate and corrosion inhibitor: experimental and theoretical studies. *Carbohydrate Polymers* 236: 116035.

143 Farahati, R., Ghaffarinejad, A., Rezania, H.J. et al. (2019). Sulfonated aromatic polyamide as water-soluble polymeric corrosion inhibitor of copper in HCl. *Colloids and Surfaces. A, Physicochemical and Engineering Aspects* 578: 123626.

144 Ruhi, G., Modi, O.P., and Dhawan, S.K. (2015). Chitosan-polypyrrole-SiO2 composite coatings with advanced anticorrosive properties. *Synthetic Metals* 200: 24–39.

145 Fares, M.M., Maayta, A.K., and Al-Qudah, M.M. (2012). Pectin as promising green corrosion inhibitor of aluminum in hydrochloric acid solution. *Corrosion Science* 60: 112–117.

146 Umoren, S.A., Obot, I.B., Madhankumar, A., and Gasem, Z.M. (2015). Performance evaluation of pectin as ecofriendly corrosion inhibitor for X60 pipeline steel in acid medium: experimental and theoretical approaches. *Carbohydrate Polymers* 124: 280–291.

147 Khairou, K.S. and El-Sayed, A. (2003). Inhibition effect of some polymers on the corrosion of cadmium in a hydrochloric acid solution. *Journal of Applied Polymer Science* 88 (4): 866–871.

148 Halambek, J., Cindrić, I., and Grassino, A.N. (2020). Evaluation of pectin isolated from tomato peel waste as natural tin corrosion inhibitor in sodium chloride/acetic acid solution. *Carbohydrate Polymers* 234: 115940.

149 Nada, C., Katarina, B., and Sandra, P. (1996). The inhibition effect of pectin and ascorbic acid on the corrosion of tin in sodium chloride solution. In: *47th Annual Meeting of the International Society of Electrochemistry* (ed. *Abstracts/ The International Society of Electrochemistry*). The Electrochemical Society.

150 Prabakaran, M., Ramesh, S., Periasamy, V., and Sreedhar, B. (2015). The corrosion inhibition performance of pectin with propyl phosphonic acid and Zn2+ for corrosion control of carbon steel in aqueous solution. *Research on Chemical Intermediates* 41 (7): 4649–4671.

151 Geethanjali, R., Sabirneeza, A.A.F., and Subhashini, S. (2014). Water-soluble and biodegradable Pectin-Grafted Polyacrylamide and Pectin-Grafted Polyacrylic Acid: electrochemical investigation of corrosion-inhibition behaviour on mild steel in 3.5% NaCl Media, Hindawi Publishing Corporation. *Indian Journal of Materials Science* 2014: 9. Article ID 356075. doi: 10.1155/2014/356075.

152 Mishra, R.K., Sutar, P.B., Singhal, J.P., and Banthia, A.K. (2007). Graft copolymerization of pectin with polyacrylamide. *Polymer-Plastics Technology and Engineering* 46 (11): 1079–1085.

153 Chauhan, K., Kumar, R., Kumar, M. et al. (2012). Modified pectin-based polymers as green antiscalants for calcium sulfate scale inhibition. *Desalination* 305: 31–37.

154 Sakai, T., Sakamoto, T., Hallaert, J., and Vandamme, E.J. (1993). [Pectin, pectinase, and protopectinase: production,[properties, and applications. *Advances in Applied Microbiology* 39: 213–294.

155 Zaafarany, I. (2012). Corrosion inhibition of aluminum in aqueous alkaline solutions by alginate and pectate water-soluble natural polymer anionic polyelectrolytes. *Electrochimica Acta* 30 (6): 419–426.

156 Zaafarany, I.A., Khairou, K.S., and Hassan, R.M. (2009). Acid-catalysis of chromic acid oxidation of kappa-carrageenan polysaccharide in aqueous perchlorate solutions. *Journal of Molecular Catalysis. A, Chemical* 302 (1–2): 112–118.

157 Chaires-Martínez, L., Salazar-Montoya, J.A., and Ramos-Ramírez, E.G., (2008). Physicochemical and functional characterization of the galactomannan obtained from mesquite seeds (Prosopis pallida). *European Food Research and Technology* 227 (6): 1669–1676.

158 Douaire, M. and Norton, I.T. (2013). Designer colloids in structured food for the future. *Journal of the Science of Food and Agriculture* 93 (13): 3147–3154.

159 Umoren, S.A., Obot, I.B., Ebenso, E.E. et al. (2006). Gum arabic as a potential corrosion inhibitor for aluminium in alkaline medium and its adsorption characteristics. *Anti-Corrosion Methods and Materials* 53: 277–282.

160 Umoren, S.A. (2009). Synergistic influence of gum arabic and iodide ion on the corrosion inhibition of aluminium in alkaline medium. *Portugaliae Electrochimica Acta* 27 (5): 565–577.

161 Verbeken, D., Dierckx, S., and Dewettinck, K. (2003). Exudate gums: occurrence, production, and applications. *Applied Microbiology and Biotechnology* 63 (1): 10–21.

162 Abdallah, M. (2004). Guar gum as corrosion inhibitor for carbon steel in sulfuric acid solutions. *Portugaliae Electrochimica Acta* 22: 161–175.

163 Biswas, A., Pal, S., and Udayabhanu, G. (2015). Experimental and theoretical studies of xanthan gum and its graft co-polymer as corrosion inhibitor for mild steel in 15% HCl. *Applied Surface Science* 353: 173–183.

164 Babaladimath, G., Vishalakshi, B., and Nandibewoor, S.T. (2018). Electrical conducting Xanthan Gum-graft-polyaniline as corrosion inhibitor for aluminum in hydrochloric acid environment. *Materials Chemistry and Physics* 205: 171–179.

165 Nawaz, M., Shakoor, R.A., Kahraman, R., and Montemor, M.F. (2021). Cerium oxide loaded with Gum Arabic as environmentally friendly anti-corrosion additive for protection of coated steel. *Materials & Design* 198: 109361.

166 Solomon, M.M., Gerengi, H., Umoren, S.A. et al. (2018). Gum Arabic-silver nanoparticles composite as a green anticorrosive formulation for steel corrosion in strong acid media. *Carbohydrate Polymers* 181: 43–55.

167 Singh, A., Liu, M., Ituen, E., and Lin, Y. (2020). Anti-corrosive properties of an effective guar gum grafted 2-acrylamido-2-methyl propane sulfonic acid (GG-AMPS) coating on copper in a 3.5% NaCl solution. *Coatings* 10 (3): 241.

168 Biswas, A., Das, D., Lgaz, H. et al. (2019). Biopolymer dextrin and poly (vinyl acetate) based graft copolymer as an efficient corrosion inhibitor for mild steel in hydrochloric acid: electrochemical, surface morphological and theoretical studies. *Journal of Molecular Liquids* 275: 867–878.

169 Devi, G.N., Unnisa, C.B.N., Roopan, S.M. et al. (2020). Floxacins: as mediators in enhancing the corrosion inhibition efficiency of natural polymer dextrin. *Macromolecular Research* 28 (6): 558–566.

170 Altin, A., Rohwerder, M., and Erbe, A. (2017). Cyclodextrins as carriers for organic corrosion inhibitors in organic coatings. *Journal of the Electrochemical Society* 164 (4): C128.

171 Amiri, S. and Rahimi, A. (2016). Anticorrosion behavior of cyclodextrins/inhibitor nanocapsule-based self-healing coatings. *Journal of Coatings Technology and Research* 13 (6): 1095–1102.

172 Khramov, A.N., Voevodin, N.N., Balbyshev, V.N., and Mantz, R.A. (2005). Sol–gel-derived corrosion-protective coatings with controllable release of incorporated organic corrosion inhibitors. *Thin Solid Films* 483 (1–2): 191–196.

173 Solomon, M.M., Umoren, S.A., Udosoro, I.I., and Udoh, A.P. (2010). Inhibitive and adsorption behaviour of carboxymethyl cellulose on mild steel corrosion in sulphuric acid solution. *Corrosion Science* 52 (4): 1317–1325.

174 Umoren, S.A., Solomon, M.M., Udosoro, I.I., and Udoh, A.P. (2010). Synergistic and antagonistic effects between halide ions and carboxymethyl cellulose for the corrosion inhibition of mild steel in sulphuric acid solution. *Cellulose* 17 (3): 635–648.

175 Solomon, M.M., Gerengi, H., and Umoren, S.A. (2017). Carboxymethyl cellulose/silver nanoparticles composite: synthesis, characterization and application as a benign corrosion inhibitor for St37 steel in 15% H2SO4 medium. *ACS Applied Materials & Interfaces* 9 (7): 6376–6389.

176 Rajendran, S., Joany, R.M., Apparao, B.V., and Palaniswamy, N. (2002). Corrosion inhibition by carboxymethyl cellulose-1-hydroxyethane-1, 1-diphosphonic acid-Zn2+ system. *Bulletin of Electrochemistry* 18 (1): 25–28.

177 El-Lateef, H.M.A., Albokheet, W.A., and Gouda, M. (2020). Carboxymethyl cellulose/metal (Fe, Cu and Ni) nanocomposites as non-precious inhibitors of C-steel corrosion in HCl solutions: synthesis, characterization, electrochemical and surface morphology studies. *Cellulose* 27 (14): 8039–8057.

178 Hasanin, M.S. and Al Kiey, S.A. (2020). Environmentally benign corrosion inhibitors based on cellulose niacin nano-composite for corrosion of copper in sodium chloride solutions. *International Journal of Biological Macromolecules* 161: 345–354.

179 Liao, S., Pan, Y., Yang, S. et al. (2022). Hydroxyethyl Cellulose Acrylamide Graft Polymer-Bentonite Nanocomposites Synthesis, Characterization and Properties. *Matéria (Rio de Janeiro)* 26.

180 Shi, S.C. and Su, C.C. (2016). Corrosion inhibition of high speed steel by biopolymer HPMC derivatives. *Materials* 9 (8): 612.

181 Muthulakshmi, L., Kumar, B.A., Rajasekar, A. et al. (2021). The benefits of k-Carrageenan-gelatin hybrid composite coating on the medical grade stainless steel (SS304) used as anticorrosive barrier. *Materials Chemistry and Physics* 258: 123909.

182 Lahrour, S., Benmoussat, A., Bouras, B. et al. (2019). Glycerin-grafted starch as corrosion inhibitor of C-Mn steel in 1 M HCl solution. *Applied Sciences* 9 (21): 4684–4702.

183 Li, X., Deng, S., Lin, T. et al. (2020). Cassava starch ternary graft copolymer as a corrosion inhibitor for steel in HCl solution. *Journal of Materials Research and Technology* 9 (2): 2196–2207.

Index

Grafted Biopolymers as Corrosion Inhibitors: Safety, Sustainability, and Efficiency, First Edition.
Edited by Jeenat Aslam, Chandrabhan Verma, and Ruby Aslam.

d

n

Printed and bound by CPI Group (UK) Ltd, Croydon, CR0 4YY

17/09/2026

14964252-0008